T0186641

FORENSIC NEUROPSYCHOLOGY

STUDIES ON NEUROPSYCHOLOGY, DEVELOPMENT, AND COGNITION

Series Editor:

Linas A. Bieliauskas, Ph.D.
University of Michigan, Ann Arbor, MI, USA

FORENSIC NEUROPSYCHOLOGY
FUNDAMENTALS AND PRACTICE

EDITED BY
JERRY J. SWEET

Psychology Press
Taylor & Francis Group
HOVE AND NEW YORK

Transferred to digital print in 2005 by Psychology Press
27 Church Road, Hove, East Sussex BN3 2FA
270 Madison Avenue, New York, NY 10016

Psychology Press is a part of the Taylor & Francis Group

© 1999 Psychology Press

First published 1999 by Swets & Zeitlinger, reprinted 2000

Cover design: Magenta Grafische Producties, Bert Haagsman
Typesetting: Red Barn Publishing, Skeagh, Skibbereen, Co. Cork, Ireland
Printed and bound in Great Britain by TJI Digital, Padstow, Cornwall

Library of Congress Cataloging-in-Publication Data

Forensic neuropsychology / edited by Jerry J. Sweet.
      p. cm. —— (Studies on neuropsychology, development, and cognition)
Includes bibliographical references and index.
ISBN 9026515448 (hardback)
1. Forensic neuropsychology.   2. Clinical neuropsychology.
I. Sweet, Jerry J. II. Series.
[DNLM: 1. Forensic Psychiatry.   2. Neuropsychology. W 740 F7143 1999]
RA1147.5.F665   1999
614.1′—dc21
DMLM/DLC
for Library of Congress   99-22363
                                        CIP

Contents

Part I Fundamentals

Part II Practice Expertise

Part III Relevant Populations

Part IV Parameters of the Forensic Arena

Dedication

*To my parents and grandparents
for their love and appreciation of learning.*

In Memoriam

Charles G. Matthews, Ph.D., Professor, Department of Neurology, University of Wisconsin, died on April 19, 1998. Dr. Matthews was well recognized in the field of clinical neuropsychology as an outstanding clinician, researcher, and teacher. He exemplified the scientist-practitioner tradition in psychology, and, in fact, mentored many young neuropsychologists who went on to become major contributors to the field. Among his many professional accomplishments, which literally are too numerous to mention here, he served as President of three of the major professional organizations for clinical neuropsychologists: International Neuropsychological Society, Division 40 (Clinical Neuropsychology) of the American Psychological Association, and the American Board of Clinical Neuropsychology. He also received a distinguished contribution award from the fourth major professional organization, the National Academy of Neuropsychology. Through professional leadership and scientific contributions Dr. Matthews influenced us all.

Dr. Matthews was to have written the Afterword for this text. Although he died before accomplishing this task, he nevertheless made a strong contribution to this text. Typical of his commitment to the field and representative of his sheer love of intellectual pursuits, before beginning to write his portion, Dr. Matthews volunteered to read each chapter carefully and provide the editor with a 'second opinion' of content and writing style. During the many months of preparation of this book he generously invested a great deal of his time and provided many very helpful suggestions. It was a distinct pleasure to work with "Chuck", as he preferred to be called, in this fashion. We shall always remember his warm and kind demeanor, extraordinary and engaging wit, and inspiring intellect.

From the series editor

Dear Reader,

This text represents the next exciting volume in the series *Studies on Neuropsychology, Development, and Cognition*, and it continues to fulfill our goal of providing a practical opening to clinical implications of contemporary scientific and professional developments in neuropsychology.

   Forensic Neuropsychology: Fundamentals and Practice, edited by Jerry Sweet, continues the tradition of the series. From a review of the basic tenets of neuropsychological testing and assessment, to the delicacies of consideration of moderator variables and the making of differential diagnoses, this text provides up-to-date coverage of issues with which the clinical neuropsychologist must deal in the forensic setting. Appropriate caveats and safeguards are stressed in line with the fundamental goal of providing information to assist with competent, thorough, and fair use of neuropsychological information for the patients and the courts when the justice system becomes involved in neurological illness or injury. Practical information is also given to familiarize the practicing clinical neuropsychologist with the nuances of delivering this information in an environment which may query their veracity and competence. The information provided by the distinguished group of contributors to this volume will provide insight and guidance to those practitioners who find themselves venturing into the exciting yet challenging forensic arena of clinical neuropsychology practice.

Linas Bieliauskas
Ann Arbor, September 1998

Foreword

The considerable growth of basic and applied knowledge in neuropsychology and the related clinical neurosciences over the past half-century has fostered the development of clinical neuropsychology as a specialty and the expansion of the scope of clinical practice. Within the intra- and interdisciplinary framework of the applied neurosciences, clinical neuropsychology has had the essentially unique role of providing quantitative measurements of behavioral consequences of various underlying neuropathological, pathophysiological and pathochemical processes that may arise across the broad spectrum of neurological and neuropsychiatric diseases. Although determination of the validity of inferences derived from neuropsychological data concerning underlying brain structure and function remains a continuing investigative challenge for clinical neuropsychologists, extant studies have demonstrated sufficient effectiveness in assisting the triers of fact in forensic arenas and earned increasing recognition of clinical neuropsychologists as expert witnesses. This is especially true in traumatic brain injury and neurotoxin exposure-related civil actions and with regard to competency determinations within criminal proceedings.

This volume represents a signal attempt to integrate relevant clinical neuropsychological knowledge and focused applications of practice methodologies that address the weighty questions posed by potential or actual litigants in judicial and other adversarial activities. Within the formal proceedings of cases litigated within the judicial system (i.e., personal injury, criminal, worker's compensation, custody, and guardianship proceedings), the trier of fact has the constitutional authority to administer the oath of truthfulness to the expert witness and to hold the witness accountable for the veracity of his or her testimony.

Unlike the routine clinical practice arena, within which the "patient" is ideally sole recipient, and the highest priority, of services provided, the expert witness usually is solicited by one or the other of the litigants. In this manner, the judicial system introduces ethical challenges unlike those encountered in traditional professional roles and indirectly forces the expert witness to address a major conflict of interest; satisfying the sworn responsibility to assist the trier of fact truthfully, while serving the interests of the soliciting litigant and involved parties.

Resolution of this ethical challenge requires the clinical neuropsychologist to render interpretations in a realistic manner, consistent with available scientific evidence of validity, as well as the relevance and appropriate use, of data selected for drawing particular inferences and conclusions under the oath of truthfulness. The authoritative information provided by the many knowledgeable contributors to this volume will help neuropsychologists at all levels of forensic experience perform more effectively in their professional roles in this interesting and important area and, thereby, enhance equitable outcomes for parties involved in adversarial proceedings.

Manfred J. Meier, Ph.D.
Professor Emeritus, University of Minnesota

Preface

Forensic neuropsychology is a relatively young practice area for many clinical neuropsychologists. In fact, when I began my career in 1979, I had no hint that my career path would involve interaction with attorneys, administrative proceedings, and courts at any degree of frequency. There was no mention of such activity from my mentors and, consequently, no didactic or supervised experiential preparation for involvement in forensic activities. As with many of my same age or older colleagues, we basically learned by doing, which is not the easiest or kindest method. It is nevertheless true that some professional 'bruising', or lessons learned 'the hard way', afford unique opportunities for unforgettable experiences that lead to forcible modification and sharpening of relevant knowledge and skills. Having been frequently asked by trainees and colleagues to explain my participation in what seems to some to be an exercise filled with anxiety and frustration, typically I have responded by describing the positive influence on clinical practice that is fostered by the realization that an exacting and demanding external system may examine critically one's daily clinical work. This kind of scenario cannot help but be intellectually stimulating. Such is the case, no matter whether retained specifically to provide expert witness services or evaluating and treating a routine clinical referral involving acquired neuropsychological dysfunction. Even the straightforward clinical evaluation of a suspected elderly Alzheimer's patient may subsequently involve legal decision making that requires the clinician's input. Such can be the case even after that patient's death, as when family members adopt adversarial positions regarding a will that was altered when the deceased arguably was impaired on formal testing that was initially ordered to assist in diagnosis and case management. While strenuous, the effect is the same as that of substantial teaching involvement with clinical trainees; there is increased motivation to perform clinical activities carefully and knowledgeably. In that sense, although unfortunately accompanied by some negative aspects, I have been grateful to have the opportunity to be involved with forensic neuropsychology.

Thanks and appreciation are extended to Linas Bieliauskas, Swets and Zeitlinger's neuropsychology book series editor, for inviting me to be editor for this topic. Also, thanks to Martha Chorney, Swets and Zeitlinger's Senior Editor in charge of this project, for her steadfast confidence, enthusiasm, and pleasant temperament. My heartfelt thanks especially to the contributors, all well qualified and recognized experts in their own right, who felt strongly enough about the value of this project to sacrifice their busy professional schedules and personal time in favor of producing quality chapters. Neuropsychology Service secretary Ann Hammond, clinical staff, and trainees deserve special thanks for enduring the added strain and stress to all when such a resource-demanding project is underway in their midst, within the already highly demanding present day healthcare environment.

Jerry J. Sweet

Introduction

It seems unlikely that any one could have predicted the explosion of interest in clinical neuropsychology from the legal community, or the resulting effects upon the profession. It was the experience of many clinical neuropsychologists, that the subspecialty was 'discovered' by the legal community in the early 1980s and has weathered increasing interest subsequently. In fact, more than merely an anecdotally- based belief, Taylor, in Chapter 14 of this text, has creatively documented that neuropsychologists had almost no impact on decisions of courts prior to 1980, but have had increasing impact subsequently, especially in the last five or six years.

Currently, the term *forensic* is associated so frequently with cases in litigation, that many have come to use it narrowly, as if it applies only to civil and criminal cases undergoing formal courtroom proceedings. Interestingly, according to Webster's Ninth New Collegiate Dictionary, forensic is derived from the Latin *forensis*, meaning public. Although the term forensic is aptly used when denoting formal legal activities, the definition also includes all activities relating to public discussion and debate, especially those involving reaching a common judgment from antagonistic positions.

At this point, neuropsychologists have been utilized in cases involving worker's compensation, disability determination, educational due process within public school systems, personal injury, criminal, child custody, impaired professional and 'fitness for duty', competency, and other cases involving adversarial administrative and judicial determinations. Truly, one must think of a larger domain for clinical neuropsychology than just those cases that are truly "litigated"in a courtroom. In fact, one might more aptly describe the cases in this arena by using the term *adversarial*, denoting involvement in adversarial administrative (e.g., disability determination, worker's compensation, due process hearings in school systems) and litigated (i.e., personal injury, child custody, divorce, criminal) cases. In most instances throughout this book, one can substitute the more encompassing term adversarial for litigated. Moreover, relating to the title of this text, all adversarial neuropsychological cases are appropriately categorized as forensic.

It can be argued cogently that the demand for involvement in adversarial activities is a natural outcome of the success of a strong scientist-practitioner orientation within the subspecialty of clinical neuropsychology. In fact, one could argue that this philosophical approach is both cause and format for clinicians engaging in forensic neuropsychology. The approach is a cause in the sense that scientific data gathering and objective behavioral methodologies created ongoing growth in an important knowledge area. Among the relevant by-products of a scientist-practitioner approach are: familiarity with disciplined scrutiny (i.e., peer review), clinical procedures emphasizing data-based decision-making (i.e., accountability), and comfort with hypothesis-testing (i.e., objective differential diagnosis). The approach is a format in the sense that related methodologies and data-based knowledge are paradigmatic in our ongoing forensic activities. While an unplanned and unforeseen outcome of a fundamentally scientific orientation, the content of this book attests to the fact that forensic neuropsychology has come to represent a growing portion of regular practice for many.

Along with forensic activities come a variety of new professional challenges for clinical neuropsychology. These challenges include the need to explicate ethical and professional issues in an arena that makes demands of clinicians that extend beyond the demands of typical practice. For example, there is a greater than normal need to share information regarding data gathering procedures, administration parameters, scoring criteria, relevance of normative comparisons, qualifications of personnel/professional credentials, and the foundations of statements contained within formal evaluation reports and treatment records. Above all, there is a unique need to assure that maximum objectivity has been used in arriving at diagnostic formulations and treatment plans. As a result, the professional literature has expanded, training programs have modified their content and relevant continuing education has been developed. Make no mistake about it, the sizeable involvement of clinical neuropsychology in forensic activities has raised the standards and resulted in more rigorous expectations with regard to efficacy and accountability. There is little doubt, for example, that the controversy that began in the late 1980's regarding the ineffectiveness of neuropsychologists in detecting neuropsychological malingering was related to neuropsychology's increased forensic involvement. Therefore, it is also these same forensic demands that should be credited, partially, with the subsequent burgeoning of methodology and literature that eventually established meaningful efficacy in detection of malingering.

Within the forensic arena, clinical neuropsychologists are asked to provide expertise regarding differential diagnosis of conditions (that may not be amenable to accurate differential by mental status or neuroimaging techniques), delineation of a broad spectrum of potential neurobehavioral dysfunctions, and, if appropriate, formulation of psychological, behavioral, and cognitive treatment plans. Neuropsychologists carry out these tasks differently than colleagues in neurology, neurosurgery, physiatry, psychiatry, speech therapy, occupational therapy, and vocational rehabilitation. In fact, much of what neuropsychologists do in the forensic arena has little or no redundancy with other health care disciplines.

The practice of clinical neuropsychology within the context of adversarial proceedings seems inherently more conceptually complex than routine clinical practice. We should not expect that neuropsychologists, including the contributors to this text, are in complete agreement with one another regarding important forensic practice issues. As the field evolves, a greater consensus among practitioners will likely follow. For the benefit of the reader, the present text on forensic neuropsychology has been organized into sections on: Fundamentals, Practice Expertise, Relevant Populations, and Parameters of the Legal Arena. By design, the text has no coverage of involvement of neuropsychologists in criminal cases, as this area currently appears to have infrequent involvement among clinical neuropsychologists. In fact, at present there does not appear to be a substantial literature within neuropsychology or a nationally recognized individual whose expertise could be drawn upon to address this subject. Perhaps the status of involvement in criminal activities will evolve and be worthy of inclusion in future writings on forensic neuropsychology.

Jerry J. Sweet

PART I

FUNDAMENTALS

There is a substantial and somewhat formidable body of technical information that forms the foundation and bases of practice in clinical neuropsychology. Before consideration of forensic practice, extensive delineation of these practice fundamentals is in order. In Chapter 1 Stanley Berent and Christine Swartz delineate essential psychometrics at the heart of formal neuropsychological assessments. W. Drew Gouvier goes on to emphasize the essential task of considering base rates, particularly low base rate conditions, in clinical decision-making in Chapter 2. Steven Putnam, Joseph Ricker, Scott Ross, and John Kurtz, in Chapter 3, discuss the omnipresent issue, and latest methods, of determining premorbid function.

Chapter 1

ESSENTIAL PSYCHOMETRICS

Stanley Berent
University of Michigan and NeuroBehavioral Resources, Inc., Ann Arbor, Michigan

Christine L. Swartz
University of Michigan and NeuroBehavioral Resources, Inc., Ann Arbor, Michigan

Detecting central nervous system dysfunction

Clinical neuropsychology is a formally recognized specialty within the field of psychology. Neuropsychologists are involved in a variety of professional activities. The majority of these activities are concerned with behavior as it relates to the central nervous system (CNS). Since the CNS regulates or is otherwise involved in virtually all human thoughts, feelings, and actions, it is not surprising that significant damage to the CNS is often reflected in behavioral symptoms. Recognizing this, a number of fields have sought to approach the problem of CNS diagnoses by 'tracing back' to the CNS from the behavioral symptoms.

---

*This work was funded in part by a SPHERE (Supporting Public Health and Environmental Research Efforts) gift award from the The DOW Chemical Foundation and NIH grants (NS 15655, AG 08671 and AG 07378).

The authors wish to thank Jacques Maurissen, Ph.D. and Jerry J. Sweet, Ph.D. for their careful review of an early version of this manuscript and their very helpful advice.

Sometimes, this may be the only way to reach a clinical diagnosis, given that the CNS is not easily accessible to direct observation. In addition, scientific investigation has not yet completely established the clinical significance of what is now able to be observed through recent technological advances. For example, techniques such as positron emission tomography (PET), electroencephalography (EEG), and magnetic resonance imaging (MRI) have allowed us to image chemical, metabolic, and other functional and structural aspects of the CNS. In some cases, we have been able to observe certain characteristics of these variables that appear to coincide with specified disease states (e.g., Adams et al., 1993; Berent et al., 1988). Nevertheless, in many instances, our understanding of the relationship between these images and a specific disease process is insufficient to make a definitive clinical diagnosis based solely on such information. In other cases, the abnormal behavioral condition appears to manifest even in the absence of underlying structural or functional abnormalities that are observable through existing techniques (e.g., Bigler & Snyder, 1995).

To effectively use behavioral information to diagnose underlying CNS damage presents noteworthy challenges as well. First, there is the problem of knowing that a given behavior is, in fact, abnormal. From a cause and effect viewpoint, the suspicion of altered behavior is raised whenever there has been some insult or possible insult to the CNS. An example is represented by the person who has been involved in a motor vehicle accident. The person may have sustained a blow to the head, a neck injury, or simply not remember the details of what happened in the collision. Has this person suffered a behavioral consequence as a result of trauma to the CNS? This is a common and reasonable question. From the same linear thinking, the identification of a behavioral problem may lead one to suspect that there may be some underlying CNS damage that is explanatory. To give a clinical example, a person with previously diagnosed mental retardation shows a dramatic change in behavior, becoming aggressive and destructive, whereas, heretofore he has shown a more docile demeanor. Does this behavior change reflect progression of neurological dysfunction? In clinical practice, such concerns translate into what are seen as complaints, or symptoms. Bothersome to the individual, or someone close to the individual, either because of their nature or because of the accompanying stress, these concerns are viewed as problems that deserve professional attention from a neuropsychologist. From a clinical diagnostic viewpoint, however, the patient's complaints alone do not provide the entire story. To understand the individual's clinical picture, it is necessary to determine objective signs that substantiate and more fully elucidate the nature of these symptoms in terms of CNS dysfunction.

The neuropsychological evaluation

In general, the neuropsychological evaluation consists of identifying the existence of an impairment, describing the problem as it relates to the nervous system, establishing the etiology when possible, and recommending further evaluations or treatment. Guided by the expressed complaints, the first question to be formally addressed by the neuropsychologist concerns the presence or absence of abnormality. While such a question is reasonable and follows logically from the tendency to think in linear, cause and effect terms, it seems deceptively simple. To answer it requires a knowledge of normal as well as abnormal behavior, including issues of individual differences. That is, people vary one from another, sometimes substantially, along a continuum for any given behavior. Once the complaint picture has been defined, a series of clinical questions is asked by the neuropsychologist. Some of these include the following:

Is the person presently functioning normally or abnormally?
What is the specific nature of the abnormality?
What is the level of severity?
Does the observed abnormality represent a change from historical baseline?
Is it acute, subacute, or chronic?
Is it progressively worsening, static, or resolving?
Does it reflect focal or general neurological involvement?
To what extent are motivation, depression, and other non-neurological factors involved?
What is the functional significance for other aspects of the individual's life?
What are the implications for diagnosis, treatment, prognosis, and etiology?

Although the neuropsychologist may employ a variety of clinical approaches to address the questions just listed, including the use of historical information and data from imaging and other methods of inquiry, a primary method rests on psychometric theory. The term *psychometrics* literally refers to the psychological theory or technique of mental measurement (Webster, 1984); neuropsychological testing is an application of such theory and technique to the clinical enterprise. The application of psychometrics enables the neuropsychologist to achieve a scientifically-based, objective and systematic approach to establishment of signs that substantiate and elucidate the patient's clinical status. Data from neuropsychological tests are thus employed to address the questions posed above.

The evaluation of an individual patient differs from experimental research, in which one could examine "experimental versus control groups", "pre-treatment versus post-treatment", "dose-response", and other scientifically-controlled manipulations or neuroepidemiological investigations that are used to identify differences between designated groups. Such group differences may be useful in establishing clinically relevant variables and in furthering our understanding of a clinically important topic area, but may be of unknown clinical significance with regard to an individual case.

For patients with suspected nervous system dysfunction, regardless of cause, the neuropsychological evaluation begins with the patient's description of the chief complaints and a review of medical, psychosocial, occupational, school and other histories. The psychometric examination follows. Information from the history (symptoms) and the examination (signs) is used to identify the abnormality and to formulate a diagnostic impression. Factual historical information (e.g., school performance records) is obtained, whenever possible. Treatment may be recommended or additional consultations suggested after the resultant information is reviewed.

A systematic and objective approach that is based on sound principles of science is critical to the clinical enterprise because of the inferential nature of the information and potential lack of precision at every level of involvement. For example, the patient may mislabel a complaint as a memory deficit, when in fact he or she is experiencing a problem with attention or some other aspect of cognition. A family history of dementia may lead a patient to wonder if he or she is suffering from a decline in memory, when that is not the case. On the other hand, problems may exist when not obvious to the patient. From the professional side, there are relatively few specific nervous system responses to trauma or other diseases, and etiology usually cannot be established by association alone without violating basic clinical and scientific principles.

The psychometric approach is also complicated by the fact that behavioral measures are not generally specific to a given kind of neurological disorder, and almost never exclusively to the CNS. For example, decreased memory may result from a toxic encephalopathy, although an identical complaint may be associated with a clinical depression or a progressive degenerative disorder, such as Alzheimer's disease. A complaint of memory impairment alone, whether made by self-report or by someone close to the patient, would be insufficient to establish a definitive diagnosis. Information obtained from the patient directly or indirectly from family members or others can be very important, but the clinical evaluation of patients with suspected neurological disease should include a balance of established clinical neuropsychological measures and lead to a process of differential diagnosis. In the evaluation of individuals exposed to potential neurotoxins, for example, the psychologist is responsible for identifying problems directly related to exposure, as well as pre-existing problems that may or may not have been aggravated by exposure. The clinician is also responsible for identifying disorders in which symptoms mimic neurotoxic damage, but are, in fact, unrelated to an environmental toxic etiology. In establishing a diagnosis of neurotoxic disease, interpretation of abnormal findings must account for findings that are unrelated to toxic exposure such as pre-existing pathology (e.g., a previous head injury) or normal physiological/psychological variation (e.g., low intelligence level or non-clinical depressed mood). It is the clinician's responsibility, in short, to arrive at an objective reality concerning the patient's symptoms – their etiology, or etiologies, the anticipated course and suggested treatment(s). Despite the cautions and limits just noted, the psychometric method lends itself well to this demand.

Assessment is a dynamic and interactive process between the examiner and the examinee. Traditionally, the procedure in quantifying behavioral aspects of human function consists of an examination that employs psychological tests, variously referred to as psychological or neuropsychological testing, or even psychometric evaluation. To quote from Anastasi (1976), "A psychological test is essentially an objective and standardized measure of a sample of behavior"

(p. 23). The term "test" has been used carelessly at times. From the psychologist's point of view a test, in its formal sense, is a task or set of tasks that has been studied and found to meet certain psychometric criteria. This is no small point. In forensic or other situations, evaluations are sometimes presented with an intimation that only "tests" have been employed, when the technical demands that define "test" have not been met by each of the procedure(s) described. That is, a procedure or device being described may have sampled behavior, but reflects neither standardization nor objectivity in a manner prescribed by the profession. These demands for standardization and objectivity have been formalized as "standards" (APA, 1985), and expectations to meet these standards have been placed by the profession on the developer of such tests and on the test user. Consumers of psychometrically-derived information should also become sophisticated about some of the basic principles, as herein presented.

Psychological and neuropsychological tests serve as tools for the examiner. Each test carries a unique set of desired attributes and capabilities, as well as limitations and sources of error. The role of the neuropsychologist as a "test user" has been noted to include the responsibility for selecting the particular test, overseeing administration and scoring of the test, and interpreting the scores within a theoretical context (Anastasi, 1992).

Psychometric properties of assessment instruments

There are basic psychometric criteria that must be met in order to formally qualify a behavioral measure as a neuropsychological test. These criteria are met through quantitative research. Before moving to a discussion of the criteria themselves, a brief overview of relevant statistical and measurement considerations is presented below.

Statistical and measurement considerations

Correlation methods are important to the establishment of psychometric criteria. Most mathematical methods for determining formal correlation values derive from the work of Sir Francis Galton and Karl Pearson in the middle to late 1890s. Correlation values differ according to the specific measurement considerations in a given test situation. While it is beyond the scope of the present chapter to go into the technical aspects of statistical tests such as correlation, it should be mentioned that correlation is usually expressed as a coefficient of correlation, with the statistical symbol, "r". The value of r represents the degree of relationship that exists between two variables (e.g., two different distributions of scores). The mathematical square of this value ("r^2") reflects the proportion of variance that is shared between the two variables. The correlation value is expressed in a numeric range from -1, through zero (0), to $+1$, where zero indicates absence of relationship, -1 indicates a perfect negative correspondence, and $+1$, a perfect positive relationship. Relative negative and positive relationships are indicated by values between the two extremes. While no causal implications derive from correlation measures, finding that two factors significantly vary together in linear or curvilinear fashion indicates an element of commonalty between the two and leaves one highly predictive of the other. These statements are true even though the correlation finding does nothing to identify the exact element of commonality.

The term, "significant", is used frequently in the field, and a number of times within this chapter. Here, we are referring to statistical significance, a technical expression used to denote that a finding is of sufficient magnitude to rule out, at specified confidence levels, chance as a probable explanation. To be specific, if a face valid item, such as "Are you depressed?", is found on formal statistical study to correlate significantly with an independent measure of depression (e.g., a history of psychiatric hospitalization for depression), evidence has been identified for a greater than face valid level of validity. On the other hand, should an insignificant correlation result from such study, validity beyond face validity has not been established. The correlation studies required to establish a test's validity, as well as other test criteria, are conducted formally in the process of test construction. Like all scientific research, the results of these studies should be replicated, published, and made available for professional scrutiny. It is unfortunate, but important to remember, that a task is sometimes presented to the consumer with the implication that the scientific and mathematical requirements for establishing psychometric criteria have been met when, in fact, they have not.

All scientific findings are probability statements, and psychological test findings are no exception. In regard to correlation, the coefficient represented by the statistic, r, will always be something less than $+1$, a perfect correspondence between two variables. This situation results from a number of factors, not the least of which relates to the topic of individual differences. Not only do people vary from one another in their performance on a given task, a given individual may vary across time. This psychological fact means that a test can meet prescribed psychometric criteria only relatively. Given this, one might ask, how strong a relationship is needed to establish the test criteria? The answer to this question is somewhat arbitrary. While statistical tests are available to determine the statistical significance of a correlation, statistical significance alone may not be sufficient to establish clinical significance. In the clinical setting, the interest is on the individual as opposed to a research setting where group differences alone may be the concern. Some workers have proposed specific levels to be used for determining that the test criterion has been reached. Rosenthal and Rosnow (1991), for example, suggest that acceptable reliability coefficients derived from criterion studies should fall above $+.85$ for the instrument to be clinically useful as a psychological test. Regardless of the level of r, the statistical information resulting from the test's development should be made available to professional test users, and is usually stated clearly in the published manual or in other readily obtainable published material. This allows the psychologist to ascertain the likely validity and reliability of the test, as well as the limits that must be placed on clinical conclusions.

The specific statistical test employed in test development depends to some extent on the nature of measurement in a given instance. Since neuropsychological tests are almost always quantitative, that is, resultant scores are expressed numerically, it is worthwhile to say a few words about the nature of measurement. The importance of quantification in neuropsychological testing can not be overstated. It is stated in this chapter that the neuropsychologist employs a clinical approach that is scientifically based, objective and systematic. Quantification helps to define this neuropsychological approach to clinical problems. Quantification is compatible with, and allows for, a scientific approach in clinical practice as well as when tests are employed for purposes of research.

Levels of measurement

In general, there are four types, or levels, of measurement – nominal, ordinal, interval, and ratio (Minium, 1970). *Nominal* measures are those that classify responses into one or more categories. The categories differ one from another only qualitatively. An example of nominal measurement would be handedness, left or right. *Ordinal* measurement allows for classification into a number of categories with a ranking in terms of order of magnitude. There is no absolute zero point to such scales, however, and only relative comparisons can be made (e.g., one ranking is higher than another). Many neuropsychological tests use an *interval* level of measurement. Here, there is a set unit of measurement and real numbers are assigned to indicate various levels that reflect relative amounts between points on a scale. An example would be a thermometer marked off in degree units. Another example is a distribution of IQ scores. With interval data, mathematical operations such as multiplication and division are not meaningful because of the absence of a true zero point. The highest level of measurement is *ratio*, which differs from interval measurement in that there is an absolute zero point. This allows for a meaningful ratio between scores. A score that is based on a ratio level of measurement is known to reflect some true multiple of another score on the same scale (e.g., a score of 100 is twice that of 50). Examples of ratio measurement are weight, height, or length. The level of measurement on a given test determines in part the types of statistical tests that can be employed in their study. The level of measurement also determines what can and cannot be concluded on the basis of test results in a given case. For example, one can conclude on the basis of an intelligence test score (an interval measure) that someone is functioning better than a specified percentage of the referent population (a percentile ranking), but one cannot conclude that a person with a given score is twice as bright as some other person.

In the following paragraphs, four essential psychometric test criterion areas will be described. These include validity, reliability, procedural standardization, and availability of normative data. At the least, a neuropsychological test must be shown to meet these basic criteria.

Validity

A test will reflect, through published research, the established "validity" of the instrument to measure what it purports to measure. Validity is a technical aspect of test development, and it is established through formal mathematical methods.

As stated in the standards, validity is the most important consideration in test evaluation as it refers to the, ". . . appropriateness, meaningfulness, and useful-ness of the specific inferences made from test scores" (APA, 1985). Validity can be examined in a number of different forms. One of the most straight-forward areas is *face validity*. Face validity refers to the overt meaningfulness or relat-edness of an instrument. For example, individual questions, such as "Are you depressed?" would reflect a high degree of face validity for the assessment of depression, in that the apparent purpose of the question is directly related to the desired inference.

Face validity can be deceptive in its simplicity and is not sufficient to quali-fy an instrument as a formal psychological test. Although the face validity of a test item may leave it a good candidate for inclusion in the final instrument, other and higher forms of validity must derive from formal statistical inquiry. Through a formal statistically-based study, an item that on the face appears to relate to an underlying theoretical construct may be found to have very little relationship to that behavior. On the other hand, some seemingly unrelated items may be found to strongly vary with the construct of interest. The face validity of an instrument can affect the respondent's motivation and interact with the patient's response set. For persons with a "malingering" and "naive" response set, particularly high inflation in symptoms and areas of deficits may be most apparent on tasks with high face validity, but less evident on tasks with low face validity. From what has just been said, it should be obvious that cor-relation methods are important to the establishment of a test's actual validity.

More complex aspects of validity are often discussed in terms of issues of content, criterion, or construct validity. *Content validity* refers to the cover-age and representativeness of a particular measure for a specified area of inter-est. In order to adequately assess an area, a test must sample relevant and representative behaviors and skills. For example, in assessing memory func-tioning, relevant content areas may include tasks of verbal memory, visual memory, recognition memory, free recall, and learning over repeated trials. During test construction, authors may request that experts generate a list of all of the domains or content areas that a test might be expected to cover. The entire test may then be later reviewed to see if it provides adequate coverage of clinically relevant areas. In contrast to the mathematical requirements for criterion and construct validity, content validity is typically established based on expert judgment.

Criterion validity of a test refers to the strength of the relationship between the test and another, independent criterion. This has often been further differ-entiated to reflect either concurrent or predictive aspects. For example, the *con-current validity* of a measure of reading comprehension might be established by correlating performance on that measure with a score on a concurrently admin-istered educational achievement test. These correlations are most often report-ed for the test in comparison to the current standard of practice. Other measures of intellectual assessment are often compared to results of Wechsler-based tests, using the latter as a "gold standard" to establish the validity of the former. As already suggested, at times a test with previously and well established validity might be used as a standard against which to compare the test under development. This is an acceptable practice, and one shared by many fields, but necessitates unequivocal scientific proof that the criterion instrument itself is valid. It should also be kept in mind that, according to Anastasi (1976), the resulting new test remains at best an approximation of the old, criterion test. In addition, there should be some compelling reason to develop the new test (e.g., it is shorter, simpler, more modern in context, or more economical).

Predictive validity refers to the relationship of the current test score to an outcome criterion in the future. Measures of intellectual performance are often correlated with future academic performances or school grades. Of note, measures of criterion validity are only as strong as the criterion against which they are measured. Therefore, a perfect relationship between school grades and earlier intellectual test performance would not be anticipated, given the multiple other factors that can affect grades.

A test is typically designed to measure a particular construct or theoretical aspect of behavior. A few examples of such constructs are anxiety, depression, intelligence, and memory. The extent to which a given test is shown to relate to such constructs is termed, its *"construct validity"*. A construct, ideally, should have clear boundaries and definitions. In theory, certain behaviors or skills would be anticipated to be related or unrelated to a selected construct. For tests, this is assessed by measures of convergent and discriminant validity. To demonstrate construct validity, test scores should be strongly associated with other measures of similar abilities and should exhibit minimal to no relationship with extraneous concepts. The multi-trait/multi-method, introduced by Campbell and Fiske (1959), is designed to examine these patterns of relationships. According to this methodology, two or more traits or behaviors are examined utilizing two or more methods (e.g., self-report inventory, observation ratings, test performance). This results in a matrix containing four types of correlation coefficients: monotrait-monomethod, monotrait-heteromethod, heterotrait-monomethod, and heterotrait-heteromethod. *Convergent validity* is established when measurements of the same trait have a high correlation (e.g., high monotrait-heteromethod coefficients), regardless of the type of method used. *Divergent validity* is established when a low correlation (e.g., low heterotrait-monomethod and low heterotrait-heteromethod coefficients) is found between different traits, even when employing the same method of measurement.

Issues of discriminant and convergent validity may also be reflected in discussions of a test's *sensitivity* and *specificity*, and may be examined with regard to a particular diagnostic question. This method has been commonly used in medicine to characterize diagnostic tests (e.g., Kramer, 1982); within psychology, sensitivity of assessment instruments also has been examined. Sensitivity refers to an instrument's ability to detect signs of behavioral dysfunction when it is present. For example, IQ measures may not be sensitive to early signs of dementia. Sensitivity has been represented mathematically as a ratio of the true positives to the sum of true positives and false negatives

(Kramer, 1982). In contrast, specificity refers to the "uniqueness" of particular scores, meaning that the test does not report an effect when it is absent. Mathematically, specificity has been represented as the ratio of true negatives to the sum of true negatives plus false positives (Kramer, 1982). For example, if Category Test scores are "sensitive" to and able to detect frontal lobe dysfunction, are they "specific" to only frontal dysfunction or can impaired test performance be related to multiple other factors? To give another example, the spelling subtest of the *Wide Range Achievement Test-Third Edition* (Wilkinson, 1993) is very sensitive in measuring a person's difficulty with spelling, but the resulting scores are not specific to any particular underlying disorder. A low score on this test could reflect neurological based disorder, but it might also reflect educational failures or even a lack of motivation. A weakness of many psychological tests is that failure on a task may often be linked to multiple possible difficulties. Issues of specificity have been less carefully examined for psychological tests. However, sensivity and specificity are interdependent concepts and cannot be dissociated. Moreover, interpretation of sensitivity and specificity rates is also dependent on knowledge about the base rate of occurance for a particular disease or symptom (Somoza & Mossman, 1990). For example, a very high false positive rate can occur when the prevalence of a particular disorder is low.

Beyond the strictly mathematical aspects of validity, Cronbach (1988) identified alternative perspectives of validity, with concepts such as *functional validity*. Others have referred to "ecological validity" in this same context (e.g., Wilson, 1993). For functional validity, the "worth" of a test would have higher value than its "truth." Even if a test can validly measure memory functioning, how do scores on that measure relate to behaviors, such as remembering to take medications or remembering a particular work-related task? The relevance of test scores to real-world behaviors and performances might carry greater weight in clinical practice than the relationship of test scores to similar theoretical constructs and theoretical measurements (see Chapter 8).

Sources of bias
Even though a test can exhibit adequate validity, individual performances on a test can be affected by multiple factors which may compromise the accuracy or meaningfulness of a particular set of test scores. These factors can include characteristics of the test, such as clarity of directions, degree of novelty, degree of cultural bias. Sources of test bias, such as cultural factors, can be particularly important issues to consider. Legislative action, as in the case of Larry P. versus Riles (1979), found that the use of standardized achievement test scores for determination of mental retardation reflected a bias against minority children. The tests were not considered to have the same level of "meaningfulness" or validity in minority cultures, given the influence of previous educational experiences and cultural factors on test performances.

These factors may also interact with characteristics of the individual test taker, such as skill level, anxiety, motivation, physical limitations, degree of bilingualism, deficiencies in educational opportunities, or unfamiliarity with testing situations (Deutsch, Fishman, Kogan, North, & Whiteman, 1964). Motivational factors of the test taker may be particularly salient in forensic neuropsychology. Persons involved in litigation may reflect a unique subpopulation with differing characteristics, including base-rates of reported symptoms (e.g., Dunn, Lees-Haley, Brown, & Williams, 1995; Lees-Haley, 1992). In addition, issues related to inappropriate standardization samples, examiner language bias, inequitable social consequences may have impact on the validity of an instrument for a given person (Reynolds & Brown, 1984). Interpersonal variables, such as differences in age, socio-economic or cultural background, may affect motivation for testing, response set, or testing taking approach. For example, some research has suggested that the "tempo of life" traditionally associated with a cultural group may affect "speed" or quickness of response to test items (Hinkle, 1994). Members of minority cultural groups may not experience the same task demands in testing situations, and this differential motivation for a task may confound the results of their performance (e.g., Brescia & Fortune, 1989; Loewenstein, Arguelles, Arguelles, & Linn-Fuentes, 1994). These moderating variables may result in a different validity coefficient for a particular subgroup and should be reported in validation studies.

Reliability

Reliability refers basically to the degree to which test scores are free from errors of measurement (APA, 1985). Put another way, reliability is an indication of the test's consistency, between items in a given administration or between two or more administrations or ratings of the same test. A test's reliability establishes the upper limit for the test's validity. In other words, an unreliable test cannot be valid. While the user of a test is not generally required to independently establish a test's reliability, the user does need to know to what extent differences between forms or administrations of a particular test reflect errors of measurement versus the effects of disease progression (e.g., metastasis of a malignant tumor), or other clinical event. It should be mentioned that every test score includes some error. This error is expressed as the error of measurement and should be specified by the test publisher in the published manual. The error of measurement is used to construct intervals for a particular score. It is the test user's responsibility to consider this error in reaching conclusions regarding a given score. For instance, the published error of measurement on a commonly used intelligence test is ± 5 points. Thus, a score (s) derived on any given occasion would actually be $s \pm 5$. If $s = 100$, it is likely that the actual score is 95–105. If the test is administered in a given case on two separate occasions, yielding scores of 95 and 102, respectively, these results would be considered in practice to be the same. As mentioned elsewhere in this chapter, all conclusions that are based on psychological test findings, as in all scientific enterprise, are probability statements, and this includes error of measurement. This error score will vary depending on the exactness required in a given instance. While professional convention leads to the use of ± 5 points in the specific example above, an error score of ± 13 points would be required to attain 99% accuracy.

Reliability of an instrument is often reported in terms of test-retest, alternate form, split-half, and other measures of internal consistency. As in the case with validity, reliability is established through formal scientific inquiry and is represented by correlation analysis. *Test-retest reliability* reflects the stability of a particular test score over time. The actual time interval between administration of the test may be particularly important. If too close together in time, artificial inflation in the correlation might be expected on the basis of practice effects (i.e., the effects of test familiarity on subsequent test performance). The relationship between performances on *alternate forms* of a test may also help assess the stability of test score, although some practice effects can be seen between alternate forms of the same test (Anastasi, 1976). If the elapsed interval between test sessions is too long, there are likely to be intervening variables, such as developmental issues, historical events, etc., which can confound the relationship between the two measurements. Methods of assessing *internal consistency* include Cronbach's coefficient alpha (Cronbach, 1951) for tests with items with multiple choice responses (e.g., "always", "sometimes", or "never") or Kuder-Richardson's formula 20 (KR-20; Kuder & Richardson, 1937) for dichotomously scored items (e.g., "true/false", "yes/no"). Split-half reliability, which examines the intercorrelations among all test items and estimates the potential effect of shortening or lengthening the instrument, is often measured with the Spearman-Brown formula, which was reported by Charles Spearman and William Brown in 1910 (Walker & Lev, 1953). On tasks that primarily assess "speed", as opposed to accuracy of response, however, measures of internal consistency are not meaningful.

Some tasks used in test construction employ items that must be judged by raters in setting the standard for test scores. For tasks with these subjectively rated criteria, *inter-rater reliability* is particularly important. For example, there may be significant variability between individual scorers' assessments on items of the Rey-Osterrieth complex figure, whereas assessment of "right" or "wrong" on a task such as the Halstead Category Test should be without disagreement. Anastasi (1976) suggests using different reliability measurements to parcel out "true" variance in the construct from test error by interpreting test-retest as variance due to time, alternate form (immediate administration) as variance due to content, and alternate form (delayed administration) as variance due to both time and content. She also suggested that split half measurements reflect variance due to content sampling, whereas KR-20 and Cronbach's coefficient alpha reflect variance due to content sampling and heterogeneity, and inter-scorer reliability reflects variance due to examiner and style of individual administration. These considerations should be attended to and resolved in the process of test development. The adequacy of the resolution should be considered by the test user in test selection.

Measurements of reliability can also be influenced by extraneous factors. For example, as the test length increases, reliability measurements often increase. Reliability can also be influenced by the variability in scores or a restriction in range of scores. When the distribution of scores is restricted in range, the reliability coefficient will be lower. As previously noted, studies of test-retest reliability are affected by the length of the interval between administration, which may be influenced by practice effects as well as by real learning or maturation. External factors, such as an individual's attempts at guessing, malingering, or other variations in the test situation, may act as confounds in assessing reliability.

Other issues in test construction and utilization

Standardization
To be formally termed a "test", a measurement device must be "standardized". That is, a procedure for administering and scoring the test needs to be specified. The concept of standardization is important in psychology as in science more generally, and its importance in the measurement of behavior can not be overestimated. Standardization applies to all aspects of the test, including such seemingly extra-test considerations as instructions and examples given, timing of stimulus presentations, how to respond to the subject's questions, the testing environment, and other details of the test situation. If the standardized procedure of administration is altered, the reliability and validity of the instrument is compromised. As already mentioned, standardization refers to uniformity in the administration of the test, but also extends to test scoring. Some tests are relatively simple instruments that require very little on the part of the examiner to insure uniform administration and scoring. In other instances, the test procedure or scoring may be complex and require substantial training for professionally acceptable use. In all cases, there are basic considerations of test administration that require formal training to be qualified as a tester. This area of practice is another that is sometimes abused, and the consumer of test information must determine that proper procedure has been followed in any given instance. Published guidelines and texts address these issues (e.g., *The TCN Guide to Professional Practice in Clinical Neuropsychology*, Adams & Rourke, 1992; Division 40's guidelines, 1989).

Normative data
Finally, a test is "normed". That is, there is some prescribed method for relating the test scores to a representative population sample. Norms do not remove the paramount importance placed on validity and reliability issues. They simply aid in understanding the meaning of a particular score. A relatively low score on a verbal learning task might be expected of someone with a life-long record of low intellect and academic achievement, for instance, whereas in another person it might reflect a cognitive impairment. Normative data are usually derived through formal research, wherein the test being developed is administered to a large group of subjects. Important factors to consider in evaluating the adequacy of a particular normative group include representativeness, size, and relevance. Results

from tests often generate standardized scores or a percentile rank, based on the normative sample, which allow comparison of an individual's performance across different tests. These standardized scores and percentile ranks are usually based on the assumption that the population's performance on this measure eventuates in a charateristic numeric distribution, usually approximating a normal distribution and creating a bell-shaped curve (see Figure 1).

The performance of the "normative" group determines the average performance anticipated for a given individual on the specified test as well as providing indication of "normally" expected variability in performance. It is through such normative testing that "raw" test scores earned by a given individual take on some clinically relevant meaning. By scaling the raw score, the psychologist is able to compare that individual performance against a meaningful referent group and determine the score's significance in the individual case. For example, most

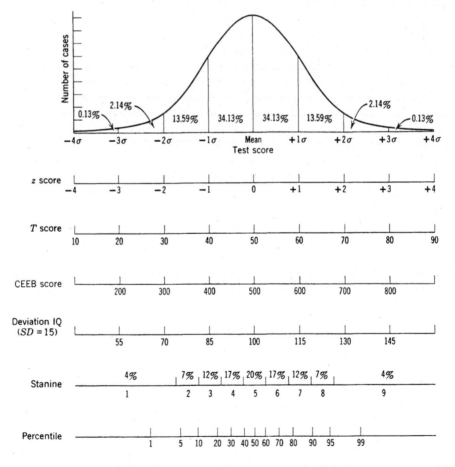

Fig. 1.   The normal distribution curve, depicting standard deviations, corresponding percentiles and standard scores. (Reprinted with permission from the publishers [Anastasi, 1976]).

commonly used tests of intelligence have been normed against a sample that reflects an average score of 100, with a standard deviation of ± 15. The psychologist who gives this test to a person who obtains a score of 85, knows that this individual falls one standard deviation below the average and is functioning at the 16th percentile in comparison to the referent population (i.e., better than only sixteen percent of the referent population). As can be seen from this description, the adequacy of the test is often determined by the adequacy of the sample against which the test has been normed. Some measurements may also generate age-equivalent or grade-equivalent scores, and not all measured variables are normally distributed (e.g., relative handedness is not normally distributed in the general population). The reference group(s) will determine the shape of the distribution, as well as its average performance level and variability.

Extra-psychometric considerations

In addition to the essential psychometric properties of a test, there are a number of what can be termed extra-psychometric considerations that impact on formal test criteria and influence the veracity as well as usefulness of the test-based neuropsychological clinical evaluation. Some of these are discussed below.

Test selection
There are many published psychological tests to choose from and a variety of non-test procedures (i.e., tasks) as well. The designation, "task", is designed to distinguish these procedures from bona fide "psychological tests" in the technical sense. Tasks are useful, however, and provide an important component to the professional's evaluation. Tasks are commonly used in research, for instance, and many automated instruments that have been introduced in recent years will fall in this category. Simple and choice reaction time procedures, for example, lend themselves to computerization and have become important in studies of neuropharmacology and behavior. In the use of such instruments, standardization of procedure becomes as important as the task itself in determining usefulness and appropriateness in a given instance.

On what basis is a given test selected to be used in a particular case? There are several factors to consider in answering such a question. A test provides a sample of behavior, usually expressed as a collection of numeric data, and the neuropsychologist must consider how the test's results will relate to the patient's complaints and symptoms as well as to the referral question(s). The technical merits of the test must also be evaluated, and the methodology required for test administration must be considered. Finally, there are factors such as customary practice, training background, professional competencies and other practical considerations. Each of these factors will be discussed more fully below.

First, the capacity of a given instrument to generate data that will address the person's complaints or symptoms must be considered. This underlies (a) the importance placed on the referral and consultation phases of the neuropsychological process, and (b) the care needed in formulating questions that will be suitable for neuropsychological inquiry. "Suitability" is used here in much the same sense as it is used in any scientific inquiry. The question posed might be as follows: Are there observable behaviors that if measured will help answer the referral question(s)? Relevant variables become "operationally defined" according to performance on a given task. In many instances, more than one kind of data may be required to answer a given question. For example, a question about a student's readiness to return to a regular classroom setting following a period of hospitalization may require test data that reflect, among other things, past school performance (e.g., reading level), general level of ability (e.g., intellect), presence or absence of cognitive dysfunction (e.g., impairment of memory), and factors that may interfere with optimal performance (e.g., depression). Particular instruments may be selected on the basis of their ability to assess multiple areas of functioning, versus a more narrow-band instrument, which would assess a single skill. Other factors to be considered may be the age of the patient and the appropriateness of a measure across the age span.

Second, the technical merits of a given test instrument, from a psychometric point of view, will be evaluated. Are the available norms suitable for comparison to the present patient? Does the test measure the theoretical construct as needed to answer the referral question? (For example, is it valid?) Will the chosen test allow for a repeat examination at a future date, or are practice effects too great to allow for such re-testing when it is needed? Other considerations might include the availability of qualitative analysis of performances, as well as the use of objective scoring criteria. Given particular referral questions, the propensity of the instrument to minimize false positives or to minimize false negatives may also be particularly important (Boll & La Marche, 1992). Further, some tests may have internal validity indicators that affect the degree to which the obtained data can be relied upon. Since not all tests have built-in validity indicators, clinicians often administer tests that are specifically aimed at determining patient effort and cooperation. For example, some measures of effort have utilized a technique based on the forced-choice procedure known as "symptom validity testing" assess signs, and make quantitative estimates of, the probability of malingering or "faking bad" (e.g., Binder & Pankratz, 1987; Faust, Hart, & Guilmette, 1988; Hiscock, Branham, & Hiscock, 1994; Hiscock & Hiscock, 1989; Pankratz, 1983), which, if present, may obviate normal interpretive approaches to test data. Evaluation of insufficient effort, and specifically malingering, is discussed in greater detail in Chapter 9.

A third area to consider in choosing a test can be termed "convention". There are certain instruments that have been used so extensively that they have become standard components in most neuropsychological test batteries. A prime example of such a "standard" is the Wechsler Adult Intelligence Scale - Revised (WAIS-R) and its recent revision (WAIS-III), or its counterpart, the Wechsler Intelligence Scale for Children - Third Edition (WISC-III; Wechsler, 1991). The Wechsler Intelligence Scales, in fact, have been translated into many languages and have been "normed" to many cultures. These scales are used by neuropsy-chologists almost everywhere.

A fourth consideration that enters into test selection relates to the training background of the person engaged in choosing. From a different perspective, one could argue that there are only two factors that enter into test selection. Listed as the first and second consideration above, these factors are represented by (a) the data needed to answer a given referral question, and (b) the technical merits of a given instrument. Using this perspective, both convention and training bias-es would be viewed as merely extensions of the two factors just mentioned. That is, one continues to use, and continues to teach others to use, those test instru-ments that meet the requirements of situation and technical suitability.

Finally, there is another source of influence on test selection represented by the realm of what can be called practical considerations. This source is listed sepa-rately because it represents a set of extra test considerations. In many instances, one might idealistically wish that this final type of influence not enter into deci-sions on test selection. Tension might arise, for example, between (a) the desires of the fiscal department of a hospital to curtail costs in delivering clinical ser-vices, and (b) the clinical need as perceived by the practitioner to administer a large battery of tests to a given patient. The potential legal and ethical implica-tions that stem from this area are many. The increase in recent years of the involvement by the neuropsychology practitioner in institutional governance (e.g., admittance of psychologists to hospital medical staff membership) provides an important tool for ensuring the delivery of appropriate care while also meet-ing institutional fiscal and other administrative demands. Other practical con-siderations may include the portability of equipment or the availability of an alternate form of the test for re-test administrations (Boll & La Marche, 1992)

Test interpretation
In addition to test selection, the examiner is also responsible for accurately inter-preting and communicating test scores. Some issues that arise in this process are related to the determinants of impairment. Use of cut-scores, which are often based on discriminant functions in empirical studies to maximize group dis-crimination, need to be used with caution, as a certain degree of false negatives and false positives is anticipated. Moreover, examiners may also struggle with the lack of established and uniform consensus of what constitutes "average" versus "low average" versus "mildly impaired" performance on a particular measure. For example, according to Wechsler's classifications (Wechsler, 1981), a score might be described as "average", whereas using the descriptive cate-gories associated with Heaton, Grant, and Matthew's normative data (Heaton et al., 1986), the same score would fall in the below average range.

Anastasi (1992) has described four "hazards" in test interpretation. The first is the "hazard of the single score". Typically, a psychological test generates a numerical score that does not reflect the degree of measurement error present for that instrument. The second is the "hazard of the single time period". Traditionally, psychological assessments have looked at a patient's functioning in one given period in time. Intervening factors, such as fatigue or mood state, at the time of the assessment may affect test performance, which, in turn, would affect the "representativeness" of that particular performance for that person. The "hazard of the single indicator" refers to the proportion of variation that may be unique to a test versus the proportion of variation that may be shared between measures. Finally, the "hazard of illusory precision" suggests that the availability of numerical scores may engender an increased sense of quantification and objectivity. This effect may be seen on measures that are scored and printed out by computer, including narrative computer-based interpretations. Matarazzo (1983) also discussed this possible "halo effect" with regard to the use of computerized testing. That is, since many people currently associate technology with conceptions of "precision" and "objectivity", the use of computers in testing may mistakenly be taken as evidence of the test's validity, reliability, and other aspects of psychometric acceptability. The incorporation of technology, no matter how advanced, into the testing enterprise holds no assurances regarding a test's usefulness. As with any other instrument, a computerized test must go through the rigorous process needed to establish the required psychometric criteria (Adams & Rourke, 1992; American Psychological Association, 1986).

In the early days of applied neuropsychology, determination of brain damage, or at least dysfunction as a sign of brain damage, was felt to be of primary importance. Arguments often centered on whether level of function, or dysfunction, on a given task was referable to a specific brain location or more generalized cerebral mediation (Russell, 1986). Today, neither a strict localizationist nor functional system point of view prevails. The brain has been accepted as an organ that has focal areas of importance for performance on specific tasks, but also one that is interactive within itself and between its various component structures and other organ systems as well. It has become increasingly clear that the information derived from the neuropsychology examination is clinically valuable on levels other than the identification of a lesion. The standardized, quantitative, and sensitive nature of the neuropsychological procedures, for example, allows for baseline comparisons with test results obtained at a later time from a given patient. This test-retest method allows for determination of progressive deterioration or even improvements in function, over time (Berent et al., 1987; Thompson & Trimble, 1981).

Assessment domains

Behavior can be divided into a number of areas, and tests have been devised to measure these various domains. These categories are not entirely arbitrary, since arguments can be made about the relative exclusiveness of each domain, in addition to their interactive qualities. In addition to an objective appraisal of the patient's history, some of the most important areas to measure include: intellect, language, cognition (e.g., learning, memory, problem solving), attention, sensation, perception, motor and psychomotor behavior, affect (e.g., anxiety, depression, etc.), and personality (e.g., coping style and effectiveness as well as presence and type of psychopathology). Advances in neuroscience in the past few decades have taught us that referral questions pertaining to establishing neurologic dysfunction are complex and often require a comprehensive examination to address. This latter statement is true for psychometric reasons, as well as because the human brain has a myriad of functions. When a different kind of referral question is posed, the nature of the behavioral data needed for its answer will also be different, as will the construction of the test battery. In this manner, the neuropsychologist examines the pattern of an individual's performance, in order to assess convergent (or supporting) findings and discriminative (or differentiating) evidence that contributes to a differential diagnosis. The domains introduced below will be discussed in greater detail in Chapter 4.

Patient history
Important aspects of the patient history include information about past scholastic performance, occupational history, family and social behavior, medical history, as well as a review of present symptoms and presenting complaints. Attempts have been made to formally standardize the patient history. Most often, however, this area of the examination takes place by interviewing the patient and the patient's family or acquaintances, when available, and, importantly, review of past records (e.g., school achievement records; records of past hospitalizations, etc.). The interview is usually semi-structured so that information that is known to be pertinent to function will be obtained, but open enough to allow for the report of unexpected items. In many instances an outline may be followed. There is no universally agreed upon format for this portion of the examination, although there is general agreement about techniques of the interview process (e.g., techniques of effective communication), rating scales to use when appropriate, and the major areas to be covered.

Intelligence
Intelligence has a relatively long history as a subject of psychological inquiry, and this aspect of behavior has become a common feature of most neuropsychological evaluations. The intellectual measurement often provides a foundation for interpretation of test results from other domains of behavior. The intellectual portion of the examination is almost always covered in a formal, psychometric way through the administration of psychological tests. The Wechsler scales of intelligence, or derivatives from these scales, are the most widely used instruments for assessing intellectual function (Harrison, Kaufman, Hickman, & Kaufman, 1988). It is well established that intelligence is multifactorial, with some factors being more susceptible to the impairing effects of damage to the brain than others. It is also accepted that general intellectual function can be reflective of educational and socio-cultural history. A further distinction is drawn between intellectual ability as generally measured by psychological tests and demands for immediate response to the environment as occurs in a cognitive act such as learning new information. This separation between intellectual ability and other cognitive functions has been well

documented in research and clinical literature, beginning with some of the earliest investigations of this sort (Berent, 1991).

Language

Language function is related to intelligence as well as to other aspects of cognition. This function is often considered in terms of expressive and receptive language. Measurement of expressive language might focus on spoken vocabulary, spelling, articulation, verbal fluency, or object naming. Receptive language may be reflected on tasks such as picture vocabulary, auditory discrimination for verbal or rhythmic sounds or comprehension of verbal directions. Vocabulary level, to give a specific instance, is widely considered to be an aspect of general intellectual ability. Indeed, the vocabulary subtest of the Wechsler Intelligence Scales (Matarazzo, 1972) has the highest correlation of all the subscales to the overall score on this test. On the other hand, verbal fluency is more likely to be measured during the cognitive part of the examination and can be found to be impaired even in the face of an adequate level of vocabulary.

Cognition

Cognition is a complex topic, but in its simplest manifestations includes measurement of learning, memory, attention, aspects of language and other related behaviors. This portion of the examination often reflects an area of primary focus for the neuropsychologist. The questions to be addressed will influence the extent to which this area is covered, but even simple questions may necessitate a relatively lengthy examination process. Certain tests have come to be widely used in effecting this part of the examination. Halstead's battery (Halstead, 1947) and Reitan's (Reitan, 1955) and its revisions by others are examples of testing approaches that devote considerable attention to measurement of cognitive function. The Wechsler Memory Scale (Wechsler, 1945) and its later revisions have come to be used extensively to quantitate as well as to make qualitative statements about learning and memory. Much has been learned about the complexities of cognition, the various manifestations of processes such as learning and memory, and their relationships to disease induced dysfunction and treatment outcome. It is well established, for instance, that verbal and visual learning are affected differently depending on the nature and location of an injury to the brain (Berent, 1991).

Sensory and motor

Many disorders can manifest in sensory or motor symptoms with little or no apparent involvement of other aspects of behavior. Motor symptoms can also interfere with adequate performance on other aspects of the examination. For example, the present author and colleagues examined a group of patients who had been diagnosed as having familial and in some cases sporadic olivopontocerebellar atrophy, a progressive disorder affecting primarily the subcortex of the brain. At first and in comparison to non-affected control participants, the patients appeared to be lower functioning on standardized measures of intellect and cognition. These differences were found to be non-significant, however, when education level and motor dysfunction were statistically considered (Berent et al., 1990).

Aspects of function that are regularly evaluated in this portion of the examination include motor strength, speed, and steadiness. Handedness and body sidedness more generally can be placed here. Adequate sensation in hearing, vision, and tactile ability is also attended to, as is evidence of imperception phenomena, often through methods similar to those of the clinical neurologic exam. Effort is made to standardize the procedure and quantify the results whenever possible. It is in this area of the functional behavioral examination that automation and computerization has found some of its most precise applications (e.g., automated reaction time devices have been commercially available for years).

Affect

The term affect is being used here to denote emotions of anxiety, depression and excitement. An effort is usually made to quantify these emotional states through the use of rating scales, self report inventories, or other psychometric devices, although in some instances data from clinical interview might be used solely in forming conclusions regarding affect. Distinctions are drawn between pathological expressions of these affects and normal variations in mood. Regarding forensic use, it should be noted that some self report inventories used to assess emotional states do not have internal validity indicators, and thus may be vulnerable to *impression management* (i.e., intentional under- or over-reporting to portray a specific clinical impression).

Relevant professional issues

The latest edition of the APA standards for testing (1985) makes note of the fact that these standards are addressing a field that is evolving. One could add that it is a field that, by design, is continually evolving. In terms of the tests themselves, the standards handbook lists a number of areas that are in need of continued work. Some of the more important of these include: (a) the further development of gender-specific and combined-gender norms, (b) an effective way of dealing with cultural bias in test outcome, (c) further development of computer-based test interpretation, (d) limits on validity generalizations from one test situation to another, and (e) differential predictions based on test results. These challenges will be among those that receive focused attention in the coming years. Dealing successfully with these challenges will require adherence to the essential psychometric principles that underlie the testing enterprise.

Summary

In this chapter, we presented essential principles of psychometric theory as they relate to the neuropsychological examination. The chapter began with a discussion of the various approaches to detecting dysfunction in the central nervous system (CNS), the rationale and some strengths and weaknesses of these methods. Because the CNS usually is not accessible to direct clinical observation, CNS dysfunction is often determined by analysis of behavior. In this realm, the neuropsychological evaluation makes a unique contribution by providing an alternative to clinical evaluation of cognitive and emotional functions and other complex behaviors that would otherwise be based solely on report of symptoms (i.e., self-complaint) or crude mental status appraisal. The neuropsychological evaluation seeks, in general, to determine the existence of an impairment, to describe the problem as it relates to the nervous system, to establish the etiology when possible, and to recommend further evaluations and/or treatment approaches. Specific and typical questions that are addressed by the evaluation were listed.

The neuropsychologist employs a variety of clinical approaches, including the use of historical information and data from imaging and other methods of inquiry. A primary method used in the neuropsychological evaluation rests on psychometric theory and involves the administration of formal tests. The application of psychometrics through testing enables the neuropsychologist to achieve a scientifically-based, objective and systematic approach that is capable of establishing signs in the form of test performances that substantiate and elucidate the patient's clinical status. Since mathematics is intimately involved in psychometric theory, a brief overview of relevant statistical and measurement principles was presented as background to introducing validity, reliability, and other essential psychometric test criteria and considerations.

This chapter reflects a philosophy that the neuropsychological enterprise depends upon its essential psychometric foundations, and the soundness of psychometrics, in turn, is based upon proper test usage. The principal issues and concepts raised within this opening chapter will be elaborated upon and receive further attention throughout the remainder of this text.

References

Adams, K.M., Gilman, S., Koeepe, R.A., Kluin, K.J., Brunberg, J.A., Dede, D., Berent, S., & Kroll, P.D. (1993). Neuropsychological deficits are correlated with frontal hypometabolism in positron emission tomography studies of older alcoholic patients. *Alcoholism: Clinical and Experimental Research, 17*, 205–210.

Adams, K. M., & Rourke, B. P. (1992). *The TCN guide to professional practice in clinical neuropsychology*. Berwyn, PA: Swets & Zeitlinger.

American Psychological Association (1985). *Standards for educational and psychological testing*. Washington, DC: Author.

American Psychological Association (1986). *Guidelines for computer-based tests and interpretations*. Washington, DC: Author.

Anastasi, A. (1976). *Psychological testing* (4th edition). New York: Macmillan Publishing Company.

Anastasi, A. (1992). What counselors should know about the use and interpretation of psychological tests. *Journal of Counseling and Development, 70*, 610–615.

Berent, S. (1991). Modern approaches to neuropsychological testing. In D. Smith, D. Treiman, & M. Trimble (Eds.), *Advances in neurology* (Vol. 55). New York: Raven Press.

Berent, S., Giordani, B., Gilman, S., Junck, L., Lehtinen, S., Markel, D. S., Boivin, M., Kluin, K., Parks, R., & Koeppe, R. A. (1990). Neuropsychological changes in olivopontocerebellar atrophy. *Archives of Neurology, 47,* 997–1001.

Berent, S., Giordani, B., Lehtinen, S., Markel, D., Penney, J. B., Buchtel, H. A., Starosta-Rubinstein, S., Hichwa, R., & Young, A. B. (1988). Positron emission tomographic scan investigations of Huntington's Disease: Cerebral metabolic correlates of cognitive function. *Annals of Neurology, 23,* 541–546.

Berent, S., Sackellares, J. C., Giordani, B., Wagner, J. G., Donofrio, P. H., & Abou-Khalil, B. (1987). Zonisamide (1,2-benzisoxazole-3-methanesfulronamide [CI-912]) and cognition: Results from preliminary study. *Epilepsia, 28,* 61–67.

Bigler, E. D., & Synder, J. L. (1995). Neuropsychological outcome and quantitative neuroimaging in mild head injury. *Archives of Clinical Neuropsychology, 10,* 159–174.

Binder, L. M., & Pankratz, L. (1987). Neuropsychological evidence of a factitious memory complaint. *Journal of Clinical and Experimental Neuropsychology, 9,* 167–171.

Boll, T. J., & La Marche, J. A. (1992). Neuropsychological assessment of the child: Myths, current status, and future prospects. In C. E. Walker & M. C. Roberts (Eds.), *Handbook of clinical child psychology* (2nd edition; pp. 133–148). New York: John Wiley and Sons.

Brescia, W., & Fortune, J. C. (1989). Standardized testing of American Indian students. *College Student Journal, 23,* 98–104.

Campbell, D. T., & Fiske, D. W. (1959). Convergent and discriminant validation by the multitrait-multimethod matrix. *Psychological Bulletin, 56,* 81–105.

Cronbach, L. J. (1951). Coefficient alpha and the internal consistency of tests. *Psychometrika, 16,* 297–334.

Cronbach, L. J. (1988). Five perspectives on validity argument. In H. Wainer & H. I. Braun (Eds.), *Test validity* (pp. 3–17). Hillsdale, NJ: Lawrence Erlbaum.

Deutsch, M., Fishman, J. A., Kogan, L., North, R., & Whiteman, M. (1964). Guidelines for testing minority group children. *Journal of Social Issues, 20,* 129–145.

Division 40 Task Force on Education, Accreditation, and Credentialing (1989). Guidelines regarding the use of nondoctoral personnel in clinical neuropsychological assessment. *The Clinical Neuropsychologist, 3,* 23–24.

Dunn, J. T., Lees-Haley, P. R., Brown, R. S., & Williams, C. W. (1995). Neurotoxic complaint base rates of personal injury claimants: Implications for neuropsychological assessment. *Journal of Clinical Psychology, 51,* 577–584.

Faust, D., Hart, K., & Guilmette, T. J. (1988). Pediatric malingering: The capacity of children to fake believable deficits on neuropsychological testing. *Journal of Consulting and Clinical Psychology, 56,* 578–582.

Halstead, W. C. (1947). *Brain and intelligence: A quantitative study of the frontal lobes.* Chicago: University of Chicago Press.

Harrison, P. L., Kaufman, A. S., Hickman, J. A., & Kaufman, N. L. (1988). A survey of tests used for adult assessment. *Journal of Psychoeducational Assessment, 6,* 188–198.

Heaton, R. K., Grant, I., & Matthews, C. G. (1986). Differences in neuropsychological test performance associated with age, education, and sex. In I. Grant & K. M. Adams (Eds.), *Neuropsychological assessment of neuropsychiatric disorders.* New York: Oxford University Press.

Hinkle, J. S. (1994). Practitioners and cross-cultural assessment: A practical guide to information and training. *Measurement and Evaluation in Counseling and Development, 27,* 103–115.

Hiscock, C. K., Branham, J. D., & Hiscock, M. (1994). Detection of feigned cognitive impairment: The two-alternative forced-choice method compared with selected conventional tests. *Journal of Psychopathology and Behavioral Assessment, 16,* 95–110.

Hiscock, M., & Hiscock, C. K. (1989). Refining the forced-choice method for the detection of malingering. *Journal of Clinical and Experimental Neuropsychology, 11,* 967–974.

Kramer, H. C. (1982). Estimating false alarms and missed events from interobserver agreement: Comment on Kaye. *Psychological Bulletin, 92,* 749–754.

Kuder, G. F., & Richardson, M. W. (1937). The theory of the estimation of test reliability. *Psychometrika, 2,* 151–160.

Lees-Haley, P.R. (1992). Neuropsychological complaint base rates of personal injury claimants. *Forensic Reports, 5,* 385–391.

Loewenstein, D. A., Arguelles, T., Arguelles, S., & Linn-Fuentes, P. (1994). Potential cultural bias in the neuropsychological assessment of the older adult. *Journal of Clinical and Experimental Neuropsychology, 16,* 623–620.

Matarazzo, J. D. (1983). *Wechsler's measurement and appraisal of adult intelligence* (5th edition). New York: Oxford University Press.

Minium, E. W. (1970). *Statistical reasoning in psychology and education.* New York: John Wiley & Sons.

Pankratz, L. (1983). A new technique for the assessment and modification of feigned memory deficit. *Perceptual and Motor Skills, 57,* 367–372.

Reitan, R. M. (1955). Investigation of the validity of Halstead's measures of biological intelligence. *Archives of Neurology and Psychiatry, 73,*28–35.

Reynolds, C. R., & Brown, R. T. (Eds.). (1984). *Perspectives on bias in mental testing.* New York: Plenum.

Rosenthal, R., & Rosnow, R. L. (1991). *Essential of behavioral research: Methods and data analysis* (2nd edition). New York: McGraw Hill.

Russell, E. W. (1986). The psychometric foundation of clinical neuropsychology. In S. B. Filskov, & T. J. Boll (Eds.), *Handbook of clinical neuropsychology* (vol. 2). New York: John Wiley & Sons.

Somoza, E., & Mossman, D. (1990). Introduction to neuropsychiatric decision making: Binary diagnostic tests. *Journal of Neuropsychiatry and Clinical Neurosciences, 2,* 297–300.

Thompson, P. J., & Trimble, M. R. (1981). Further studies on anticonvulsant drugs and seizures. In M. Karl-Axel (Ed.), *Second workshop on memory functions* (pp. 51–57). Copenhagen: Munksgaard.

Walker, H. M., & Lev, J. (1953). *Statistical inference.* New York: Holt.

Webster, A.M. (1984). *Webster's new collegiate dictionary.* Springfield, MA: Merriam Press.

Wechsler, D. (1945). A standardized memory scale for clinical use. *Journal of Psychology, 19,* 87–95.

Wechsler, D. (1981). *Manual for the Wechsler Adult Intelligence Scale - Revised.* San Antonio, TX: The Psychological Corporation.

Wechsler, D. (1991). *Manual for the Wechsler Intelligence Scale of Children - third edition.* New York: New York Psychological Corporation.

Wilkinson, G. S. (1993). *Administration Manual for the Wide Range Achievement Test.* Wilmington, DE: Wide Range.

Wilson, B. A. (1993). Ecological validity of neuropsychological assessment: Do neuropsychological indexes predict performance in everyday activities? *Applied and Preventive Psychology, 2,* 209–215.

Chapter 2

BASERATES AND CLINICAL DECISION MAKING IN NEUROPSYCHOLOGY

Wm. Drew Gouvier
Louisiana State University, Baton Rouge, Louisiana

In the ivory tower of academic clinical psychology, teaching examples often start with a set of hypothetical circumstances designed to highlight a single point by simplifying or dismissing the effect of other relevant variables that confuse or confound the issue. Teachers have a habit of saying "all other things being equal", without offering the caveat that, typically speaking, all other things are *never* equal unless artificially set that way, as is true in the type of controlled experimental research that constitutes the core of much that we purport to know. Clinicians trained under such circumstances are prone to view their cases accordingly, acting as if they actually believe that each case being presented for evaluation of disorder X has a 50–50 chance of having the disorder or not. Such thinking is dangerous, in that it is quite likely to create circumstances in which a well meaning, but ill informed professional will assert what he or she knows to be true, when in fact, the matter at issue is most likely false. When this problem arises, one of the most likely culprits is the clinician's failure to consider the influence of base rates in affecting the accuracy of our clinical decision making (Meehl & Rosen, 1955).

Unlike the epidemiologist's incidence and lifetime prevalence statistics (Bar-Hillel, 1980; Gordon, 1977), base rates refer to the *current population prevalence* of some phenomena of interest. This phenomena could be a clinical diagnosis, a diagnostic test sign or score, or merely a symptom complaint. The base rate for a particular clinical disorder is the most commonly studied facet of the base rate problem, and base rate research has been conducted primarily in the context of interpreting the significance of diagnostic test findings. Base rate information is essential in establishing the validity of clinical assessment procedures (Duncan & Snow, 1987; Faust & Nurcombe, 1989; Meehl, 1954; Willis, 1984).

Base rate information is *a priori*; a person's behavior cannot change this information. The only influence an individual can have on base rate information is by providing sufficiently detailed background demographic information to allow more precise definition of the sub-class within the population to which the individual belongs, and therefore a more accurate specification of which set of base rate information should apply in interpreting the significance of his or her diagnostic findings.

According to classical test theory, with any test or psychometric instrument, the obtained score is equal to the true score plus the error score. Test theorists and pundits emphasize the core constructs of reliability and validity in psychological testing. Stated simply, reliability represents the degree to which the obtained score approximates the true score, while validity reflects the degree to which the obtained score represents relevant aspects of the phenomena of interest. For most of us, our training taught us that "If a test is reliable and valid, it is good", and for many psychologists, the matter ends there (Faust, 1986). Equally important, but neglected by most is a third concept needed to support our evaluation endeavor. This is the matter of *effectiveness* (Faust & Nurcombe, 1989). It is quite possible for a test to be a reliable and valid indicator of some phenomenon of interest, but still have nothing to offer in terms of improving our diagnostic accuracy. Meehl and Rosen (1955) asserted that for a test to be clinically useful, it's classification accuracy must exceed the accuracy of classifications made using base rate prediction alone.

In order to examine this, one must construct a 2 by 2 contingency table which lays out the hits, misses, false alarms, and correct rejections for each test score, diagnostic sign, or symptom report of interest. A sample contingency table is shown in Figure 1.

Overall, the diagnostic accuracy of the score, sign, or symptom can be computed by summing up hits plus correct rejections. The combined error rate is represented by the sum of misses plus false alarms. Reliability and validity are the sole determinants of diagnostic accuracy when 50% of the subjects are "true state positive" and 50% are "true state negative". Under this circumstance, in the hypothetical ivory tower where $p = q$, conditions allow examination of how accurate a test score, sign, or symptom can be in classification where "all other things are equal". These hypothetical conditions are rarely seen however, and even when they are, we are often at a loss to recognize them.

STATE INDICATOR

	+	−
+	HIT SENSITIVITY VALID POSITIVE TRUE POSITIVE 1	MISS FALSE NEGATIVE TYPE II ERROR 2
−	3 FALSE ALARM FALSE POSITIVE TYPE I ERROR	4 CORRECT REJECT SPECIFICITY VALID NEGATIVE TRUE NEGATIVE

TRUE STATE

Fig. 1.   Sensitivity, specificity and error type contingency table.

What happens when the base rate for the true condition begins to skew toward 0 or 1? How do base rates influence the diagnostic accuracy of any given score, sign, or symptom report? The clinical utility of any diagnostic indicator varies as a function of the base rate of the condition being diagnosed. This introduces how the concept of "effectiveness" (Faust & Nurcombe, 1989) is a key to enhancing the diagnostic decision making process. A test can be reliable and valid, but still not contribute to diagnostic accuracy beyond what we could accomplish by using base rates alone. Faust and Nurcombe (1989, p 200) put it as follows:

> A sign cannot improve the accuracy achieved by playing the baserates if its error rate (the combined frequency of false positive and false negative errors) exceeds the frequency of the condition to be identified or predicted. For example, if a sign yields a 20% error rate in diagnosing a condition that occurs 5% of the time, the use of the sign will not enhance base rate judgment. Stated in its reverse form, the rule is as follows: A sign can increase accuracy only if the frequency of the condition exceeds the sign's error rate. This rule is ironclad.

The computation of base rates is a simple matter using the following formula:

$$\text{Base rate} = \frac{\text{\# of cases with the condition}}{\text{\# of cases in the population}}$$

In order to better understand how to compute and use base rate information, variants of the *Cab Problem* (Kahneman & Tversky, 1982; Lyon & Slovic, 1976) will be used. Two cab companies operate in Fairyland. There are 85 Blue cabs and 15 Green cabs. A witness to a hit and run accident last night claimed that it was a Green cab in the accident. The Judge showed wisdom in taking the witness into chambers and testing the witness' ability to distinguish photographs of Blue and Green cabs taken at night. Viewing pictures of 5 Blue and 5 Green cabs, the witness correctly identified 4/5 for each. Repeating this exercise with 10 additional photographs, the witness replicated this original performance exactly. Clearly then, this witness is a valid and reliable detector of cab color. Had this witness been tested identifying each of the 100 cabs in the city, 80% of the 15 Green cabs would be correctly identified as Green ($N = 12$), and 20% of the 85 Blue cabs would be incorrectly identified as Green as well ($N = 17$). We can now ask: "What is the true probability that the hit and run cab was Green?". This can be computed by dividing the number of correct Green calls over the total number of Green calls: $12/(12+17) = .414$. Take note that even though the witness is a reliable and valid indicator, and even though the witness was sure the hit and run cab was Green, more probably than not, the cab was really Blue. Even though this witness is 80% accurate in his judgments, only when the base rates for Blue cabs and Green cabs are equal will 80% overall diagnostic accuracy be obtained.

This example shows one of the ways base rate information can affect diagnostic accuracy. Because of the relative rarity of Green cabs, the witness was more likely to misidentify a Blue cab as Green than to actually see a Green cab involved in the accident. But base rate information can have an opposite effect too. Had the witness said the cab was Blue, there would be a significantly enhanced probability of an accurate identification. Using the same formula of correct Blue calls divided by total number of Blue calls, we compute $68/(68+3) = .958$. Thus, had the witness seen a Blue cab, there would be a 96% chance

WITNESS SAYS
(Test Data, Symptoms, or Signs)

		Blue	Green
	Blue 85	68	17
TRUE COLOR (Disease State)			
	Green 15	3	12

Fig. 2.   Cab problem contingency table: Observer accuracy = 80%, Baserate for Green cabs = 15%.

that that the cab involved was actually Blue. Thus, whenever the base rates for the two conditions are not equal, base rates consistently affect diagnostic accuracy by reducing the accuracy of prediction of the rarer category (which is usually the clinical diagnosis), and boosting the accuracy of prediction of the more common category (Gordon, 1977).

The degree to which this base rate influence affects diagnostic accuracy is controlled by at least 2 variables. The first influence is the degree of base rate skew. The effect becomes increasingly more pronounced the further the base rates of the two conditions skew away from 50/50. Thus, had there been 95 Blue cabs and only 5 Green cabs in Fairyland, the odds of the witness' Green call being correct would be only 17%, while the odds of a Blue call being correct would be 99%.

The second influence is observer accuracy. If the witness had correctly identified only 3 of each 5 pictures in chambers, but had done so consistently under conditions where $p = q$, the witness would remain a reliable and valid indicator of cab color. With the base rate for Green and Blue cabs equated, the observer accuracy would be 60% for each color, and the witness' testimony would more probably than not be true. However, as the base rate for Green is reduced to .15, the likelihood of a Green call being correct drops to 21%, and at a base rate for Green of .05, the chance of a Green call being correct drops to only 7%. With an extremely accurate witness, capable of consistently making 9 out of 10 correct observations for either color, the influence of base rates becomes less pronounced. For this highly accurate witness, observing in the condition with a Green base rate of .15, the chances are, more probably than not, that the witness's Green call is now correct (61% probability). The effect of these manipulations of base rate and observer accuracy on overall diagnostic accuracy are depicted in Table 1.

With the base rate for Green cabs set at .15, the 80% accurate witness was most likely wrong when making a Green call, but the 90% accurate witness was most likely correct. The testimony of a witness with 85% accuracy is likely to match the diagnostic accuracy attained by playing base rates alone. Although

Table 1. Cab Problem Outcomes across Differing Base Rates and Observer Accuracies.

Base Rate for Green Cabs				
Observer Accuracy 60%		.5	.15	.05
Likelihood of correct identification	Green	60%	21%	7%
	Blue	60%	89%	97%
Observer Accuracy 80%		.5	.15	.05
Likelihood of correct identification	Green	80%	41%	17%
	Blue	80%	96%	99%
Observer Accuracy 90%		.5	.15	.05
Likelihood of correct identification	Green	90%	61%	32%
	Blue	90%	98%	99%

the overall accuracy for each witness is reasonably good, in large part because of their base rate induced inflated accuracy of Blue calls, none of the witnesses is particularly good at making accurate Green calls. If maximizing overall diagnostic accuracy is the goal, one would be better off saying that all cabs are Blue rather than relying on the observations of the 80% accurate witness. This demonstrates the principle of effectiveness (Faust & Nurcombe, 1989); this 80% witness is simply not an effective test of cab color under the circumstance outlined in the 15% Green base rate scenario, but becomes as effective as not when accuracy goes to 85%, and becomes clearly effective when witness accuracy jumps to 90%.

We have seen that observations can be reliable and valid but still lead to wrong conclusions, and that the likelihood of this happening varies as a function of base rate of the condition and observer accuracy in detecting it. Valid tests can be ineffective, as we saw in the case of our 80% accurate observer operating in a field where only 15% of the cabs were Green. Formulae for computing whether a particular test or observation is effective can be derived from Faust and Nurcombe (1989), and are as follows (Gouvier, Hayes, & Smiroldo, 1998):

For baserates below 50%, a test is effective when BR > FP + FN errors
For baserates above 50%, a test is effective when 1−BR > FP + FN errors

Valid tests can be ineffective; effectiveness is determined by the base rates of the condition and the combined error rate of the test. All tests are inherently imperfect, and the effectiveness formula can be used to define the range of baserates within which imperfect tests can be effective, or as Meehl and Rosen (1955) stated, "clinically useful". The only time base rates do not affect diagnostic accuracy is when p=q at 50% for either diagnostic condition.

In the case of our Cab Problem witness with 80% accuracy, this witness becomes effective only when the base rates for Blue or Green cabs range between .21 and .79.

Most of us are oblivious to the importance of base rates in everyday practice, even though the effect size for baserates may be much larger than the effect size for positive test findings. This is what Bar-Hillel (1980) refers to as the "base-rate fallacy". Bar-Hillel (1980) and Arkes (1981) have suggested a number of reasons for the widespread adoption of this fallacy among professionals. These reasons include our failure to report base rate data (Meehl & Rosen, 1955), the lack of salience of this type of information in contrast to juicy clinical vignettes, our lack of "experience" with the relation between base rate information and diagnostic accuracy (Christensen-Szalawski & Beach, 1982), our relative cockiness and overconfidence in our clinical acumen (Chapman & Chapman, 1967, 1969), laziness (Davidson & Hirtle, 1990), and our fear of having the responsibility for being fluent in this area, but not being comfortable in our knowledge (Gouvier et al., 1997). It is the sort of thing that makes you want to hide your head in the sand. Watterson (1992), the creator of *Calvin and Hobbes* describes a similar situation in the cartoon depicted in Figure 3.

Meehl and Rosen (1955, p. 194) offered a partial explanation for the disuse of base rate information, saying that "base rates are virtually never reported", and that "our ignorance of base rates is nothing more subtle than our failure to compute them" (p. 213). Thirty years later, Duncan and Snow (1987) reaffirmed this observation. While it is still relatively infrequent to see reports include base rate information about the frequency of diagnoses, the recent interest in malingering detection *has* spurred an interest in ascertaining the base rates for this diagnosis among different population subgroups, and therefore a renewed attention to the issue of base rate information in diagnostic judgments in general. Much of this new information has focused on reporting baserates for symptom complaints and test failures among various groups of subjects (Gouvier, Cubic, Jones, Brantley, & Cutlip, 1992; Gouvier, Uddo-Crane, & Brown, 1988; Hilsabeck, Gouvier, & Bolter, 1998; Lees-Haley & Brown, 1993; Lees-Haley, Williams, & English, 1996; Martin, Hayes, & Gouvier, 1996; Roberts et al., 1990; Wong, Regennitter, & Barrios, 1994). This latter set of research reports is useful for helping parse out the relative contributions of type of injury, effects of stress, participation in litigation, and coaching instructions in contributing to symptom self report and test failure, but does not specifically address the more fundamental issue of the effect of base rates on diagnostic test accuracy. Regarding this rather more fundamental issue, for how many of us did the Watterson (1992) phrase "willfully stupid" in the cartoon ring a resonant chord?

Fig. 3.   Hobbes *is* smarter! (Calvin and Hobbes © 1992 Watterson. Dist. by Universal Press Syndicate. Reprinted with permission. All rights reserved).

Enhanced overall diagnostic accuracy is a laudable goal, but "base rates should guide our practice, not rule it" (Faust & Nurcombe, 1989, p. 201), and the Meehl and Rosen (1955) dictum about the clinical usefulness of tests being dependent on surpassing baserate classification alone fails to consider the consequences and acceptability of different error types. Accuracy must be weighed against utility and the implications of different error types (Gordon, 1977; MacLennan, 1988). If only one of a hundred cafeteria workers has typhoid (Hailey, 1959), Dr. Smith, who relies on baserates is 99% accurate when he gives all workers a clean bill of health. Dr. Jones, who has a typhoid test with perfect sensitivity, but only 90% specificity, is only 91% accurate overall. But accuracy be damned, which doctor would you have supervising the health of the cafeteria workers? Dr. Jones is assuredly going to catch the one typhoid infected worker, but he does so at a cost of inconveniencing 9 healthy workers, and subjecting them to additional testing to rule out the diagnosis of typhoid. Even though Dr. Jones makes nearly 10 times more errors than Dr. Smith, this is probably acceptable. It does not take much imagination to envision circumstances that strain the limits, however. As Dr. Jones' specificity drops to 80% or 60%, there is a point at which following up on his diagnoses becomes an exercise in "hollering wolf", and the costs of following up on false positives may begin to outweigh the benefits of the occasional hit.

These considerations apply in the diagnosis of clinical pathology, but they also apply to the diagnosis of malingering. There are important differences between these two types of diagnoses, however. In clinical diagnosis of a true pathology, there are naturally occurring checks and balances that let us catch our mistakes much of the time. Sick patients who are misclassified as healthy will typically continue to present with symptoms enough to warrant reexamination. Well patients wrongly labeled as sick continue to present for evaluation as their response to treatment is followed and their healthy status will likely be recognized in these follow up visits.

No such safeguards occur when the diagnosis being considered is malingering, however, as this diagnosis is most often made in a "one-shot" IME/IPE (independent medical/psychological exam) type evaluation in which the evaluator is not regarded as a treating practitioner. As such, this represents an area of significant liability exposure for practitioners who make this diagnosis (Hayes, Hilsabeck, & Gouvier, in press), because once an individual is labeled as a malingerer, professionals are less likely to take seriously any complaints that are raised by that individual, and may be inclined to write off the purported malingerer altogether. In civil cases in which the diagnosis of malingering is incorrect, the net result can be undue hardship or deprivation being wrongly applied to an undeserving soul. In criminal forensic arenas, such misdiagnosis could make a life or death difference in sentencing judgments.

While the emphasis in test development often seems to value sensitivity over specificity (Lyon & Slovic, 1976), the latter construct is far more important in malingering detection, if one's primary concern is protecting the rights of the incorrectly diagnosed or unjustly accused. If our malingering detection strategy is perfect at identifying normals (non-malingerers), we never have to worry about damning a patient unjustly, even though some malingerers would successfully slip through the cracks (being the false negatives from our malingering testing). The protection of the innocent (in civil as well as criminal arena) is a fundamental tenant of our justice system, and as such, it is encouraging to see efforts to develop sophisticated techniques to augment our traditional malingering detection strategies and enhance their ability to detect normals. For example, Rosenfeld, Sweet, Chuang, Ellwanger, and Song (1996) have introduced a procedure to augment forced choice malingering testing with event related potential recording measurements. This use of physiological as well as behavioral indices for detecting normal and deviant response sets offers promise toward our goal of enhanced detection of normalcy as well as improved detection of pathology in our diagnostic endeavors (see Chapter 10 within this book). Sadly, however, this is but an early step come lately; the idea of combining physiological and behavioral indices to augment psychological testing for anxiety disorders was a novel idea decades ago. Our assessment technology still lags behind the goal of perfect specificity in malingering detection, but efforts to improve our identification of normalcy *as well as* to improve our ability to detect malingering are both laudable and important steps toward the development of more effective procedures (see Chapter 9 within this book).

Other approaches to managing the problem of baserates and misdiagnosis in malingering detection strategies can be applied using tests with imperfect specificity and sensitivity in a multi-tiered assessment process requiring a criterion of multiple test failures as a minimum standard upon which to base a diagnosis of malingering (Gouvier et al., 1998). Such an approach knowingly lets more malingerers slip though the cracks than it catches, but does so in order to ensure that when a diagnosis of malingering is made, the recipient is more probably than not a real malingerer, and not some hapless person who was wrongly classified as a malingerer because of the test's inherent false positive error rate and the relative scarcity of real malingerers.

The net result of all of this is that there is a reasonably high likelihood that well meaning, but misguided clinicians offer sworn testimony about what they believe to be true, even when the baserate analysis shows that their testimony is unknowingly, but more probably than not, false. How much does such a scenario resemble our individual clinical practice and forensic testimony? A lot more than most of us want to acknowledge. After all, admitting this vulnerability forces a re-examination of the entire manner in which psychologists (and all health care diagnosticians for that matter) conduct their business of making clinical and malingering diagnoses. We can no longer function in a state of ignorance in which we delude ourselves into believing that our patients have a 50–50 chance of having the disorder and our tests are the *sine qua non* of diagnostic accuracy.

In *Daubert v. Merrell Dow Pharmaceuticals, Inc.* (1993) the U.S. Supreme Court ruled that to be admissible as "scientific knowledge", expert testimony must be derived by accepted scientific methods, and such testimony must be based on evidence that is reliable and valid (Cecil, Drew, Cordisco, & Miletich, 1992). By taking the bull by the horns now, our profession has the opportunity to improve forensic practice and prevent self-annihilation by the Daubert ruling or future related rulings. We should take every measure to attempt to stop misguided forensic practice; putting our heads in the sand won't protect us from a strong Daubert 'junk science' challenge that could be readily grounded in base rate analysis.

References

Arkes, H. (1981). Impediments to accurate clinical judgment and possible ways to minimize their impact. *Journal of Consulting and Clinical Psychology, 49,* 323–330.

Bar-Hillel, M. (1980). The base-rate fallacy in probability judgments. *Acta Psychologica, 44,* 211–233.

Cecil, J.S., Drew, C.E., Cordisco, M., & Miletich, D.P. (1992). *Reference manual on scientific evidence,* Washington, DC: Federal Judicial Center.

Chapman, L., & Chapman, J. (1967). Genesis of popular, but erroneous psychodiagnostic observations. *Journal of Abnormal Psychology, 72,* 193–204.

Chapman, L., & Chapman, J. (1969). Illusory correlation as an obstacle to the use of valid psychodiagnostic signs. *Journal of Abnormal Psychology, 74,* 271–280.

Christensen-Szalawski, J., & Beach, L. (1982). Experience and the base rate fallacy. *Organizational Behavior and Human Performance, 29,* 270–278.

Daubert v. Merrell Dow Pharmaceuticals, Inc., 113 S. Ct. 2786 (1993).

Davidson, D., & Hirtle, S. (1990). Effects of nondiscrepant and discerpant information on the use of base rates. *American Journal of Psychology, 103,* 343–357.

Duncan, D., & Snow, W.G. (1987). Base rates in neuropsychology. *Professional Psychology: Research and Practice, 18,* 368–370.

Faust, D. (1986). Learning and maintaining rules for decreasing judgement accuracy. *Journal of Personality Assessment, 50,* 585–600.

Faust, D., & Nurcombe, B. (1989). Improving the accuracy of clinical judgement. *Psychiatry, 52,* 197–208.

Gordon, N. (1977). Base rates and the decision making model in neuropsychology. *Cortex, 13,* 3–10.

Gouvier, W., Cubic, B., Jones, G., Brantley, P., & Cutlip, Q. (1992). Postconcussion symptoms and daily stress in normal and head injured college populations. *Archives of Clinical Neuropsychology, 7,* 193–211.

Gouvier, W., Hayes, J., & Smiroldo, B. (1998). The significance of baserates, test sensitivity, test specificity, and subjects' knowledge of symptoms in assessing TBI sequelae and malingering. In C. Reynolds (Ed.), *Detection of malingering in head injury litigation.* New York: Plenum Press.

Gouvier, W., Uddo-Crane, M., & Brown, L. (1988). Base rates of postconcussional symptoms. *Archives of Clinical Neuropsychology, 3,* 273–278.

Hailey, A. (1959). *The final diagnosis.* New York: Bantam Books.

Hayes, J., Hilsabeck, R., & Gouvier, W. (in press). Malingering traumatic brain injury: Current issues and caveats in assessment and classification. In R. Robert & N. Varney (Eds.), *Neurobehavioral dysfunction following mild head injury: Mechanisms, evaluation, and treatment.*

Hilsabeck, R., Gouvier, W., & Bolter, J. (1998). Reconstructive memory processes in recall of neuropsychological symptomology. *Journal of Clinical and Experimental Neuropsychology, 20,* 328–338.

Kahneman, D., & Tversky, B. (1982). On the psychology of prediction. In D. Kahneman, P. Slovic, & A. Tversky (Eds.), *Judgement under uncertainty: heuristics and biases.* New York: Cambridge University Press.

Lees-Haley, P., & Brown, R. (1993). Neuropsychological complaint base rates of 170 personal injury claimants. *Archives of Clinical Neuropsychology, 8,* 203–209.

Lyon, D., & Slovic, P. (1976). Dominance of accuracy information and neglect of base rates in probability estimation. *Acta Psychologica, 40,* 287–298.

Lees-Haley, P., Williams, C., & English, L. (1996). Response bias in self-reported history of plaintiffs compared with nonlitigating patients. *Psychological Reports, 79,* 811–818.

MacLennan, R. (1988). Correlation, base-rates, and the predictability of behaviour. *Personality and Individual Differences, 9,* 675–684.

Martin, R., Hayes, J., & Gouvier, W. (1996). Differential vulnerability between postconcussion self-report and objective malingering tests in identifying simulated mild head injury. *Journal of Clinical and Experimental Neuropsychology, 18,* 265–275.

Meehl, P. (1954). *Clinical vs. statistical prediction.* Minneapolis: University of Minnesota Press.

Meehl, P., & Rosen, A. (1955). Antecedent probability and the efficiency of psychometric signs, patterns, or cutting scores. *Psychological Bulletin 52,* 194–216.

Roberts, R., Varney, N., Hulbert, J., Richardson, E., Springer, J., Shepherd, J., Swan, C., Legrand, J., Harvey, J., & Struchen, M. (1990). The neuropathology of everyday life: The frequency of partial seizure symptoms among normals. *Neuropsychology, 4,* 65–85.

Rosenfeld, J.P., Sweet, J.J., Chuang, J., Ellwanger, J., & Song, L. (1996). Detection of simulated malingering using forced choice recognition enhanced with event related recording. *The Clinical Neuropsychologist, 10,* 163–179.

Watterson, W. (1992). *Calvin and Hobbes,* Universal Press Syndicate, May 17.

Willis, W. (1984). Reanalysis of an actuarial approach to neuropsychological diagnosis in consideration of base rates. *Journal of Consulting and Clinical Psychology, 52,* 567–569

Wong, J., Regenniter, R., & Barrios, F. (1994). Base rates and simulated symptoms of mild head injury among normals. *Archives of Clinical Neuropsychology, 9,* 411–425.

Chapter 3

CONSIDERING PREMORBID FUNCTIONING:
BEYOND COGNITION TO A CONCEPTUALIZATION OF PERSONALITY IN POSTINJURY FUNCTIONING

Steven H. Putnam
Rehabilitation Institute of Michigan, Wayne State University, Novi, Michigan

Joseph H. Ricker
Rehabilitation Institute of Michigan, Wayne State University, Novi, Michigan

Scott R. Ross
DePauw University, Greencastle, Indiana

John E. Kurtz
Villanova University, Villanova, Pennsylvania

Introduction

Involvement in forensic neuropsychology offers substantial financial rewards and, in accord, appears to have attracted the involvement of an increasing number of practitioners in recent years (Putnam, DeLuca, & Anderson, 1994). The principal question to be addressed (perhaps after proximal cause), is whether there has been any substantive or *bona fide* change in an individual's neuropsychological or psychological status vis-à-vis some injury or traumatic event.

For this reason the topic of estimating premorbid intellectual and cognitive functioning is one of particular relevance to the neuropsychologist performing forensic examinations. Several excellent reviews of the topic are available in the literature, but most are not written in the context of the forensic neuropsychological evaluation. As noted by Lees-Haley (1992) "litigation is a special context that appears to have different base rates and different evaluation requirements than traditional therapeutic or clinical environments" (p. 387). Therefore, the present chapter will review several issues that are pertinent to neuropsychological assessment, but will focus predominantly on those details that arise in the context of forensic neuropsychological evaluation involving personal injury claims such as head injury. However, the ideas presented in this chapter are equally relevant to personal injury litigation associated with toxic exposure, medical malpractice, and other foci of litigation.

Despite the requirement that neuropsychologists infer premorbid cognitive and psychological status, at present there is no uniformly agreed upon method of achieving this aim. Of course, a systematic and thoughtful review of all available school, medical, occupational, etc., records continues to be elemental in arriving at estimations in the individual case. Unfortunately, standardized or established procedures for fully utilizing these data across cases have yet to be developed and are not available. With regard to assessing a patient's current clinical status Matarazzo (1990) emphasizes:

> The assessment of intelligence, personality, or type or level of impairment is a highly complex operation that involves extracting diagnostic meaning from an individual's personal history and objectively recorded test scores. Rather than being totally objective, assessment involves a subjective component... In the hands of a good clinician, the results of an examination of intelligence or personality, correlated with information from the person's history, are as useful as analogous information would be in the hands of a good surgeon, internist, accountant or plumber (pp. 1000–1001).

Of course, the complexity of conceptualizing the patient's premorbid status increases exponentially. Sbordone (1996) asserts that certain patients with neurological disorders, such as head injury, may be unreliable historians and recommends seeking collateral sources of information. He offers that a careful review of the patient's background information, "will frequently uncover a history of pre-existing neurological disease, psychological problems, prior head trauma, drowning episodes, seizures, hyperactivity, dyslexia, left-handedness, stuttering, sexual molestation and rape, alcoholism, criminal behavior, poor academic performance, or drug abuse" (p. 27). Such information can be crucial in making necessary discriminations between pre-existing functional and adaptive limitations and those directly associated with head injury or some other ictus. Indeed, identifying the primary causes and mitigants of current symptomatology and cognitive change is of particular interest to attorneys and third party payers responsible for negotiating issues of liability. It is this collection of archival information and pursuit of corroboration from third-parties that distinguishes the forensic context from the clinical or therapeutic context.

In a forensic context, relying solely on self-reported symptoms and history as evidence of brain dysfunction is *not* recommended (Lees-Haley & Brown, 1993). Within a clinical concept, the establishment of trust and rapport with the patient takes precedence over seeking the factual basis of the patient's claims (Melton, Petrila, Poythress, & Slobogin, 1997).

The chapter is divided into two major sections. The first covers some of the rather substantial empirical literature related to estimating premorbid intellectual and cognitive functioning. The second section is concerned with conceptualizing premorbid personality as it may relate to current symptomatic status and psychosocial functioning. This section begins with an extrapolation of theoretical literature pertaining to personality formation and constructs and is followed by coverage of relevant empirical studies, including several clinical measures of personality.

Intellectual functioning and cognition

Actuarial approaches to assessing premorbid cognitive status

Several approaches have been developed to estimate premorbid cognitive functioning, most of which have dealt with the mathematical estimation of measured intellectual functioning (i.e., an intelligence quotient (IQ) or composite score). In general, empirically or actuarially-based approaches to estimating premorbid intellectual functioning may be more compelling than approaches that rely strictly on subjective human judgment (e.g., clinical interview), given the problems that clinicians have been reported to demonstrate when relying on judgment to estimate presence or extent of cognitive decline (Kareken, 1997; Kareken & Williams, 1994). In spite of these findings a recent survey based on a modest sized sample of clinical psychologists found that the clinical interview was the most routinely utilized method for assessing premorbid status (Smith-Seemiller, Franzen, Burgess, & Prieto, 1997). In any event, various actuarial procedures have been developed in an effort to provide some estimation of premorbid cognitive functioning. These procedures include mathematical modeling of sociodemographic variables, as well as current measure estimates, such as vocabulary based word reading. In addition to the summary that follows, the interested reader may wish to refer to the recent review by Graves, Carswell, and Snow (1999).

Demographically-based approaches to estimating premorbid IQ

Mathematical estimation of premorbid IQ is a common point of departure in the process of inferring premorbid status. Although IQ is representative of certain aspects of cognitive functioning, clinicians must be attentive to the fact that overall IQ and specific cognitive functions are not synonymous concepts (e.g., Ruff, Mueller, & Jurica, 1996). There can be tremendous variation in specific cognitive skills even within the same level of intellectual classification (e.g., Low

Average). This appears to be particularly true for individuals towards the higher end of the IQ range. Possessing a higher than average IQ should not be viewed as tantamount to demonstrating comparably elevated performances on measures of specific neuropsychological abilities (Dodrill, 1997).

Regression modeling of relevant sociodemographic variables in postdicting WAIS-R IQ values (Barona, Reynolds, & Chastain, 1984) is useful in providing gross parameter estimates. Perhaps the most frequently used actuarial approach for estimating premorbid Wechsler Adult Intelligence-Scale Revised (WAIS-R) IQ scores is that of Barona and colleagues (Barona et al., 1984). The Barona regression equation (BRE) is derived from ordinal scaling of the following demographic variables: age, education, gender, race, geographic region, and occupation. The regression model produces an approximation of (premorbid) Verbal (VIQ), Performance (PIQ), and Full Scale IQ (FSIQ), but it frequently generates values within a restricted range complicated further by large standard error of estimates (12–13 IQ points). For example, even if a given patient presents with demographic characteristics which are assigned the highest ordinal rating, the maximum premorbid FSIQ predicted by the BRE algorithm is 120.66. Conversely, the lowest possible predicted premorbid FSIQ is 69.4. In cases where actual premorbid FSIQ is above 115 or below 85 (i.e., one plus or minus standard deviation from the mean) the use of this formula will result in artifactual ceiling/floor effects. However, the inclusion of certain performance variables (WAIS-R Vocabulary and Picture Completion subtests) has been reported to attenuate the range problem while increasing the predictive validity of the BRE (Krull, Scott, & Sherer, 1995). Perez, Schlottman, Holloway, and Ozolins (1996) examined the use of the BRE in comparison with the Intellectual Correlates Scale (ICS), a self-rating scale that is comprised of interests, attitudes, and beliefs thought to be associated with intellectual functioning (Schlottmann & Johnsen, 1991). The BRE was reported to be more accurate in estimating IQ than the ICS. Sweet, Moberg, and Tovian (1990), reported that the Barona formula did not predict better than chance when applied to neurologically normal psychiatric patients and a heterogeneous sample of brain damaged patients. They concluded:

> Although the formulas appear to be fairly reliable for *groups* of patients, there is no assurance that the estimates produced by the regression formulas will be accurate in individual cases . . . In general, these formulas may be useful in research (i.e., matching subjects) or when cautiously used in conjunction with past records (e.g., school or military). However, they should not be used in isolation with individual patients (p. 43).

While limited, this approach does offer the advantage of being applicable to a variety of patients, and, in fact, one need not actually examine a patient directly to apply the index. Given the weight assigned to educational attainment in these models and unreliability of patient reports (Johnson-Greene & Binder, 1995), corroboration should be pursued. Furthermore, it should be emphasized that attainment of a certificate (e.g., GED) resulting from passing an equivalency exam is not tantamount to completing 12 years of education.

Performance-based approaches

Early on, Babcock (1930) and later Yates (1956) reported that certain psychological tasks, such as vocabulary, were relatively resilient to the effects of cerebral damage, whereas performance on other tests, most notably memory, learning, and motor speed, were frequently diminished. These findings became a basis for the belief that premorbid functioning could, in fact, be estimated and used as a comparison with currently obtained assessment results. Such performance based approaches, of course, have the advantage of providing a criterion-related index in estimating premorbid functioning. There is the additional advantage of being able to more accurately account for individual differences in outcome (e.g., premorbid verbal learning disabilities).

Procedures requiring the correct pronunciation of irregularly spelled words (e.g., colonel, debris) which have little relationship to common phonetic rules can provide another estimate of a literate patient's premorbid intellectual functioning. This is based on two principal assumptions: (1) accurate reading of such words requires *previous* familiarity (recognition) which is based on rote knowledge and past educational exposure rather than *current* ability to phonetically decode (process) the words, and (2) vocabulary based word reading ability in general is fairly resilient to intellectual decline (Blair & Spreen, 1989; Nelson & McKenna, 1975).

One measure in widespread use is the National Adult Reading Test (NART; Nelson, 1982), and its North American revision, the North American Adult Reading Test (NAART; Blair & Spreen, 1989). Unfortunately, these procedures also produce a restricted range of scores with overestimates and underestimates. However, the inclusion of gender and education as predictor variables provided a superior estimate to either word reading measured with the National Adult Reading Test (NART) or demographic variables alone (Willshire, Kinsella, & Prior, 1991). Race and parental education, when combined with performance on the Wide Range Achievement Test-Revised (WRAT-R) expanded the range of prediction of this method which does not rely foremost on irregularly spelled words (Kareken, Gur, & Saykin, 1995). The NAART produces much smaller standard errors of estimate than the Barona index, and a discrepancy of 15 points between estimated and obtained VIQ or FSIQ identifies intellectual deterioration at the 95% confidence level (Blair & Spreen, 1989). Obviously such procedures cannot be used with patients who are aphasic, have a pre-existing reading disorder, or have a disturbance of visual acuity. Its application with people whose native language is not English needs to be done very cautiously, if done at all.

The WAIS-R Vocabulary subtest is the most highly correlated with FSIQ (Wechsler, 1981) and is often considered to be somewhat robust in the presence of cerebral compromise (Larabee, Largen, & Levin, 1985), thus making it an adequate index of general intellectual functioning. Given its strong verbal response component, however, performance on the Vocabulary subtest may be lowered by left hemisphere injury, which, of course, could lead to potential underestimation of premorbid ability in such cases. In fact, all Wechsler scales have been reported to be somewhat susceptible to cerebral compromise (Russell, 1972). WAIS-R Vocabulary may also be discrepant from overall intellectual functioning in cases of verbal learning disability, inadequate educational opportunity, or impoverished quality of the individual's early educational and home environment.

A primary shortcoming of performance-based approaches is that they are often highly dependent upon adequately developed reading and language dependent skills. While these skills may be expected to remain relatively intact in mild to moderate diffuse brain injuries, they may be compromised in the presence of severe head injury, certain focal lesions (e.g., penetrating TBI, CVA), or certain neurodegenerative conditions (e.g., Alzheimer's disease).

Analysis of IQ test variance, which was discussed by Thorp and Mahrer (1959) in the context of the Wechsler Adult Intelligence Scale (WAIS), is a more general performance-based approach. This approach makes the assumption that greater variability among subtest scores may indicate the presence of intellectual decline. A similar approach, sometimes referred to as the "best performance method" (Lezak, 1995), assumes that an individual's highest performance level (in this situation, highest WAIS-R subtest) is the best estimate of premorbid functioning and is therefore the benchmark against which other tests and inferences should be gauged. This type of inference makes the questionable assumption that intellectual abilities are uniformly developed, however, and may tend to *over*estimate general premorbid abilities (Mortensen, Gade, & Reinisch, 1991). Wilson, Rosenbaum, and Brown (1979) compared demographic versus performance-based strategies in estimating premorbid WAIS IQ in two samples (neurologic patients and non-neurologic controls). They found that when a 10 point ordinally scaled occupation rating was substituted for Wechsler's "hold tests" in a regression based deterioration equation, the prediction of premorbid IQ improved by 11%. These results may suggest that utilizing the "hold" tests may not be accurate in neurologic and psychiatric populations, and that demographic factors may, in fact, be of greater predictive value. It remains to be seen if this would generalize to the WAIS-R or WAIS-III, however. In a recent application of best-performance equations developed by Vanderploeg and Schinka (1995), regression functions based on "hold tests" overestimated IQ by only 9 to 5 points. Further, best-performance estimates were significantly more highly correlated (median $r = .85$) than the BRE (median $r = .60$) with actual IQ scores (Vanderploeg, Schinka, & Axelrod, 1996). These results suggest that the best-performance method for estimating IQ is promising in its clinical application with neurologic samples.

Inspection of intrasubtest scatter in the WAIS-R subtest profile has also been suggested as an indirect means of assessing intellectual decline. Although it is true that research studies comparing *groups* of head injury survivors with groups of non-injured controls demonstrate a statistically greater amount of variance in the subtest profiles of individuals with head injury (Mittenberg, Hammeke, & Rao, 1989), there is *no* compelling evidence that intratest scatter can be used to reliably distinguish the presence of brain injury from its absence in a given individual.

Johnstone et al. (1997) evaluated demographic (years of education) versus performance (WRAT-R and WRAT-3 Reading) approaches in estimating premorbid cognitive functioning in a sample of individuals with medically substantiated head injury. They report that each approach yielded different findings regarding patients, and suggested that the use of reading performance was of greater assistance than years of education in estimating premorbid functioning and actual decline in neuropsychological status.

In recent years, investigators have begun to expand the scope of prediction from simply estimating premorbid IQ. Crawford et al. (1992) and colleagues have used the NART to predict FAS (Crawford et al., 1992), and Paced Auditory Serial Addition Test (PASAT) performance (Crawford et al., in press). Hawkins et al. (1993) demonstrated that reading ability was strongly correlated with performance on the Boston Naming Test in both psychiatric and non-psychiatric samples. It appears that confrontation naming ability is not a unitary construct but rather may, to some extent, be dependent upon the premorbid factor of linguistic ability. These investigations provide evidence that reading skill, as indexed by tests such as the NART, may be employed to estimate specific cognitive abilities beyond summary IQ alone. However, Williams (1997) recently attempted to predict performance on the Memory Assessment Scales from demographic variables and concluded; "demographic variables have low and significant relationships to memory scores but do not predict memory scores at the same magnitude that they predict IQ" (p. 752).

Combined demographic and performance-based approaches
Clinicians have sought to capitalize upon both demographic and performance-based strategies. Crawford, Parker, and Stewart (1989) combined NART performance with demographic variables to increase accuracy of predicting premorbid WAIS IQ.

A contemporary approach is that of the Oklahoma Premorbid Intelligence Estimate (OPIE). The OPIE uses both demographic variables (age, education, race, occupation) and current performance (WAIS-R Picture Completion and Vocabulary raw scores). Using the WAIS-R standardization sample, Krull et al. (1995) demonstrated the accuracy of the OPIE in estimating WAIS-R IQ's. The OPIE is also useful in the estimation of premorbid Full Scale IQ using only Vocabulary or Picture Completion performances. Scott, Krull, Williamson, Adams, and Iverson (1997) demonstrated that the OPIE is accurate with clinically referred patients and does not demonstrate the same restriction of range or over-estimation of premorbid status present in certain other approaches.

Kareken, Gur, and Saykin (1995) examined the utility of considering reading performance combined with demographic variables (parental education and race) in estimating premorbid IQ. In a sample of adults without neurologic or psychiatric disturbances these researchers developed a series of regression equations that predicted WAIS-R IQ's from WRAT-R Reading, average number of years of parental education, and race.

Contributions of personality variables to estimating premorbid intelligence
Wrobel and Wrobel (1996) utilized Minnesota Multiphasic Personality
Inventory (MMPI) variables in an attempt to improve BRE estimates of IQ in
psychiatric patients. They found that MMPI indices such as the Infrequency (F)
and Psychopathic Deviate (Pd) scales demonstrated particular promise in estimating FSIQ, lowering BRE overestimates of IQ and adding to the overall predictive utility of the BRE in estimating premorbid WAIS-R IQ. Additionally,
zero-order correlations also indicated that the Lie (L) scale was significantly,
though modestly, related to VIQ, PIQ, and FSIQ in their psychiatric sample.
Other studies (Ackerman & Heggestad, 1997; Costa & McCrae, 1992b;
McCrae, 1987) have found substantial, resilient relationships between Openness
to experience and task-based measures of intelligence and cognitive performance. Especially where such broad traits would not be expected to change (as
in mild head injury), self- or observer-reports of normatively stable personality
dispositions such as Openness may incrementally add to the estimation of premorbid cognitive ability over traditional demographic and performance-based
approaches. Although recent studies have focused on the role of Openness and
Conscientiousness in measured intellectual ability (Costa & McCrae, 1992b),
Neuroticism and Extraversion are broad traits that have been studied more
extensively in relation to cognitive ability in normal persons (Ackerman &
Heggestad, 1997). Meta-analytic results reported by Ackerman and Heggestad
(1997) suggest that Neuroticism is uniformly negatively related to most cognitive domains, including general intelligence. Extraversion was found to be uniformly positively related to test performance on measures of general intelligence
and other cognitive domains. Ackerman and Heggestad's (1997) results are also
consistent with findings and contentions by McCrae (1987) and Costa and
McCrae (1992b) that Openness is related to general intellectual ability and
achievement.

Although these results may implicate the role of core personality features in
the assessment of cognitive abilities in normal persons, it is unclear to what extent
these results would generalize to clinical samples of persons reporting with symptoms of brain-behavior dysfunction or in what precise manner they would be
applied in clinical assessment. However, in a large sample ($N = 176$) of patients
with presumed head injury representing a wide range of injury severity, Ross and
Putnam (1998) have also found that Openness to experience is moderately related to VIQ, PIQ, and FSIQ (mean $r = .44$, $p < .01$) as expected. The Revised
NEO Personality Inventory (NEO PI-R) (Costa & McCrae, 1992a) measures of
Extraversion and Neuroticism were also significantly related to measured intellectual functioning as predicted by Ackerman and Heggestad's (1997) findings in
normal samples. In particular, Extraversion was positively related to VIQ, PIQ,
and FSIQ (mean $r = .23$, $p < .01$); whereas Neuroticism was negatively related
to VIQ, PIQ, and FSIQ (mean $r = -.22$, $p < .01$. Further, in this sample, self-
reported standing on these traits significantly added to the incremental prediction
of VIQ, PIQ, and FSIQ scores over NAART estimates of premorbid functioning
in hierarchical multiple regression models. Provided that self-reported Openness

remains stable after head injury (Kurtz, Putnam, & Stone, 1998) and is relatively insensitive to social desirability (Ross, Bailley, & Millis, 1997), trait Openness may hold some promise in the estimation of preinjury intellectual ability.

In summary, our assessment of the various *psychometric* approaches to estimating premorbid functioning is mixed, but promising. Regression-based approaches appear to be useful, particularly those that incorporate both demographic and current performance-based methodologies. In addition, personality dispositions may also hold some promise in the estimation of premorbid ability. These approaches are often limited, however, by their focus on estimating WAIS-R composite scores rather than specific areas of neurocognitive functioning. Obviously, such approaches can be used when attempting to compare obtained WAIS-R performance with premorbid estimates, but generalizations about specific areas of functioning may, at present, be more speculative in the absence of additional information.

School records, academic data, and standardized achievement testing

Years of formal education completed is a variable that is almost invariably collected and given thoughtful consideration by clinical neuropsychologists. In many clinical situations this information is likely to suffice for routine purposes. However, a simple self-report of the number of years of formal education may be insufficient when a more detailed estimate of premorbid functioning is sought, such as in a forensic neuropsychological assessment.

It is quite possible that the student with 12 years of formal education who received mostly A's may be cognitively and psychosocially different from the student with 12 years of education who earned marginal grades. The correlation between past academic attainment and currently measured cognitive performance is impressive. For each year of missed or abbreviated schooling there is an accompanying decline in IQ scores (Ceci, 1991). This finding relates to the quantity of schooling completed although the quality of the educational environment (e.g., class size, per pupil expenditures, etc.), which is more difficult to assess, would seem to be similarly contributory in producing individual differences in cognitive ability. In any case, formal education introduces the individual to modes of cognizing such as conceptual organization, taxonomic sorting, and hypothetical thinking, skills emphasized on most conventional intelligence tests in popular use (Ceci, 1991).

Particular strengths or weaknesses in academic performance should be thoughtfully considered. Many neuropsychological tests that are assumed to measure one particular domain may actually be heavily influenced by acquired core academic abilities. For example, the PASAT (Gronwall, 1977) is often considered to be a measure of sustained attention, but performance is correlated with mathematical ability (Sherman, Strauss, & Spellacy, 1997). An individual who failed to develop requisite mathematical skills in school, independent of cerebral dysfunction, may display performance inefficiencies on such a speeded central processing measure. Reading proficiency has been shown to be related to performance on the Stroop Color and Word Test (Cox et al., 1997) suggesting that poor performance is not exclusively pathognomonic for deficient selective response inhibition. With reference to verbal learning tasks, such as the California Verbal Learning Test (CVLT), both vocabulary (Keenan, Ricker, Lindamer, Jiron, & Jacobson, 1996) and verbal IQ (Rapport, Axelrod, Theisen, Brines, Kalichstein, & Ricker, 1997) have been demonstrated to be correlated with performance on this measure. The Keenan et al. (1996) study also raises an interesting issue regarding the extent to which the ordinal variable of education adequately predicts performance. In that study, it was noted that although there was a strong relationship between performance on the CVLT and general fund of lexical knowledge (WAIS-R Vocabulary subtest), years of education *per se* were not significantly statistically correlated with performance. Thus, although a variable that is clearly influenced by education (WAIS-R Vocabulary) was strongly related to verbal list learning, a direct statistical relationship between education and list learning was not in evidence, suggesting perhaps that vocabulary may be best conceptualized as a moderator variable. This does not, however, argue against the use of demographic considerations, but rather suggests that certain factors in addition to years of education, may need to be considered in evaluating the relationship between premorbid abilities and current neuropsychological functioning.

School records, standardized test scores, grades, teachers' progress reports and impressions, may yield information suggestive of suboptimal learning status or emotional problems (e.g., "does not try hard enough", "emotionally immature"). It is necessary to request corroborating data, however; such comments may be viewed as indicia of inchoate cognitive or emotional dysfunction. Another potential "red flag" is reference to an individual being "advanced" in school for reasons other than substantive academic progress (e.g., passed into the next grade because of size, age, pressure from parents, etc.).

Another problematic issue is that of comparability across schools. That is to say, the variability across schools and school districts in degree of rigor, evaluation, and quality of environment is a commonly accepted phenomenon. The abilities or effort that would result in a superior grade in one school, or for that matter with a particular teacher, might earn a more modest grade in another setting. Standardized achievement tests based on national or regional standards are likely to be more reliable, at least in terms of a more uniform metric, and provide a more robust standard of comparison. Of course, caution must be exercised given the fact that some students may not be proficient test-takers or, conversely, may be very clever test takers, particularly on multiple choice tests administered in a group-format.

One of the authors recalls a litigated case wherein two neuropsychologists (one of whom was suggested by the plaintiff's attorney, but not officially retained as an expert, while the other was expressly enlisted as an expert for the plaintiff) and a neurologist (also hired as an expert for the plaintiff) unquestioningly accepted the assertion by the plaintiff's mother that her teenage daughter's bicycle accident had transformed her from an "A" student to a "D" student. This assertion was uncontested, and many subsequent opinions and attributions about the plaintiff's cognitive status were gauged against this pre-supposed baseline. For example, the only notable findings from either of the very comprehensive neuropsychological evaluations were a mild retrieval impairment on one verbal list learning task, and rather dramatically below-aver-age performances on the Wide Range Achievement Test-Revised (WRAT-R), most notably on the language dependent subtests (Reading and Spelling). Both neuropsychological examiners interpreted this performance as indicative of an acquired cognitive disorder that was causally related to the putative brain injury (even though there was no actual injury to the head, no loss or alteration of con-sciousness, and no post-fall confusion). One of the present authors was retained to conduct an independent neuropsychological evaluation of the plaintiff. Upon reviewing the actual academic records, it was ascertained that the plaintiff had never, in fact, been an "A" student, nor had she been reduced to a "D" student. Her grades remained consistently in the "C" and "D" range throughout her entire academic career. Moreover, she had also been formally identified as hav-ing a language-based learning disability and subsequently enrolled in a special education curriculum from elementary school through high school.

Military records
Military records can provide a potential source of information pertaining to pre-morbid background. Documentation of conduct problems, conflicts with authorities, substance abuse, or psychiatric or neurologic illness or injury may contribute to a more accurate baseline estimation vis-à-vis the individual's cur-rent status. In certain cases previous psychological or neuropsychological test-ing may have been performed which, of course, is particularly valuable as a baseline reference.

Previous litigation
Some litigants who present for neuropsychological examination have been involved in previous or concurrent matters of litigation. Multiple law suits may reflect a pattern of coping with stressors wherein an external cause is sought and "exoneration" is achieved through convincing authorities that one is not respon-sible or capable of change. Perhaps a less psychologically elaborated but cer-tainly face valid interpretation of repetitive personal injury litigation is that the individual is motivated, at least in part, by secondary gain.

Premorbid factors that may affect related neurodiagnostic tests
Individuals who present for forensic neuropsychological evaluations may have prior medical histories that contribute to, or were exacerbated by, the event that initiated the current litigation. Although the potential contributions of certain medical or neurological issues will be apparent to the forensic neuropsycholog-ical examiner (e.g., history of previous head injury, cerebrovascular disease, sub-stance abuse, mental retardation, etc.), there are certain issues involved in forensic assessment in which the contribution of premorbid factors may be as potent, but not as obvious.

The forensic neuropsychologist must be very sensitive to premorbid factors that can affect the results of subsequent medical tests, even those presumed to be objective. While many physiological tests are likely to have fewer sources of error variance (e.g., complete blood count), some medical tests that have gained increasing use in clinical and forensic contexts can have more potential sources of error that may not be apparent to the non-specialist professional consumer. Single photon emission computed tomography (SPECT) is one such procedure with numerous potential sources of measurement error (Zhang, Park, & Kim, 1994). SPECT technology, which was originally developed in the context of investigating cerebrovascular disease, is based on the finding that regional changes in brain activity or chemistry can be indirectly measured via externally placed gamma cameras which detect the regional accumulations of tracer flow or receptor-binding isotopes. SPECT is primarily used as a measure of regional cerebral blood flow (rCBF), although some specific neuroreceptor imaging studies can also be accomplished (Holman & Devous, 1992).

SPECT has been shown to be of use in research studies following head injury (e.g., Ichise, Chung, Wang, Wortzman, Gray, & Franks, 1994), but at present there is no particular SPECT profile that is pathognomonic or reliable for brain injury (Herscovitch, 1996). Clinically, the literature does not support the routine use of SPECT in head injury or postconcussion syndrome, and the Therapeutics and Technology Subcommittee of the American Academy of Neurology (1996a) has rated SPECT as an *investigational* procedure for the study of head trauma. (In contrast, this same committee has rated neuropsychological assessment as an *established* procedure in the clinical evaluation of head injury and postconcussive symptoms; 1996b). In spite of these professional cautions and the absence of scientific support for the routine use of SPECT in head injury, SPECT appears to be used frequently in clinical and forensic contexts as a means of "diagnosing" such.

SPECT is affected by many factors other than a primary neurologic illness or injury, and can be affected by factors that, in fact, pre-date a putative head injury (Juni, 1994). For example, SPECT and related rCBF investigations have been shown to be abnormal in *psychological disorders*, such as depression (Sackeim, Prohovnik, & Moeller, 1990), manic-depressive disorder (Iidaka, Nakajima, Ogikubo, & Fukuda, 1995), obsessive-compulsive disorders (Adams, Warneke, McEwan, & Fraser, 1993), generalized anxiety disorders (Uchiyama, Sue, Fukumitsu, Mori, & Kawakami, 1997), postraumatic stress disorder (Semple, McCormick, Goyer, Compton-Toth, 1996), and schizophrenia (Paulman, Devous, & Gregory, 1990). SPECT frequently demonstrates abnormalities in dementia (Read, Miller, Mena, & Kim, 1995) and cerebrovascular disease (Masdeu & Brass, 1995), and also suggests decreased tracer uptake in normal aging (Mozley et al., 1996). Of additional relevance to forensic neuropsychological assessment are studies that have demonstrated unusual functional neuroimaging findings in domains that frequently covary with (or are mistaken for) head injury, such as learning disabilities (Wood, Flowers, Buschbaum, & Tallal, 1991), pain (Derbyshire, Jones, Davani, & Friston,

1994), and somatization disorder (Lazarus & Cotterell, 1989). Finally, SPECT often demonstrates abnormalities in aggression (Amen, Stubblefield, Carmichael, & Thisted, 1996), alcohol dependence (Modell & Mountz, 1995), alcohol withdrawal (Mampunza, Verbanck, Verhas, & Martin, 1990), opiate use (Krystal, Woods, Kosten, & Rosen, 1995), hallucinogenic drug use (Hertzman, Reba, & Kotlyarov, 1990) and cocaine abuse (Miller, Mina, Giombetti, & Villanueva-Meyer, 1992). Alcoholics in long-term abstinence with full behavioral remission of drinking also continue to demonstrate SPECT abnormalities (Dupont et al., 1996).

While it may not be within the direct purview of the forensic neuropsychologist to address issues related to SPECT (or other neuroimaging procedures) in detail, it is important that neuropsychologists who encounter "positive" SPECT findings in forensic cases consider the multitude of conditions and factors that can influence the results, and not simply accept the findings as unequivocal evidence of head injury and proceed to make assumptions regarding causality vis-à-vis neuropsychological test performance.

Another neuromedical assessment procedure that has found increasing use in forensic contexts is that of quantitative or digital electroencephalagram (EEG). As with SPECT, digital EEG is affected by certain factors other than neurotrauma. The Therapeutics and Technology Assessment Subcommittee of the American Academy of Neurology, along with the American Clinical Neurophysiology Society (formerly the American Electroencephalographic Society), have published a report on the use of digital EEG which discusses the insufficiency of evidence for the use of digital EEG in the diagnosis and assessment of postconcussion syndrome, or minor or moderate head injury (Nuwer, 1997). In addition, these committees also recommend against the use of digital EEG in civil or criminal judicial proceedings, and consider the clinical use of digital EEG by non-physicians to be unacceptable.

Personality

Consideration of premorbid personality
Unlike the quantitative summary yielded by an IQ test, personality traits certainly represent a more variegated and complex construct that is difficult to generically quantify or operationally define. Arguably, this has hindered clinical utility in some applied health care settings in which neuropsychologists are likely to practice. Of course, the ease with which certain constructs can be measured has little or no bearing on their importance. Thus, a determination of premorbid personality traits are perhaps less amenable to the accepted methodologies reviewed in the first section of this chapter. Why should the forensic examiner be concerned with premorbid personality traits? Alterations in so-called personality are commonly cited complaints among patients and family members following traumatic injuries. In cases where medical substantiation of

injury is equivocal or absent, these reported changes may become the focus of litigation. While an impressive store of knowledge exists about the development of personality traits and their behavioral manifestations, the nature of personality change in response to life events, whether traumatic or normal developmental processes, has become a recently burgeoning area of research. Given the relative shortage of empirical data and the difficult conceptual ventures involved for personality theorists, monumental challenges in addressing these issues in the forensic examination exist.

The belief that psychological or personality factors may influence the subsequent expression of physical symptoms and the course of illness has been a basic assumption inherent in the practice of health care since the time of Hippocrates. Indeed, we propose that lifelong personality traits serve as a substrate and potent context for understanding response to stress, injury, and various expressions of psychopathology and, to some extent, neuropathology. Symptom reporting does not invariably represent a one-to-one correspondence with *bona fide* physiological injury, but rather, represents the perception of the individual, a perception that is mediated through the individual's core personality disposition. Machulda, Berquist, Ito, and Chew (1998) reported that an individual's perception of stress, independent of the frequency of actual stressful events, was significantly related to the level of symptom reporting in mild head injury. These authors assert; "normal variance in individual's response to stress in most environmental conditions is exaggerated when they are faced with more challenging, and often confusing, environmental conditions, such as MTBI (mild traumatic brain injury)" (p. 421). Therefore, it is necessary for clinicians to conscientiously integrate the potential contributions of premorbid personality with the patient's *present* clinical status.

Personality development and stability
Millon and Davis (1996) have offered the following working definition of personality:

> Personality is seen today as a complex pattern of deeply embedded psychological characteristics that are largely non-conscious and not easily altered, expressing themselves automatically in almost every facet of functioning. Intrinsic and pervasive, these traits emerge from a complicated matrix of biological dispositions and experiential learnings, and ultimately comprise the individual's distinctive pattern of perceiving, feeling, thinking, coping, and behaving (p. 4).

Such a conceptualization of personality assumes that each person's biophysical disposition influences his or her life experience, and the range and frequency of experiences to which the individual is subsequently exposed throughout life. These biogenic and psychogenic factors strongly influence the development of personality as individuals develop unique sets of inherent behaviors, cognitions, and affects clearly distinguishing them from others with different dispositional backgrounds. Temperamental dispositions in the child, in turn, evoke counterreactions from others that may strongly reinforce these initial dispositions. Over time these influences shape the development of abiding personality traits (Millon & Davis, 1996). Tellegen (1988) has defined a trait as "a psychological (therefore organismic) structure underlying a relatively enduring behavioral disposition, i.e., a tendency to respond in certain ways under certain circumstances. In the case of a personality trait, some of the behaviors expressing the disposition have substantial adaptational implications" (p. 622). A person's genetic proclivities and the continuity of environments that comprise one's life history jointly create, develop, and maintain these structures that with time and experience produce enduring personality traits.

Costa and McCrae (1994) have emphasized that, "Stable personality traits affect characteristic adaptations, but so do external influences. The result is that behaviors, attitudes, skills, interests, roles, and relationships change over time, but in ways that are consistent with the individual's underlying personality... Stability in personality thus gives rise to continuity in life" (p. 35). These authors offer the example of individuals who demonstrate a high degree of openness to experience as perhaps more inclined to initiate midcareer shifts than individuals closed to experience. In their view, this actually underscores the predictability of personality; "Such a midlife change requires new habits and skills . . . but the change itself is characteristic because the need for variety is part of what it means to be open to experience . . . the new adaptations are also likely to be characteristic of the basic tendencies of the individual" (Costa & McCrae, 1994, p. 35).

According to Millon and Davis (1996):

> The importance of early learning cannot be overstated for creatures that continue to live in the same environments as their ancestors... Experiences in early life are not only ingrained more pervasively and forcefully, but their effects tend to persist and are more difficult to modify than later experiences. . . early events occur at a presymbolic level and cannot easily be recalled and unlearned; they are reinforced frequently as a function of the child's restricted opportunities to learn alternatives; they tend to be repeated and perpetuated by the child's own behavior (pp. 120–121).

For example, in relation to the parental relationship, the isolated child may achieve some sense of recognition in the family order by assuming a role that the family system has not filled. The chronically sick child may assume a role in which parental attention focuses upon and reinforces maladaptive ways of coping with life circumstances. What was perhaps a competing system of relationships can now recognize a child whose importance was at issue before illness behaviors evolved (often in a non-deliberate manner) and, as a result, introduced change. To generalize from this model, a minority of individuals with chronic complaints of attentional inefficiency, dysfunction of memory processes, and a host of nonspecific complaints often associated with head injury, may find a role in life which has meaning and purpose, even if defective and incomplete. The importance of appreciating enduring personality in this regard is that it offers a backdrop upon which misattributions of cognitive abnormalities can be understood conceptually.

The foregoing discussion is not intended to suggest that personality change is not possible, but rather, that it may be predictable on the basis of a priori dispositions and no more likely following trauma than some other developmental event. Costa and McCrae (1994) assert; "In many fundamental ways adults remain the same over periods of many years and their adaptation to life is profoundly shaped by their personality. People surely grow and change, but they do so on the foundation of enduring dispositions" (pp. 35–36). The basic stability of individual differences in personality provides a particularly valuable framework from which to assess an individual's response to the experience of injury, illness, or trauma.

Personality traits mediate the experience of injury or illness

Individual differences in personality may be neglected unless the patient presents with a behavioral disturbance or is highly symptomatic. The meaningfulness of personality variables in evaluating and treating patients may be poorly appreciated by some in the neuropsychological community. Adams (1989) believes the role of personality variables have been disregarded in contemporary clinical neuropsychological practice and facetiously proposes the historical perspective that even "the famous cases of aphasias, alexias, apraxias, aphemia without aphonia and a-*this*-ias without a-*that*-ias did not occur in people without premorbid personalities, developmental histories, defense mechanisms, hopes, fears, or inner emotional life!"

When individuals with particular personality characteristics, even if not formally identified as a personality disorder *per se*, sustain a head injury, the manner of response and the course of recovery will vary considerably. Although a variety of somatic and functional symptoms are characteristic of normal human experience, certain individuals seem more disposed to assign meaning and significance to such symptoms. The insidious role of expectancy in the etiology of somatic complaints after mild head injury has been addressed by Mittenberg, DiGiulio, Perrin, and Bass (1992). These investigators found that patients having sustained mild head injury consistently underestimated the normal prevalence of various somatic symptoms in their retrospective accounts when compared with the base rate established by normal controls. Mittenberg et al. (1992) suggested that "patients may reattribute benign emotional, physiological and memory symptoms to their head injury" (p. 203). On an *a priori* basis this tendency would appear to be further accentuated in cases involving potential financial gain such as personal injury litigation. Indeed, Lees-Haley and Brown (1993) have suggested that the litigation process itself is likely to increase the rate of endorsement of traditional post-concussional symptoms. Mittenberg et al. (1992) argue that such circular reinforcement of expectations (i.e., the individual simply "discovers" what he or she is expecting to find), may, in fact, account for the persistence of these symptoms in the absence of objective indices of impairment. In other words, individuals can be implicitly conditioned to interpret commonly occurring premorbid sensations as manifestations of altered brain functions following an injury event.

With respect to head injury in particular, a problematic factor is the generality of assumptions by some clinicians that personality will be altered by head injury. Considering the paucity of compelling empirical evidence for this postulate, clinicians are at a loss to accurately predict specifically what changes (i.e., which traits, in which direction, and to what extent?) are expected from the various severities and myriad types of injury that can be suffered by the brain. Changes in symptomatic status should not be construed as reflecting changes in stable personality traits. The majority of clinicians are unacquainted with the patient previous to the injury and inclined to operate on vague or mistaken assumptions that may lead to the belief that symptomatic manifestations represent changes in core personality. Such thinking may further contribute to or reinforce perceptions of personality change and initiate a continuing confirmatory bias among other providers. The role of such iatrogenic influences is beyond the scope of this chapter, but particularly relevant to this discussion.

Personality contributions to the development of medical disorders

Personality traits have been associated with certain medical conditions (e.g., asthma, rheumatoid arthritis, irritable bowel syndrome, nonulcerative dyspepsia, inflammatory bowel disease, and dermatoses; Folks & Kinney, 1992a, 1992b). However, there has been no compelling evidence to date that any specific type of personality trait alone can account for the development of a particular physical illness, with perhaps the exception of coronary artery disease. Elements of hostility are believed to be cardiotoxic components of the Type A personality style which may expedite the development of arteriosclerosis. Heightened physiologic reactivity in such individuals, resulting in acute and chronic activation of the sympathetic nervous system, is one mechanism in which arteriosclerosis and cardiac arrhythmias are believed to be more likely to develop.

Several conditions in which the individual's response to stress has been associated with heightened physiologic reactivity leading to the expression or exacerbation of disease symptoms include urticarial reactions, eczema, atopic dermatitis, and pruritus. Exacerbations of asthmatic reactions in particular have been associated with stress reactions, although the mechanisms involved are, at present, not entirely clear (Folks & Kinney, 1992a, 1992b). Evidence for a personality style affecting the onset or course of cancer is less clear, although some studies seem to show that personal determination may be associated with a more auspicious outcome. Of course, attributing etiologic importance to certain personality factors that may, in fact, be complications rather than causes of a particular disease must be avoided. The degree of psychological adaptation in response to illness, particularly psychological defenses against anxiety, may positively influence outcomes in certain conditions.

Personality disorder and characterological styles

Although individuals are likely to display many attributional styles and expectancies across life situations and contexts, a person's overreliance on certain behaviors may constitute an ingrained pattern and psychological rigidity. Where this inflexibility interferes with appropriate adaptation to new situations and challenges, the personality style may represents a pathologic condition. Thus, the notion of a personality disorder is a construct employed to represent specific styles or patterns in which the personality system behaves rigidly and maladaptively in relation to the environment. Millon and Davis (1996) emphasize:

> From a developmental perspective, maladaptive personality characteristics result from the same influences shaping the development of normal personality. Important differences in the character, timing, and intensity of these influences contributes to individual differences in personality. When an individual displays the capacity to respond to the environment in a flexible manner, and his or her perceptions and behaviors facilitate personal satisfaction, then the person is viewed as possessing a normal or healthy personality (p. 13).

Among the most interesting and relevant characterological styles involved in forensic assessment of personality is a disorder which does not specifically fall within the axis II offerings for the DSM-IV (American Psychiatric Association, 1994). However, because it represents a particular, personological style of interpreting and attributing physiologic symptoms to physical causes and conditions that is stable over time, some have argued that somatization may represent a form of personality disorder (Stern, Murphy, & Bass, 1993). Somatization appears to be highly associated with personality disorder, especially among cluster B (flamboyant) and C (anxious/fearful) subtypes (Tyrer, Gunderson, Lyons, & Tohen, 1997). In most studies, the majority of patients with somatization disorder also have personality disorder (Fink, 1995; Rost, Atkins, Brown, & Smith, 1992; Stern et al., 1993). Thus, understanding premorbid personality styles and dispositions that may interact with unexpected traumatic events is an important step in evaluating the psychological importance of patient claims regarding post-injury functioning.

Premorbid emotional and adverse life events

Of particular relevance to the neuropsychologist is the work of Fenton, McClelland, Montgomery, MacFlynn, and Rutherford (1993). These investigators found among a sample of individuals having sustained a mild head injury that those who demonstrated a persisting pattern of symptoms also reported twice as many adverse life events or social difficulties occurring in the year preceding the injury as compared to individuals whose symptoms had resolved. The authors suggested, "The emergence and persistence of the postconcussional syndrome are associated with social adversity before the accident. . . . A holistic approach which takes account of people's physical nature and in which meaning and relationships matter is essential for a full understanding of the complex evolving sequelae of minor head injury" (pp. 493, 496). In an earlier investigation, patients having sustained mild head injury with reported pre-injury emotional problems scored higher on a scale of post-concussive cognitive symptoms and had disproportionately higher scores on an emotional-vegetative scale than a mild head injury subgroup not reporting such pre-injury complications (Bohnen, Twijnstra, & Jolles, 1992).

Premorbid vulnerabilities and propensities

Substance abuse is of particular importance in addressing a patient's premorbid personality. The risk of developing a substance abuse disorder is not equally distributed across the general population and the incidence rates for such disorders among head injury samples has been reported to be remarkably high (National Head Injury Foundation, 1988; Putnam & Adams, 1992). Graham and Strenger (1988) reviewed the literature pertaining to MMPI characteristics of alcoholics and identified several distinct pathological-characterological alcoholic profile types based on factor- and cluster-analytic studies. They concluded, "As a group, alcoholics have in common a tendency to be impulsive, to resent authority, to have low frustration tolerance, and to have poorly controlled anger" (p. 202). These findings are consistent with other studies suggesting that Cluster B personality disorders which are characterized by a flamboyant interpersonal style (e.g., antisocial and borderline personality disorders) have a high comorbidity with alcohol abuse (Helzer & Pryzbeck, 1988), cocaine (Barber et al., 1996) and opiod abuse (Brooner, Herbst, Schmidt, Bigelow, & Costa, 1993; Darke, Hall, & Swift, 1994; Links, Heslegrave, Mitton, van Reekum, & Patrick, 1995). Along with cluster B variants, there are many other comorbid personality disorders present in patients with polysubstance abuse (Tyrer et al., 1997). In fact, the comorbidity of many mixed personality and substance abuse disorders may represent a diagnostically separate group (Brooner et al., 1993). For example, in a study of polysubstance abusers by DeJong, Van den Brink, Harteveld, and van der Wielen (1993), identifying a particular set of personality disorders that was associated with abuse was difficult because of the co-occurrence of several additional personality disorders among individuals who composed this sample. In discussing the clinical implications of a meta-analytic review of mild head injury and the contributions of premorbid influences on outcome, Binder (1997) asserted, "A history of alcohol abuse could explain some symptoms because of the toxic effects of alcohol or the presence of psychological problems associated with alcohol abuse" (p. 441). His contention implies that clinicians need to be particularly careful to avoid the potential confounds of chronic alcohol abuse and issues of individual differences in personality that often co-occur with significant levels of alcohol abuse.

Failure to return to previous levels of activity and involvement in life following very mild gradients of head injury may result from a number of conditions related to "secondary gain". Although monetary compensation associated with personal injury litigation is often an entry point for discussing this topic, more implicit aspects of secondary gain may play an important role in the persistence of symptoms after injury in some patients. In this regard, one might consider the work of Zuckerman, Kieffer, and Knee (1998) on the issue of self-handicapping. Self-handicaps are defined as obstacles to successful performance that are constructed by a person to protect or enhance self-esteem (Berglas & Jones, 1978; Jones & Berglas, 1978). Arkin and Baumgardner (1985) have classified various forms of self-handicaps into *acquired obstacles* (obstacles that actually lower the likelihood of success) and *claimed obstacles* (obstacles that people claim to have). For some, a mild head injury (acquired obstacle) may potentially become a basis for claimed deficits (claimed obstacle). Zuckerman et al. (1998) show how a propensity toward high self-handicapping results in poorer adjustment which, in turn, results in higher prospective self-handicapping. Such a model addressing a self-perpetuating cycle may be useful in explaining a subset of individuals with mild head injury who not only do not recover following injury, but seem to progressively deteriorate. The cognitive and psychiatric handicaps putatively produced by the injury serve the purpose of providing an "explanation" for their misfortunes and failures.

While psychological premorbidities are known to contribute to outcome from mild head injury, we propose that enduring personality traits not necessarily deemed psychopathological, similarly shape an individual's interpretation of the injury event, adaptive or maladaptive response to symptoms, and extent to which resolution is accomplished and forward momentum is resumed.

Base rates of commonly reported neuropsychological symptoms
Setting aside the issue of personality traits in symptom expression and maintenance, it is important to empirically consider the actual occurrence of certain behaviors and symptoms in the general population. Dikmen and Levin (1993) recognize that "A common mistake in clinical practice is automatically to attribute the cause of the difficulties observed in patients seen long after the injury to the head injury . . . morbidities seen after mild head injury may have predated the injury, or it is possible that preexisting conditions may compound the effects of the mild head injury" (pp. 33, 35). This introduces the important issue of symptom base rates and their application to patient assessment. The base rate issue is well illustrated by a recent study which reported that 52% of a sample of asymptomatic patients had various lumbar spine abnormalities and that only 36% demonstrated a normal disc at all levels (Jensen et al., 1994). The authors concluded "The discovery of a bulge or protrusion on an MRI scan in a patient with low back pain may frequently be coincidental . . . Abnormalities of the lumbar spine by MRI examination can be meaningless if considered in isolation" (p. 72). In a study by Newman (1992) on the rates of endorsement of PCS symptoms in normal college students, he found a high level of endorsement of headaches (41%), memory problems (32%), word-finding difficulty (42%), depression (47%), and poor concentration (41%), among others. These findings led Newman (1992) to conclude, "It would appear that many of the symptoms characteristic of the post-concussional syndrome are not unique to head injury" (p. 3). In fact, Iverson and McCracken (1997) also reported that

chronic pain patients *without* head injury endorsed common PCS symptoms at a high rate. In this study, approximately two thirds of the pain patients endorsed three or more PCS symptoms.

It is of interest to also consider the percentage of normal subjects in the MMPI-2 standardization sample endorsing certain items possibly suggestive of deficient physical and emotional health (Butcher, Dahlstrom, Graham, Tellegen, & Kaemmer, 1989). Certain items illustrate the extent to which functional complaints seem to be commonly reported by normal adults. For instance, in response to the item "I feel tired a good deal of the time", 34% and 25% of the females and males, respectively, positively endorsed the item. The item "I have few or no pains" was endorsed "false" by almost 20% of the sample and the item "I forget where I leave things" was endorsed affirmatively by 40% of the standardization sample. The essential point is that such symptoms are not diagnostically definitive or exclusive, but rather, appear to be commonly experienced by non-patient adults.

Further, in a more direct comparison by Fox, Lees-Haley, Earnest, and Dolezal-Wood (1995), the base-rates of PCS symptoms (e.g., headache, memory, dizziness, concentration, and fatigue) for neurology patients was not significantly different from patients receiving psychotherapy. Nonetheless, patients who reported loss of consciousness from head injury also reported significantly higher levels of PCS symptoms than patients who reported simply a bump on the head, were in litigation, or were apparently normal. Thus, the symptoms reported by patients with head injury involving loss of consciousness are not qualitatively idiosyncratic to head injury but may only be significant from a quantitative standpoint. One possible conclusion derived from the Fox et al. (1995), Lees-Haley and Brown (1993), and Mittenberg et al. (1992) investigations is that while PCS-type symptoms appear to be common even in the absence of CNS injury, the retrospective self-analysis performed by mild head injury patients often underestimates or dismisses the existence of the pre-injury symptoms they may have experienced. Specifically, Mittenberg et al.'s (1992) study suggests that mild head injury patients tend to underestimate preinjury cognitive faults and difficulties. Thus, the comparison of pre- and post-injury levels of functioning may contribute to a contrast effect in which post-injury cognitive faults are notably salient to patients with presumed mild head injury. Consequently, the underestimation of pre-injury cognitive faults in conjunction with the salience of the injury event may contribute to the re-attributing of cognitive faults and symptoms predominantly to the head injury. This failure to accurately interpret present difficulties in the context of pre-injury problems and functioning often reflects individual differences in perception associated with core personality traits.

The prevalence of sleep disturbances has ranged from 10.2% to 52% in several community surveys with the prevalence of psychiatric disorders reported to be much higher among individuals with sleep complaints (Ford & Kamerow, 1989). The National Comorbidity Survey (Kessler et al., 1994), based on a stratified multistage area probability sample of 8,098 individuals, found that about

50% of the subjects reported one or more lifetime psychiatric disorders. Approximately 30% reported one or more twelve-month disorder. However, less than 40% of subjects with a lifetime disorder had received any formal treatment. The most prevalent psychiatric disorders found were major depression, alcohol dependence, social phobia, and simple phobia. Clinicians should be mindful of these epidemiological data to avoid presuming that a patient, even one without a known or reported history, was exempt from psychological or psychosocial disturbance prior to the injury or traumatic event.

Psychiatric disorders not only result in social and occupational disability, but some have also been associated with neuropsychological impairment (Klonoff & Lamb, 1998; Veiel, 1997; Winokur & Clayton, 1994). Of course, patients with affective disorders may demonstrate slowing in the rate of information processing, poor concentration, and memory deficits (Johnson & Magaro, 1987). Depression has similarly been associated with a diminution in the ability to sustain effort in formal assessment (Cohen, Weingartner, Smallberg, Pickar, & Murphy, 1982). Not surprisingly patients with anxiety disorders also report memory inefficiencies (Noyes & Holt, 1994). In general, psychopathology has been associated with retardation of the encoding and storage processes and a disruption of trace storage and recall resulting from a disorganizing effect (Johnson & Magaro, 1987). Premorbid learning disabilities also warrant careful evaluation because they may be over-represented in the mild head injury population and can be associated with pre-existing neuropsychological impairment. Dicker (1992) reported that 50% of the mild head injury subjects in her sample described a premorbid history of learning disability or poor academic attainment. Haas, Cope, and Hall (1987) reported that 40% of their severe head injury sample had a positive history of learning disability.

Post-Traumatic Stress Disorder (PTSD): An exemplar of premorbid vulnerabilities

The post-traumatic reaction to injury or a stressor event is a critical concern in forensic assessment, especially when the event appears to have exacerbated premorbid vulnerabilities and is associated with a maladaptive response to the event. When the patient is able to recall the injury event as in cases of presumed MHI, the development of post-traumatic stress disorder (PTSD) and other post-traumatic reactions may be a valid concern (Bryant & Harvey, 1995). In a sample of 158 victims of motor vehicle accident, Blanchard and Hickling (1995) reported that approximately 33% met criteria for PTSD one to four months after the injury event. More than half of those who were diagnosed with PTSD also met criteria for current major depression. Although most of those with major depression appeared to have developed depressive symptoms subsequent to the accident, a prior history of major depression was identified as a significant risk factor for developing PTSD. In a recent review, McDonald and Davey (1996) indicate that psychiatric disorders are more prevalent in accident victims as compared to the general population. In particular, they note that the frequency of characterological disorders and alcohol abuse is higher in accident victims. Thus, it would appear that an individual's response to the injury event may, in some cases, be influenced by pre-existing psychiatric vulnerabilities that facilitate the development and maintenance of symptoms.

The symptom syndrome that has defined PTSD was primarily conceived as a set of normal reactions to outstandingly abnormal circumstances. Recently, however, investigators are beginning to focus on the sometimes abnormal reactions to commonly occurring but unfortunate incidents such as motor vehicle accidents (Mayou, Bryant, & Duthie, 1993), medical procedures (Shalev, Schreiber, Galai, & Melmed, 1993), or myocardial infarctions (Kutz, Shabtai, Solomon, Neumann, & David, 1994). This approach of identifying individual difference variables among persons who have suffered a trauma has been further supported by the results of the National Vietnam Veterans Readjustment Study (Kulka et al., 1990) which found that only 15.2% of male Vietnam veterans suffer from prolonged symptoms of PTSD. These findings suggest that the contention that trauma *always* results in psychopathology is inaccurate. Persons exposed to the same environmental stressors appear to be differentially at risk for the development of later traumatic reactions. In fact, a number of pretrauma vulnerability factors have already been identified in the development of PTSD. For example, pre-injury socioeconomic factors including poverty and lower educational attainment, as well as early parental separation and negative parenting behaviors have been useful in predicting the onset of PTSD upon exposure to later traumatic events (Davidson, Hughes, Blazer, & George, 1991; Green, Grace, Lindy, Gleser, & Leonar, 1990; McCranie, Hyer, Boudewins, & Woods, 1992). Further, the presence of early traumatic experiences (e.g., childhood sexual and physical abuse) and exposure to traumas similar to the precipitating traumatic event also appear to place individuals at increased risk (Shalev, 1996). From a study of Viet Nam veterans, King et al. (1996) have reported that prior trauma, however is most strongly related to PTSD development in veterans who had been exposed to a high degree of war combat. Thus, premorbid vulnerabilities seem to be least important when the traumatic event proximal to the development of symptoms is highly significant and traumatic. Foy, Resnick, Sipprelle, and Carroll (1987) and Davidson, Smith, and Kudler (1989) have reported that familial psychiatric disorders (particularly alcohol abuse) are related to the development of PTSD under conditions of low combat exposure in veterans. In addition, normatively stable personality traits such as neuroticism and introversion also place individuals at particular risk for developing PTSD. Peri, Beh-Shachar, and Shalev (1994) have reported that increased conditioning susceptibility, which Eysenck (1967) hypothesizes is a mark of high neuroticism and low extraversion, appears to be a significant personal vulnerability factor for developing PTSD.

An especially potent variable contributing to the development of PTSD is an individual's propensity to (premorbidly) engage in dissociative coping or experiencing. In a prospective study of injured trauma survivors, Shalev, Peri, Caneti, and Schreiber (1996) found that after controlling for the effects of gender, education, age, event severity, depression, and intrusive and avoidant anxiety, reports of dissociative experiencing of the trauma at one week after the event predicted 30% of the variance in PTSD symptoms 6-months later. Likewise, Koopman, Classen, and Spiegel (1994) found that dissociative symptoms reported by trauma survivors early after the event predicted PTSD symptoms at seven month follow-up. These results are consistent with earlier findings by Bremner, et al. (1992). which also support the importance of dissociative coping in response to the trauma vis-à-vis later development of PTSD.

The literature pertaining to PTSD is particularly relevant for clinicians evaluating the contribution of premorbid personality to the development of subjectively reported cognitive symptoms (e.g., misplacing car keys, forgetting appointments) following an injury event. Cognitive inefficiencies of persons with suspected head injury seem also to be shared by persons with high levels of posttraumatic stress, anxiety, and depressive disorders. Richardson and Snape (1984) reported that orthopedic trauma controls who had *not* sustained head injury, demonstrated impaired performances on free recall memory tasks. These investigators interpreted their findings to suggest that state anxiety associated with traumatic injury may lead to an impairment of cognitive functions by reducing the capacity of working memory. The diagnosis of PTSD per se, may not specifically include instances of cognitive changes. However, the dissociative aspects of PTSD (e.g., forgetfulness, impaired memory for the traumatic event, poor concentration) in addition to the long-term manifestation of post-traumatic stress (which may include depressive symptoms such as fatigue and anhedonia), may be potent contributors to cognitive disruption and inefficiency. As noted, a high premorbid propensity toward dissociation may be a strong etiologic contributor to the development of PTSD. Indeed, the relationship between dissociation and PTSD is a resilient finding in the literature. Further, given the high association between PTSD and somatization (McFarlane, Atchinson, Rafalowicz, & Papay, 1994; Saxe et al., 1994; Walker, Katon, Neraas, Jemelka, & Massoth 1992), the assessment of PTSD may be all the more important. That is, persons who are at risk for developing post-traumatic symptoms may also be at parallel risk for developing psychogenically-based somatic symptoms. Thus, premorbid cognitive inefficiencies may be exacerbated under high levels of stress and may result in the reattribution of cognitive faults to the injury event in question.

In addition to confirming the link between dissociation and PTSD, other studies point to the relationship between somatization and dissociative experiences. For example, in the DSM-IV field trials, Pribor, Yutzy, Dean, and Wetzel (1993) reported that more than 90% of 100 women with somatization disorder had reported some type of abuse, with 80% reporting a history of sexual abuse. Although dissociation does not necessarily follow from previous physical or sexual abuse, childhood traumatic experiences that include abuse seem to place individuals at high risk for the development of PTSD (including dissociative symptoms; King et al., 1996). Consequently, a positive history of physical or sexual abuse may be an important marker and prognostic indicator of traumatic reaction and somatization as well as dissociation following an event such as mild head injury.

Conceptually, the role of premorbid vulnerabilities and individual differences in personality in the later development of PTSD may bear a resemblance to the development of the protracted post-concussive syndrome following mild head injury. Among persons who experience lower levels of objective trauma, premorbid individual differences appear to be of critical importance in predicting the development of PTSD. Similarly, this may provide a model for investigating the disproportionate response to mild or equivocal head injury observed among certain individuals (Millis & Putnam, 1996; Putnam, Millis, & Adams, 1996). It may be hypothesized that the role of somatization and expectancy in mild head injury is similar to that of dissociation in the development of PTSD. In addition, mild head injury and PTSD, respectively, share certain symptoms in common with depression including sleep disturbance, attentional difficulties, psychomotor slowing, etc. Indeed, Blanchard, Buckley, Hickling, and Taylor (1998) argue that symptom overlap between depression and PTSD is sufficiently high that the threshold for diagnosing comorbid depression should be raised. They found that of 62 motor vehicle accident victims who met the criteria for PTSD one to four months post-accident, 33 also met the criteria for major depression. Further, comorbid depression predicted higher levels of subjective distress, greater major role impairment, and poorer remission rates for PTSD over the first 6 months after the occurrence of the accident.

However, the similarities between mild head injury and PTSD may extend even further. Various studies have been directed at examining the neuropsychological sequelae of both conditions. Overall, studies investigating cognitive functioning in PTSD samples find impairment of attention, learning, and memory functions (Vasterling, Brailey, Constans, & Sutker, 1998). Of course, disrupted functioning in these domains is often reported following head injury. Thus, there appears to be overlap in reported functional impairment between these separate diagnostic groups of patients. However, in a meta-analysis of studies investigating neuropsychological impairment in representative cases of mild head injury, Binder (1997a) found that only attention was robustly found to be impaired. Thus, one must wonder to what extent factors such as PTSD may play a role in the assessment of persons sustaining a mild head injury with questionable loss of consciousness who demonstrate poor performance on neuropsychological measures of memory, attention, and executive functioning.

An objective measure of PTSD that may hold promise in civilian populations is the Clinician-Administered PTSD Scales (CAPS-1 and CAPS-2; Blake et al., 1990; Nagy et al., 1991). The CAPS are structured interviews which allow for the rating of 17 DSM-III-R symptoms of PTSD, as well as 8 PTSD-associated symptoms. Further, because the frequency and intensity of each symptom is rated on 5-point Likert scales, the continuous nature of symptom expression is preserved. Since their introduction, the CAPS interviews have gained relatively wide acceptance as a reliable and valid measure of PTSD symptoms (Blake, 1994).

Measurement and inference of premorbid personality
As noted earlier, there are a number of ways to conceptualize basic personality. Nonetheless, the measurement of personality almost always involves some inference of broad-based traits that are thought to color the patient's behavior across different contexts and environmental demands. Probably the three most influential models of basic personality are the three-factor theory proposed by Eysenck (1952, 1967), the Circumplex model of interpersonal behavior proposed by Leary (1957), and the Five-Factor model (FFM) recently revived by Costa and McCrae (1992a). An explicit comparison and critique of these three theories is outside the bounds of this manuscript. However, we will focus primarily on the FFM because it has gained a prominent position in the field of personality theory (McAdams, 1994) and because a valid and reliable measure of the FFM, the NEO PI-R (Costa & McCrae, 1992a), allows for standardized measurement of personality traits based on reports of significant others or observers. In addition, we will also briefly review the importance of the MMPI-2 and Personality Assessment Inventory (PAI) in the assessment of stable personality dispositions. Although the PAI does not have the long history of clinical usage and research of the MMPI-2, preliminary findings indicate that the PAI may also be of use in clinical and forensic settings (Morey, 1991). However, because the veridicality of self-report after a severe injury event may be suspect as a result of possible personality change, observer- or significant other- reports of personality dispositions may be a necessary step in the evaluation of premorbid personality dispositions in the absence of personality assessment before the injury event.

Stability of measured normal dispositional traits
The factors of the FFM have been variously named by investigators as Neuroticism (or Emotional-Stability), Extraversion (also named "Surgency" by Goldberg, 1981), Openness (or Intellect, Culture), Agreeableness, and Conscientiousness (Costa & McCrae, 1992a). Of central importance to the clinician evaluating premorbid personality traits is the fact that cross-sectional investigations in most persons who are aged 30 or older, suggest that basic personality traits are stable over time (McCrae & Costa, 1990). However, although longitudinal investigations of self-reported standing on FFM measures over intervals varying from three to 30 years also demonstrate adequate stability coefficients (median = .65), individual standing on the FFM may change substantially (Costa & McCrae, 1994). Specifically, Costa and McCrae (1988) report six-year test-retest correlations of .83 for Neuroticism, .82 for Extraversion, and .83 for Openness. However, less stability was found for Agreeableness (r = .63) and Conscientiousness (r = .79) over three years with earlier versions of the NEO. Investigations of the stability of FFM constructs in the MMPI have also been impressive. In a 30 year study of MMPI factor scales, Finn (1986) reports correlation coefficients of .56 for Neuroticism, .56 for Extraversion, .62 for Intellectual Interests (representing Openness), and .65 for Cynicism (representing Agreeableness). An adequate representation of

Conscientiousness was not reported in Finn's study. Further, heritability estimates for FFM traits have been reported to be as high as .73 and .63 in the case of Extraversion and Neuroticism (Heath, Neale, Kessler, Eaves, & Kendler, 1992). Based only on self-report measures of FFM traits, heritability estimates reportedly range from .28 to .49 (Loehlin, 1992). Thus, normative personality traits appear relatively stable over time and are likely constitutional determinants of behavior. However, it should be noted that the younger the patient at the time of injury, the more likely it is that estimates of premorbid personality will inaccurately represent what the patient would have been like if he or she had not incurred an injury. This contention is supported by the observation that test-retest coefficients are substantially lower for younger individuals. In the 30 year follow-up reported by Finn (1986), the stability coefficients for men aged 43 to 53 were substantially higher (median =.53) in comparison to college students aged 17 to 25 (median =.38). Likewise, Helson and Moane (1987) demonstrated higher coefficients of stability for the interval from age 27 to 43 than for the much shorter interval from age 21 to 27. Thus, the fact that basic personality remains stable over time seems to be moderated by age effects on retest. Provided that males in their twenties are at the greatest risk for head injury, the lower stability estimates in younger individuals may be a mitigating factor in drawing conclusions about premorbid personality. This point is especially potent considering that younger patients are also more likely to have suffered more severe head injuries that are thought to sometimes involve significant personality change.

It has been postulated that psychological dispositions, including some aspects of personality, may change as a result of major life trauma (Heatherton & Weinberger, 1994). However, Caspi and Moffitt (1993) report that personality at the trait level in normal individuals remains largely stable, even across major changes in life circumstances. Bagby, Joffe, Parker, Kalemba, and Harkness (1995) investigated the stability of self-reported FFM standing on the NEO PI-R in persons experiencing an acute major depressive episode. Patients completed the NEO PI-R at the time of initial presentation for therapy and at remission of symptoms (five weeks to three months) after first assessment. Bagby et al. (1995) found significant differences in self-reported Neuroticism and Extraversion from pre-treatment to post-treatment scores. Neuroticism decreased and Extraversion increased at post-treatment which also suggests that acute symptomatology can influence self-reported standing on some broad traits.

Knowledge regarding the stability of personality dispositions in the general population over time will aid the task of premorbid personality assessment. In the case of head injury, claims of marked changes in personality are often made by patients, family members, and clinicians. The often cited case of Phineas Gage (Harlow, 1868) is a widely-known and vivid example, although accounts of change in personality associated with neurotrauma are found in more recent literature as well (Blumer & Benson, 1975; Damasio, Tranel, & Damasio, 1990; Meyers, Berman, Scheibel, & Hayman, 1992; Saver & Damasio, 1990). This literature typically relies on clinical descriptions and single case studies, but most notable is the absence of premorbid assessment data from which to document the direction and degree of these putative changes in dispositional traits. Most evidence for personality changes depends on the accuracy of retrospective accounts of premorbid characteristics that may be biased by the passage of time, exposure to post-injury sequelae, and personal needs of the patient or others to view circumstances as having changed. Of course, the cases most frequently bringing attorneys and neuropsychologists together involve milder gradients of head injury in which the issue of alterations in brain functioning is far less certain. A head trauma producing little or no significant loss of consciousness would not appear, from the standpoint of a neurobehavioral substrate model, to significantly influence fundamental personality functioning (Binder, 1997).

One practical and feasible solution to this dilemma is to collect "quasi-premorbid" data on personality traits or other psychosocial variables during the acute phase following the injury event. Timely assessments of premorbid status, such as those gathered before the patient is discharged from the hospital, can circumvent the biases and distortions that are inherent in retrospective accounts collected after much exposure to the patient post-injury. This approach has been used effectively to study changes in children's behavior following head injury (Donders, 1992; Fletcher et al., 1992; Rivara et al., 1993) and in community adjustment among adults (Corrigan & Deming, 1995; Oddy & Humphrey, 1980). Kurtz, Putnam, and Stone (1998, p. 12) collected ratings of head injured patients from relatives and other well-known significant others at an average of 13 days post-injury using the NEO PI-R. The injuries sustained were moderate to severe by conventional standards (e.g., Glasgow Coma Scale) and most of these "premorbid" assessments were obtained while the patient remained hospitalized; in several cases, the patient had not yet regained full consciousness.

Six months later, these raters were re-contacted and asked to complete a follow-up NEO PI-R describing the patient's personality during the post-injury period. Apart from a moderate decline in mean Extraversion to a more average level, the data revealed little in the way of systematic changes in trait levels for these patients compared to six-month retest ratings of a non-injured control group from the same urban community. However, the measured changes in these NEO ratings revealed little correspondence with the raters' subjective and retrospective sense of personality change. After the follow-up NEO ratings were obtained, all raters were asked to judge the degree of change (i.e. "more", "less", or "no change") they had observed in various FFM characteristics of the patients compared to the pre-injury period. These data, which are more representative of that which typically informs routine evaluations of head injury patients, deviated from the serial assessments provided by the same raters in both the overall extent of change and in the trait domains most affected. This discordance between retrospective estimations of change and change assessed using objective serial assessments is consistent with the hypothesis that new attributions are being made for old dispositions by observers of the patients' behavior. Moreover, the unsystematic changes in trait levels seen in the

NEO PI-R data were uncorrelated with several indicators of head injury severity. The study's authors interpret these results "as a cautionary statement against presuming changes in specific traits or even change at all in the personality of any individual TBI patient, even following severe injury" (Kurtz, Putnam, & Stone, 1998, p. 12). Despite the preliminary nature of these findings, the direct implications for forensic neuropsychological practice are indisputable. While they place perhaps some extra burden of proof upon litigants and evaluators contending personality changes due to neurological injury, they also provide an objective and defensible means of providing evidence for such changes when employing psychological measurement data.

Stability of measured psychopathological dispositional traits

Part of the difficulty in estimating pre-injury personality dispositions involves the stability or instability of personality scales and measures across normative samples and over time. That is, a measure that is unstable in normal populations would be assumed to remain unstable when applied clinically. The converse of this proposition, however, may not be accurate. That is, a scale or measure that is stable in normal populations may not be stable from pre- to post-injury if a neurological event has occurred in the interim. As noted earlier, severe disruption of the brain has been clinically (or, anecdotally) related to changes in personality (Damasio, Tranel, & Damasio, 1990; Meyers, Berman, Scheibel, & Hayman, 1992; Saver & Damasio, 1990). However, these cases often represent *severe* neurobehavioral dysfunction *unlike* that noted in mild head injury.

The most influential and widely used self-report inventory for the measurement of psychopathology and personality has been the MMPI and its revision, the MMPI-2 (Butcher et al., 1989). The original test-retest studies of the MMPI-2 involved rather short intervals of 7 to 10 days (Butcher et al., 1989). However, later studies by Putnam, Kurtz, and Houts (1996) and Spiro, Butcher, Levenson, Aldwin, and Bosse (1993) have reported stability estimates on Basic, Supplementary, and Content scales at intervals of four months and five years, respectively. Although a thorough review of these studies can be found in Putnam et al. (1996), we report some of their findings here. Of the Basic scales, Social Introversion is most stable (.92–.85) whereas Paranoia (.67–.55) and Hysteria (.72–.51) were least stable across the three studies. Also, at longer retest intervals, K-correction (.83–.75) and Depression (.74–.71) were reliably most stable over the Putnam et al. (1996) and Spiro et al. (1993) studies. Overall stability of the Basic scales was generally adequate across the three studies (*r*'s ranging from .51–.92). Review of the Supplementary scales indicate that measures of noncontent-based responding, VRIN (.60–.54) and TRIN (.34–.31), were most changeable at retest. However, among content-based scales, Overcontrolled-Hostility (.64–.68) was least stable; the Welch Anxiety (.91–.85) and Schlenger's Post-Traumatic Stress Disorder Scale (PS;.85–.92) were most stable. Again, overall retest reliability of the MMPI-2 Supplementary scales was .76–.81. Of the Content scales, Social Discomfort (.83–.91) and Work Interference (.82–.90) were most reliable followed by Low Self-Esteem (.85–.78) and Anxiety (.91–.76). Bizarre Mentations was least stable (.78–.63) which may be a function of the low base-rate of such reported characteristics in the normal population. Overall, the stability of the Content scales was excellent (.76–.84). In terms of broad traits, neuroticism (Depression, Welch Anxiety, Content Anxiety) and extraversion (Social Introversion, Social Discomfort) seem to be particularly stable constructs as measured by the MMPI-2.

Although the PAI (Morey, 1991) has a much shorter history, data on the short-term test-retest reliability of this measure is available. Utilizing a community sample ($N = 75$) tested an average of 24 days apart and a college sample ($N = 80$) tested exactly 28 days apart, test-retest reliability correlations for PAI Full Scales suggest that Alcohol Problems (.92) is most stable followed by Antisocial Features (.89), Anxiety (.88), Depression (.87), and Borderline Features (.86). Of scales measuring stylistic tendencies in the interpersonal circumplex, Warmth (.77) maintained an acceptable degree of stability, whereas Dominance (.68) was the lowest of the nonvalidity-based scales. Although these results are encouraging, longer term studies of the stability of the PAI are in order. However, if the scales of the PAI function similarly to those of the MMPI-2 over time, most of these findings based on shorter intervals would also apply to longer test-retest intervals.

Contextual issues in trait assessment: ratings from observers

The assessment of broad-based traits and psychopathology has typically been achieved via self-report (McAdams, 1994). However, when changes in personality are hypothesized to have occurred from pre-injury levels, observer reports from spouse, relatives, or friends may be an important source of information regarding premorbid personality. Friends or relatives, having been familiar with the patient's pre-injury behavior and characteristics, would appear to be in an excellent position to observe changes. However, judgments of premorbid personality made by friends and family may be subject to many of the same information processing biases found in social judgment in general. Specifically, context effects such as assimilation, contrast, and mood congruency may lead persons to either over- or under-estimate the patient's premorbid level of personality adjustment. The influence of these types of situational priming on social judgments has been well documented (e.g., Herr, 1986; Higgins, Rholes & Jones, 1977; Lombardi, Higgins, & Bargh, 1987; Martin, Seta, & Crelia, 1990; Srull & Wyer, 1979; Wegener & Petty, 1995). Given the interest in formulating accurate estimates of premorbid personality functioning, it is important to identify possible sources of bias in these estimates, and to clarify administration instructions in order to control for these biases.

With respect to relatives' judgments of current personality, situational factors surrounding the patient's illness may influence relatives' perceptions of the patient's personality. For example, when making personality ratings of the patient, relatives are likely to also consider the patient's physical symptoms following the head injury. These types of considerations could create a primed context which may, in turn, influence their judgments of the patient's personality.

Specifically, contemplating the severity of the patient's memory and psychomotor difficulties could result in an overestimate of the patient's current personality problems. Even in the absence of clear objective changes in functional status post-injury, perceptions of cognitive failures in memory and attention may foster expectations of change in basic personality and disposition. Judgments of the patient's current functional problems could in turn create a context for judgments of patient's premorbid personality. One condition that appears to give rise to contrast effects is extreme primes (Herr, 1986). Where relatives have already judged the patient's current personality adjustment to be extremely poor, they are likely to underestimate premorbid levels of personality pathology (e.g., "Well, compared to how difficult he is now, Henry was an angel before the car wreck!"). Given that this sequence of contextual effects withstands the test of empirical investigation, one practical implication is that family and friends' judgments of current and premorbid personality functioning must be counterbalanced. Specifically, relatives should be asked to rate premorbid levels of personality functioning *before* rating the patient's present, post-injury personality status. In this manner, the clinician may circumvent the biasing effect of postmorbid perceptions of change when rating pre-injury personality. A second method of decreasing the biasing effects of situational priming is to make raters aware of these effects. Research on correction-based processes (Martin et al., 1990; Wegner & Petty, 1995) suggests that increasing awareness of these potential biases leads participants to mentally correct for them. Given that participants do not "overcorrect", judgments of targets may be more accurate. When asking informants to provide ratings of patients' personality, instructions such as the following could be employed:

> I'm going to ask you to make a series of judgments about your family member's current personality. When making these judgments, many people are tempted to assume that because their relative may have experienced physical changes as a result of the injury, that personality has also changed as well. However, even if you have noticed physical changes as a result of the accident, personality may or may not have changed. Please consider carefully before answering each question and be forthright in your answers.

Additional instructions such as these may assist the rater in decreasing the biasing effects of context and further elucidate possible interactions between pre-injury personality and post-injury functioning and adaptation.

Conclusions

The importance of estimating a premorbid cognitive baseline from which to interpret current neuropsychological assessment data is readily apparent and continues to be addressed by the neuropsychological community. Unfortunately, there has not been a corresponding growth in the development and application of ideas regarding abiding personality dispositions and their role in conceptualizing a patient's neuropsychological status and symptom expression. Formal assessment of premorbid personality characteristics is essential for several reasons. First, premorbid personality traits correlate with features of postmorbid emotional behavior and adaptation (Chatterjee, Strauss, Smyth, & Whitehouse, 1992). An individual with prominent premorbid histrionic or narcissistic personality features will likely respond to an injury event in a qualitatively different manner than an individual without such attributes, and, accordingly, be placed in a higher risk category for development of certain Axis I disorders (APA, 1994). Likewise, the role of personal injury litigation may be perceived differently by individuals with certain personality traits and the likelihood of malingering may similarly increase. Premorbid behavior may also contribute to the prediction of difficulties during the recovery process and offer a framework for conceiving of strategic interventions and management. For example, premorbid difficulty associated with substance abuse, aggressive behavior, conflicts with authority figures, noncompliance and social nonconformity, and poor motivation obviously have important implications for treatment response and outcome (Prigatano, 1987). Dicker (1992) has asserted that the personality factors that increase a patient's risk of sustaining a head injury may also complicate recovery and subsequent adaptation following injury. Second, knowledge of premorbid functioning provides a partial basis for estimating potential strengths as well as realistic parameters for behavioral improvement during periods of recovery from injury or illness. Third, premorbid estimation is necessary given that significant change in interpersonal behavior may have an impact on the patient's continuing role within the family as well as the family's ability to respond adaptatively.

The method of acute assessment of premorbid personality used in the Kurtz, Putnam, and Stone (1998) study has utility for research purposes and could be employed clinically as well. *If* these data are collected from informants in close temporal proximity to the injury event, their memory for this information may be relatively free of interference from exposure to the patient during post-injury adjustment or litigation stages. By doing so, the perceptions of informants should be less affected by expectancy and confirmatory biases, and thus, the data may be more valid for later comparisons. Of course, the technique is best suited for changes that putatively arise from an acute event, accident or trauma, rather than conditions that may have evolved over time, such as chronic exposure to toxins. Personality data need not be limited to third-party sources. Self-reports provided by the patient represent additional means for measuring changes relative to premorbid assessments obtained during the acute stage. Research comparing self- and informant-based personality assessments of head injury patients (Kurtz & Putnam, 1996) has demonstrated that such data, even from patients who have suffered severe injuries, display levels of reliability and validity comparable to that observed for the same instrument in the general population.

References

Ackerman, P. L., & Heggestad, E. D. (1997). Intelligence, personality, and interests: Evidence for overlapping traits. *Psychological Bulletin, 121,* 219–245.

Adams, K.M. (1989, December). *The role of emotional factors in neuropsychological assessment.* Paper presented at the meeting of the Northwest Neuropsychological Society.

Adams, B.L., Warneke, L.B., McEwan, A.J.B., & Fraser, B.A. (1993). Single photon emission computerized tomography in obsessive compulsive disorder: A preliminary study. *Journal of Psychiatry and Neuroscience, 18,* 109–112.

Amen, D.G., Stubblefield, M., Carmichael, B., & Thisted, R. (1996). SPECT findings and aggressiveness. *Annals of Clinical Psychiatry, 8,* 129–137.

American Psychological Association (1994). *Diagnostic and statistical manual of mental disorders* (4th ed. rev.). Washington, DC: American Psychiatric Press.

Arkin, C. M., & Revenson, T. A. (1987). Does coping help? A reexamination of the relation between coping and mental health. *Journal of Personality and Social Psychology, 53,* 337–348.

Babcock H. (1930). An experiment in the measurement of mental deterioration. *Archives of Psychology,117,* 312–316..

Bagby, R. M., Joffe, R. T., Parker, J. D. A., Kalemba, V., & Harkness, K. L. (1995). Major depression and the five-factor model of personality. *Journal of Personality Disorders, 9,* 224–234.

Barber, J. P., Frank, A., Weiss, R. D., Blaine, J., Siqueland, L., Moras, K., Calvo, N., Chittams, J., Mercer, D., & Salloum, I. M. (1996). Prevalence and correlates of personality disorder diagnoses among cocaine dependent outpatients. *Journal of Personality Disorders, 10,* 297–311.

Barona, A., Reynolds, C.R., Chastain, R.A. (1984). A demographically based index of premorbid intelligence for the WAIS-R. *Journal of Clinical Psychology, 52,* 885–887.

Binder, L.M. (1997). A review of mild head trauma. Part II: Clinical implications. *Journal of Clinical and Experimental Neuropsychology, 19,* 432–457.

Berglas, S., & Jones, E. E. (1978). Drug choice as a self-handicapping strategy in response to noncontingent success. *Journal of Personality and Social Psychology, 36,* 405–417.

Blair, J.R., & Spreen, O. (1989). Predicting premorbid IQ: A revision of the National Adult Reading Test. *The Clinical Neuropsychologist, 3,* 129–136.

Blake, D. D., Weathers, F. W., Nagy, L. M., Kaloupek, D. G., Klauminzer, G., Charney, D. S., & Keane, T. M. (1990). A clinician rating scale for assessing current and lifetime PTSD: The CAPS-1. *Behavior Therapist, 13,* 187–188.

Blake, D. D. (1994). Rationale and development of the Clinician-Administered PTSD Scales. *PTSD Research Quarterly, 5,* 1–2.

Blanchard, E. B., Hickling, E. J., Taylor, A. E., & Loos, W. (1995). Psychiatric morbidity associated with motor vehicle accidents. *Journal of Nervous and Mental Disease, 183,* 495–504.

Blanchard, E. B., Heckling, E. J., Taylor, A. E., Loos, W. R. et al. (1996). Who develops PTSD from motor vehicle accidents? *Behavior Research and Therapy, 34,* 1–1 0.

Blanchard, E. B., Buckley, T. C., Hickling, E. J., & Taylor, A. E. (1998). Posttraumatic Stress Disorder and comorbid Major Depression: Is the correlation an illusion? *Journal of Anxiety Disorders, 12,* 21–37.

Blumer, D., & Benson, D.F. (1975). Personality changes with frontal and temporal lobe lesions. In D. Blumer & D.F. Benson (Eds.), *Psychiatric aspects of neurological disease* (pp. 151–170). New York: Grune and Stratton.

Bohnen, N., Twijnstra, A., & Jolles, J. (1992). Post-traumatic and emotional symptoms in different subgroups of patients with mild head injury. *Brain Injury, 6,* 481–487.

Breier, A., Kelsoe, J.R., Kirwin, P.D., Beller, S.A., Wolkowitz, O.M., & Pickar, D. (1988). Early parental loss and development of adult psychopathology. *Archives of General Psychiatry, 45,* 987–993.

Bremner, J. D., Southwick, S., Brett, E., Fontana, A., Rosenbeck, R., & Charney, D. S. (1992). Dissociation and posttraumatic stress disorder in Vietnam combat veterans. *American Journal of Psychiatry, 149,* 328–332.

Brooner, R. K., Herbst, J. H., Schmidt, C. W., Bigelow, G. E., & Costa, P. T. J. (1993). Antisocial personality disorder among drug abusers. Relations to other personality diagnoses and the five-factor model of personality. *Journal of Nervous and Mental Disease, 181,* 313–319.

Bryant, R. A., & Harvey, A. (1995). Acute stress response: A comparison of head injured and non-head injured patients. *Psychological Medicine, 25,* 869–873.

Buss, D.M. (1994). Personality evoked: The evolutionary psychology of stability and change. In T. F. Heatherton & J.L. Weinberger (Eds.), *Can personality change?* (pp. 41–58). Washington, DC: American Psychological Association.

Butcher, J. N., Dahlstrom, W. G., Graham, J. R., Tellegen, A., & Kaemmer, B. (1989). *Minnesota Multiphasic Personality Inventory-2 (MMPI-2): Manual for administration and scoring.* Minneapolis: University of Minnesota Press.

Caspi, A., & Moffitt, T. E. (1993). When do individual differences matter? A paradoxical theory of personality coherence. *Psychological Inquiry, 4,* 247–271.

Ceci, S.J. (1991). How much does schooling influence intellectual development and its cognitive components? A reassessment of the evidence. *Developmental Psychology, 27,* 703–722.

Chatterjee, A., Strauss, M. E., Smyth, K. A., & Whitehouse, P. J. (1992). Personality changes in Alzheimer's disease. *Archives of Neurology, 49,* 486–491.

Cohen, R.M., Weingartner, H., Smallberg, S.A., Pickar, D., & Murphy, D. L. (1982). Effort and cognition in depression. *Archives of General Psychiatry, 39,* 593–597.

Corrigan J.D., & Deming R. (1995). Psychometric characteristics of the Community Integration Questionnaire: Replication and extension. *Journal of Head Trauma Rehabilitation, 10,* 41–53.

Costa, P.T., & McRae, R.R. (1994). Set like plaster? Evidence for the stability of adult personality. In T. F. Heatherton & J.L. Weinberger (Eds.), *Can personality change?* (pp. 21–40). Washington, DC: American Psychological Association.

Costa, P. T., Jr., & McRae, R. R. (1992a). *Revised NEO Personality Inventory (NEO PI-R) and NEO Five-Factor Inventory (NEO-FFI) professional manual.* Odessa, FL: Psychological Assessment Resources.

Costa, P. T., & McCrae, R. R. (1992b). Normal personality assessment in clinical practice: The NEO Personality Inventory. *Psychological Assessment: A Journal of Consulting and Clinical Psychology, 4,* 5–13.

Costa, P. T., Jr., & McCrae, R. R. (1988). Personality in adulthood: A six-year longitudinal study of self-reports and spouse ratings on the NEO Personality Inventory. *Journal of Personality and Social Psychology, 54,* 853–863.

Cox, C.S., Chee, E., Chase, G.A., Baumgardner, T.L., Shuerholz, L.J., Reader, M.J., Mohr, J., & Denckla, M. (1997). Reading proficiency affects the construct validity of the Stroop Test interference score. *The Clinical Neuropsychologist, 11,* 105–100.

Crawford, J.R., Parker, D.M., & Stewart, L.E. (1989). Prediction of WAIS IQ with the National Adult Reading Test: Cross-validation and extension. *British Journal of Clinical Psychology, 28,* 267–273.

Crawford, J., Obousawin, M., & Allan, K. (1998). PASAT and components of WAIS-R performance: Convergent and discriminative validity. *Neuropsychological Rehabilitation, 8,* 255–272.

Crawford, J., Moore, J., & Cameron, I. (1992). Verbal fluency: A NART-based equation for the estimation of premorbid performance. *British Journal of Clinical Psychology, 31,* 327–329.

Damasio, A. R., Tranel, D. & Damasio, H. (1990). Individuals with sociopathic behavior caused by frontal damage fail to respond autonomically to social stimuli. *Behavioural Brain Research, 41,* 81–94.

Darke, S., Hall, W., & Swift, W. (1994). Prevalence, symptoms and correlates of antisocial personality disorder among methadone maintenance clients. *Drug and Alcohol Dependence, 34,* 253–257.

Davidson, J. R. T., Smith, R., & Kudler, H. (1989). Familial psychiatric illness in chronic posttraumatic stress disorder. *Comprehensive Psychiatry, 30,* 339–345.

Davidson, J. R. T., Hughes, D., Blazer, D. G., & George, L. K. (1991). Posttraumatic stress disorder in the community: An epidemiological study. *Psychological Medicine, 21,* 713–721.

DeJong, C. A., Van den Brink, W., Harteveld, F. M., & van der Wielen, E. (1993). Personality disorders in alcoholics and drug addicts. *Comprehensive Psychiatry, 34,* 87–94.

Derbyshire, S.W.G., Jones, A.K.P., Davani, P., Friston, K.J. (1994). Cerebral responses to pain in patients with atypical facial pain measured by positron emission tomography. *Journal of Neurology, Neurosurgery, and Psychiatry, 57,* 1166–1172.

Dicker, B. G. (1992). Profile of those at risk for minor head injury. *Journal of Head Trauma Rehabilitation, 7,* 83–91.

Dikmen, S.S., & Levin, H.S. (1993). Methodological issues in the study of mild head injury. *Journal of Head Trauma Rehabilitation, 8,* 30–37.

Dodrill, C.B. (1997). Myths of neuropsychology. *The Clinical Neuropsychologist, 11,* 1–17.

Donders J. (1992). Premorbid behavioral and psychosocial adjustment of children with traumatic brain injury. *Journal of Abnormal Child Psychology, 20,* 233–246.

Dupont, R.M., Rourke, S.B., Grant, I., Lehr, P.P., Reed, R.J., Challakere, K., Lamoureux, G., & Halpern, S. (1996). Single photon emission computed tomography with iodoamphetamine-123 and neuropsychological studies in long-term abstinent alcoholics. *Psychiatry Research, 67,* 99–111.

Engdahl, B. E., Eberly, R. E., & Blake, J. D. (1996). Assessment of Posttraumatic Stress Disorder in World War II Veterans. *Psychological Assessment, 8,* 445–449.

Eysenck, H. J. (1952). *The scientific study of personality.* London: Routledge & Kegan Paul.

Eysenck, H. (1967). *The biological basis of behavior.* Springfield, IL.: Charles C. Thomas.

Fenton, G., McClelland, R., Montgomery, A., MacFlynn, G. & Rutherford, W. (1993). The postconcussional syndrome: Social antecedants and psychological sequelae. *British Journal of Psychiatry, 162,* 493–497.

Fink, P. (1995). Psychiatric illness in patients with persistent somatization. *British Journal of Psychiatry, 166,* 93–99.

Fletcher J.M., Ewing-Cobbs L., Miner M.E., Levin H.S., & Eisenberg H.M. (1990). Behavioral changes after closed head injury in children. *Journal of Consulting and Clinical Psychology, 58,* 93–98.

Finn, S. E. (1986). Stability of personality self-ratings over 30 years: Evidence for an age/cohort interaction. *Journal of Personality and Social Psychology, 50,* 813–818.

Folks, D. G., & Kinney, F. C. (1992a). The role of dermatologic conditions. *Psychosomatics, 33,* 45–54.

Folks, D. G., & Kinney, F. C. (1992b). The role of psychological factors in gastrointestinal conditions: A review pertinent to DSM-IV. *Psychosomatics, 33,* 257–270.

Ford, D.E., & Kamerow, D.B. (1989). Epidemiologic study of sleep disturbances and psychiatric disorders. *Journal of the American Medical Association, 262,* 1479–1484.

Fox, D. D., Lees-Haley, P. R., Earnest, K., & Dolezal-Wood, S. (1995). Base rates of post-concussive symptoms in health maintenance organization patients and controls. *Neuropsychology, 9,* 606–611.

Foy, D. W., Resnick, D. B., Sipperelle, R. C., & Carroll, E. M. (1987). Premilitary, military, and postmilitary factors in the development of combat-related posttraumatic stress disorder. *The Behavior Therapist, 10,* 3–9.

Gaston, L., Brunet, A., Koszycki, D., & Bradwejn, J. (1996). MMPI profiles of acute and chronic PTSD in a civilian sample. *Journal of Traumatic Stress, 9,* 817–832.

Goldberg, L. R. (1981). Language and individual differences: The search for universals in personality lexicons. In L. Wheeler (Ed.), *Review of personality and social psychology* (Vol. 2; pp. 141–166). Beverly Hills, CA: Sage.

Graham, J. R. (1990). *MMPI-2: Assessing personality and psychopathology.* New York: Oxford University Press.

Graham, J.R., & Strenger, V.E. (1988). MMPI characteristics of alcoholics: A review. *Journal of Consulting and Clinical Psychology, 56,* 197–205.

Graves, R., Carswell, L., & Snow, G. (1999). An evaluation of the sensitivity of premorbid IQ estimators for detecting cognitive decline. *Psychological Assessment, 11,* 29–38.

Green, B. L., Grace, M. C., Lindy, J. D., Gleser, G. C., & Leonard, A. (1990). Risk factors for PTSD and other diagnoses in a general sample of Vietnam veterans. *American Journal of Psychiatry, 147,* 729–733.

Gronwall, D. (1977). The Paced Auditory Serial Addition Task: A measure of recovery from concussion. *Perceptual and Motor Skills, 44,* 367–373.

Haas, J.F., Cope, D.N., & Hall, K. (1987). Premorbid prevalence of poor academic performance in severe head injury. *Journal of Neurology, Neurosurgery, and Psychiatry, 50,* 52–56.

Harlow, J.M. (1868). Recovery from the passage of an iron bar through the head. *Publication of the Massachusetts Medical Society, 2,* 327–347.

Hawkins, K.A., Sledge, W.H., Orleans, J.F., Quinlan, D.M., Rakfeldt, J., & Hoffman, R.E. (1993). Normative implications of the relationship between reading vocabulary and Boston Naming Test performance. *Archives of Clinical Neuropsychology, 8,* 525–537.

Harkness, A. R., McNolty, J. L., & Ben-Porath, Y. S. (1995). The Personality Psychopathology Five (PSY-5): Constructs and MMPI-2 scales. *Psychological Assessment, 7,* 104–114.

Heath, A. C., Neale, M. C., Kessler, R. C., Eaves, L. J., & Kendler, K. S. (1992). Evidence for genetic influences on personality from self-reports and informant ratings. *Journal of Personality and Social Psychology, 44,* 1198–1213.

Heatherton, T. F., & Weinberger, J. L. (1994). *Can personality change?* Washington, DC: American Psychological Association.

Helson, R., & Moane, G. (1987). Personality change in women from college to midlife. *Journal of Personality and Social Psychology, 53,* 176–186.

Helzer, J. E., & Pryzbeck, T. R. (1988). The co-ocurrence of alcoholism with other psychiatric disorders in the general population and its impact on treatment. *Journal of Studies on Alcohol, 49,* 219–224.

Herscovitch, P. (1996). Functional brain imaging: Basic principles and application to head trauma. In M. Rizzo & D. Tranel (Eds.), *Head injury and postconcussive syndrome* (pp. 89–118). New York: Churchill Livingstone.

Hertzman, M., Reba, R.C., & Kotlyarov, E.V. (1990). Single photon emission computed tomography in phencyclidine and related drug abuse. *American Journal of Psychiatry, 147,* 255–256.

Higgins, E.T., Rholes, W.S., & Jones, C.R. (1977). Category accessibility and impression formation. *Journal of Experimental Social Psychology, 13,* 141–154.

Holman, B.L., & Devous, M.D. (1992). Functional brain SPECT: The emergence of a powerful clinical method. *Journal of Nuclear Medicine, 33,* 1888–1904.

Herr, P. M. (1986). Consequences of priming: Judgment and behavior. *Journal of Personality and Social Psychology, 51,* 1106–1115.

Hyer, L., Walker, C., Swanson, G., Sperr, S., Sperr, E., & Blount, J. (1992). Validation of PTSD measures for older combat veterans. *Journal of Clinical Psychology, 48,* 579–588.

Hyer, L. A., Albrecht, J. W., Boudewyns, P. A., Woods, G. M., & Brandsma, J. (1993). Dissociative experiences of Vietnam veterans with chronic posttraumatic stress disorder. *Psychological Reports, 73,* 519–530.

Ichise, M., Chung, D.G., Wang, P., Wortzman, G., Gray, B.G., & Franks, W. (1994). Technetium-99m-HMPAO SPECT, CT, and MRI in the evaluation of patients with chronic traumatic brain injury: A correlation with neuropsychological performance. *Journal of Nuclear Medicine, 35,* 1217–1226.

Iidaka,T., Nakajima, T., Ogikubo, T., & Fukuda, H. (1995). Correlations between regional cerebral blood flow and the Hamilton Rating Scale for Depression in mood disorders: A study using 123-iodoamphetamine single photon emission computerized tomography. *Clinical Psychiatry, 37,* 951–958.

Iverson, G., & McCracken, L. (1997). 'Postconcussive' symptoms in persons with chronic pain. *Brain Injury, 11,* 783–790.

Jensen, M.C., Brant-Zawadzki, M.N., Obuchowski, N., Modic, M.T., Malkasian, D., & Ross, J.S. (1994). Magnetic resonance imaging of the lumbar spine in people without back pain. *New England Journal of Medicine, 331,* 69–73.

Johnson, M. H., & Magaro, P. A. (1987). Effects of mood and severity on memory processes in depression and mania. *Psychological Bulletin, 101,* 28–40.

Johnson-Greene, D., & Binder, L.M. (1995). Evaluation of an efficient method for verifying higher educational credentials. *Archives of Clinical Neuropsychology, 10,* 251–253.

Johnstone, B., Slaughter, J., Schopp, L., McAllister, J., Schwake, C., & Luebbering, A. (1997). Determining neuropsychological impairment using estimates of premorbid intelligence: Comparing methods based on level of education versus reading score. *Archives of Clinical Neuropsychology, 12,* 591–601.

Jones, E. E., & Berglas, S. (1978). Control of attributions about the self through self-handicapping strategies: The appeal of alcohol and the role of underachievement. *Personality and Social Psychology Bulletin, 4,* 200–206.

Juni, J.E. (1994). Taking brain SPECT seriously: Some reflections on recent clinical reports in the Journal of Nuclear Medicine. *Journal of Nuclear Medicine, 35,* 1891–1895.

Kareken, D.A. (1997). Judgment pitfalls in estimating premorbid intellectual function. *Archives of Clinical Neuropsychology, 12,* 701–709.

Kareken, D.A., & Williams, J.M. (1994). Human judgment and estimation of premorbid intellectual function. *Psychological Assessment, 6,* 83–91.

Kareken, D.A., Gur, R.C., & Saykin, A.J. (1995). Reading on the Wide Range Achievement Test-Revised and parental education as predictors of IQ: Comparison with the Barona formula. *Archives of Clinical Neuropsychology, 10,* 147–157.

Keane, T. M., Malloy, P. F., & Fairbank, J. A. (1984). Empirical development of an MMPI subscale for the assessment of combat-related posttraumatic stress disorder. *Journal of Consulting and Clinical Psychology, 52,* 888–891.

Keenan, P.A., Ricker, J.H., Lindamer, L.A., Jiron, C.C., & Jacobson, M.W. (1996). Relationship between WAIS-R Vocabulary and Performance on the California Verbal Learning Test. *The Clinical Neuropsychologist, 10,* 455–458.

Kessler, R. C., McGonagle, K. A., Zhao, S., Nelson, C. B., Hughes, M., Eshleman, M. A., Wittchen, H. U., & Kendler, K. S. (1994). Lifetime and 12-month prevalence of DSM-III-R psychiatric disorders in the United States: Results from the national comorbidity survey. *Archives of General Psychiatry, 51,* 8–19.

King, D.W., King, L.A., Foy, D.W., & Gudanowski, D.M. (1996). Prewar factors in combat-related post-traumatic stress disorder: Structural equation modeling wiht a national sample of female and male Vietnam veterans. *Journal of Consulting and Clinical Psychology, 63,* 520–531.

Klonoff, P.S., & Lamb, D.G. (1998). Mild head injury, significant impairment on neuropsychological test scores, and psychiatric disability. *The Clinical Neuropsychologist, 12,* 31–42.

Koopman, C., Classen, C., & Speigal, D. (1994). Predictors of posttraumatic stress symptoms among survivors of the Oakland/Berkeley, California firestorm. *American Journal of Psychiatry, 151,* 888–894.

Krull, K.R., Scott, J.G., Sherer, M. (1995). Estimation of premorbid intelligence from combined performance and demographic variables. *The Clinical Neuropsychologist, 9,* 83–88.

Krystal, J.H., Woods, S.W., Kosten, T.R., & Rosen, M.I. (1995). Opiate dependence and withdrawal: Preliminary assessment using single photon emission computerized tomography (SPECT). *American Journal of Drug and Alcohol Abuse, 21,* 47–63.

Kulka, R. A., Schlenger, W. E., Fairbank, J. A., Hough, R. L., Jordan, B. K., Marmar, C. R., & Weiss, D. S. (1990). *Trauma and the Vietnam War generation: Report of findings from the National Vietnam Veterans Readjustment Study.* New York: Brunner/Mazel.

Kurtz, J. E., Putnam, S.H., & Stone, C. (1998). Stability of normal personality traits after traumatic brain injury. *Journal of Head Trauma Rehabilitation, 13,* 1–14.

Kurtz, J.E., & Putnam, S.H. (1996, April). *Concordance of self-report and informant-based personality assessments in traumatic brain injury.* Paper presented at the meeting of the National Center for Medical Rehabilitation Research, Bethesda, MD.

Kutz, I., Shabtai, H., Solomon, Z., Neumann, M., & David, D. (1994). Posttraumatic stress disorder in myocardial infarction patients: Prevalence study. *Israel Journal of Psychiatry and Related Science, 31,* 48–56.

Labouvie, E.W., & McGee, C.R. (1986). Relation of personality to alcohol and drug use in adolescence. *Journal of Consulting and Clinical Psychology, 54,* 289–293.

Larrabee, G.J., Largen, J.W., & Levin, H.S. (1985). Sensitivity of age-decline resistant ("Hold") WAIS subtests to Alzeimer's disease. *Journal of Clinical and Experimental Neuropsychology, 7,* 497–504.

Lazarus, A., & Cotterell, K.P. (1989). SPECT scan reveals abnormality in somatization disorder patient. *Journal of Clinical Psychiatry, 50,* 475–476.

Leary, T. (1957). *Interpersonal diagnosis of personality.* New York: Ronald Press.

Lees-Haley, P. R. (1992). Neuropsychological complaint base rates of personal injury claimants. *Forensic Reports, 5,* 385–391.

Lees-Haley, P., & Brown, R. (1993). Neuropsychological complaint base rates of 170 personal injury claimants. *Archives of Clinical Neuropsychology, 8,* 203–209.

Lezak, M.D. (1995). *Neuropsychological assessment* (pp. 106–108). New York: Oxford University Press.

Links, P. S., Heslegrave, R. J., Mitton, J. E., van Reekum, R., & Patrick, J. (1995). Borderline personality disorder and substance abuse: Consequences of comorbidity. *Canadian Journal of Psychiatry, 40,* 9–14.

Loehlin, J. C. (1992). *Genes and environment in personality development.* Newbury Park, CA: Sage.

Lombardi, W.J., Higgins, E.T., & Bargh, J.A. (1987). The role of consciousness in priming effects on categorization: A function of awareness of the priming task. *Personality and Social Psychology Bulletin, 13,* 411–429.

Machulda, M.M., Berquist, T.F., Ito, V., & Chew, S. (1998). Relationship between stress, coping, and postconcussion symptoms in a healthy adult population. *Archives of Clinical Neuropsychology, 13,* 415–424.

Mampunza, S., Verbanck, P., Verhas, M., Martin, P. (1995). Cerebral blood flow in just detoxified alcohol dependent patients: A 99 m Tc-HMPAO-SPECT study. *Acta Neurologica Belgica, 95,* 164–169.

Martin, L.L., Seta, J.J., & Crelia, R.A. (1990). Assimilation and contrast as a function of people's willingness and ability to expend effort in forming an impression. *Journal of Personality and Social Psychology, 59,* 27–37.

Masdeu, J.C., & Brass, L.M. (1995). SPECT imaging of stroke. *Journal of Neuroimaging, 5 (Suppl. 1),* 14–22.

Matarazzo J.D. (1990). Psychological assessment versus psychological testing: Validation from Binet to the school, clinic, and courtroom. *American Psychologist, 45,* 999–1017.

Mayou, R., Bryant, B., & Duthie, R. (1993). Psychiatric consequences of road traffic accidents. *British Medical Journal, 307,* 647–651.

McAdams, D. P. (1994). Can personality change? Levels of stability and growth in personality across the life span. In T. F. Heatherton & J. L. Weinberger (Eds.), *Can personality change?* (pp. 299–313). Washington, DC: American Psychological Association.

McCrae, R. R., & Costa, P. T., Jr. (1990). *Personality in adulthood.* New York: Guilford Press.

McCranie, E. W., Hyer, L. A., Boudewins, P. A., & Woods, M. G. (1992). Negative parenting behavior, combat exposure, and PTSD symptom severity: Test of the person-event interaction model. *Journal of Nervous and Mental Disease, 180,* 431–438.

McDonald, A. S., & Davey, G. C. L. (1996). Psychiatric disorders and accidental injury. *Clinical Psychology Review, 16,* 105–127.

McFarlane, A. C., Atchinson, M., Rafalowicz, E., & Papay, P. (1994). Physical symptoms in posttraumatic stress disorder. *Journal of Psychosomatic Research, 38,* 715–726.

Melton, G.B., Petrila, J., Poythress, N.G., & Slobogin, C. (1997). *Psychological evaluations for the courts: A handbook for mental health professionals and lawyers* (2nd ed.) New York: Guilford.

Meyers, C. A., Berman, S. A., Scheibel, R. S., & Hayman, A. (1992). Case report: Acquired antisocial personality disorder associated with unilateral left orbital frontal lobe damage. *Journal of Psychiatry and Neuroscience, 17,* 121–125.

Miller, B.L., Mina, I., Giombetti, R., & Villanueva-Meyer, J. (1992). Neuropsychiatric effects of cocaine: SPECT measurements. *Journal of Addictive Diseases, 11,* 47–58.

Millis, S.R., & Putnam, S.H. (1996). Detection of malingering in postconcussive syndrome. In M. Rizzo & D. Tranel (Eds.), *Head injury and the postconcussion syndrome,* (pp. 481–498). New York: Churchilll Livingstone.

Millon, T., & Davis, R.D. (1996). *Disorder of personality: DSM-IV and beyond* (2nd ed.). New York: Wiley Interscience.

Mittenberg, W., Hammeke, T.A., & Rao, S.M. (1989). Intrasubtest scatter on the WAIS-R as a pathognomonic sign of brain injury. *Psychological Assessment, 1,* 273–276.

Mittenberg, W., Dibuilio, D. V., Perrin, S., & Bass, A. E. (1992). Symptoms following mild head injury: Expectation as aetiology. *Journal of Neurology, Neurosurgery, and Neuropsychiatry, 35,* 200–204.

Modell, J.G., & Mountz, J.M. (1995). Focal cerebral blood flow change during craving for alcohol measured by SPECT. *Journal of Neuropsychiatry and Clinical Neurosciences, 7,* 15–22.

Morey, L. C. (1991). *The Personality Assessment Inventory: Professional manual.* Odessa, FL: Psychological Assessment Resources.

Mortensen, E., Gade, A., & Reinisch, J. (1991). A critical note on Lezak's 'best performance method' in clinical neuropsychology. *Journal of Clinical and Experimental Neuropsychology, 13,* 361–371.

Mozley, P.D., Kim, H.J., Gur, R.C., Tatsch, K., Muenz, L.R., McElgin, W.T., Kung, M.P., Mu, M., Myers, A.M., & Kung, H.F. (1996). Iodine-123-iPT SPECT imaging of CNS dopamine transporters: Nonlinear effects of normal aging on striatal uptake values. *Journal of Nuclear Medicine, 37,* 1965–1970.

Nagy, L. M., Blake, D. D., Dan, E., Riney, S., Mangine, W., Southwick, S. M., Gusman, F., & Charney, D. S. (1991, October). Clinician Administered PTSD Scale - Weekly Version (CAPS-2): Reliability, validity, and sensitivity to change. In D. D. Blake (Chair), *An update on the Clinician Administered PTSD Scales (CAPS-1 and CAPS-2).* Symposium paper presented at the 7th Annual Conference of the International Society for Traumatic Stress Studies, Washington, DC.

National Head Injury Foundation (1988). *National Head Injury Foundation: Substance Abuse Task Force: White Paper on Substance Abuse.* Southborough, MA: Author.

Nelson, H.E. (1982). *The National Adult Reading Test.* Windsor-Berks, UK: NFER-Nelson.

Nelson, H.E., McKenna, P. (1975). The use of current reading ability in the assessment of dementia. *British Journal of Social and Clinical Psychology, 14,* 259–267.

Newman, B. (1992). Research in mild traumatic brain injury. *New York University Medical Center Research Training Center News, 4,* 3.

Nuwer, M. (1997). Assessment of digital EEG, quantitative EEG, and EEG brain mapping. Report of the American Academy of Neurology and the American Clinical Neurophysiology Society. *Neurology, 49,* 277–292.

Oddy, M., & Humphrey, M. (1980). Social recovery during the year following severe head injury. *Journal of Neurology, Neurosurgery, and Psychiatry, 43,* 798–802.

Paulman, R.G., Devous, M.D., & Gregory, R.R. (1990). Hypofrontality and cognitive impairment in schizophrenia: dynamic single photon tomography and neuropsychological assessment of schizophrenic brain function. *Biological Psychiatry, 27,* 377–399.

Perez, S.A., Schlottman, R.S., Holloway, J.A., & Ozolins, M.S. (1996). Measurement of premorbid intellectual ability following brain injury. *Archives of Clinical Neuropsychology, 11,* 491–501.

Peri, T., Beh-Shachar, G., & Shalev, A. (1994, May). *Heightened conditionability in PTSD and panic disorder.* Paper presented at the 147th Annual Meeting of the American Psychiatric Association, Philadelphia, PA.

Pribor, E. F., Yutzy, S. H., Dean, T., & Wetzel, R. D. (1993). Briquet's syndrome, dissociation, and abuse. *American Journal of Psychiatry, 150,* 1507–1511.

Prigatano, G. P. (1987). Psychiatric aspects of head injury: Problem areas and suggested guidelines for research. In H.S. Levin, J. Grafman, & H.M. Eisenber (Eds.) *Neurobehavioral recovery from head injury* (pp. 215–231). New York: Oxford University Press.

Putnam, S.H., & Adams, K.M. (1992). Regresssion-based prediction of long-term outcome following multidisciplinary rehabilitation for traumatic brain injury. *The Clinical Neuropsychologist, 6,* 383–405.

Putnam, S.H., DeLuca, J., & Anderson, C. (1994). The Second TCN salary survey: A survey of neuropsychologists, Part 1. *The Clinical Neuropsychologist, 8,* 3–37.

Putnam, S.H., Millis, S.R., Adams, K.M. (1996). Mild traumatic brain injury: Beyond cognitive assessment. In I. Grant, & K.M. Adams (Eds.), *Neuropsychological assessment of neuropsychiatric disorders* (2nd ed.; pp. 529–551). New York: Oxford University Press.

Putnam, S.H., Kurtz, J.E., & Houts, D.C. (1996). Four-month test-retest reliability of the MMPI-2 with normal male clergy. *Journal of Personality Assessment, 67,* 341–353.

Query, W. T., Megran, J., & McDonald, G. (1986). Applying posttraumatic stress disorder MMPI subscale to World War II POW veterans. *Journal of Clinical Psychology, 42,* 315–317.

Rapport, L.J., Axelrod, B.N., Theisen, M., Brines, B., Kalechstein, A., & Ricker, J.H. (1997). The relationship of IQ to verbal learning and memory: Test and retest. *Journal of Clinical and Experimental Neuropsychology, 19,* 655–666.

Read, S.L., Miler, B.L., Mena, I., & Kim, R. (1995). SPECT in dementia: Clinical and pathological correlation. *Journal of the American Geriatrics Society, 42,* 1243–1247.

Richardson, J.T.E., & Snape, W. (1984). The effects of closed head injury upon human memory: An experimental analysis. *Cognitive Neuropsychology, 7,* 217–231.

Rivara J.B., Jaffe K.M., Fay G.C., Polissar N.L., Martin K.M., Shurtleff H.A., & Liao S. (1993). Family functioning and injury severity as predictors of child functioning one year following traumatic brain injury. *Archives of Physical Medicine and Rehabilitation, 74,* 1047–1055.

Ross, S. R., & Putnam, S. H. (1998, August). *Estimating premorbid intelligence in head injury: The contributions of personality.* Poster session presented at the 106th Annual Meeting of the American Psychological Association, San Francisco, CA.

Ross, S. R., Bailley, S. E., & Millis, S. R. (1997). Positive self-presentation and the detection of defensiveness on the NEO PI-R. *Assessment, 4,* 395–408.

Rost, K. M., Akins, R. N., Brown, F. W., & Smith, G. R. (1992). The comorbidity of DSM-III-R personality disorders in somatization disorder. *General Hospital Psychiatry, 14,* 322–326.

Ruff, R.M., Mueller, J., & Jurica, P. (1996). Estimation of premorbid functioning after traumatic brain injury. *Neurorehabilitation, 7,* 39–53.

Russell, E. (1972). WAIS factor analysis with brain-damaged subjects using criterion measures. *Journal of Consulting and Clinical Psychology, 39,* 133–139.

Sackeim, H.A., Prohovnik, I., & Moeller, J. (1990). Regional cerebral blood flow in mood disorders, I: Comparison of major depressives and normal controls at rest. *Archives of General Psychiatry, 47,* 60–70.

Saver, J. L., & Damasio, A. R. (1991). Preserved access and processing of social knowledge in a patient with acquired sociopathy due to ventromedial frontal damage. *Neuropsychologia, 39,* 1241–1249.

Saxe, G. N., Chinman, G., Berkowitz, R., Hall, K., Lieberg, G., Schwartz, J., & van der Kolk, B. A. (1994). Somatization in patients with dissociative disorders. *American Journal of Psychiatry, 151,* 1329–1335.

Sbordone R.J. (1996). Ecological validity: Some critical issues for the neuropsychologist. In R.J. Sbordone & C.J. Long (Eds.), *Ecological validity of neuropsychological testing* (pp. 15–41). DelRay Beach, FL: GR Press/St. Lucie Press.

Schlottman, R.S., & Johnsen, D.H. (1991). The Intellectual Correlates Scale and the prediction of premorbid intelligence in brain damaged adults. *Archives of Clinical Neuropsychology, 6,* 363–374.

Scott, J.G., Krull, K.R., Williamson, D.J.G., Adams, R., & Iverson, G. (1997). Oklahoma Premorbid Intelligence Estimation (OPIE): Utilization in clinical sample. *The Clinical Neuropsychologist, 11,* 146–154.

Semple, W.E., McCormick, R., Goyer, P.F., & Compton-Toth, B. (1996). Attention and regional cerebral blood flow in posttraumatic stress disorder patients with substance abuse histories. *Psychiatry Research: Neuroimaging, 67,* 17–28.

Shalev, A. Y. (1996). Stress versus traumatic stress: From acute homeostatic reactions to chronic psychopathology. In B. A. van der Kolk, A. C. McFarlane, & L. Weisaeth (Eds.). *Traumatic stress: The effects of overwhelming experience on mind, body, and society,* (pp. 77–101). New York: Guilford Press.

Shalev, A. Y. Schreiber, S., Galai, T., & Melmed, R. (1993). Post traumatic stress disorder following medical events. *British Journal of Clinical Psychology, 32,* 352–357.

Shalev, A. Y., Peri, T., Caneti, L., & Schreiber, S. (1996). Predictors of PTSD in injured trauma survivors. *American Journal of Psychiatry, 53,* 219–224.

Sherman, E.M.S., Strauss, E., & Spellacy, F. (1997). Validity of the Paced Auditory Serial Addition Test (PASAT) in adults referred for neuropsychological assessment after head injury. *The Clinical Neuropsychologist, 11,* 34–45.

Smith-Seemiller, L., Franzen, M.D., Burgess, E.J., & Prieto, L.R. (1997). Neuropsychologists' practice patterns in assessing premorbid intelligence. *Archives of Clinical Neuropsychology, 12,* 739–744.

Spiro, A., Butcher, J., Levenson, M. Aldwin, C., & Bosse, R. (1993, August). *Personality change and stability over five years: The MMPI-2 in older men.* Paper presented at the meeting of the American Psychological Association, Toronto, Canada.

Spiro, A., Schnurr, P. P., & Aldwin, C. A. (1994). Prevalence of combat-related PTSD in older men. *Psychology and Aging, 9,* 17–26.

Srull, T. K., & Wyer, R. S. (1979). The role of category accessibility in the interpretation of information about persons: Some determinants and implications. *Journal of Personality and Social Psychology, 38,* 1660–1672.

Stern, J., Murphy, M., & Bass, C. (1993). Personality disorders in patients with somatization disorder: A controlled study. *British Journal of Psychiatry, 163,* 785–789.

Sweet, J.J., Moberg, P.J., & Tovian, S.M. (1990). Evaluation of the Wechsler Adult Intelligence Scale-Revised premorbid IQ formulas in clinical populations. *Psychological Assessment, 2,* 41–44.

Talbert, F. S., Albrecht, N. N., Albrecht, J. W., Boudewyns, P. A., Hyer, L. A., Touze, J. H., & Lemmon, C. R. (1994). MMPI profiles in PTSD as a function of comorbidity. *Journal of Clinical Psychology, 50,* 529–537.

Tellegen, A. (1988). The analysis of consistency in personality assessment. *Journal of Personality, 56,* 621–663.

Thorp, T.R., & Mahrer, A.R. (1959). Predicting potential intelligence. *Journal of Clinical Psychology, 15,* 286–288.

Therapeutics and Technology Subcommittee of the American Academy of Neurology (1996a). Assessment of brain SPECT. *Neurology, 46,* 278–285.

Therapeutics and Technology Subcommittee of the American Academy of Neurology (1996b). Assessment: Neuropsychological testing of adults. *Neurology, 47,* 592–599.

Tyrer, P., Gunderson, J., Lyons, M., & Tohen, M. (1997). Special feature: Extent of comorbidity between mental state and personality disorders. *Journal of Personality Disorders, 11,* 242–259.

Uchiyama, M., Sue, H., Fukumitsu, N., Mori, Y., & Kawakami, K. (1997). Assessment of cerebral benzodiazepine receptor distribution in anxiety disorders by 123-i iomazenil SPECT. *Nippon Acta Radiologica, 57,* 41–46.

Vanderploeg, R. D., Sison, G. F. P., & Hickling, E. J. (1987). A reevaluation of the use of the MMPI in the assessment of combat-related postraumatic stress disorder. *Journal of Personality Assessment, 51,* 140–150.

Vanderploeg, R. D., & Schinka, J. A. (1995). Predicting WAIS-R IQ premorbid ability: Combining substest performance and demographic variable predictors. *Archives of Clinical Neuropsychology, 10,* 225–239.

Vanderploeg, R. D., Schinka, J. A., & Axelrod, B. N. (1996). Estimation of WAIS-R premorbid intelligence: Current ability and demographic data used in a best-performance fashion. *Psychological Assessment, 8,* 404–411.

Vasterling, J. J., Brailey, K., Constans, J. I., & Sutker, P. B. (1998). Attention and memory dysfunction in Posttraumatic Stress Disorder. *Neuropsychology, 12,* 125–133.

Veiel, H. O. F. (1997). A preliminary profile of neuropsychological deficits associated with major depression. *Journal of Clinical and Experimental Neuropsychology, 19,* 587–603.

Walker, E. A., Katon, W. J., Neraas, K., Jemelka, R. P., & Massoth, D. (1992). Dissociation in women with chronic pelvic pain. *American Journal of Psychiatry, 149,* 534–537.

Wechsler, D. (1981). *The Wechsler Adult Intelligence Scale-Revised.* New York: Psychological Corporation.

Wegner, D. T., & Petty, R. E. (1995). Flexible correction processes in social judgment: The role of naive theories in corrections of perceived biases. *Journal of Personality and Social Psychology, 68,* 36–51.

Williams, J.M. (1997). The prediction of premorbid memory ability. *Archives of Clinical Neuropsychology, 12,* 745–756.

Willshire, D., Kinsella, G., Prior, M. (1991). Estimating WAIS-R IQ from the National Adult Reading Test: A cross-validation. *Journal of Clinical and Experimental Neuropsychology, 13,* 204–216.

Wilson, R.S., Rosenbaum, G., & Brown, G. (1979). The problem of premorbid intelligence in neuropsychological assessment. *Journal of Clinical Neuropsychology, 1,*

Wise, E. A. (1996). Diagnosing posttraumatic stress disorder with the MMPI Clinical Scales: A review of the literature. *Journal of Psychopathology and Behavioral Assessment, 18,* 71–82.

Wrobel, N. H., & Wrobel, T. A. (1996). The problem of assessing brain damage in psychiatric samples: Use of personality variables in prediction of WAIS-R scores. *Archives of Clinical Neuropsychology, 11,* 625–635.

Wood, F., Flowers, L., Buchsbaum, M., & Tallal, P. (1991). Investigation of abnormal left temporal functioning in dyslexia through rCBF, auditory evoked potentials, and positron emission tomography. *Reading & Writing, 3,* 379–393.

Yates A.J. (1956). The use of vocabulary in the measurement of intellectual deterioration: A review. *The Journal of Mental Science, 102,* 409–440.

Zhang, J.J., Park, C.H., & Kim, S.M. (1994). Brain SPECT artifact in multidetector SPECT systems. *Clinical Nuclear Medicine, 19,* 789–791.

Zuckerman, M., Kieffer, S. C., & Knee, R. C. (1998). Consequences of self-handicapping: Effects on coping, academic performance, and adjustment. *Journal of Personality and Social Psychology, 74,* 1619–1628.

PART II

PRACTICE EXPERTISE

Effectiveness in the forensic arena requires a wide range of clinical skills and familiarity with a diverse and rapidly growing knowledge base. This section is intended to assist clinicians with acquisition of relevant information on practice expertise. Chapter 4, by Eugene Rankin and Russell Adams, describes the clinical and scientific foundations of formal neuropsychological assessment with forensic cases. In Chapter 5, Linas Bieliauskas delineates the assessment of personality and emotional functioning relevant to these cases. Thomas Kay, in Chapter 6, elaborates the issues of differential diagnosis of neuropsychological impairments in forensic cases, for which the question is "what's really wrong?". In Chapter 7, David Osmon discusses theoretical, functional neuroanatomical and practical issues underlying complexities of evaluating executive functions, as well as relevant experimental research. Chapter 8, by Robert Sbordone and Thomas Guilmette, details issues pertaining to the pervasive issue of ecological validity of neuropsychological assessment, with special consideration to predictions of daily living and vocational activities. The essential question of how to determine validity of test performances is the focus of the discussion on differential diagnoses involving neuropsychological malingering found in Chapter 9 by Jerry Sweet. Finally, to close this section, in Chapter 10, J. Peter Rosenfeld and Joel Ellwanger chronicle an innovative and methodical research program that has resulted in a combination of psychophysiological and neuropsychological assessment procedures for the detection of malingering. This latter chapter may be prefatory to a new technological age of neuropsychological malingering detection that could affect forensic activities profoundly.

Chapter 4

THE NEUROPSYCHO-LOGICAL EVALUATION:
CLINICAL AND SCIENTIFIC FOUNDATIONS

Eugene J. Rankin
Younker Memorial Rehabilitation Center, Iowa
Methodist Medical Center, Des Moines, Iowa

Russell L. Adams
University of Oklahoma Health Sciences Center,
Oklahoma City, Oklahoma

Introduction

Within the context of legal proceedings, the neuropsychological evaluation must maintain the same high level of commitment to the clinical and scientific foundations of the field, as would be expected within the context of a non-legal referral. Given this opening statement, the philosophical emphasis of this chapter will be on the scientist-practitioner model. The chapter begins with a brief history of contemporary neuropsychology, followed by information on the general purpose of the evaluation, background review of clinical information, and interpretation of test results from both quantitative and clinical approaches. In addition, this chapter will discuss practical and statistical considerations of neuropsychological test interpretation. The chapter concludes with a discussion of ethical issues involved in the release of raw neuropsychological test data to non-psychologists, and possible resolutions of such issues.

In general, the overriding theme throughout the chapter is that the neuropsychological evaluation is more than administering tests to persons with brain injury, but rather is a complicated process requiring the integration of multiple sets of data. As stated by the Therapeutics and Technology Assessment

Subcommittee of the American Academy of Neurology (1996), "neuropsycho-logical information is subject to intensive scrutiny in forensic proceedings and can be successfully challenged if it is over interpreted, obtained during the acute phase of an injury or when the patient is taking medications that might affect performance, ignores the presence of depression or anxiety when the tests were performed, or fails to take premorbid characteristics, developmental irregulari-ties, and substance abuse into account" (p. 596).

Brief history of contemporary neuropsychology

Historical background

Cognitive neurosciences at the end of the nineteenth century believed that cog-nitive processes could be localized to specific brain regions. The localizationist-connectionist approach is generally considered to have had its start with the work of Broca (1865), Bastian (1869), and Wernicke (1874), all of whom attempted to demonstrate the connection between "cortical centers" and cog-nitive processes. More specifically, Broca (1865) reported on several patients manifesting right hemiplegia and expressive speech deficits, who on postmortem examination demonstrated large lesions in the left frontal brain region. Bastian (1869) implicated separate "centers" for word knowledge and motor knowl-edge. Wernicke's (1874) work focused on the identification of three separate types of language impairments, implicating not only language "centers," but commissural connections between these "centers," thus accounting for motor, conduction, and sensory aphasias.

The strict localizationist-connectionist approach fell from favor for several reasons. Karl Lashley discovered through experimental animal investigations that these proposed "centers" were not localized, but rather diffusely repre-sented within the brain. This holistic approach became known as the mass action doctrine. More specifically, the mass action doctrine considered that the consequences of brain lesions were a direct function of the amount, rather than the location, of tissue removed (Lashley, 1929, 1938).

Current understanding

In the latter half of the twentieth century came the advent of new computer technology that allowed neuroscientists access to the interactive workings of the living human brain. This technology facilitated our current understanding of the relationship between neuropsychological impairment and brain injury. No longer did neuroscientists have to conjecture about the inner workings of the human brain through post-mortem examinations; there was now technology that allowed access to *in vivo* examination of the brain. However, and unfortu-nately, this access did not come without its share of methodological shortcom-ings. More specifically, the majority of the neuroscience investigations that characterize this period examined persons who sustained brain injury, identified through neuroimaging, and then examined the behavioral consequences of their injury. A major methodological shortcoming for early neuroscience in general, and neuropsychology in particular, was that of inferring normal mechanisms of brain functioning from ablation or lesion paradigms. For example, a brain lesion may result in a generalized behavioral *disinhibition*, in addition to a discrete loss of function. This is related to concepts such as *diaschisis*, among others (Heilman & Valenstein, 1993). More specifically, overestimation of cognitive function to a lesioned area of brain can occur not because that area of the brain "contains" that cognitive function, but because that lesioned area of the brain somehow disturbed functioning important to other nearby brain regions (i.e., the concept of diaschisis), and hence other cognitive functions, thereby resulting in a generalized behavioral disinhibition. As a result, researchers sought "double dissociations," referring to a need to demonstrate that comparable lesions in other brain regions did not cause similar behaviors, in order to rule out diaschisis effects. Brain stimulation paradigms have a similar problem in that such experimental manipulation may *overinhibit* normal brain functioning. These arguments are particularly salient with regard to renewed attempts at correlating cognitive processes observed with neuropsychological tests and metabolic differences observed with modern brain imaging techniques, chiefly functional MRI, positron emission tomography (PET) and single photon emission computed tomography (SPECT). These techniques evaluate brain metabolism rather than brain structure, and secondarily can be used to correlate behavioral deficits with metabolic deficits. Such investigations frequently operate under the assumption that cognitive operations are localizable to focal brain regions or systems (Sarter, Berntson, & Cacioppo, 1996), quite similar to the assumptions of the initial localizationist-connectionist approach. Unfortunately, there is often not a one-to-one correspondence between deficits observed (or not observed) on neuroimaging and deficits observed (or not observed) on neuropsychological testing. Contemporary neuropsychological theories suggest that a brain region does not act on its own, but interacts with other regions and at multiple levels of processing in producing behavior. In other words, while it is tempting to conclude that region X causes behavior Y, what we actually can conclude is that region X and behavior Y are somehow related. Such explanations help to explain the frequent absence of a one-to-one correspondence between brain lesion and behavioral deficit. Consequently, as Lezak (1995) says, "the uncertain relation between brain activity and human behavior obligates the clinician to exercise care in observation and caution in prediction, and to take nothing for granted when applying the principles of functional localization to diagnostic problems." It is at this point that our discussion of contemporary neuropsychology begins.

Development of the specialty

The development of the science of neuropsychology stemmed from two very different methodological approaches to evaluating cognitive processes. These approaches are often considered to be at opposite ends of the same continuum:

an actuarial approach, often attributed to American schools of thought (Reitan & Davison, 1974; Reitan & Wolfson, 1993), and a clinical-theoretical approach attributed to the Russian neuropsychologist, A.R. Luria (1966, 1973). An example of the actuarial approach is Ward Halstead's attempts at investigating the concept of biological intelligence, which was believed to be affected by injury to the prefrontal cortex. Halstead's theory was psychometrically operationalized by Ralph Reitan through the investigation of various clinical populations. Through these investigations, the Halstead-Reitan Neuropsychological Test Battery was developed, constituting a major scientific attempt at linking theory and clinical practice (Reitan & Wolfson, 1996). In contrast, the clinical-theoretical approach of A.R. Luria was an attempt at integrating localizationistic-connectionistic thinking with mass action doctrine, resulting in a theory of higher cortical functioning by analyzing various behavioral syndromes. In general, these methodological approaches can be distinguished by an emphasis on quantification of performance decrements using an actuarial approach versus an emphasis on the demonstration of the cognitive processes underlying dysfunctional performance using a clinical-theoretical approach.

The methodological approaches described above allow for both quantification of cognitive deficit as well as qualification of the processes underlying the cognitive deficit. From these two divergent approaches, contemporary clinical neuropsychology as a psychological specialty developed. Today, many neuropsychologists ascribe to an interpretive approach that integrates both an actuarial approach and a clinical-theoretical approach, which appears desirable.

These methodological approaches are reflected in the neuropsychologist's training experiences. Because of the pivotal role of the training background on both administration and interpretation, standards have subsequently been adopted by neuropsychologists to ensure adequate training for those persons who call themselves neuropsychologists (International Neuropsychological Society-Division 40 Task Force on Education, Accreditation, and Credentialing, 1987). These standards are described in more detail below. As with most training guidelines, these training suggestions may be modified in future years.

Training in neuropsychology

According to Division 40 of the American Psychological Association (APA), the definition of a clinical neuropsychologist is as follows:

> A clinical neuropsychologist is a professional psychologist who applies principles of assessment and intervention based upon the scientific study of human behavior as it relates to normal and abnormal functioning of the central nervous system. The clinical neuropsychologist is a doctoral level psychology provider of diagnostic and intervention services who has demonstrated competence in the application of such principles for human welfare following: a) successful completion of systematic, didactic, and experiential training in neuropsychology and neurosciences at a regionally accredited university, b) two or more years of appropriate supervised training applying neuropsychological services in a clinical setting, c) licensing and

certification to provide psychological services to the public by the laws of the state or providence in which he or she practices, d) review by one's peers as a test of these competencies. Attainment of the ABCN-ABPP diplomate in clinical neuropsychology is the clearest evidence of competence as a clinical neuropsychologist assuring that all these criteria have been met. This statement reflects the official position of the Division of Clinical Neuropsychology and should not be construed as either contrary to or subordinate to the policies of the APA at large (Adams & Rourke, 1992, p. 5).

Training in clinical neuropsychology takes place at the doctoral, internship and post-doctoral level. Standards for these various levels of training have been published by the International Neuropsychological Society-Division 40 Task Force on Education, Accreditation, and Credentialing (1987). Briefly, they involve the awarding of a Ph.D. degree from a regionally accredited university, with the student having been prepared for health service delivery, basic clinical research, teaching, and consultation. A clinical neuropsychology internship should devote "at least 50% of a one-year full-time training experience to neuropsychology", with at least 20% of the training experience devoted to general clinical training to ensure a competent background in clinical psychology. In recent years, postdoctoral training has become an expected accomplishment in the career path of a clinical neuropsychologist. It is expected that postdoctoral training will be directed by a board-certified clinical neuropsychologist, extending over a two-year period. Exceptions to this recommendation include those individuals who completed a specific clinical neuropsychology specialization in their graduate programs and/or a clinical neuropsychology internship. Specific training guidelines at each of these levels are provided within the document.

Finally, attainment of diplomate status in clinical neuropsychology, indicating Board Certification through the American Board of Professional Psychology, represents the highest level of competence in the field. Since 1982, this specialty board has established eligibility criteria, a written examination, and a competency-based oral examination for the field. Although not sufficient by itself, board certification can be an important method of determining competency in neuropsychology.

The described guidelines provide a standard to compare a psychologist's training in neuropsychology. The guidelines are explicit and clear-cut. As stated by Satz (1988), "psychologists (including clinical) who practice clinical neuropsychology outside these standards are operating in violation of APA ethical and practice standards" (p. 99).

General purpose of neuropsychological evaluations

The neuropsychological evaluation is a psychometric investigation of the behavioral manifestation of brain dysfunction. The evaluation compares the patient's performance *across* (i.e., against similar individuals) age and education, as well as comparing performance, for example with estimated premorbid performance, *within* the same individual (Adams & Rankin, 1996). If the case is likely to be adversarial, as in application to forensic cases, the evaluation will most often need to be comprehensive, employing multiple measures of various cognitive domains. Reasons for comprehensiveness include the complexity of the issues (e.g., etiology, damages, prognosis, treatment, etc.) and the standard of evidence that neuropsychologists are held to as expert witnesses.

As cited in the 1981 Report of the Task Force on Education, Accreditation and Credentialing of the International Neuropsychological Society,

> referrals for clinical neuropsychological assessment typically consist of, but may not be limited to, the following: 1) differential diagnosis between psychogenic and neurogenic syndromes (e.g., depression vs. dementia); 2) differential diagnoses between two or more suspected etiologies of cerebral dysfunction (e.g., neoplasm vs. cerebral vascular disorder); 3) delineation of spared and impaired functions secondary to an episodic event (e.g., cerebrovascular accident, head trauma, infection); 4) establishment of baseline measures to monitor progressive cerebral disease (e.g., neoplasms, demyelinating disease); 5) comparison of pre- and post-pharmacologic, surgical and behavioral interventions (e.g., drug trails, tissue excision, shunts, vascular repair and language or cognitive therapy); and 6) assessment of cognitive functions for the formulation of rehabilitation strategies (p. 5).

With an understanding of reasons for referral, an appropriate and responsive evaluation typically can be planned. It can be argued that the actual testing portion of the examination is an essential, but small, part of the evaluative process. The larger part of the evaluation is *how* the data is interpreted - idiographically within the context of the individual from which the data were derived, and nomothetically within the context of neurological and neuropsychological commonalities among similar persons. It can also be argued that the process of the evaluation, the tests administered and their interpretation, is greatly dependent upon the training background of the examiner. Table 1 provides a partial listing of the types of litigation-related cases referred for neuropsychological evaluation.

Background preparation

The following section will discuss the necessary preparatory work involved prior to the actual evaluation. More specifically, this section will discuss the need to review academic and vocational records, prior medical and psychiatric records, and prior neuropsychological testing data (if available). In general, as stated by the Therapeutics and Technology Assessment Subcommittee of the American Academy of Neurology (1996), "like other tests, neuropsychological assessments are of limited usefulness by themselves and must be interpreted in conjunction with other clinical, imaging, and laboratory information." (p. 592). In other words, neuropsychological tests are no different than most medical tests in terms of needing a context for accurate interpretation. Neuropsychological test data must be interpreted with due consideration of the patient's premorbid academic and vocational, premorbid clinical, and premorbid medical histories. Table 2 provides a list of moderator variables routinely considered in the neuropsychological interpretive process.

Table 1. Types of Cases Referred for Neuropsychological Evaluations.

Personal injury cases
Motor vehicle accidents in which the victim with a traumatic brain injury was a passenger, driver, or pedestrian;
Falls or fights in which traumatic brain injury occurred or is suspected of having occurred;
Head injury caused by a faulty piece of equipment.

Product liability cases
Patients exposed to certain neurotoxic chemicals (e.g., perchlorethylene, trichloroethylene, hydrogen sulfide, solvents);
Carbon monoxide poisoning caused by a faulty heating system; Seatbelt or safety equipment malfunction resulting in brain injury.

Medical malpractice suits
Brain injury suffered at birth;
Problems occurring during the surgery itself that resulted in brain dysfunction (i.e., an otolaryngologist penetrating the roof of the ethmoid sinus during sinus surgery and creating an encephalocele involving the frontal lobe).

Workers' compensation cases
Client suffered a traumatic brain injury as a direct result of his or her employment;
Chronic exposure to neurotoxic chemicals in the workplace.
Social security determination and other disability assessment
Client suffered some presenile dementia and was claiming inability to be gainfully employed;
Client claimed inability to work due to some type of brain injury.
Inability to make a contract or will
Family members or other claimed that the patient was not competent to make a will or enter into a contract and an attempt to void a will in which he or she gave all of his or her assets to a nonfamily member;
An elderly patient who was demented sold her house at a price well below market value.

Note. From "A practical guide to forensic neuropsychological evaluations and testimony" by R.L. Adams and E.J. Rankin, 1996, In R.L. Adams, O.A. Parsons, J.L. Culbertson, & S.J. Nixon (Eds.), *Neuropsychology for clinical practice: Etiology, assessment and treatment of common neurological disorders* (p. 457), Washington, DC: American Psychological Association. Copyright 1996 by the American Psychological Association. Adapted with permission of the publisher.

Academic and vocational records

Access to academic and vocational records may be essential in order to assist in determining the patient's premorbid level of ability. In particular, educational attainment is highly correlated with neuropsychological test performance (Finlayson, Johnson, & Reitan, 1977; Leckliter & Matarazzo, 1989; Vega & Parsons, 1967). Educational attainment also has positive prognostic value for many neuropsychological conditions (Grafman et al., 1988). However, merely knowing how many years a person was at school is not sufficient, nor is the patient's self-report of academic performance (which is typically stated by the patient as "average"). In many cases, standardized academic measures have

Table 2.   Partial Listing of Moderator Variables or Alternative Hypotheses Commonly
           Considered in the Neuropsychological Interpretive Process Pertaining to
           Primary Brain Dysfunction.

Age
Gender
Handedness
Emotional states (e.g., depression, anxiety)
Suboptimal motivation and effort
Poor physical stamina or fatigue
Low education
Congenital learning disorder, pre-morbid cognitive or intellectual deficiency
Significant psychiatric disorder (e.g., Bipolar Disorder, Schizophrenia)
Peripheral injury or other non-central nervous system medical disorder
Acute effects of substances (recreational, prescription or non-prescription)
External distraction in the testing environment
Internal distractions (e.g., pain or other internal bodily discomforts)
English as a second language
Sociocultural background
Non-neurological medical disorders that may affect the central nervous system(e.g., liver
     disease causing hepatic encephalopathy)
Deliberate attempts to feign deficit (i.e., malingering)

Note. From *Psychological assessment in medical settings* (p. 152) by R.H. Rozensky, J.J.
Sweet, & S.M. Tovian, 1997, New York: Plenum. Copyright 1997 by Plenum. Reprinted
with permission of publisher.

been administered at some point during the patient's academic career and can
serve as useful and more valid benchmarks for premorbid intellectual abilities.

Vocational history is another important variable to review prior to the neuropsychological evaluation. In particular, employment complexity (e.g., type of
occupation), number of employment positions over the course of one's life, reasons for changing jobs, duties involved at each employment position, and vocational performance can be important variables to review in order to assist in
determining premorbid intellectual abilities. Vocational history can be best
investigated in a chronology from first employment to current (or last) employment, in order to determine whether expected (promotions, pay raises, etc.) or
unexpected (demotions, terminations, etc.) patterns exist.

Ascertaining employment complexity (e.g., surgeon versus laborer) helps in
gaining an understanding of premorbid abilities. Determining number of
employment positions and reasons for job changes can also be helpful in gaining an understanding of possible authority and interpersonal conflicts, particularly with persons who were terminated or simply left several positions for
unexplained reasons. If available, information pertaining to performance at
each position can facilitate an understanding of whether the person received
promotions, or frequently ran into conflicts with supervisors, resulting in the
patient's termination, for example, persons who were employed initially at a
higher vocational level and subsequently drifted to employment at a lower
vocational level (e.g., a patient who was employed initially as an accountant,

but now works as a food service worker at a food store.) Reasons for downward drift require investigation to determine whether the job change was due to legal causes (losing one's license to practice, etc.), substance abuse, medical disability (prior head trauma with loss of intellectual capacity), or free choice. It is our experience that the latter cause is an infrequent occurrence at best.

The workforce of today more frequently comes into contact with toxic chemical substances than in the past. While not related to premorbid ability, vocational information gathering may also reveal whether toxic exposure in the workplace may have occurred. If relevant, it is important to know the chemicals, as well as frequency of contact and duration of contact to which the worker was exposed. (The reader is referred to Chapter 12 on neurotoxin-related cases for more information.)

One final note regarding vocational history has to do with prior job satisfaction. Specifically, the willingness of a head injured worker to return to work after a head injury is significantly impacted by previous job satisfaction. If the job was not satisfying to the patient prior to the incident, it is unlikely that the patient will want to return to that job. Two important moderator variables to consider in predicting return to work are accommodation to residual cognitive and behavioral deficits (Bruckner & Randle, 1972), and level of education (Rimel et al., 1981).

Medical and psychiatric records
In addition to reviewing academic and vocational records, the patient's medical and psychiatric records require review, including pertinent neurological, neuroradiological, neurosurgical and psychiatric data. Although not within the purview of the neuropsychologist to provide either neurological, neuroradiological, or neurosurgical opinions, it is necessary for the neuropsychologist to have at least a reasonable understanding of these areas of investigation in order to assist in the confirmation or refutation of generated neuropsychological hypotheses. These areas of investigation are briefly described within this section.

Neurological investigation involves formulation of four key areas: symptoms, signs, site of involvement (localization) and probable cause (differential diagnosis) (Kaufman, 1990). In providing this formulation, a neurological examination is conducted. This examination typically involves assessment of six basic areas of neurological functioning: mental status, cranial nerves, cerebellar functioning, motoric integrity, integrity of the sensory functions, and reflexes (deep tendon, pathologic, and sensory) (DeMyer, 1994; Kaufman, 1990; Weiner & Levitt, 1983). From this information, inferences regarding whether the patient's presentation is secondary to cortical, subcortical, brainstem, spinal cord, peripheral nervous system, muscle dysfunction, or psychogenic factors are made by the neurologist. The latter inference may be made if either the neurological examination is within normal limits, or the elicited neurological signs and symptoms violate the laws of neuroanatomy (Kaufman, 1990). However, it is noteworthy, as with neuropsychological misdiagnosis, that neurological misdiagnosis can occur when inferences of psychogenic causes are given automatically to clinical presentations simply because they are: unique or bizarre, severity is greater than expected, or there are no accompanying objective physical abnormalities (Kaufman, 1990).

Neuroradiological procedures are often employed to assist in the diagnostic process. Briefly, neuroradiological procedures can be divided into those involving assessment of brain structure (CT scanning and MR imaging), and those involving brain function (positron emission tomography [PET], single photon emission computed tomography [SPECT], and functional MRI [fMRI]). Briefly, CT scanning works by using exogenous X-ray penetration that identifies tissue density changes, particularly at the interface between tissue and fluid or tissue and bone. MR imaging works by detecting radiofrequency energy absorbed by brain tissue after the tissue is subjected to a strong magnetic field. In contrast to structural neuroimaging approaches, PET, SPECT, and fMRI examine changes in cellular metabolism in the brain. While functional neuroimaging techniques, such as SPECT, are becoming increasingly useful tools to aid in the diagnosis of a number of neuropsychiatric disorders (Holman & Devous, 1992), there is a consensus at present of limitations that preclude forensic applications (Mayberg, 1996). For example, SPECT is often abnormal in severely emotionally disturbed individuals without structural abnormality (American Academy of Neurology, 1996; Ritchie et al., 1997).

One final diagnostic procedure that neuropsychologists need to be familiar with is the electroencephalogram (EEG). Specifically, the EEG is a device that records electrical activity of the brain. Although the EEG is considered to be a highly sensitive test of abnormal electrical activity in the brain, in neurological conditions such as epilepsy, abnormalities on EEG are not proof positive of abnormal brain functioning. In fact, abnormal EEG patterns are observed in approximately 3% of normal (i.e., seizure free) individuals (Fenton, 1982), and approximately 20% of epilepsy patients have completely normal interictal EEG patterns (Kaufman, 1990). One final note is that the use of quantitative EEG (e.g., "brain mapping") in the clinical diagnosis of conditions such as traumatic brain injury is considered controversial (American Academy of Neurology, 1989), and has *not* been widely accepted in the neurological community.

Review of neurosurgical data prior to neuropsychological evaluation may allow an understanding of injury severity in cases of traumatic brain injury. Specifically, pertinent information leading up to neurosurgical intervention includes the condition of the patient at the time of the injury, condition at the time of admission to the hospital, coma scale scores both immediately and subsequently, and additional physical injuries. When applicable, review of the method of neurosurgical intervention is also important with regard to complexity of the neurological condition, medical complications, and course of recovery.

Past psychiatric history would include prior history of clinical depression and mental stability, in that psychiatric disorders may affect neuropsychological performance significantly (e.g., Sweet, Newman, & Bell, 1992). Several investigators have found, for example, that clinical depression is commonly accompanied by psychomotor slowing, impaired attention, decreased cognitive flexibility, and poor retrieval memory (Bornstein & Kozora, 1990; Caine 1986; Nelson et al., 1993). Substance abuse may also significantly affect performance on neuropsychological tests. For example, chronic alcoholism is associated with deterioration in abstraction, visuospatial skills, and problem-solving abilities (Parsons, 1987). Persons abusing cocaine may demonstrate brain perfusion anomalies and persistent neurocognitive deficits (Strickland & Stein, 1995), although variables such as intensity of use have not been fully investigated. This history must be sought and if present, integrated into the interpretation of the neuropsychological test data.

Prior neuropsychological records

An ethical approach to obtaining prior neuropsychological test records (the actual test protocols) is discussed at the end of this chapter. In general, if a prior neuropsychological evaluation was performed, test data and protocols should be examined for accuracy in scoring, and the report examined to determine whether the conclusions were sound, based on current scientific understanding and literature. The credentials of the neuropsychologist can also be reviewed.

Clinical interview

The clinical interview allows for additional data gathering prior to performing neuropsychological testing. Interviews are essential in ascertaining perceived changes from pre-injury state to post-injury condition. In general, the interview should cover the following areas of functioning: demographic information, history of present illness, current complaints, past medical history, past surgical history, past psychiatric history, current medications, educational history, work history (including military history, if applicable), social history, substance abuse history, and legal history. In addition, it is helpful to ascertain whether current life stressors exist, as well as a typical day in the life of the patient. The interviewer should also be attentive to nonspecific factors, such as inconsistencies between the patient's verbal self-report and facts obtained through review of records, as well as inconsistencies between the patient's statements and behavior, and finally, between information obtained from the patient report and that obtained from significant others (e.g., spouse, child, close friend).

Testing methods may require adaptation if it becomes apparent that an individual does not have the necessary stamina for a complete evaluation, either due to physical and/or mental impairment. To supplement the interview, some clinicians find it helpful to have the patient complete a history questionnaire prior to the evaluation appointment. This document can serve as a written record, in the patient's own writing, of the course of their illness, whether there were pertinent developmental milestone inconsistencies, other illnesses or injuries sustained prior to the present illness (which may or may not impact the present illness), psychiatric history, educational and occupational histories, legal history including prior arrests and convictions, as well as a description of their current complaints. However, it is noteworthy that current complaints of individuals claiming mild traumatic brain injury, particularly when obtained through symptom checklists, are fraught with difficulties, given recent findings that seemingly naïve college students are able to accurately endorse over half the symptoms associated with mild brain injury (Lees-Haley & Dunn, 1994). Symptom checklists suffer from an inherent weakness of not having validity indicators, making their use in forensic cases difficult, unless the information is verified by independent means.

At the outset, it is important to determine whether the person is aware of the reason for the evaluation, which is an indirect test of orientation and awareness. Once ascertained, we describe the reason for the evaluation, as well as a brief description of what the patient is likely to experience throughout the evaluation process. Again, as noted above, sensitivity to inconsistencies between the patient's self-report and their behavior is also of importance during the clinical interview (e.g., a patient who reports severe memory impairment, yet describes in detail all aspects of the injury, or on the other hand, a patient who reports no memory dysfunction but is unable to remember why he or she was brought to the evaluation).

In order to prevent deception on our part, given that several of the tests administered will be tests sensitive to malingering, we typically tell the patient that ". . . some of the tests we administer will be testing motivation, but you will not know which tests these are. Therefore, we ask you to try to do your best on all the tests given to you today." Methods for evaluation of insufficient effort and malingering are presented in chapters 9 and 10.

Interviewing patients in a forensic neuropsychological evaluation can be challenging, especially when the patient is defensive, or does not give candid responses. The patient may have already given his deposition in the case in question, and that experience for some patients can be an unpleasant one. In some instances, the patient may view the clinical interview as another deposition. If possible to obtain, interviews of collateral sources such as spouses, significant others, parents or children, can be essential. These interviews may reveal inconsistencies in what the patient says or may confirm the patient's self-report.

Neuropsychological test interpretation

This section discusses neuropsychological test interpretation from both the quantitative and clinical perspective. Today's modern neuroimaging technology, despite shortcomings from a cognitive science perspective, decreases the need to localize dysfunction by neuropsychological tests (Trenerry, 1996). Rather, the continued importance of neuropsychology resides in its ability to quantify a patient's brain-dependent cognitive abilities (Trenerry, 1996), as well as its ability to identify bases for dysfunctional performance.

This chapter section initially discusses neuropsychological test interpretation by comparing a patient's performance to persons with similar demographic backgrounds, and also to his or her own expected premorbid performance. This approach takes into account both quantification methods and qualification or process methods. Kaplan (1988) describes this different methods as a focus on global achievement versus a focus on observation of behavior enroute to a solution (process). The quantification approach, as exemplified by Reitan (1967), includes the following methods of interpretation: level of performance, pattern of performance, right-left differences, and pathognomonic signs. This chapter section discusses the analysis of test data within the context of various domains of higher cognitive functioning in an attempt to understand the cognitive processes underlying dysfunctional performance.

A review of the use of specific neuropsychological tests with specific neuropsychological conditions is beyond the scope of this chapter. The interested reader is referred to either the work of Heilman and Valenstein (1993), Lezak (1995), or Spreen and Strauss (1991). Additionally, a compilation of specific neuropsychological test batteries for specific neuropsychological conditions is also beyond the scope of this chapter. Here, the reader is referred to the edited text by Adams, Parsons, Culbertson, and Nixon (1996).

Quantification of neuropsychological deficits
The interpretation of neuropsychological test results using quantitative comparative methods follows a process of examining test data along several interpretive steps. These interpretive steps are interdependent; false positive or false negative results may occur *if* one step is relied upon to the exclusion of the others. The following discussion is taken largely from the work of Reitan (Reitan, 1967; Reitan & Wolfson, 1996).

Level of performance
Initially, a patient's neuropsychological test data are interpreted by comparing that person's data to persons with similar demographic backgrounds. Specifically, a patient earns a score on a test, and, based on estimated premorbid levels, that patient's performance is compared to others of similar age, education and gender, as a general expected normative comparison standard. Estimates of premorbid level of functioning are obtained in several fashions, many of which are described within this chapter, and also in detail in chapter 3 of this text. Certainly, when evaluating whether a patient's current test performance represents a decline from a higher level of functioning, adequate normative information is required. These comparative normative data are available from several sources, typically the test manuals, as well as subsequently published information, such as the demographically corrected normative data on the Halstead-Reitan battery developed by Heaton, Grant, and Matthews (1991).

The *level of performance* approach in quantifying neuropsychological deficits is not sufficient in itself, but requires combinations with other approaches. When used without consideration of other interpretive approaches, clinical utility typically suffers. For example, considering only performance below a certain cut-off score (i.e., impaired range) can result in false positive results for persons with a low level of premorbid functioning, and when considering only performance above a certain cut-off score (i.e., non-impaired range) can result in false negative results for persons with a high level of premorbid functioning. Emotional dysfunction can also result in this kind of misclassification, as can malingering.

Pattern of performance

The second step in test interpretation is determining whether a recognized pattern of deficits emerge from the test data. More specifically, the evaluator examines those tests that were performed poorly, and those tests performed at a level consistent with premorbid estimates. Common areas of deficit are identified, and attempts are then made to determine whether these deficits are associated with known patterns of cognitive dysfunction evident in brain injured individuals. Obviously, a pattern analysis approach assumes some degree of neuroanatomical homogeneity of neuropsychological functions. For example, in the majority of right handed individuals, language and praxis functions are lateralized to the left hemisphere (Rasmussen & Milner, 1977); whereas visuospatial skills are often associated with right hemisphere functioning (Kirk & Kertesz, 1993). As most right handed persons are left hemisphere dominant for language and praxis, in persons with a language-related disorder and ideomotor apraxia a hypothesis of left hemisphere dysfunction would be considered. As another example, deficient motor functioning in the presence of normal somatosensory functioning, may implicate anterior brain dysfunction, relative to intact posterior brain function. However, these associations are not absolute findings, and may differ within individual cases for a variety of reasons. A pattern approach can be helpful with localizing lesions to specific areas of the brain, if needed.

Right-left differences

This interpretive approach examines neuropsychological test data by comparing sensory and motor performances on one side of the body with those of the other side. This is often performed using motor tests of speed, strength, and dexterity, and comparing the motoric performances against normative data based on estimates of premorbid functioning. Deficient performance may manifest as an atypical superiority of the nondominant hand, implying dysfunction to the hemisphere contralateral to the affected extremity. Sensory impairment in one extremity may also imply contralateral brain dysfunction, although such a finding is not always supported. Right-left differences on these types of neuropsychological tests can be compared to findings from the neurological examination, within which demonstrations of lateralized patterns of weakness or reflex asymmetries are sought in order to implicate brain dysfunction. However, one strong caveat is required here. In many individuals, there may be peripheral (non central nervous system) explanations for the neuropsychological findings, as would be the case for neurological findings. For example, peripheral neuropathy may limit sensory integrity, leading to false positive results. Another example would be degenerative arthritis, which may limit motor functioning in terms of diminished strength, speed, and dexterity, without affecting brain integrity, and possibly lead to a false positive result.

Pathognomonic signs

Pathognomonic signs are considered absolute (or approaching absolute) indications of brain injury. Examples of pathognomonic signs in a previously literate and non-symptomatic adult include: aphasia, apraxia, sensory neglect, visual field cuts, acalculia, and alexia, among others. Although these signs are almost always, in and of themselves, indications of brain dysfunction, the utility of this approach is reduced by the rarity with which these signs are observed in general clinical practice.

Qualification of cognitive processes

Neuropsychological tests are variable in their sensitivity and specificity to brain dysfunction. Factor analytic studies demonstrate that some of these tests intercorrelate better than others, and hence, are considered to represent theoretical constructs of cognition. How these theoretical constructs relate, not only to each other, but to brain dysfunction is the focus of this chapter section. More specifically, within this section, commonly discussed theoretical domains of cognitive functioning are presented, including (1) attention, learning, and memory, (2) speech and language, (3) visuospatial and constructional skills, and (4) executive functioning and concept formation. Obviously, given space limitations, a complete discussion of these topics is not attempted.

Attention, learning and memory

Measures of attention, learning, and memory represent overlapping constructs on a continuum from initial registration of information, improvement with repetition, storage, and finally, retrieval from storage. These constructs are discussed separately.

In consideration of an information processing approach, attention would represent the basic foundation for adequate registration of to-be-remembered information. Attention can be conceptualized both in terms of attentional capacity (how much information the system can process at once) and attentional efficiency (how fast the attentional system can operate) (Lezak, 1995). Within each of these dimensions, it is possible to further subdivide attention into several components, including sustained attention, selective attention, and divided attention.

Sustained attention capacity is often evaluated by supraspan procedures, such as digit span forward and initial word recall on serial word learning lists; sustained attention efficiency is often evaluated by tests of vigilance (Strub & Black, 1985). Selective attention efficiency can be evaluated by one of the various Stroop procedures, as well as by symbol substitution procedures (e.g., Digit Symbol or Symbol Digit Modalities Test). Divided attention efficiency can be evaluated by tests such as Part B of the Trail Making Test and Paced Auditory Serial Addition Test (PASAT).

Learning capacity is considered part of a rehearsal process involved in memory functioning. Evaluation of learning capacity typically involves performance on serial word learning lists, although, as would be expected, some tests may confound the information processes of registration, storage and retrieval. There are numerous list learning procedures available to the examiner, for which the choice depends on patient capacity, the kind of information required, the ease of administration and scoring, and available norms. The benefit of these measures includes the ability to assess initial recall (described above), capacity to assess the development of a learning curve with repeated trials, interference effects, short- and long-term retention, and improvement with cues and/or recognition formats. Factor analytic investigations of many of the list learning procedures (e.g., Rey Auditory Verbal Learning Test [RAVLT], Selective Reminding Procedures) generally demonstrate three basic factors of acquisition, storage and retrieval (Vakil & Blachstein, 1993). Clinically, these list learning procedures are beneficial in delineating expected performance decrements. For example, traumatic brain- injured patients typically demonstrate a less than normal learning curve across trials, reduced free recall on delay, but improvement with recognition cues, suggesting memory retrieval rather than memory consolidation deficits with the RAVLT (Bigler, Rosa, Schultz, et al., 1989). Using the California Verbal Learning Test, Crosson, Novack, Trenerry, and Craig (1989) demonstrated three types of memory deficits among traumatically brain injured patients (i.e., encoding deficits, consolidation deficits, and retrieval deficits) similar to findings observed with the RAVLT. In contrast, pervasive dementing disorders like Alzheimer's disease typically demonstrate little to no improvement with repetition, limited improvement with recognition cues, and a higher incidence of intrusions (Bigler et al., 1989).

Memory is a complex process, involving the acquisition and storage of information (i.e., learning) and the retrieval of stored information (Thompson, 1988). Frameworks for understanding human memory have been developed with varying degrees of clinical utility. Some of these frameworks distinguish between episodic and semantic memory (Kinsbourne & Wood, 1975; Tulving, 1972, 1983), reference and working memory (Olton, Becker, & Handelmann, 1979), vertical and horizontal associative memory (Wickelgen, 1979), and declarative and procedural memory (Cohen & Squire, 1980). One framework proven useful in understanding memory dysfunction (Craik, 1977; Erickson, 1978) is a three-stage model of memory functioning developed by Atkinson and Shiffrin (1968). Briefly, this model, based on information processing theory, conceptualizes memory functioning in terms of different stages, (i.e., sensory, primary [short-term memory] and secondary [long-term memory]), and different processes, (i.e., encoding or registration, storage/rehearsal, and retrieval).

The utility of this theoretical model lies in its ability to allow for the systematic assessment of memory dysfunction at one of several points along the information processing sequence. Memory dysfunction can be conceptualized as a breakdown in information processing at one or several points along the sequence. Notably, a majority of the research conducted on memory impairment has utilized an information processing approach (Bauer, Tobias, & Valenstein, 1993).

Consistent with this information processing theory is the notion that the brain does not function as a whole unit; rather, as discussed earlier, the brain appears to function concertedly in localized functions or systems. Anatomical correlates to the hypothesized three-stage information processing model of memory functioning are supported by clinical literature on the effects of brain surgery on human memory functioning (Milner, 1966), as well as by the reported effects of various neuropathological processes on human memory functioning (Cermak, 1976; Damasio, Eslinger, Damasio, Van Hoesen, & Cornell, 1985). Such evidence suggests that memory is not a unitary ability. Notably, Russell's (1975) adaptation to the original Wechsler Memory Scale (WMS; Wechsler, 1945) was one of the first clinical demonstrations that memory is not a unitary process. Based on his research with patients exhibiting unilateral and diffuse brain damage, Russell concluded that at least three types of memory can be measured clinically, which he termed "immediate memory, short-term memory, and long-term memory" (Russell, 1975).

Additionally, memory for certain types of information appears to be lateralized in the brain (Milner, 1968, 1970). Specifically, in right-handed individuals, memory for verbal information appears to be located primarily in the left hemisphere, and memory for figural information appears to be localized primarily in the right hemisphere. Although not universally supported by researchers, these findings have been validated in the assessment of modality-specific memory impairment associated with unilateral brain damage, using the Wechsler Memory Scale-Revised (e.g., Chelune & Bornstein, 1988). Specifically, subjects with left hemisphere damage were more proficient with their retention of non-verbal/visual material than of comparable verbal material; subjects with right hemisphere impairment demonstrated the opposite pattern.

Pervasive memory impairment is a hallmark of dementing disorders (Moss & Albert, 1988). Typically, this decline in memory functioning begins insidiously as increased forgetfulness. At this stage, such a decline is often difficult to differentiate from memory changes associated with normal aging (Hartley, Harker, & Walsh, 1980). In fact, the predictive power of individual neuropsychological tests, such as tests of memory functioning, to differentiate dementia from depression in a single patient is modest, with 10 to 30% of depressed patients classified as demented (Harper, Chacko et al., 1992; Lamberty & Bieliauskus, 1993). At more advanced stages of dementia, assessment of memory disturbances provides benchmarks for staging the level of severity of the disorder. However, emotional disturbances, most notably depression (Cronholm & Ottosson, 1961; Miller & Lewis, 1977; Sternberg & Jarvik, 1976; Stromgren, 1977; Walton, 1958), have also been associated with concomitant memory dysfunction. Memory dysfunction secondary to depression has been observed in populations both below and beyond age sixty-five. In populations below age sixty-five, Cronholm and Ottosson (1961) suggested that memory deficits caused by depression are limited to impaired short-term memory abilities. Sternberg and Jarvik (1976) supported Cronholm and Ottosson's (1961) findings, and found that depressed patients manifested significant impairment in short-term memory without a concomitant impairment in long-term memory. Using the WMS, Stromgren (1977) found that depression affected performance on several of the subtests, most notably those assessing attentional or registration capacities. This investigation was subsequently replicated by Breslow, Kocsis, and Belkin (1980). In populations age sixty-five and above, Walton (1958) demonstrated that depressed elderly patients obtained significantly lower scores on the WMS compared to control subjects. Miller and Lewis (1977) demonstrated that depressed elderly subjects actually remembered the same amount of information as normal elderly control subjects, but the depressed subjects appeared to adopt a conservative response strategy.

Tools used in evaluating memory deficits are varied, and include single tests and battery approaches. Unfortunately, memory batteries, although seemingly comprehensive, typically lack theoretical sophistication in assessing memory disorders. Naïve examiners may feel that by using a prescribed battery of memory tests, a comprehensive assessment of memory functions is possible, when in fact, important aspects concerning the reason for referral were actually not assessed (i.e., attentional contributions to dysfunctional memory). For example, the original WMS is predominantly verbal, and thus an adequate assessment of memory may be compromised in those persons with speech and language deficits. The original WMS also does not assess either delayed memory or recognition memory, and its normative base is deficient. While some of these problems were rectified with the Wechsler Memory Scale-Revised (Wechsler, 1987), it too suffers from numerous confounds, including the finding that the various memory indices do not measure the factors they purport to measure, with the exception of the Attention-Concentration Index (Chelune, Bornstein, & Prifitera, 1990). Thus, at present, there is no single test of attention, learning and memory that alone will suffice. Therefore, experienced neuropsychologists use a variety of attention, learning, and memory tests with appropriate normative data in order to evaluate the diverse components and modality specific nature of these abilities, within an information processing interpretive approach.

Speech and language
Language refers to a symbol system used to exchange information (Benson, 1993). The disorder of aphasia is an acquired condition of language, and therefore refers to the loss or impairment of language use caused by brain damage (Benson, 1993). Included in this definition are the language subcomponents of reading and writing, stressing the point that aphasia is not simply a disorder of speech, but a disorder of language use. Except under unusual circumstances (e.g., aphemia, pure word deafness, and alexia without agraphia), dissociation

of writing and speech, or of reading and auditory comprehension, does not occur without a localized lesion.

Evaluation of language disorders essentially involves assessment of language expression, comprehension and repetition. Within the domain of language expression, assessment of oral speech and graphomotor output is required; within the domain of language comprehension, assessment of auditory comprehension and reading is required. Language repetition links the domains of language comprehension with language expression. Language can be further subdivided into gestural and prosodic components (disorders of which are typically associated with damage to either hemisphere), semantic and syntactic components (disorders of which are typically associated with damage to the left hemisphere), and pragmatic components (which refers to a hierarchical processing of language involving the additional cognitive processing of memory and executive functioning) (Benson, 1993).

As described above, the semantic and syntactic components of language are essentially left-hemisphere dominant activities in the majority of right handed individuals (Goodglass & Quadfasel, 1954; Roberts, 1969), as well as in most left handed individuals (Benson, 1985). It is from an understanding of how semantic and syntactic components of language are disordered in persons with left-hemisphere injury that conceptualizations of various aphasic syndromes have developed. Benson (1993) describes a superordinate method of understanding the various aphasic syndromes in terms of an anatomic division of perisylvian versus extrasylvian involvement. Essentially, the aphasic syndromes of Broca's (or expressive), Wernicke's (comprehension), conduction (repetition), and global aphasia are typically observed with injury to the perisylvian region of the left hemisphere, while the aphasic syndromes of transcortical motor, transcortical sensory, transcortical mixed and anomic are typically observed with injury to the extrasylvian region of the left hemisphere. The key to this classification system is the relative preservation of the capacity to repeat: repetition capacity is typically preserved in the extrasylvian aphasias. Finally, with the exception of the anomic variety of aphasia, and in combination with associated physical signs, general correlations between dysfunctional anatomy and aphasic syndromes are possible with good reliability.

Given this outline of language functions, it stands to reason that assessment of language for determination of specific aphasia follows an orderly process of assessing language expression (both written and oral modalities), language comprehension (both auditory and reading modalities), language repetition, and naming skills. Assessment of these functions can be cursory, such as at the bedside during brief evaluations, or comprehensive, such as in the laboratory. With varying degrees of clinical utility, a variety of testing strategies have been proposed to assess these language domains.

Visuospatial and constructional skills
Visuospatial dysfunction has been thought to result from sole injury to the right hemisphere (Kirk & Kertesz, 1993). However, the work of Kaplan (1988) and

colleagues has demonstrated that this is not always the case. By analyzing how dysfunctional test performance occurs, these researchers suggest that the strategies used by patients with injury anywhere in the brain reveal more about the behavioral expression of an injury than does focus on a global achievement score. In no other area of cognitive functioning does this qualitative approach demonstrate its strength more so than in the area of visuospatial and constructional skills.

Initially, it can be conceptualized that visuospatial skills require perception of stimuli, followed by conceptualization of structure to order the stimuli into meaningful information, and then motoric output. Breakdown of visuospatial skill can occur anywhere along this process prior to motoric output, as can breakdown of constructional skill. Analysis of dysfunction through appropriate neuropsychological testing can help discern at which point along this process the deficiency occurs.

Visual perception of stimuli requires an intact visual system from the level of the retina, through the optic tracts and lateral geniculate nucleus of the thalamus. Visual processing advances through the optic radiations, and finally terminates at the occipital cortex. Perception of stimuli also requires coordination of extraocular muscles to ensure coordination of gaze. This coordination is achieved through a coordinated system involving both involuntary feedback from the superior colliculus, medial longitudinal fasciculus, and cranial nerves which innervate the eye muscles, and voluntary feedback from the frontal eye fields. Injury at any point within these systems can result in dysfunction in visual perception. For example, injury to the retina can result in blindness in one eye; injury to the superior portion of the optic radiations on the left can result in right inferior quadrantopsia; injury to the frontal eye field can result in incoordination of the eye muscles resulting in blurred vision, as can injury to one or more of the cranial nerve nuclei innervating the various eye muscles.

Injury to the posterior portions of the right hemisphere can typically result not only in visual field defects, but also in inattention phenomena known as visual neglect. A sensory inattention phenomenon, such as visual neglect, is considered the result of deficits in awareness, not a deficit in sensation. In the majority of these cases, sensory capabilities are intact. The disorder frequently accompanies injury to posterior portions of the right hemisphere (Heilman, Watson, & Valenstein, 1993; Lezak, 1995). Manifestations of this disorder are frequently observed in deficient performance on line cancellation tests, line bisection tests, and double simultaneous stimulation. As noted above, this inattention phenomenon is considered pathognomonic of brain injury.

Once the integrity of the visual system at its basic level has been ascertained, differences in information processing can be manifested in dysfunctional performance on constructional tasks. Patients with right hemisphere injury typically make configural errors characterized by loss of gestalt and/or hemi-inattention characterized by failure to reproduce stimuli located in the left hemispace. Patients with left hemisphere injury typically make featural errors, characterized by a loss of details in their reproductions of test stimuli. According to Kaplan and colleagues (1991), these solution strategy differences help to distinguish between injuries to either the left- or right-hemisphere.

In addition, information processing differences can be observed when evaluating deficient performance along an anterior-posterior axis. Patients with frontal lobe injuries tend to make errors secondary to program organizational difficulties, while patients with parietal lobe injuries tend to make errors secondary to spatial organizational difficulties (Pillon, 1981). Improvements were observed in the Rey Complex Figure productions of patients with parietal injuries when they were provided spatial references, but not program references, while patients with frontal injuries demonstrated the reverse pattern of improvement with the opposite instructional set (Pillon, 1981). In keeping with similar clinical descriptions of frontal involvement, the degree of structure inherent in the task can also distinguish between injury to the anterior or posterior portions of the brain, a quality Lezak (1995) refers to as "concrete-mindedness".

Discerning the locus of dysfunction requires knowledge not only of the neuroanatomical bases of vision, but also knowledge of neuropsychological consequence of that injury. For example, the ability to perceive angular relationships, such as assessed with the Judgment of Line Orientation test, is considered primarily a visuoperception test, sensitive to posterior brain functioning. On the other hand, the ability to reproduce designs with blocks (e.g., Block Design test), requires not only visuoperception and visuospatial integrity, sensitive to posterior brain functioning (Chase et al., 1984), but also the capacity to motorically integrate the various stimuli into a conceptual whole, sensitive to anterior brain functioning.

Executive functioning and concept formation
Dysfunction of executive skills is frequently associated with injury to the frontal lobes, in particular the prefrontal cortex (Damasio, 1991). However, due to the complexity of interactions between brain regions, such dysfunction can also be associated with injury within other brain regions (Goldberg & Bilder, 1987). More specifically, in routine tasks, posterior brain systems can integrate and work in concert with the prefrontal cortex quite competently. However, in novel tasks, such as those tasks requiring solutions to previously unencountered problems, initial attempts at coordination between these brain regions becomes inefficient compared to routine tasks (Williamson, Scott, & Adams, 1996). For example, Stuss (1987) describes executive functioning as the ability "to extract and use information from the posterior brain systems, and to anticipate, select, plan, experiment, modify, and act on such information in novel situations" (p. 175). In other words, executive functioning is the cognitive capacity to decide, plan, implement and evaluate goal-directed behavior based on available information.

Lezak (1995) conceptualizes executive functioning in terms of volition, planning, purposive action, and effective performance. Dysfunction can occur at any point along this sequence, requiring careful evaluation at each step. Unfortunately, much of what is considered to be executive skill is difficult to measure using traditional neuropsychological assessment tools (Damasio &

Anderson, 1993; also see Chapter 7 for detailed discussion of executive functions). One reason for this state of affairs is that executive dysfunction secondary to brain injury is most pronounced only in novel settings. Therefore, the standardized nature of the assessment environment, while necessary to compare performance of one individual to a normative population, hinders assessment of executive functions because of the high level of structure imposed within the testing setting. In fact, patients with known frontal lobe disease may score entirely adequately on neuropsychological testing, yet be dysfunctional within their daily setting. Consequently, a qualitative approach to evaluating executive function is typically required, although even this approach may fail to detect impairment.

It is necessary to first ascertain a patient's capacity for volitional behavior. Dysfunction of volition typically presents as an abulic state, as altered self-awareness, or with disconnection between understanding and action. A patient may "look" clinically depressed, but not report depression. Dysfunctional self-awareness may present as inability to profit from social cues for appropriate behavior, or as anosognosia (neurological denial of illness), or anosodiaphoria (neurological unconcern of illness), among others. Disconnection between understanding and action may be seen when asking a patient to perform an activity; the patient acknowledges the request, but does not subsequently engage in the requested task.

Planning requires several subcomponents, including breaking a complex action into smaller steps, evaluating performance, and subsequently modifying performance if and when necessary. Assessment of planning ability can include tasks such as the Tower of London or Porteus Mazes. Asking significant others about how the patient functions during novel activities may also elicit evidence of substantial planning deficits. Purposive action refers to the ability to intentionally carry out a series of programmed steps. Deficits that reduce one's ability to demonstrate purposive action include: decreased speed of information processing, decreased ability to establish a mental set, cognitive inflexibility, and decreased capacity to shift mental set (i.e., perseveration). Failure to inhibit responses may be observed in injury to the frontal lobes, and in essence is a type of behavioral release phenomenon, manifested as a executive planning deficit. Such deficits may be manifested during the clinical interview or though various test procedures, including go/no-go tasks, and tasks assessing utilization behaviors (Lhermitte, 1983). Finally, effective performance refers to the capacity to self-monitor and self-correct.

Deficits of concept formation can also result from injury to various locations in the brain. Like executive functioning, concept formation deficits are more likely to manifest in terms of the quality of the response, rather than the content of the response. Verbal concept formation tests include measures of logical abstract reasoning, such as the Abstraction subtest of the Shipley Institute of Living Scale, the Similarities subtest from the Wechsler intelligence series, and responses to proverbs. Nonverbal concept formation tests include the Category Test from the Halstead-Reitan Neuropsychological Test Battery and the Wisconsin Card Sorting Test.

Interpretive considerations

Neuropsychological test interpretation: practical considerations
This section will discuss practical considerations of neuropsychological test interpretation, including appropriate times to evaluate post injury, temporal order of tests, practice effects, effects of age, education, and gender, and briefly, assessment of premorbid functioning.

When to evaluate post injury
Extensive neuropsychological assessment is not normally undertaken during the acute or post-acute stages of an illness. Disorientation and anterograde amnesia typically accompany the patient's behavioral presentation during the acute and subacute stages of traumatic brain injury. Levin, O'Donnell and Grossman (1979) recommend that formal assessment not take place until such acute effects have subsided, as measured by Galveston Orientation and Amnesia Test scores falling within the normal range. Additional reasons for not evaluating during this period include the rapid recovery often observed in such patients (thus making obtained data invalid soon after being collected), and fatigue, which further threatens the validity and reliability of the data. As a general rule, the earliest point in time for formal comprehensive neuropsychological evaluation post injury would be three to six months in severe cases, or possibly as early as one month in cases showing with more dramatic than normal recovery or with milder injury. After six months, rapid early improvement typically slows and gradually plateaus in subsequent months for both traumatic brain injured patients (Jennett, Snoek, Bond, & Brooks, 1981; Jennett & Teasdale, 1981), and cerebrovascular accidents (Kertesz, 1993; Kertesz & McCabe, 1977).

However, this is not to say that brief bedside evaluations or other less comprehensive evaluations should not be undertaken. In fact, such preliminary evaluations can be quite important in ascertaining basic level of impairment, and capacity and direction of rehabilitative efforts, cognitive competency to make decisions, and capacity for independent living prior to discharge from the hospital. Clinical approaches to evaluation (as opposed to quantitative approaches) warrant greater importance in these situations, in order to provide coverage of basic areas of cognition, even if in a rudimentary fashion.

Temporal order of tests
Previous investigations have demonstrated that the order of test presentation generally has no appreciable effect on performance (Cassel, 1962), although throughout a long day of evaluation, fatigue may play a factor in poorer performance during the latter half of an evaluation session. In one study demonstrating order effects (Neuger et al., 1981), adverse effects of fatigue were generally considered to be the cause of lowered Finger Tapping performance. Clinicians may find it prudent to document test order with individual patients, thus allowing discussion of this issue in depositions or at trial. In contrast,

although order of test presentation generally has no effect on performance other than possible effects of fatigue, temporal order of tests may make a difference when attempting to detect malingering. Specifically, testing for malingering early in the session may be more effective than testing later, due to an individual's initial lack of data with which to compare the relative difficulty of different measures (Guilmette, Whelihan, Hart, Sparadeo, & Buongiorno, 1996; also see Chapter 9 for more on malingering).

Practice effects
The issue of practice effects is of particular concern within the context of the forensic neuropsychological evaluation, as many patients routinely undergo two or more neuropsychological evaluations during the course of their litigation. The reasons for repeated testings can include the desire to measure change (or lack of change) over time, or to either confirm or refute the findings of an earlier neuropsychological evaluation. Essentially, after several testings in non-neurological controls, some degree of improvement in test scores would be expected; test-retest variability in this normal population may be the result of: low reliability of the measure, low reliability of the construct measured, changes in moderator variables (e.g., emotional state, medication), or learning from prior exposure to the task. In brain-injured persons, some degree of improvement would also be expected, although the cause for this improvement could reflect true change in the underlying neurological condition (e.g., recovery or deterioration) (Dikmen, Machanmer, Temkin, & McLean, 1990), practice effects, or all. Dikmen et al. (1990) in their study of neuropsychological outcome two years post traumatic brain injury suggest that in repeated testings, the magnitude of change as a result of practice is initially pronounced, and diminishes with additional exposure due to ceiling effects. Use of normal comparison subjects by these investigators to control for practice effects assumed that the effect of practice on performance was equal in the two groups, and statistically may have underestimated actual improvements in the traumatically brain injured group two years post due to "overestimation of the magnitude of practice effects." (Dikmen et al., 1990).

Neuropsychological tests that have been shown to demonstrate significant practice effects include timed tests, tests with easily conceptualized single solutions (e.g., Bornstein, Baker & Douglass, 1987; Matarazzo, 1990; Matarazzo & Herman, 1984), tests involving learning (Ryan, Morris, Yaffa, & Peterson, 1981), and manual dexterity tasks (McCaffrey et al., 1992). In examining data from the Wechsler Adult Intelligence Scale-Revised (WAIS-R) standardization sample, Matarazzo and colleagues (1990; Matarazzo & Herman, 1984) reported that for some individuals, scores obtained on re-evaluation will be considerably different, both higher and lower. Consequently, there are two issues to consider when considering practice effects: the stability of the construct (expressed in terms of test-retest reliability) and improvements or decrements in scores from first to second testing. Moore and colleagues (1990) discuss these issues in terms of psychometric retest reliability and clinical retest reliability, suggesting that greater temporal stability of scores is expected on measures dependent upon acquired knowledge and lesser temporal stability of scores is expected on measures dependent upon reasoning skills.

Assessment of premorbid functioning
Chapter 3 thoroughly reviews this topic, and aspects of this topic have been briefly reviewed earlier in this chapter under the heading of "Background preparation". However, a more general, brief review of this area is in order, given the importance of the topic with regard to interpretation of neuropsychological test data. It is important to note that knowledge of the patient's premorbid level of functioning is a necessity in the evaluator's ability to accurately interpret the obtained neuropsychological test results. This is a frequent point of contention within the legal arena. Therefore, it is also essential to have an appreciation of the various methods of estimating premorbid abilities and the strengths and weaknesses of these methods.

Currently, there are several methods of estimating premorbid intelligence. These include methods utilizing demographic variables (Barona, Reynolds, & Chastain, 1984), measures that are believed resistant to the effects of both brain injury and advancing age (Blair & Spreen, 1989; Johnstone & Wilhelm, 1996), best performance methods (Vanderploeg, Schinka, & Axelrod, 1996), and combined demographic and best performance methods (Krull, Scott, & Sherer, 1995; Scott, Krull, Williamson, Adams, & Iverson, 1997). Many of the premorbid estimation approaches yielded systematic under- or overestimation of IQ, have range restriction problems, and are of limited clinical utility (Eppinger, Craig, Adams, & Parsons, 1987; Scott et al., 1997; Sweet, Moberg, & Tovian, 1990). Recent research (Ritchie, Lam, & Rankin, 1996) using the combined approach (i.e., Krull et al., 1995; Scott et al., 1997) in neurological and non-neurological control patients suggests that this approach provides an adequate estimate of premorbid intelligence compared to other approaches, as it approximates the mean and standard deviation of the Wechsler Adult Intelligence Scale-Revised standardization sample, is not susceptible to the range restriction problem inherent in other approaches, and does not systematically under- or overestimate intelligence.

Neuropsychological test interpretation: statistical considerations

This section will discuss statistical considerations in neuropsychological test interpretation, specifically threats to clinical utility, consideration of the standard error of measurement, and construct validity of various cognitive domains.

Threats to clinical utility
A necessary assumption in the practice of neuropsychology is that the tests administered are sensitive in various degrees to the effects of brain injury. The outcome of this assumption is commonly referred to as deficit measurement, whereby a person's current level of functioning is compared to a normative ideal or estimated prior level of functioning. In making inferences of impairment, the

concepts of sensitivity and specificity of the tests become important for decision making. This topic has been thoroughly discussed earlier in chapter 2 of this text, but is briefly revisited here to affirm its' importance in test interpretation.

Briefly, *sensitivity* refers to the degree to which if a disease is present, its presence will manifest as an abnormal test score. *Specificity* refers to the degree to which an absence of disease is reflected in a normal test score. Within the context of neuropsychological assessment, there are situations in which tests with high sensitivity are preferred, such as screening for abnormal performance, and situations in which tests with high specificity are preferred, such as examining specific neuropsychological functions in order to delineate specific deficits (Teng, Wimer, Roberts, et al., 1989).

Building on this concept, Retzlaff and Gibertini (1994) state that knowledge of a test's sensitivity or specificity is not enough, but rather it is equally important to have knowledge of the *positive predictive power* and *negative predictive power* of the tests. In other words, given a positive (abnormal) or negative (normal) test result, what is the likelihood that a disease state is present or absent, respectively? These concepts become particularly important when neuropsychological tests results are impaired and neuroradiological and neurological studies are negative (Bauer, 1997), discussed in more detail below. More specifically, positive and negative predictive powers of tests directly relate to the commission of errors in decision-making (i.e., erroneously concluding normality or brain-injury); there are several situations in which errors in interpretation can be made. These errors can be discussed in the decision-theory language of false positive errors (i.e., concluding someone is brain injured, when they actually are not), and false negative errors (i.e., concluding someone is not brain injured, when they actually are). Of importance for neuropsychologists, Bauer (1997) further subdivides false positive errors into those errors in which abnormal tests results truly reflect neurogenic causes versus psychogenic, socioeconomic, or motivational causes, and those errors in which normal test results are interpreted as abnormal. The clinical implication of these false positive and false negative errors can be great, and can include inappropriate treatment of a condition that does not exist, or failure to treat a condition that does exist, respectively. The legal implications of these errors can include eggregious granting of monetary rewards for an alleged permanent condition that does not exist, or failure to grant monetary rewards for an actual permanent condition that does indeed exist, respectively.

Possible causes of false positive errors when inferring brain injury include failing to take into account the effects of: age, the effects of educational level, a person's social and cultural background, premorbid intelligence level, acute effects of alcohol, street drugs, and prescription drugs, emotional factors such as anxiety and depression, fatigue, poor motivation, and malingering. Possible causes of false negative errors include some of the same conditions (e.g., premorbid intelligence, and educational level), but also can include frontal lobe dysfunction that is not identified by traditional testing approaches (Damasio & Anderson, 1993; also see Chapter 7 in this text for detailed discussion), chronicity of the condition, as well as development of compensatory strategies to overcome the primary deficits.

Standard error of measurement. Standard error of measurement refers to a band of error around each obtained score, within which a person's true score is believed to exist. Matarazzo (1990) cautions that all too frequently, the standard error of measurement is not taken into account when interpreting neuropsychological test results. More specifically, Matarazzo (1990) states that intrasubtest scatter among test scores "cannot be used ipso facto to either (a) estimate (using the highest score) the examinee's supposed premorbid level of cognitive function or (b) identify areas (using the lowest scores) of current cognitive impairment." Rather, psychometric information obtained during evaluation must be contextualized within relevant background information to include prior test results, academic history, medical history, and other relevant supplementary information (Matarazzo, 1990).

Construct validity. A construct is simply a word used to represent an idea. In the case of neuropsychological assessment, words such as *intelligence, attention,* and *memory* are labels for ability constructs. The degree to which a construct has biological or neurological significance reflects the degree that the construct actually represents a neurological event. Neuropsychological tests attempt to measure the neurological event in question, and do so with greater or lesser precision. This precision is construct validity. Previous research has shown that no neuropsychological test is a pure measure of a neurological event in questions. Factor analytic studies of neuropsychological tests (Goldstein, 1984; Heilbronner, Buck, & Adams, 1989; Matarazzo, 1990) have demonstrated that a test that purports to measure the construct *memory* may not only measure memory, but may also measure associated constructs, such as language use. In other words, a so-called memory test will likely share statistical variance with other constructs, such as language use (Zihl, 1989). Construct validity must be kept in mind when interpreting neuropsychological test data, because it indicates that cognitive processes are multiply determined, and may help to explain performance variability across tests that purport to measure the same construct. For discussion of issues related to ecological validity, see chapter 9.

Release of neuropsychological test materials

When performing neuropsychological evaluations in the context of litigation, neuropsychologists are often confronted with the request to release under subpoena or court order not only the evaluation report, but raw neuropsychological test data and test forms (i.e., test protocols). At this point, this creates a major ethical dilemma for the neuropsychologist regarding releasing such information to non-psychologists. The dilemma rests upon two major principles: (a) the test data may be misused by those unqualified to use such information, and

(b) the integrity and security of the test and its questions and answers may be compromised (American Psychological Association, 1992). For our purposes regarding adversarial cases, the issue is not that other clinical psychologists or neuropsychologists are the only persons qualified to form an opinion about the work in question, as discussed by Lees-Haley (1995); the issue is one of test security. Note that there is a difference between test scores (which are considered releasable, attested to by the fact that such scores are frequently referenced in, or appended to, reports) and test protocols (which are copyrighted, and considered not releasable). Nevertheless, the neuropsychologist is confronted with the requirements of the legal system when served with either a subpoena or court order requiring such release.

In an effort to mitigate this dilemma of releasing raw neuropsychological test data, it has been suggested (Ackerman & Kane, 1993; Frumkin, 1995; Shapiro, 1991), and we agree, that if the neuropsychologist is in receipt of a subpoena, it is prudent to first contact the requesting attorney to explain the ethical conflict pertaining to "test security" issues. The neuropsychologist can recommend either (1) sending copies of the test protocols directly from one psychologist's office to another psychologist's office, thereby guaranteeing test security, or alternatively, (2) entering a special order of protection into the legal case at hand, such that the test materials will be guaranteed to be secure by the law firms during the course of the legal proceedings, *and* will either be destroyed or returned to the psychologist from whom they were obtained at the end of the court proceeding. In situations in which one legal party does not wish to disclose a consulting expert to another, a third psychologist can transfer the test protocols between the experts, thereby maintaining the right of having an undisclosed consultant.

Summary

The focus of this chapter has been on the clinical and scientific foundations of the neuropsychological evaluation. The goal was to convey that the practice of neuropsychology is much more complicated than merely administering a number of tests, and determining whether a particular score falls above or below a predetermined cut-off score. Rather, the neuropsychological evaluation, which is central to the practice of neuropsychology, requires an understanding of a variety of conditions that may affect test results other than brain injury, in order to avoid false positive and false negative errors, as well as an understanding of neuroanatomical and neuropathological conditions in order to place in context the obtained test results. Within the context of an adversarial proceeding, the neuropsychological evaluation requires: judicious selection of well normed and validated tests, a level of comprehensiveness equal to that of scientific investigations, an appreciation of biological and behavioral variables affecting test performance, and an unbiased, dispassionate interpretation of the data.

References

Ackerman, M.J., & Kane, A. (1993). *Psychological experts in divorce, personal injury, and other civil actions* (Vol. 1, 2nd ed.). New York: Wiley Law Publications.

Adams, K.M., & Rourke, B.P. (Eds.). (1992). *The TCN guide to professional practice in clinical neuropsychology.* Lisse, The Netherlands: Swets & Zeitlinger.

Adams, R.L., Parsons, O.A., Culbertson, J.L., & Nixon, S.J. (Eds.). (1996). *Neuropsychology for clinical practice: Etiology, assessment, and treatment of common neurological disorders.* Washington, DC: American Psychological Association.

Adams, R.L., & Rankin, E.J. (1996). A practical guide to forensic neuropsychological evaluations and testimony. In R.L. Adams, O.A. Parsons, J.L. Culbertson, & S.J. Nixon (Eds.), *Neuropsychology for clinical practice: Etiology, assessment, and treatment of common neurological disorders.* (pp. 455–488). Washington, DC: American Psychological Association.

American Academy of Neurology, Therapeutics and Technology Assessment Subcommittee (1989). Assessment: EEG brain mapping. Neurology, 39, 1100–1101.

American Academy of Neurology (Therapeutics and Technology Assessment Subcommittee (1996). Assessment of brain SPECT: Report of the Therapeutics and Technology Assessment Subcommittee of the American Academy of Neurology. *Neurology, 46,* 278–285.

American Academy of Neurology (Therapeutics and Technology Assessment Subcommittee) (1996). Assessment: Neuropsychological testing of adults: Consideration for neurologists. *Neurology, 47,* 592–599.

American Psychological Association (1992). Ethical principles of psychologists and code of conduct. *American Psychologist, 47,* 1597–1611.

Atkinson, R.C., & Shiffrin, R.M. (1968). Human memory: A proposed system and its control processes. In K.W. Spence & J.T. Spence (Eds.), *The psychology of learning and motivation,* II (pp. 89–195). New York: Academic Press.

Bastian, H.C. (1869). On the various forms of loss of speech in cerebral disease. *British Foreign Medico-Surgical Review, 43,* 470–492.

Barona, A., Reynolds, C.R., & Chastain, R. (1984). A demographically based index of premorbid intelligence for the WAIS-R. *Journal of Consulting and Clinical Psychology, 52,* 885–887.

Bauer, R.M. (1997, February). *Brain damage caused by collision with forensic neuropsychologist.* Paper presented at the meeting of the International Neuropsychological Society, Orlando, FL.

Bauer, R.M., Tobias, B., & Valenstein, E. (1993). Amnesic disorders. In K.M. Heilman & E. Valenstein (Eds.), *Clinical neuropsychology* (pp. 523–602). New York: Oxford University Press.

Benson, D.F. (1993). Aphasia. In K.M. Heilman & E. Valenstein (Eds.), *Clinical neuropsychology* (pp. 17–36). New York: Oxford University Press.

Benson, D.F. (1985). Language in the left hemisphere. In D.F. Benson & E. Zaidel (Eds.), *The dual brain* (pp. 193–203). New York: Guilford Press.

Bigler, E.D., Rosa, L., Schultz, F., et al. (1989). Rey-Auditory Verbal Learning and Rey-Osterrieth Complex Figure Design performance in Alzheimer's disease and closed head injury. *Journal of Clinical Psychology, 45,* 277–280.

Blair, J.R., & Spreen, O. (1989). Predicting premorbid IQ: A revision of the National Adult Reading Test. *The Clinical Neuropsychologist, 3,* 129–136.

Bornstein, R.A., Baker, G.B., & Douglass, A.B. (1987). Short-term retest reliability of the Halstead-Reitan battery in a normal sample. *Journal of Nervous and Mental Disease, 175,* 229–232.

Bornstein, R.A., & Kozora, E. (1990). Content bias of the MMPI Sc scale in neurological patients. *Journal of Neuropsychiatry, Neuropsychology, and Behavioral Neurology, 3,* 200–205.

Breslow, R., Kocsis, J., & Belkin, B. (1980). Memory deficits in depression: Evidence utilizing the Wechsler Memory Scale. *Perceptual and Motor Skills, 51*, 541–542.

Broca, P. (1865). Sur la faculté du langage articulé'. *Bulletin of Soc. Anthropol. Paris, 6*, 337–393.

Bruckner, F., & Randle, A. (1972). Return to work after severe head injuries. *Rheumatology and Physical Medicine, 11*, 344–348.

Caine, E.D. (1986). The neuropsychology of depression: the pseudodementia syndrome. In I. Grant & K. Adams (Eds.), *Neuropsychological assessment of neuropsychiatric disorders*. (pp. 221–243). New York: Oxford University Press.

Cassel, R.H. (1962). The order of the tests in the battery. *Journal of Clinical Psychology, 18*, 464–465.

Cermak, L.S. (1976). The encoding capacity of patients with amnesia due to encephalitis. *Neuropsychologia, 14*, 311–326.

Chase, T.N., Fedio, P., Foster, N.L., et al. (1984). Wechsler Adult Intelligence Scale performance: Cortical localization by flurodeoxyglucose F18-positron emission tomography. *Archives of Neurology, 41*, 1244–1247.

Chelune, G.J., & Bornstein, R.A. (1988). Wechsler Memory Scale-Revised patterns among patients with unilateral brain lesions. *The Clinical Neuropsychologist, 2*, 121–132.

Chelune, G.J., Bornstein, R.A., & Prifitera, A. (1989). The Wechsler Memory Scale-Revised: Current status and applications. In J. Rosen, P. McReynolds, & G.J. Chelune (Eds.), *Advances in psychological assessment*. New York: Plenum Press.

Cohen, N.J., & Squire, L.R. (1980). Preserved learning and retention of pattern analyzing skill in amnesia: Dissociation of knowing how and knowing that. *Science, 210*, 207–209.

Craik, F.I.M. (1977). Age differences in human memory. In J.E. Birren & K.W. Schaie (Eds.), *Handbook of the psychology of aging* (pp. 384–420). New York: Van Nostrand Reinhold.

Cronholm, B., & Ottosson, J. (1961). Memory functions in endogenous depression. *Archives of General Psychiatry, 5*, 193–197.

Crosson, B., Novack, T.A., Trenerry, M.R., & Craig, P.L. (1989). Differentiation of verbal memory deficits in blunt head injury using the recognition trial of the California Verbal Learning Tests: An exploratory study. *The Clinical Neuropsychologist, 3*, 29–44.

Damasio, A.R. (1991). Concluding comments. In H.S. Levin, H.M. Eisenberg & A.L. Benton (Eds.), *Frontal lobe function and dysfunction* (pp. 401–407). New York: Oxford University Press.

Damasio, A.R., & Anderson, S.W. (1993). The frontal lobes. In K.M. Heilman & E. Valenstein (Eds.), *Clinical neuropsychology* (3rd ed.; pp. 409–460). New York: Oxford University Press.

Damasio, A.R., Eslinger, P.J., Damasio, H., Van Hoesen, G.W., & Cornell, S. (1985). Multimodal amnesic syndrome following bilateral temporal and basal forebrain damage. *Archives of Neurology, 42*, 252–259.

DeMyer, W.E. (1994). *Technique of the neurologic examination: A programmed text.* New York: McGraw-Hill.

Dikmen, S., Machamer, J., Temkin, N., & McLean, A. (1990). Neuropsychological recovery in patients with moderate to severe head injury: Two year follow-up. *Journal of Clinical and Experimental Neuropsychology, 12*, 507–519.

Eppinger, M., Craig, P., Adams, R., & Parsons, O. (1987). The WAIS-R index for estimating premorbid intelligence: Cross-validation and clinical utility. *Journal of Consulting and Clinical Psychology, 55*, 86–90.

Erickson, R.C. (1978). Problems in clinical assessment of memory. *Experimental Aging Research, 4*, 255–272.

Fenton, G.W. (1982). Hysterical alterations of consciousness. In A. Roy (Ed.), *Hysteria* (pp. 229–246). New York: Wiley.

Finlayson, M.A., Johnson, K.A., & Reitan, R.M. (1977). Relationship of level of education to neuropsychological measures in brain-damaged and non-brain-damaged adults. *Journal of Consulting and Clinical Psychology, 45,* 536–542.

Frumkin, L.B. (1995). How to handle attorney requests for psychological test data. In L. VandeCreek, S. Knapp, & T.L. Jackson (Eds.), *Innovations in clinical practice: A source book* (Vol. 14; pp. 275–291). Sarasota, FL: Professional Resource Press.

Goldberg, E., & Bilder, R.M., Jr. (1987). The frontal lobes and hierachical organization of cognitive control. In E. Perecman (Ed.), *The frontal lobes revisited*. New York: The IRBN Press.

Goldstein, G. (1984). Comprehensive neuropsychological assessment batteries. In G. Goldstein & M. Hersen (Eds.), *Handbook of psychological assessment* (pp. 181–210). New York: Pergamon Press.

Goodglass, H., & Quadfasel, F. (1954). Language laterality in left-handed aphasics. *Brain, 77,* 521–548.

Grafman, J., Jonas, B.S., Martin, A., et al. (1988). Intellectual function following penetrating head injury in Vietnam veterans. *Brain, 111,* 169–184.

Guilmette, T.J., Whelihan, W.M., Hart, K.J., Sparadeo, F.R., & Buongiorno, G. (1996). Order effects in the administration of a forced-choice procedure for detection of malingering in disability claimants' evaluation. *Perceptual and Motor Skills, 83,* 1007–1016.

Harper, R.B., Chacko, R.C., Kotic-Harper, D., et al. (1992). Comparison of two cognitive screening measures for efficacy in differentiating dementia from depression in a geriatric inpatient population. *Journal of Neuropsychiatry and Clinical Neurosciences, 4,* 178–184.

Hartley, J.T., Harker, J.O., & Walsh, D.A. (1980). Contemporary issues and new directions in adult development of learning and memory. In L.W. Poon (Ed.), *Aging in the 1980s: Psychological issues* (pp. 239–252). Washington, DC: American Psychological Association.

Heaton, R.K., Grant, I., & Matthews, C.G. (1991). *Comprehensive norms for an expanded Halstead-Reitan battery: Demographic corrections, research findings, and clinical applications*. Odessa, FL: Psychological Assessment Resources.

Heilbronner, R.L., Buck, P., & Adams, R.L. (1989). Factor analysis of verbal and non-verbal clinical memory tests. *Journal of Clinical Neuropsychology, 4,* 299–309.

Heilman, K.M., & Valenstein, E. (1993). Introduction. In K.M. Heilman & E. Valenstein (Eds.), *Clinical neuropsychology* (3rd ed.; pp. 3–16). New York: Oxford University Press.

Heilman, K.M., Watson, R.T., & Valenstein, E. (1993). Neglect and related disorders. In K.M. Heilman & E. Valenstein (Eds.), *Clinical neuropsychology* (pp. 279–336). New York: Oxford University Press.

Holman, B.L., & Devous, M.D., Sr. (1992). Functional brain SPECT: The emergence of a powerful clinical method. *Journal of Nuclear Medicine, 33,* 1888–1904.

International Neuropsychological Society, Division 40 Task Force on Education, Accreditation and Credentialing (1987). Guidelines for doctoral training programs, neuropsychology internships, and post-doctoral training in clinical neuropsychology. *The Clinical Neuropsychologist, 1,* 29–34.

Jennett, B., Snoek, J., Bond, M.R., & Brooks, N. (1981). Disability after severe head injury: Observations on the use of the Glasgow Outcome Scale. *Journal of Neurology, Neurosurgery, and Psychiatry, 44,* 285–293.

Jennett, B., & Teasdale, G. (1981). *Management of head injuries*. Philadelphia: F.A. Davis Co.

Johnstone, B., & Wilhelm, K.L. (1996). The longitudinal stability of the WRAT-R reading subtest: Is it an appropriate estimate of premorbid intelligence? *Journal of the International Neuropsychological Society, 2,* 282–285.

Kaplan, E. (1988). A process approach to neuropsychological assessment. In T. Boll & B.K. Bryant (Eds.), *Clinical neuropsychology and brain function: research, measurement and practice* (pp. 129–167). Washington, DC: American Psychological Association.

Kaplan, E., Fein, D., Morris, R., & Delis, D. (1991). *WAIS-R as a neuropsychological instrument*. San Antonio, TX: The Psychological Corporation.

Kaufman, D.M. (1990). *Clinical neurology for psychiatrists*. Philadelphia: Harcourt Brace Jovanovich.

Kertesz, A. (1993). Recovery and treatment. In K.M. Heilman & E. Valenstein (Eds.), *Clinical neuropsychology* (3rd ed.; pp. 647–674). New York: Oxford University Press.

Kertesz, A., & McCabe, P. (1977). Recovery patterns and prognosis in aphasia. *Brain, 100*, 1–18.

Kinsbourne, M., & Wood, F. (1975). Short-term memory processes and the amnestic syndrome. In D. Deutsch & J.A. Deutsch (Eds.), *Short-term memory* (pp. 258–291). New York: Academic Press.

Kirk, A., & Kertesz, A. (1993). Subcortical contributions to drawing. *Brain and Cognition, 21*, 57–70.

Krull, K.R., Scott, J.G., & Sherer, M. (1995). Estimation of premorbid intelligence from combined performance and demographic variables. *The Clinical Neuropsychologist, 9*, 83–88.

Lamberty, G.J., & Bieliauskus, L.A. (1993). Distinguishing between depression and dementia in the elderly: A review of neuropsychological findings. *Archives of Clinical Neuropsychology, 8*, 149–170.

Lashley, K.S. (1929). *Brain mechanisms and intelligence: A quantitative study of injuries to the brain*. Chicago: University of Chicago Press.

Lashley, K.S. (1938). Factors limiting recovery after central nervous lesions. *Journal of Nervous and Mental Diseases, 888*, 733–755.

Leckliter, I.N., & Matarazzo, J.D. (1989). The influence of age, education, IQ, gender, and alcohol abuse on Halstead-Reitan Neuropsychological Test Battery performance. *Journal of Clinical Psychology, 45*, 484–512.

Lees-Haley, P.R. (1995). Neurobehavioral assessment in toxic injury evaluations. *Toxicology Letters, 82/83*, 197–202.

Lees-Haley, P.R., & Dunn, J.T. (1994). The ability of naïve subjects to report symptoms of mild brain injury, post-traumatic stress disorder, major depression, and generalized anxiety disorder. *Journal of Clinical Psychology, 50*, 252–256.

Levin, H.S., O'Donnell, V.M., & Grossman, R.G. (1979). The Galveston Orientation and Amnesia Test: A practical scale to assess cognition after head injury. *Journal of Nervous and Mental Disease, 167*, 675–684.

Lezak, M.D. (1995). *Neuropsychological assessment* (3nd ed.). New York: Oxford University Press.

Lhermitte, F. (1983). 'Utilization behaviour' and its relation to lesions of the frontal lobes. *Brain, 106*, 237–255.

Luria, A.R. (1966). *Higher cortical functions in man*. New York: Basic Books.

Luria, A.R. (1973). *The working brain: An introduction to neuropsychology* (trans. B. Haigh). New York: Basic Books.

Matarazzo, J.D. (1990). Psychological assessment versus psychological testing: Validation from Binet to the school, clinic, and courtroom. *American Psychologist, 45*, 999–1017.

Matarazzo, J.D., & Herman, D.O. (1984). Base rate data for the WAIS-R: Test-retest stability and VIQ-PIQ differences. *Journal of Clinical Neuropsychology, 6*, 351–366.

Mayberg, H.S. (1996). Medical-legal inferences from functional neuroimaging evidence. *Seminars in Clinical Neuropsychiatry, 1*, 195–201.

McCaffrey, R.J., Ortega, A., Orsillo, S.M., et al. (1992). Practice effects in repeated neuropsychological assessments. *The Clinical Neuropsychologist, 6*, 32–42.

Miller, E., & Lewis, P. (1977). Recognition memory in elderly patients with depression and dementia: A signal detection analysis. *Journal of Abnormal Psychology, 86,* 84–86.

Milner, B. (1966). Amnesia following operation on the temporal lobes. In C.W.M. Whitty & O.L. Zangwill (Eds.), *Amnesia* (pp. 109–133). London: Butterworths.

Milner, B. (1968). Visual recognition and recall after right temporal lobe excision in man. *Neuropsychologia, 6,* 191–209.

Milner, B. (1970). Memory and the medial temporal regions of the brain. In K.H. Pribram & D.E. Broadbent (Eds.), *Biology of memory* (pp. 29–50). New York: Academic Press.

Moore, A.D., Stambrook, M., Hawryluk, G.A., Peters, L.C., Gill, D.D., & Hymans, M.M. (1990). Test-retest stability of the Wechsler Adult Intelligence Scale-Revised in the assessment of head-injured patients. *Psychological Assessment: A Journal of Consulting and Clinical Psychology, 2,* 98–100.

Moss, M.B., & Albert, M.S. (1988). Alzheimer's disease and other dementing disorders. In M.S. Albert & M.B. Moss (Eds.), *Geriatric neuropsychology* (pp. 145–178). New York: Guilford Press.

Nelson, D.V., Harper, R.G., Kotik-Harper, D., et al. (1993). Brief neuropsychologic differentiation of demented versus depressed elderly patients. *General Hospital Psychiatry, 15,* 409–416.

Neuger, G.J., O'Leary, D.S., Fishburne, F. et al. (1981). Order effects on the Halstead-Reitan Neuropsychological Test Battery and allied procedures. *Journal of Consulting and Clinical Psychology, 49,* 722–730.

Olton, D.S., Becker, J.T., & Handelmann, G.E. (1979). Hippocampus, space and memory. *Behavioral Brain Science, 2,* 313–365.

Parsons, O.A. (1987). Neuropsychological consequences of alcohol abuse: Many questions—some answers. In O.A. Parsons, N. Butters, & P.E. Nathan (Eds.), *Neuropsychology of alcoholism: Implications for diagnosis and treatment* (pp. 153–175). New York: The Guilford Press.

Pillon, B. (1981). Troubles visuo-constructifs et methodes de compensation: Resultats de 85 patients atteints de lesions cerebrales. *Neuropsychologia, 19,* 375–383.

Rasmussen, T., & Milner, B. (1977). The role of early left-brain injury in determining the lateralization of cerebral speech functions. *Annals of New York Academy of Sciences, 299,* 355–369.

Reitan, R.M. (1967). Psychological assessment of deficits associated with brain lesions in subjects with normal and subnormal intelligence. In J.L. Khanna (Ed.), *Brain damage and mental retardation: A psychological evaluation* (pp. 44–93). Springfield, IL: Charles C. Thomas.

Reitan, R.M., & Davison, L.A. (1974). *Clinical neuropsychology: Current status and applications.* New York: Winston/Wiley.

Reitan, R.M., & Wolfson, D. (1993). *The Halstead-Reitan neuropsychological test battery: Theory and clinical interpretation.* Tucson, AZ: Neuropsychology Press.

Reitan, R.M., & Wolfson, D. (1996). Theoretical, methodological, and validation bases of the Halstead-Reitan Neuropsychological Test Battery. In I. Grant & K.M. Adams (Eds.), *Neuropsychological assessment of neuropsychiatric disorders* (pp. 3–42). New York: Oxford University Press.

Report of the Task Force on Education, Accreditation, and Credentialing (1981). *INS Bulletin,* 5–10.

Retzlaff, P.D., & Gibertini, M. (1994). Neuropsychometric issues and problems. In R. Vanderploeg (Ed.), *Clinician's guide to neuropsychological assessment* (pp. 185–209). Hillsdale, NJ: Lawrence Erlbaum.

Rimel, R., Giordani, B., Barth, J., Boll, T., & Jane, J. (1981). Disability caused by minor head injury. *Neurosurgery, 9,* 221–228.

Ritchie, A.J., Lam, M., & Rankin, E.J. (1996, November). *Estimating premorbid intelligence: Comparison of the Barona Formula, North American Adult Reading Test*

(NAART), Wide Range Achievement Test-3 Reading subtest (WRAT-3), and Oklahoma Premorbid Intelligence Estimate (OPIE). Paper presented at the Annual Meeting of the National Academy of Neuropsychology, New Orleans, LA.

Ritchie, A.J., Terryberry-Spohr, L., Lam, M., et al. (1997, February). Concordance of neuropsychological and psychiatric diagnosis with SPECT in a geropsychiatric population. Paper presented at the Annual Meeting of the International Neuropsychological Society, Orlando, FL.

Roberts, L. (1969). Aphasia, apraxia and agnosia in abnormal states of cerebral dominance. In P.J. Vinken & G.W. Bruyn (Eds.), Handbook of clinical neurology, Vol. 4 (pp. 312–326). Amsterdam: North Holland.

Rozensky, R.H., Sweet, J.J., & Tovian, S.M. (1997). Psychological assessment in medical settings. New York: Plenum Press.

Russell, E.W. (1975). A multiple scoring method for the assessment of complex memory functions. Journal of Consulting and Clinical Psychology, 43, 800–809.

Ryan, J.J., Morris, J., Yaffa, S., & Peterson, L. (1981). Test-retest reliability of the Wechsler Memory Scale, For I. Journal of Clinical Psychology, 3, 847–848.

Sarter, M., Berntson, G.G., & Cacioppo, J.T. (1996). Brain imaging and cognitive neuroscience: Toward strong inference in attributing function to structure. American Psychologist, 51, 13–21.

Satz, P. (1988). Neuropsychological testimony: Some emerging concerns. The Clinical Neuropsychologist, 2, 89–100.

Scott, J.G., Krull, K.R., Williamson, D.J.G., Adams, R.L., & Iverson, G.L. (1997). Oklahoma Premorbid Intelligence Estimation (OPIE): Utilization in clinical samples. The Clinical Neuropsychologist, 11, 146–154.

Shapiro, D. (1991). Forensic psychological assessment: An integrative approach. Boston, MA: Allyn and Bacon.

Spreen, O., & Strauss, E. (1991). A compendium of neuropsychological tests. New York: Oxford University Press.

Sternberg, D.E., & Jarvik, M.E. (1976). Memory functions in depression: Improvement with antidepressant medication. Archives of General Psychiatry, 33, 219–224.

Strickland, T.L., & Stein, R. (1995). Cocaine-induced cerebrovascular impairment: challenges to neuropsychological assessment. Neuropsychology Review, 5, 69–79.

Stromgren, L.S. (1977). The influence of depression on memory. Acta Psychiatrica Scandinavica, 56, 109–128.

Strub, R.L., & Black, F.W. (1985). Mental status examination in neurology (2nd ed.). Philadelphia: F.A. Davis.

Stuss, D.T. (1987). Contributions of frontal lobe injury to cognitive impairment after closed head injury: Methods of assessment and recent findings. In H.S. Levin, J. Grafman, & H.M. Eisenberg (Eds.), Neurobehavioral recovery from head injury (pp. 166–177). New York: Oxford University Press.

Sweet, J.J., Moberg, P.J., & Tovian, S.M. (1990). Evaluation of Wechsler Adult Intelligence Scale-Revised premorbid IQ formulas in clinical populations. Psychological Assessment: A Journal of Consulting and Clinical Psychology, 2, 41–44.

Sweet, J.J., Newman, P., & Bell, B. (1992). Significance of depression in clinical neuropsychological assessment. Clinical Psychology Review, 12, 21–45.

Teng, E.L., Wimer, C., Roberts, E., et al. (1989). Alzheimer's dementia: Performance on parallel forms of the Dementia Assessment Battery. Journal of Clinical and Experimental Neuropsychology, 11, 899–912.

Thompson, R.F. (1988). Brain substrates of learning and memory. In T. Boll & B.K. Bryant (Eds.), Clinical neuropsychology and brain function: research, measurement and practice (pp. 57–84). Washington, DC: American Psychological Association.

Trenerry, M.R. (1996). Neuropsychologic assessment in surgical treatment of epilepsy. *Mayo Clinic Proceedings, 71*, 1196–1200.

Tulving, E. (1972). Episodic and semantic memory. In E. Tulving and W. Donaldson (Eds.), *Organization of memory* (pp. 381–403). New York: Academic Press.

Tulving, E. (1983). *Elements of episodic memory.* New York: Oxford University Press.

Vakil, E., & Blachstein, H. (1993). Rey Auditory Verbal Learning Test: Structure analysis. *Journal of Clinical Psychology, 49*, 883–890.

Vanderploeg, R., Schinka, J., & Axelrod, B. (1996). Estimation of WAIS-R premorbid intelligence: Current ability and demographic data used in a best-performance fashion. *Psychological Assessment, 8*, 404–411.

Vega, A., Jr., & Parsons, O.A. (1967). Cross-validation of the Halstead-Reitan tests for brain damage. *Journal of Consulting Psychology, 31*, 619–625.

Walton, D. (1958). The diagnostic and predictive accuracy of the Wechsler Memory Scale in psychiatric patients over sixty-five. *Journal of Mental Science, 104*, 1111–1116.

Wechsler, D.A. (1945). A standardized memory scale for clinical use. *Journal of Psychology, 19*, 87–95.

Wechsler, D. (1987). *Wechsler Memory Scale-Revised Manual.* San Antonio, TX: The Psychological Corporation.

Weiner, H.L., & Levitt, L.P. (1983). *Neurology for the house officer* (4th ed.). Baltimore: Williams & Wilkins.

Wernicke, C. (1874). *Des Aphasische Symptomenkomplex.* Breslau: Cohn and Weigart.

Wickelgen, W.A. (1979). Chunking and consolidation: A theoretical synthesis of semantic networks, configuring in conditioning, S-R v. cognitive learning, normal forgetting, the amnestic syndrome and the hippocampal arousal system. *Psychological Review, 86*, 44–60.

Williamson, D.J.G., Scott, J.G., & Adams, R.L. (1996). Traumatic brain injury. In R.L. Adams, O.A. Parsons, J.L. Culbertson, & S.J. Nixon (Eds.), *Neuropsychology for clinical practice: Etiology, assessment and treatment of common neurological disorders* (pp. 9–64). Washington, DC: American Psychological Association.

Zihl, J. (1989). Cerebral disturbances of elementary visual functions. In J.W. Brown (Ed.), *Neuropsychology of visual perception.* New York: IRBN Press.

Chapter 5

THE MEASUREMENT OF PERSONALITY AND EMOTIONAL FUNCTIONING

Linas A. Bieliauskas
University of Michigan, Ann Arbor, Michigan

Introduction

Defining what psychologists mean when they speak of "personality and emotional functioning" is not a simple task. Hall and Lindzey (1970) basically assert that "no substantive definition of personality can be applied with any generality" and that how one defines personality will depend on the viewpoint of "the theory of personality employed by the observer" (p. 9). Emotional functioning is not necessarily easier to define, the term being generally used to describe "feeling" and "mood" (Hinsie & Campbell, 1970, p. 261). For the purpose of this chapter, we shall consider personality as an organizing construct, a way of describing consistency of behavior for a given individual (borrowing from Allport, 1961) in such a way that an individual's organization of behavior is unique from that of other persons (borrowing from Guilford, 1959). This position is similar to Prigatano's (1987, p. 217) conceptualization: "Personality is defined as patterns of emotional and motivational responses that develop over the lifetime of the organism, are highly sensitive to biological and environmental contingencies . . . and are resistant to change but are, nevertheless, modifiable. . ." The tests used to describe personality will be considered a means of measuring this consistent and unique behavioral organization in a given individual. Emotions, for the purposes of this chapter, will be considered descriptors of the expression of various feelings and moods and the tests used to describe them, measures of the degree of that expression.

In the context of forensic evaluations, a description of personality, or the organization of individual behavioral consistency, is important for two primary determinations. These determinations are represented in the following questions:

(a) *Are manifest behaviors related to an individual's personality?* In other words, is the behavior which one is observing, be it violence, passivity, forgetfulness, obtuseness, etc., consistent with what would be expected given this individual's historical pattern of behavioral organization? Do the behaviors represent a deviation from what would be expected from the individual when faced with a particular set of circumstances?

(b) *Is the individual's personality the same as before?* Is the individual's pattern of behavioral organization consistent with what it has always been, or has it changed due to some circumstance or injury?

Emotional functioning, in the forensic setting, generally refers to the degree to which feelings and moods are characteristic (i.e., distinctive and long standing) of an individual vs. whether the degree and nature of their expression has changed due to incident or illness. When emotions are of sufficient degree, their expression can be maladaptive and lead to harm to the individual or others. In this situation, emotional functioning may be addressed as a measure of the degree to which expression of feeling or mood has clouded judgment and interfered with reason, or even interfered with the individual's perception of responsibility for their own behavior.

Scientific underpinnings of measurement and clinical judgment

In the forensic setting, the testimony of experts has long been used to discern whether given observations are true, likely to be true, unusual, causally related to another observation, untrue, or exist with some other likelihood. Traditionally, courts have admitted testimony into evidence if ". . . the principle underlying scientific evidence . . . (is) sufficiently established to have gained general acceptance in the particular field in which it belongs" (Ayala & Black, 1993, pp. 231–232). These rules were generally related to the case *Frye v. United States*, decided 75 years ago, and generally "qualified experts . . . to testify about 'scientific, technical or other specialized knowledge' if their testimony will assist [in understanding] the evidence or to determine a fact or issue" (Ayala & Black, 1993, p. 232). As one might surmise, there remains a fair degree of latitude under such rules to argue whether or not certain principles or facts have *gained general acceptance* in a particular field.

More recently, the case of *Daubert v. Merrell Dow Pharmaceuticals, Inc* has clarified a scientific basis as necessary for the admission of testimony into evidence. Referred to simply as *Daubert*, this case and its appeals established the precedent that scientific knowledge implies a grounding in the methods and procedures of science, that testimony should be based on the scientific method, that knowledge is more than subjective belief, and that theories or techniques have been tested and subject to peer review (Rotgers & Barrett, 1996).

These evolving standards regarding admission of testimony directly affect psychological testimony on the issues of measurement of personality and emotional functioning. They indicate that psychologists offering expert testimony must be "qualified" and offer testimony that is scientifically based. In addition to the discussions already cited, Goodman-Delahunty and Foote (1995) offer a detailed examination of how the Daubert ruling may impact testimony regarding psychological injury, life stressors, stress disorders, emotional sequelae of life stressors, and opinion regarding causation of injuries.

For the purposes of this chapter, we will look at measures of personality and emotional functioning in light of the Daubert ruling and how these measures can be employed in the forensic setting in cases involving neuropsychological evidence. We will consider the qualifications of the individual clinical neuropsychologist, the identification of long-standing personality traits, and the use of some appropriate measures of personality and emotional functioning. Finally, brief reviews will be offered of circumscribed clinical syndromes that often present in the forensic setting.

Who should render clinical judgment?

In any state or province, psychologists who hold a valid license are free to practice their profession independently. In the forensic setting, the license for independent practice is also the primary credential to recognize independent competence to render psychological judgment. Psychologists are licensed, however, as generic practitioners in almost all jurisdictions and there is little regulation beyond a licensed individual's adherence to ethical principles to prevent practice outside of one's realm of competency.

The profession of psychology, however, has identified specialties, including Clinical Psychology, Counseling Psychology, and Clinical Neuropsychology. The best recognition of competence within any specialty is, as in medicine, legitimate *board certification*, meaning that an individual has undergone a thorough review of credentials and rigorous examination by one's peers and been found competent to practice. Though there are multiple groups claiming to offer such external verification in any profession, there is generally a recognized umbrella organization that oversees recognized and valid specialty board certification. In medicine, this umbrella organization is The American Board of Medical Specialties (ABMS, 1996) and is comprised of the relevant medical specialty boards with which most of us are familiar (American Board of Psychiatry and

Neurology, American Board of Family Practice, American Board of Radiology, etc.). Similarly, in professional psychology, the umbrella organization is The American Board of Professional Psychology (ABPP, 1997) which is comprised of the relevant psychological specialty boards such as the American Board of Clinical Psychology, the American Board of Counseling Psychology, and the American Board of Clinical Neuropsychology, among others. In psychology, such board certification indicates an individual has undergone peer examination and can be expected to have been tested on the foundations of their knowledge and are thus "qualified" in the Daubert sense. Of course, logically, the presence or absence of a specific professional credential or accomplishment neither guarantees nor precludes a valid opinion in any specific forensic case. Beyond such general qualifications, specific relevant experience in the differential diagnosis, or treatment, if relevant, of the psychiatric and neurological conditions germane to the case at hand will be expected of a credible expert witness.

While the specialty of Clinical Neuropsychology has been fortunate enough to have specialty board certification through ABPP since 1981, not all individuals who practice the specialty are so qualified, again, a situation not unlike that in medicine. The Division of Clinical Neuropsychology of the American Psychological Association (Division 40) has issued a standing "Definition of a clinical neuropsychologist" (Division 40, 1989) in an attempt to offer guidance for identification of specialty education and training for those who do not yet hold the board credential. Nevertheless, while this definition has served the profession well during the past ten years, the field has continued to progress and in September of 1997, a conference was held in Houston, Texas, in which an integrated model of education and training in Clinical Neuropsychology was developed. Reference to the Houston Conference document (Hannay et al., 1998) will provide the most contemporary review of *current* training expectations for specialty practice in clinical neuropsychology.

Determination of long-standing personality traits

It almost seems too obvious to mention, but as its violation is so commonly encountered, we shall again blow the horn of an old rule in psychology: *The best predictor of future behavior is past behavior*. It is amazing how often previous testing and history, relevant to the behaviors being measured at hand, are ignored in the forensic setting. One of the clearest examples that comes to mind is a report we reviewed, attesting that a 45 year old male, referred for testing "to delineate possible effects of closed head trauma to the front of the head, . . . revealed drastic and unexpected decrements in overall cognitive abilities given his education and employment history. While premorbid cognitive functioning is estimated to be in the high average to the superior range . . . (he) is currently unable to function at such levels." The report goes on to suggest frontal lobe dysfunction that will permanently prevent the patient from again functioning as a mechanical engineer. The patient's history was rendered as unremarkable, consistent with a college education, and appropriate employment. On our examination, with thorough questioning, it was revealed that the patient had a long-standing history of treatment for schizophrenia, with multiple hospitalizations in the past. He had completed his education at a community college, in a sporadic fashion, and had never held steady employment. The actual scores, obtained from a less-commonly used battery of tests, generally reflected mild weaknesses in writing and visual functioning, with most other scores in the Low Average Range. In general, the scores, were not at all unusual for someone with a long-standing history of schizophrenia, with recurrent exacerbations.

The determination of prior psychological and psychiatric history is critical in terms of addressing the initial points made in this chapter (i.e., whether the patient's current behavior is consistent with his or her personality and whether that personality is the same as it always was). While clear measures or indications of long standing personality traits are not always available, there is no excuse for not inquiring about them. A positive psychiatric history for the individual or his/her family involving formal treatment or specific outstanding symptoms (e.g., suicide attempts, run-ins with the law, "driving while intoxicated" arrests, disciplinary actions) may prove crucial in understanding the organizational principles of behavior for a given individual. Of course, if prior formal measurement of personality and emotional functioning exists, the results should be obtained as well. The point to be made here is that long-standing personality traits will exert a moderating influence on any current personality assessment and these must be considered in arriving at conclusions related to current trauma or disease. Discussion of how personality and other moderating variables can affect the results of cognitive testing during the neuropsychological evaluation can also be found in Chapter 3, by Putnam, Ricker, Ross and Kurtz and in Chapter 4, by Rankin and Adams, in this volume.

It is important to be careful in obtaining the history of a patient directly from the patient himself or from relatives and friends. On the face of it, these individuals should know the patient's situation the best, and some (e.g., Lezak, 1995) even disavow the appropriateness of evaluating a patient without such full information being obtained. However, it is also the case that the veracity of interview information is open to question in forensic evaluation. The same can be said for obtaining symptom information via self-report checklists (e.g., of cognitive or personality symptoms) that do not contain measures of validity (Lees-Haley, 1997). As an example, after a minor head injury, by history, a young man presented to the author as being unable to perform *any* formal neuropsychological testing. On interview, he had difficulty giving his name; he could not give any other orienting information and seemed unable to comprehend the instructions to any task given to him. In an hour-long interview with his wife, the examiner listened with compassion as she related the devastating effects of the head injury on her life and that of her husband. She elaborated how he had changed from a vibrant, gainfully employed individual to the pathetic shadow of a man we now saw. Even though the degree of deficit was out of proportion to the history of the injury (see Chapter 9), an experienced clinician could not help but be impressed at the apparent genuiness of the description of the dev-astating consequences of the patient's injury. About a month later, when video-taped surveillance of the patient showed him having a good time at a party, freely conversing, joking, and drinking, faith in the personally obtained inter-view information was soundly shaken.

Review of medical records from other professionals reveals similar opportu-nities for the influence of slanted symptom reporting on professional judgment. Often, reports of medical examination, neurological examination, psychiatric examination, etc., contain only normal objective findings (in terms of signs), but nevertheless conclude with diagnoses of neurological conditions based solely on patient self-report. Negative reports of prior psychiatric history must also be considered cautiously. The advantage of clinical neuropsychological opinion in the forensic setting rests on a dependence on measurement of behaviors, not just self-report, in reaching conclusions about personality organization and emo-tional functioning. Available history in this regard should be externally verifiable to the extent possible.

Measures of personality and emotional functioning

MMPI & MMPI-2 (prototype of objective tests)

While there are many available instruments for the measurement of personality and emotional functioning, those that are commonly seen in the forensic setting are relatively limited. I will review some of those most often encountered in forensic cases. Additional description of relevant tests can be found in Sweet and Westergaard (1997). By far, the most commonly used are the Minnesota Multiphasic Personality Inventory (MMPI; Hathaway & McKinley, 1943) and the Minnesota Multiphasic Personality Inventory-2 (MMPI-2; Butcher, Dahlstrom, Graham, Tellegen, & Kaemmer, 1989). There is no need to extol the virtues of the MMPI or its solid scientific base, most aspects of which are covered thoroughly and excellently covered with regard to use in the forensic setting by Pope, Butcher, and Seleen (1993). Suffice it to say that the MMPI and MMPI-2 (referred to hereafter collectively as MMPI/2) are extensively validat-ed on large samples and have good test-retest reliability. Added to this is the extensive literature documenting MMPI/2 profile types in numerous conditions, summaries of which can be found in Butcher and Williams (1992), Graham (1990), and Greene (1991).

The MMPI/2 has also been extensively used in characterizing the tendency of individuals to produce exaggerated or false patterns of psychological and cog-nitive symptoms associated with neurological conditions. For example, MMPI profiles tend to be exaggerated in groups of individuals involved in litigation when compared to individuals with genuine injury or disability who are not seeking financial compensation (Pollack & Grainey, 1984; Shaffer, Nussbaum, & Little, 1972) and when general issues of chronic pain are involved (see Keller

& Butcher, 1991, and Pope et al., 1993 for a more thorough review). The same seems to be true of chronic neurological conditions, such as mild head injury (e.g., Dikmen et al., 1992). When litigation is not considered, some report greater emotional distress after more serious head injury (Dikmen & Reitan, 1977), while others report the opposite (Leininger, Kreutzer, & Hill, 1991). The general topic of exaggeration and malingering of symptoms in neurological conditions, especially head injury, will not be covered here as the reader can refer to Chapter 9 and other chapters in this text, as well as other extensive reviews (Millis & Putnam, 1996; Putnam, Millis & Adams, 1996).

Other objective tests that are used frequently in forensic evaluations include: Beck Depression Inventory (Beck, 1987), Millon Clinical Multiaxial Inventory-III (Millon, Millon, & Davis, 1994), Profile of Mood States (McNair et al., 1981), Symptom Check List-90-r (Derogatis, 1994), and the Self-Rating Depression Scale (Zung, 1965). These and other objective tests can be reviewed elsewhere for their particular characteristics (Lezak, 1995; Sweetland & Keyser, 1991).

Rorschach (prototype of projective tests)
While the MMPI/2 and other "objective" tests (i.e., tests that permit only limited answers to a question such as "True" or "False") are effective at comparing an individual's responses to particular groups of normal or impaired individuals, such tests alone may not permit a clear characterization of the complete uniqueness of behavioral organization to a particular individual. Objective tests are very useful in forensic evaluations for assessing validity of responding, degree of distress, somatic overconcern, or presence of presence of unusual notions. The clinician, however, is cautioned that elevation on a given clinical scale, such as represented in the MMPI/2, is *not* equivalent to a psychological diagnosis. Unfortunately, it is not uncommon for a psychologist to state that an individual is "paranoid" based purely on the fact that the individual has an elevated score on the *Paranoia* scale of the MMPI. Of course, the well-trained psychologist would not assume so simplistic a score-diagnosis relationship, knowing full well that other factors may relate to the elevated score, such as general suspiciousness of health care systems based on past experience or limited educational background.

Under circumstances in which diagnosis is crucial, the use of a projective test (i.e., a test that permits virtually unlimited responses by an individual) is desirable to provide confirmation of particular personality characteristics. For example, without going into a course on the Rorschach (1942), comprehensive psychological training will permit the clinician to discern test patterns that are associated with paranoia, thought disorder, specific somatic concerns, and depression among other emotional disturbances. Specific patterns of response on the Rorschach have been associated with head injury and its recovery (Ellis & Zahn, 1985; Exner, Colligan, Boll, & Stischer, 1996).

While perhaps not as psychometrically aesthetic as the objective tests, use of a test such as the Rorschach is effective in the hands of a knowledgeable clinician, and often can add to information obtained from objective measures (Ganellen, 1996). It is most important in the forensic setting, however, that the test be given in a way that takes advantage of consensually validated scoring systems (Beck et al., 1961; Exner, 1986). This permits reference to available empirical research and ensure that test interpretation is not limited to purely unique clinical impressions.

Other projective procedures occasionally seen in the forensic setting include the Thematic Apperception Test (Murray, 1943) and the Incomplete Sentences Blank (Rotter, 1950). Again, these and other projective measures can be reviewed elsewhere for their particular characteristics (Lezak, 1995; Sweetland & Keyser, 1991).

Katz Adjustment Scale (functional rating scales)
Functional rating scales provide quantitative information about personal, inter-personal, and environmental adjustment of patients in daily life situations. The Katz Adjustment Scale (Katz & Lyerly, 1963) is one such scale that has been shown to be effective in characterizing daily life adjustment of patients with head trauma (Fordyce, Roueche, & Prigatano, 1983; Goodman, Ball, & Peck, 1988; Hinkeldey & Corrigan, 1990). The advantage of such scales is that they provide indicators of personality and emotional functioning that are supported by research as to their reliability and validity (Lezak, 1996). There is some dis-advantage, of course, if potentially nonobjective raters provide ratings of patient behavior.

Other rating scales used with some frequency in forensic situations include the Neurobehavioral Rating Scale (NBRS; Levin, High, Goethe, et al., 1987), and the Portland Adaptability Inventory (PAI; Lezak, 1995). The NBRS was developed primarily for patients with head trauma and focuses on cognitive symptoms, whereas the PAI concentrates on more general behavioral outcome measures for patients with head trauma and includes motor and sensory ratings, as well as functional categories of cognitively-related capacities (i.e., initiative, social contacts, etc.). As with the Katz scale, these scales have solid reliability and are sensitive to behavioral change.

Major relevant diagnostic entities encountered in forensic settings

Postconcussion syndrome
Postconcussion syndrome (PCS) refers to the myriad of symptoms often report-ed by individuals after experiencing a blow to the head resulting in a concussion. Concussion generally refers to a "brief loss of consciousness after head injury, with no immediate or delayed evidence of structural brain damage" (Lewis & Sciarra, 1989, p. 370). Gennarelli (1987) reflects a commonly accepted classifi-cation among neuropathologists of concussion as reflecting reversible neuro-logical deficiency caused by loss of consciousness of less than six hours. As

described by Rizzo and Tranel (1996, p. 3) "The effects of an uncomplicated concussion are usually transient. In most cases they resolve over weeks or months." Levin, Benton, and Grossman (1982) describe the symptoms associated with PCS as:

> ... a constellation of somatic and psychological symptoms including headaches, dizziness, fatigue, diminished concentration, memory deficit, irritability, anxiety, insomnia, hypochondriacal concern, hypersensitivity to noise, and photophobia. (p. 181)

However, Levin et al. (1987) also suggest that the most common persisting PCS complaints are headache, fatigue, and dizziness.

As it directly relates to concussion, which relates to a shock or shaking of the brain with effects that are generally transient (Rizzo & Tranel, 1996), PCS would seem to be a temporary phenomenon by definition. Potential persistence of PCS symptoms, however, has been a subject of debate. Levin et al. (1982) suggest that postconcussional symptoms extending up to two to three months likely reflect some subtle neurologic dysfunction, while more prolonged symptoms likely reflect a preexisting psychiatric condition or personality disorder that may have been intensified by the trauma. Similarly, most patients with mild head injuries (as defined by loss of consciousness of 20 minutes or less) are reported to have resumed work and leisure activities, at least to the same levels as control groups, within one year of injury (Dikmen, Temkin, & Armsden, 1989). Nevertheless, there are reports of some persisting symptoms at least up to 12 months (Rutherford, 1989). Most recently, Binder (1997) suggests that about 7–8% of patients with mild head injury remain chronically symptomatic, though without evidence that there is a neurological basis for their complaints. DSM-IV (1994) has a proposed diagnostic category for "Postconcussional disorder" which would include many of the symptoms noted above along with neuropsychological assessment evidence of in attention or memory. The DSM-IV proposed classification, however, suggests that symptoms must persist *at least* three months for the diagnosis to be made, a time frame somewhat at variance with the findings summarized above. From this author's perspective, it would seem prudent to err on the conservative side of allowing the possibility of PCS during a time frame up to 18 months post injury. This would encompass the commonly described early periods of symptom presentation, while addressing the possibility that some longer recovery, up to a year or more, may indeed be taking place. Consideration of PCS after longer post injury periods does not seem prudent, as at 18 months post injury, whatever symptoms are present will most likely be a permanent feature of an individual's behavioral presentation, similar to Rutherford's (1989) findings of striking similarities between symptom reports one year and 15 years post head injury in certain groups of patients. For interested readers, Couch (1995) presents a recent review of treatment issues associated with PCS.

Nevertheless, as one can surmise from the range of symptoms represented in PCS, their occurrence may not be particularly diagnostic of nervous system dysfunction. As already noted by Levin et al. (1982), after three months, symptoms are regarded as primarily psychiatric. In addition, Fox, Lees-Haley, Earnest, and Dolezal-Wood (1995a, 1995b) have found that symptoms associated with postconcussive syndrome were reported at high prevalence rates both by samples of medical patients in health maintenance organizations and in groups of patients seeking psychotherapy, suggesting that variables other than brain trauma are often related to the PCS symptom complex.

In terms of the patient's personality and emotional functioning, it is important for the neuropsychologist to assess: (1) whether the symptoms reported by the patient are significantly different from the general pattern of behavior before a given head trauma, (2) whether symptoms reported are different from relevant peer groups, and (3) whether the symptoms are more likely to reflect an emotional reaction than a genuine concussive phenomenon.

These distinctions are not always easy. The best approach to understanding pre-injury behavior is, of course, an accurate personal history, though the objectivity of this information can be subject to question if obtained from the patient or close friends or relatives, as already indicated. When available, descriptions of patient behavior from pre-injury medical records or behavioral observations obtained from school or work records can sometimes provide valuable clues to tendencies toward PCS-like behaviors. And finally, the personality measures used, and the formulation of their results arrived at by the clinician, can give important clues as to the long-standing vs. more acute behavioral patterns being exhibited by the patient.

Peer groups would include fellow students, family, and social and working peers. It is important for the neuropsychologist to determine whether the patient is part of a group undergoing social or occupational stress or general medical conditions. For example, PCS symptoms would not be unexpected around the time of final exams, in the context of divorce or loss of job, or as part of an outbreak of flu symptoms in a dormitory. Ruling out such corollary conditions is important to assessing the probability that PCS symptoms are genuinely related to a head trauma itself.

Finally, and perhaps with most difficulty, individuals can react psychologically to any situation with emotional reactivity, which can include PCS symptoms. The distinction here requires the establishment of a psychological diagnosis that can account for the symptoms reported by the patient, apart from a direct influence of head trauma. Such other diagnoses will be addressed in the rest of this chapter.

Posttraumatic Stress Disorder (PTSD)

Wilson (1994) argues that it was Freud's original conceptualization of traumatic neurosis that first was written into the DSM-I (American Psychiatric Association, 1952) diagnosis for an entity called Gross Stress Reaction, a category under Transient Situational Personality Disorders. It was described as reflecting overwhelming fear after exposure to severe physical demands or extreme emotional stress such in combat or civilian catastrophe. DSM criteria for PTSD were initially largely based on experience in combat and later extrapolated to civilian group and individual experience (Sparr & Boehnlein, 1990). By the time of DSM-IV (American Psychiatric Association, 1994), this concept had evolved to what we now regard as PTSD, which simply involves exposure to a traumatic event which is life or health threatening, following which, the event is re-experienced, stimuli (including thoughts) associated with the event are avoided, and symptoms of increased arousal persist.

PTSD often enters the forensic arena in the wake of physical and emotional trauma as a persisting harmful syndrome to the individual who experienced the trauma. It has even been suggested that PTSD may be a longer-lasting reaction to PCS in head injured patients (Davidoff, Kessler, Laibstain, & Mark, 1988). Leigh (1979) also suggests that PCS and PTSD can be distinguished, with PCS occurring immediately after an injury, with symptoms such as headaches and dizziness, and PTSD occurring later in some patients with symptoms such as anxiety, intrusive thoughts, and nightmares.

Determination of PTSD in forensic settings, however, is not without controversy. Sparr (1995) points out that PTSD has come to represent a bifurcated entity (1) a formal psychiatric diagnosis and (2) a legal-lay classification for disability and personal injury claims. The credibility of the diagnosis in the courtroom has come into question, with Slovenko (1994, p. 439) phrasing the observation "A lot of distressed people are feeling better these days – thanks to the courts." Authors have pointed out the likelihood of malingering and misuse associated with the use of the diagnosis of PTSD in forensic settings (Lees-Haley, 1997; Rosen, 1996). Slovenko (1994) notes that while psychiatric diagnosis associated with stress disorders is not essential for a particular legal action, the listing of stress disorders in the psychiatric nomenclature has tended to provide a sense of legitimacy on purely a descriptive basis. Sparr and Boehnlein (1990) similarly stress the value of DSM diagnoses as vehicles of classification for PTSD, rather than diagnostic justification per se; in the hands of inexperienced individuals, the apparent ease of use of DSM criteria may give a false sense of security to the diagnosis. They also, however, decry the lack of adherence to DSM criteria that often accompanies use of the diagnosis in the courtroom setting. As stated at the beginning of this chapter, there is a strong need for historical data regarding the patient, including prior psychiatric illness and potential predisposing characteristics to the symptoms observed when PTSD is being inferred (Marciniak, 1986), as well as other life-event and lifestyle factors, other than the matter of immediate concern in the courtroom.

It also needs to be realized that PTSD has sub-classifications, including acute vs. delayed and acute vs. chronic – generally revolving around the occurrence of symptoms within or after six months of the identified event (Peterson, Prout, and Schwarz, 1991). With regard to delayed onset, an inference of relationship between a given event and symptoms of PTSD can be particularly difficult. PTSD may also represent a cumulation of multiple events that eventually lead to the recognized onset of symptoms (Slovenko, 1994).

In this author's experience, PTSD is a diagnosis that has been used more often than is justified; this diagnosis requires that DSM-IV (1994) criteria be strictly met. The prevalence of PTSD among individuals exposed to violent crime, death, or accident is estimated at 7–11% (Norris, 1992), thus occurring in a minority of individuals. Assertion of the presence of PTSD for forensic purposes requires positive evidence, rather than argument based on natural probability. For forensic purposes, it is thus recommended that DSM-IV (1994) criteria be closely followed and that clinicians be familiar with the proposed variants of PTSD. Peterson et al. (1991) provide a good general summary of PTSD symptoms, variants, theory, and assessment practices.

More cogent to the forensic setting, the reader is referred to Simon (1995), wherein clear guidelines are offered for forensic dealings with PTSD, such as the need for use of official diagnostic manuals, current literature and research, and the use of standard assessment tools. For psychological assessment, Keane (1995) emphasizes the explicit specification of social and psychiatric history, the use of standardized assessment tools, and measures of social functioning.

A summary of supported use of psychological assessment tools in the evaluation of PTSD is provided by Allen (1994), and includes objective test instruments and functional ratings as indicated above, as well as test measures specifically developed for identification of PTSD (e.g., PTSD subscale of the MMPI; Keane, Malloy, & Fairbank, 1984 and the Mississippi PTSD scale; Keane, Caddell, & Taylor, 1988). Caution needs to be exercised with these MMPI scales, however, as high false-positive rates for anxious patients have been reported (Cannon, Bell, Andrews, & Finkelstein, 1987), as well as for patients seeking compensation based on PTSD symptoms (Frueh, Gold, & Arellano, 1997) and subjects instructed to fake symptoms of PTSD (Wetter & Deitsch, 1996).

Among other scales used to measures subjective stress following untoward life events is the Impact of Events Scale (Horowitz, Wilner, & Alvarez, 1979) Though potentially subject to self-report bias and tied directly to a particular formulation of response to stress (Bieliauskas, 1982), it has been cross-validated for assessing particular characteristics of stress disorders (Zilberg, Weiss, & Horowitz, 1982), including psychological sequelae of combat (Schwarzwald, Solomon, Weisenberg, & Mikulince, 1987). A longer, more recently developed stress response inventory, the Penn Inventory for Postraumatic Stress Disorder (Hammarberg, 1992) has also been extensively validated and should be given serious consideration in assessment of PTSD. Finally, the Davidson Trauma Scale (DTS; Davidson et al., 1997) is a brief questionnaire answered by the patient which also has solid test-retest reliability and internal consistency. A structured interview approach which produces very reliable DSM diagnosis for PTSD is represented by the Clinician-Administered PTSD Scale (CAPS; Blake et al., 1990). Finally, use of projective testing techniques for identification of PTSD, such as the Rorschach, has also been supported, especially for veteran populations (Frueh & Leverett, 1995; Hartman et al., 1990; Swanson, Blount, & Bruno, 1990).

Compensation neurosis

The term "compensation neurosis" must first be placed in the context of "neurosis," a term not used since the adoption of DSM-III (1980). Neuroses previously were defined as disorders that represented efforts

> to deal with specific private, internal psychological problems and stressful situations that the patient is unable to master without tension or disturbing psychological devices caused by the anxiety aroused...The symptoms of these disorders consist either of a manifestation of anxiety as it is directly felt and expressed or of automatic efforts to control it by such defenses as conversion, dissociation, displacement . . . (Kolb, 1974, p. 399).

By definition, then, an individual fended off anxiety through neurotic disorders in a manner that disavowed personal responsibility for the anxiety. As such, the behaviors that guarded against the anxiety became part of the personality makeup of the individual and were not necessarily under voluntary control.

Compensation neurosis was described as a neurotic disorder that encompassed

> . . . not only a desire for financial gain but the conviction on the part of the patient that he has the right to expect indemnification. . . . Social custom and public opinion tend to encourage the development of accident neuroses by their attitude toward the question of responsibility. (Kolb, 1974, p. 420).

The symptoms that were attributed to compensation neurosis, by now, are probably becoming familiar to the reader: The symptoms include "irritability, stubbornness, argumentativeness, crying spells, anxiety, depression, sleeplessness, headache, and dizziness. The patient is garrulous in describing his feelings and may complain of poor memory and inability to concentrate." (Kolb, 1974, p. 420).

In other words, the patient may be demonstrating symptoms that one might variously attribute to entities that we have already discussed, such as postconcussive syndrome or posttraumatic stress disorder. In this case, however, the syndrome that we observe is motivated by a desire for, or feeling of entitlement to, compensation related to a given incident, albeit the motivating desire may not be readily conscious to the patient.

The term "compensation neurosis" generally fell out of favor after the view was put forth that individuals involved in accident or trauma should have their legal process finished as quickly as possible and that convalescence would then dramatically accelerate, whether they were compensated or not (Miller, 1961); others actively attacked its validity (Mendelson, 1985; Modlin, 1986). Nevertheless, compensation neurosis appears to be re-emerging as a useful concept (Levy, 1992) that encompasses: a distorted sense of justice, victim status, and lack of an honorable response to an event (Bellamy, 1997), the strong influence of secondary gain (Modlin, 1986), and potential homeostatic (adjustment) response of the family or social environment to stress or crisis (Rickarby, 1979).

In this author's opinion, compensation neurosis is a useful conceptualization in an individual for whom compensation appears to be a determinant factor for their symptoms, yet for whom malingering or factitious disorder (see Chapter 9) does not appear to be present. Frequently, these individuals will present with strong financial incentive for compensation related to accident or injury, such as lifestyle history of marginal financial adjustment, deterioration in financial security *prior* to the incident that brought them to the courtroom, or evidence of persistent frustration at financially or status-driven aspirations.

Somatoform disorder

DSM-IV (1994) describes Somatoform disorders as those that suggest a general medical condition, but are not fully explained on a medical basis. Symptoms must cause impairments in functioning and the physical symptoms must not be intentional (i.e., under volitional control of the patient). The disorders include:

Somatization disorder (a.k.a. hysteria)
Undifferentiated somatoform disorder
Conversion disorder
Pain disorder
Hypochondriasis
Body dysmorphic disorder
Somatoform disorder not otherwise specified

The reader is referred to DSM-IV (1994) for specific diagnostic criteria for each of the disorders listed above.

Somatoform disorders generally are not specified diagnostically as related to specific incidents or events and will thus generally be brought into the forensic environment as explanations of symptom behaviors exclusive of the trauma-related events. Emotional factors are considered as significant contributors to the conditions and these disorders are frequently encountered in medical settings. Nevertheless, somatoform disorders have been described as potentially associated with psychologically traumatic life events (Waitzkin & Magana, 1997) and there are scattered reports of somatoform disorders occurring after rape, industrial disasters, and assault, though mostly in European literature sources and usually involving retrospective investigation in individuals with psychiatric disorders (Darves-Bornoz, Benhamou-Ayache, Degiovanni, Lepine, & Gaillard, 1995; Darves-Bornoz, Berger, Degiovanni, Soutoul, & Gaillard, 1994; Darves-Bornoz, Delmotte, Benhamou, Degiovanni, & Gaillard, 1996; Schnyder, 1996).

At this time, it appears that somatoform disorders relate to long-standing personality organization, rather than to individual traumatic incidents or events. It is not inconceivable that a predisposition to experience a somatoform disorder would be triggered by a specific incident in a given individual, but far more empirical evidence connecting somatoform disorders to real-life events needs to be gathered before such relationships can be regarded as having reasonable degrees of probability. An individual with an identifiable long-standing somatoform disorder may incorporate a given injury or event into his or her symptom organization, though any cause-effect relationships are speculative at best. In such instances, when faced with the question about cause and effect relationships, this author has found it appropriate to distinguish *necessary* vs. *sufficient* cause. In the case of a somatoform disorder possibly relating to a given injury or event, an appropriate position might be to indicate that while the injury or event could be sufficient to cause the symptoms, it is not necessary for such symptom expression; in other words, other meaningful events could also have caused a similar symptom expression because of the patient's predisposition.

DSM-related diagnoses

As noted by Yates (1996), the DSM classification of psychiatric syndrome related to organic factors changed significantly with DSM-IV. Psychiatric and psychological disorders due to physical causes are now listed with primary psychiatric syndromes. Thus, disorders in DSM-IV, such as delirium, dementia, amnestic disorders, and "other cognitive disorders", (American Psychiatric Association, 1994, p. 163) are listed with likely contributing etiologies (such as head trauma), while others such as psychoses, mood, and anxiety disorders also have sub-classifications as to their potential etiology (e.g., Anxiety Disorder due to . . . *specify general medical condition* (American Psychiatric Association, 1994, p. 436).

Although such a classification scheme specifies particular criteria for arriving at diagnosis of the disorder, the causal attribution is generally left to the clinician. Thus, DSM-IV specifies criteria for dementia (Criteria A: "The development of multiple cognitive deficits manifested by. . ." and Criteria B: "The cognitive deficits . . . cause significant impairment in social or occupational functioning. . ." (p. 142)), but attributing the dementia to causes other than Alzheimer's or cerebrovascular disease is generally left to the clinician to determine on a "rule out" basis. In other words, the clinician determines whether a particular disease is sufficiently proximal in time or similar in nature to be related to the symptoms. Of course, one can see how this can be misapplied in trauma situations, if it is believed that *any* cognitive change can be related to *any* environmental event. The description of Dementia Due to Head Trauma (American Psychiatric Association, 1994, p. 148) is particularly troublesome as it translates a descriptive term (dementia) historically used to refer to progressive cognitive deterioration associated with diseases of aging and/or neurodegenerative conditions to instances of trauma or other medical conditions in younger individuals. This author has seen dementia diagnosed in cases of relatively mild head trauma by clinicians who strictly follow the nomenclature of DSM-IV classification, with little appreciation of the severity of its implication.

For psychiatric disorders, this weakness in classification is further compounded. Thus, while DSM-IV provides clear criteria for diagnoses of mood disorders, anxiety disorders, sleep disorders, sexual disorders, etc., the classification of "due to (*Indicate the General Medical Condition*)" leaves open the interpretation of causality without strict guidelines. This is not to criticize the DSM criteria, most of which it is wise to follow. Rather, it is to caution against the cavalier use of causality when using a diagnosis that is frankly at the discretion of the individual cataloguing the symptoms, rather than strict diagnostic rules. As will be reiterated later, it is important in the forensic setting to follow diagnostic criteria strictly for the primary diagnosis. As noted below, however, determining implications for etiology requires far more careful clinical formulation than "bean-counting" of symptoms.

Psychopathology due to accident or trauma

In general, the most commonly reported psychiatric sequelae after injury are depression and anxiety (Cartlidge, Shaw, & Kalbag, 1981; Lezak, 1995; Tuokko, Vernon-Wilkinson, & Robinson, 1991). Factors that affect the degree of personality change, for example those associated with head trauma, include the length of post-traumatic amnesia, physical disfigurement, disruption of social and occupational roles, and financial factors (Yates, 1996). Psychoses are much more rare and are generally associated with personal or family history of psychiatric disorder (Yates, 1996). Nevertheless, there is a report of manic episodes occurring after head trauma involving the basal region of the right temporal lobe (Starkstein et al., 1990), suggesting the possibility of psychotic-like behaviors after serious brain injury. It seems generally the case that greater levels of psychopathology are associated with more serious head injuries, being correlated with serious neuropsychological deficits, and that emotional difficulties tend to decline over time (Dikmen & Reitan, 1977).

Nevertheless, it should be noted that diagnosis of emotional disturbance in forensic cases can be influenced by numerous factors, including psychosocial difficulties, reaction to other injuries (i.e., disfigurement, loss of a limb, etc.), or the litigation process (Dikmen et al., 1992). Youngjohn, Davis, and Wolf (1997) have shown that that elevations on the MMPI-2 were related to litigation in moderate/severely injured patients, with mild head injury patients having even higher elevations on neurotic clinical scales than those who were injured more severely. It should also be noted, though perhaps too briefly in the context of this chapter, that interactions with ongoing treatment (especially medications) are often overlooked in judging a patient's emotional reaction to trauma or accident. For example, a patient's experience of pain can be expected to affect his or her lifestyle significantly, as well as ability to experience pleasure and even to concentrate and remember effectively. Yet, chronic headaches that commonly occur after trauma to head and neck, for example, may very well be influenced to a greater degree by the quantity and nature of analgesic medications given to the patient than to trauma. In fact, it is estimated that the majority of patients with chronic post-traumatic headaches are experiencing rebound headaches as a result of persistent use of analgesics (Warner & Fenichel, 1996). If a patient's degree of distress, anxiety, and depression are related to their experience of pain, one must also consider the relative influence of pain treatment on the patient's symptoms and emotional reaction.

As a final caveat, the kind of symptoms being discussed here are often part of a universal reaction that can accompany any mental disorder, be it primarily psychological or trauma/accident-related. Hoch (1972) points out that: "The basic clinical manifestations which can occur in any mental disorder are *anxiety, depression,* and *paranoid reactions.* And at times one may want to include elation and obsessive reactions." (p. 502).

Thus, the very symptoms to which we continually refer (i.e., anxiety, depression, and potentially paranoia, mania, and obsessive reactions), may very well be general manifestations of cognitive and emotional disruption after a trauma, incident, or accident, rather than specific diagnostic entities.

With the more serious psychopathological entities it seems safest to conclude the following:

(1) More serious emotional disturbances, especially psychotic symptoms, very seldom occur in the absence of a prior or predisposing history.

(2) More serious emotional disturbances seem to inevitably occur in the context of more serious injury, especially with more significant neuropsychological deficits.

(3) Emotional disturbances related to more serious injuries appear to decline over time.

To the extent that a serious personality or emotional disturbance occurs in a forensic case in which accident-related injury is potentially responsible, the further the measured behavior deviates from a relationship to injury severity or a decline over time, the less likely it is due to the accident at issue. In all cases, prior history needs to be ruled out and predisposition considered.

Summary and conclusions

From the foregoing, we can see that measurement of personality in the forensic setting is complex and requires simultaneous consideration of multiple factors. The following may best summarize the points we have been discussing:

(1) Personality reflects an organizational concept for behavior. In the forensic setting, the main question generally asked is whether the personality organization seen at time of measurement reflects long-standing characteristics or more recent changes due to an incident, accident or injury. It is imperative, therefore, to obtain as thorough a psychological/behavioral/psychiatric history as possible to best characterize these long-standing personality traits. Measurement of personality and emotional functioning will give one a "snapshot in time" at the very best; judgment as to its static vs. changing quality requires an historical basis.

(2) In light of the Daubert ruling, psychologists who testify in an expert witness role should embody a solid background in the foundations of knowledge germane to the profession. As in medicine, legitimate board

certification reflects examination of competency by one's peers and is one means of establishing a foundation as an expert witness, given relevant experience to the case at hand. Psychologists who wish to participate in expert witness roles, and thus also participate on a stage with their medical peers, are urged to obtain board certification in their appropriate specialty through the American Board of Professional Psychology.

(3)   Measurement of personality and emotional functioning includes use of tests and techniques that are empirically validated and have support in the professional literature regarding their application to the psychological entities to which they are applied. Similarly, when forming diagnostic impressions, it is important to be aware of limitations in diagnosing certain syndromes and to adhere to sound diagnostic criteria when available. This latter point is especially important in categories such as PCS and PTSD, which have gained popular exposure in forensic settings.

(4)   Several general guidelines concerning measurement of personality and emotional functioning should be considered, including: (1) anxiety and depression are frequently observed symptoms in situations causing mental distress and are not necessarily diagnostic in and of themselves; (2) emotional reactions are multiply determined (e.g., one's response to situational, family, and physical changes that may have little to do with the trauma or accident in question, as well as medical treatment for the complaint that gains the forensic spotlight); (3) more serious emotional disturbance generally occurs in the context of a positive psychiatric history or predisposition, which needs to be ruled out or considered, if one is to link the disturbance to an event at hand; and, (4) more serious emotional disturbance is generally seen with more serious injury, the typical pattern of which is to subside, at least somewhat, over time.

References

ABMS (1996). *The Official ABMS Directory of Board Certified Medical Specialists*. New Providence, NJ: Reed Reference Publishing.

American Board of Professional Psychology (1997). *Directory of Diplomates*. Columbia, MO: Author.

Allen, S.N. (1994). Psychological assessment of post-traumatic stress disorder. *Psychiatric Clinics of North America, 17*, 327–349.

Allport, G.W. (1961). *Pattern and growth in personality*. New York: Holt, Rinehart & Winston.

American Psychiatric Association (1952). *Diagnostic and statistical manual, Mental disorders* (1st edition). Washington, DC: Author.

American Psychiatric Association (1980). *Diagnostic and statistical manual of mental disorders* (3rd edition). Washington, DC: Author.

American Psychiatric Association (1994). *Diagnostic and statistical manual of mental disorders* (4th edition). Washington, DC: Author.

Ayala, F.J., & Black, B. (1993). Science and the courts. *American Scientist, 81*, 230–239.

Beck, A.T. (1987). *Beck Depression Inventory*. San Antonio, TX: The Psychological Corporation.

Beck, S.J., Beck, A.G., Levitt, E.E., & Molish, H.B. (1961). *Rorschach's test. I: Basic processes* (3rd edition). New York: Grune & Stratton.

Bellamy, R. (1997). Compensation neurosis: Financial reward for illness as nocebo. *Clinical Orthopaedics and Related Research, 336*, 94–196.

Bieliauskas, L.A. (1982). *Stress and its relationship to health and illness*. Boulder, Colorado: Westview Press.

Binder, L.M. (1997). A review of mild head trauma. Part II: Clinical implications. *Journal of Clinical and Experimental Neuropsychology, 19*, 432–457.

Blake, D.D., Weathers, F.W., Nagy, L.M., Kaloupek, D.G., Klauminzer, G., Charney, D.S., & Keane, T.M. (1990). A clinician rating scale for assessing current and lifetime PTSD: The CAPS-1. *The Behavior Therapist, 13*, 187–188.

Butcher, J.N., & Williams, C.L. (1992). *Essentials of MMPI-2 and MMPI-A interpretation*. Minneapolis: University of Minnesota Press.

Butcher, J.N., Dahlstrom, W.G., Graham, J.R., Tellegen, A., & Kaemmer, B. (1989). *Minnesota Multiphasic Personality Inventory-2 (MMPI-2): Manual for administration and scoring*. Minneapolis, MN: University of Minnesota Press.

Cannon, D.S., Bell, W.E., Andrews, R.H., & Finkelstein, A.S. (1987). Correspondence between MMPI PTSD measures and clinical diagnosis. *Journal of Personality Assessment, 51*, 517–521.

Cartlidge, N.E.F., Shaw, D.A., & Kalbag, R.M. (1981). *Head injury*. Philadelphia: Saunders.

Couch, J.R. (1995). Post-concussion (post-trauma) syndrome: Special issue: Neurorehabilitation of the head injured patient. *Journal of Neurologic Rehabilitation, 9*, 83–89.

Darves-Bornoz, J.M., Benhamou-Ayache, P., Degiovanni, A., Lepine, J.P., & Gaillard, P. (1995). *Annales Medico-Psichologiques, 153*, 77–80.

Darves-Bornoz, J.M., Berger, C., Degiovanni, A., Soutoul, J.H., & Gaillard, P. (1994). Treating psychic traumas: A psychiatric emergency. *Annales Medico-Psichologiques, 152*, 649–652.

Darves-Bornoz, J.M., Delmotte, I., Benhamou, P., Degiovanni, A., & Gaillard, P. (1996). Syndrome secondary to post-traumatic stress disorder and addictive behaviors. *Annales Medico-Psichologiques, 154*, 190–194.

Davidson, J.R.T. et al. (1997). Assessment of a new self-rating scale for post-traumatic stress disorder. *Psychological Medicine, 27*, 153–160.

Dcrogatis, L.R. (1994). *SCL-90-r: Symptom Checklist-90-r*. Minneapolis, MN: National Computer Systems.

Davidoff, D.A., Kessler, H.R., Laibstain, D.F., & Mark, V.H. (1988). *Cognitive Rehabilitation, 6*, 8–13.

Dikmen, S.S., Temkin, N., & Armsden, G. (1989). Neuropsychological recovery: Relationship to psychosocial functioning and postconcussional complaints. In H.S. Levin, H.M. Eisenberg, & A.L. Benton (Eds.), *Mild head injury* (pp. 229–241). New York: Oxford University Press.

Dikmen, S.S., & Reitan, R.M. (1977). Emotional sequelae of head injury. *Annals of Neurology, 2*, 492–494.

Dikmen, S.S., Reitan, R.M., Temkin, N.R., & Machmer, J.E. (1992). Minor and severe head injury emotional sequelae. *Brain Injury, 6*, 477–478.

Division 40 (1989). Definition of a clinical neuropsychologist. *The Clinical Neuropsychologist, 3*, 22.

Ellis, D.W., & Zahn, B.S. (1985). Psychological functioning after severe closed head injury. *Journal of Personality Assessment, 49*, 125–128.

Exner, J.E. (1986). *The Rorschach: A comprehensive system* (Vol. 1, 2nd. edition). New York: Wiley.

Exner, J.E., Colligan, S.C., Boll, T.J., & Stischer, B. (1996). Rorschach findings concerning closed head injury patients. *Assessment, 3,* 317–326.

Fordyce, D.J., Roueche, J.R., & Prigatano, G.P. (1983). Enhanced emotional reactions in chronic head trauma patients. *Journal of Neurology, Neurosurgery, and Psychiatry, 46,* 620–624.

Fox, D.D., Lees-Haley, P.R., Earnest, K., & Dolezal-Wood, S. (1995a). Base rates of post-concussive symptoms in health maintenance organization patients and controls. *Neuropsychology, 9,* 606–611.

Fox, D.D., Lees-Haley, P.R., Earnest, K., & Dolezal-Wood, S. (1995b). Post-concussive symptoms: Base rates and etiology in psychiatric patients. *The Clinical Neuropsychologist, 9,* 89–92.

Frueh, B.C., & Leverett, J.P. (1995). Interrelationship between MMPI-2 and Rorschach variables in a sample of Vietnam veterans with PTSD. *Journal of Personality Assessment, 54,* 312–318.

Frueh, B.C., Gold, P.B., & de Arellano, M.A. (1997). Symptom overreporting in combat veterans evaluated for PTSD: Differentiation on the basis of compensation seeking status. *Journal of Personality Assessment, 68,* 369–384.

Ganellen, R.J. (1996). *Integrating the Rorschach and the MMPI-2 in personality assessment.* Mahwah, NJ: Lawrence Erlbaum.

Gennarelli, T.A. (1987). Cerebral concussion and diffuse brain injuries. In P.R. Cooper (Ed.), *Head injury* (2nd ed.). Baltimore: Williams & Wilkins.

Goodman, W.A., Ball, J.D., & Peck, E. (1988). Psychosocial characteristics of head-injured patients: A comparison of factor structures of the Katz Adjustment Scales. (Abstract). *Journal of Clinical and Experimental Neuropsychology, 10,* 42.

Goodman-Delahunty, J., & Goote, W.E. (1995). Compensation for pain, suffering, and other psychological injuries: The impact of *Daubert* on employment discrimination claims. *Behavioral Sciences and the Law, 13,* 183–206.

Graham, J.R. (1990). *MMPI-2. Assessing personality and psychopathology.* New York: Oxford University Press.

Greene, R.L. (1982). *The MMPI-2/MMPI: An interpretive manual.* Needham Heights, MA: Allyn & Bacon.

Guilford, J.P. (1959). *Personality.* New York: McGraw-Hill.

Hall, C.S., & Lindzey, G. (1970). *Theories of personality.* New York: Wiley.

Hammarberg, M. (1992). Penn inventory for Postraumatic Stress Disorder: Psychometric properties. *Psychological Assessment, 4,* 67–76.

Hannay, H.J., Bieliauskas, L.A., Crosson, B.A., Hammeke, T.A., Hamsher, K. deS., & Koffler, S.P. (1998). Proceedings: The Houston Conference on Specialty Education and Training in Clinical Neuropsychology. *Archives of Clinical Neuropsychology, 13.*

Hartman, W.L., Clark, M.E., Morgan, M.K., Dunn, V.K., Fine, A.D., Perry, G.G., & Winsch, D.L. (1990). Rorschach structure of a hospitalized sample of Vietnam veterans with PTSD. *Journal of Personality Assessment, 54,* 149–159.

Hathaway, S.R., & McKinley, J.C. (1943). *Manual for administering and scoring the MMPI.* Minneapolis, MN: University of Minnesota Press.

Hinkeldey, N.S., & Corrigan, J.D. (1990). The structure of head-injured patients' neurobehavioral complaints: A preliminary study. *Brain Injury, 4,* 115–134.

Hinsie, L.E., & Campbell, R.J. (1970). *Psychiatric dictionary* (4th ed.). New York: Oxford University Press.

Hoch, P. (1972). *Differential diagnosis in clinical psychiatry.* New York: Jason Aronson.

Horowitz, M., Wilner, N., & Alvarez, W. (1979). Impact of event scale: A measures of subjective stress. *Psychosomatic Medicine, 41,* 209–218.

Keane, T.M. (1995). Guidelines for the forensic psychological assessment of posttraumatic stress disorder claimants. In R.I. Simon (Ed.), *Postraumatic stress disorder in litigation. Guidelines for forensic assessment.* Washington, DC: American Psychiatric Press.

Keane, T.M., Caddell, J.M., & Taylor, K.L. (1988). Mississippi Scale for Combat-Related Posttraumatic Stress Disorder: Three studies in reliability and validity. *Journal of Consulting and Clinical Psychology, 56,* 85–90.

Keane, T.M., Malloy, P.F., & Fairbank, J.A. (1984). Empirical development of an MMPI subscale for the assessment of combat-related post-traumatic stress disorder. *Journal of Consulting and Clinical Psychology, 5,* 888–891.

Keller, L.S. & Butcher, J.N. (1991). *Assessment of chronic pain patients with the MMPI-2.* Minneapolis: University of Minnesota Press.

Kolb, L.C. (1973). *Modern clinical psychiatry.* Philadelphia: Saunders.

Lees-Haley, P.R. (1997). MMPI-2 base rates for 492 personal injury plaintiffs: Implications and challenges for forensic assessment. *Journal of Clinical Psychology, 53,* 745–755.

Leigh, D. (1979). Psychiatric aspects of head injury. *Journal of Continuing Education in Psychiatry, 40,* 21–33.

Leininger, B.E., Kreutzer, J.S., & Hill, M.R. (1991). Comparison of minor and severe head injury emotional sequelae using the MMPI. *Brain Injury, 5,* 199–205.

Levin, H.S., Benton, A.L., & Grossman, R.G. (1982). *Neurobehavioral consequences of closed head injury.* New York: Oxford University Press.

Levin, H.S., Gary, H.E., High, W.M., Mattis, S., Ruff, R.M., Eisenberg, H.M., Marshall, L.F., & Tabaddor, K. (1987). Minor head injury and the postconcussional syndrome: Methodological issues in outcome studies. In H.S. Levin, J. Grafman, & H.M. Eisenberg (Eds.), *Neurobehavioral recovery from head injury* (pp. 262–275). New York: Oxford.

Levin, H.S., High, W.M., Goeth, K.E., Sisson, R.A., Overall, J.E., Rhoades, H.M., Eisenberg, H.M., Kalisky, Z., & Gary, H.E. (1987). The Neurobehavioral Rating Scale assessment of the behavioural sequelae of head injury by the clinician. *Journal of Neurology, Neurosurgery, and Psychiatry, 50,* 183–193.

Levy, A. (1992). Compensation neurosis rides again. *Brain Injury, 6,* 401–410.

Lezak, M.D. (1995). *Neuropsychological assessment* (3rd ed.). New York: Oxford.

Marciniak, R.D. (1986). Implications to forensic psychiatry of post-traumatic stress disorder. *Military Medicine, 151,* 434–437.

McNair, D.M., Lorr, M., & Droppleman, L.F. (1981). *EDITS manual for the Profile of Mood States.* San Diego, CA: Educational and Industrial Service.

Mendelson, G. (1985). "Compensation neurosis". An invalid diagnosis. *Medical Journal of Australia, 142,* 561–564.

Miller, H. (1961). Accident neurosis. *British Medical Journal, 51,* 919–925, 992–998.

Millis, S.R., & Putnam, S.H. (1996). Detection of malingering in postconcussive syndrome. In M. Rizzo & D. Tranel (Eds.), *Head injury and postconcussive syndrome* (pp. 481–498). New York: Churchill Livingstone.

Millon, T., Millon, C., & Davis, R. (1994). *Millon Clinical Multiaxial Inventory-III manual.* Minneapolis, MN: National Computer Systems.

Modlin, H.C. (1986). Compensation neurosis. *Bulletin of the American Academy of Psychiatry and the Law, 14,* 263–71.

Murray, H.A. (1943). *Thematic Apperception Test.* Cambridge, MA: Harvard University Press.

Norris, F.H. (1992). *Journal of Consulting and Clinical Psychology, 60,* 409–418.

Peterson, K.C., Prout, M.F., & Schwarz, R.A. (1991). *Post-traumatic stress disorder. A clinician's guide.* New York: Plenum.

Pollack, D.R., & Grainey, T.F. (1984). A comparison of MMPI profiles for state and private disability insurance applicants. *Journal of Personality Assessment, 48,* 121–125.

Pope, K.S., Butcher, J.N., & Seleen, J. (1993). *The MMPI, MMPI-2, & MMPI-A in court.* Washington, DC: American Psychological Association.

Prigatano, G.P. (1987). Psychiatric aspects of head injury: Problem areas and suggested guidelines for research. In H.S. Levin, J. Grafman, & H.M. Eisenberg (Eds.), *Neurobehavioral recovery from head injury* (pp. 215–231). New York: Oxford.

Putnam, S.H., Millis, S.R., & Adams, K.M. (1996). Mild traumatic brain injury: Beyond cognitive assessment. In I. Grant & K.M. Adams (Eds.), *Neuropsychological assessment of neuropsychiatric disorders* (2nd ed.). (pp. 529–551). New York: Oxford.

Rickarby, G.A. (1979). Compensation-neurosis and the psychosocial requirements of the family. *British Journal of Medical Psychology, 52,* 333–338.

Rizzo, M., & Tranel, D. (1996). Overview of head injury and postconcussive syndrome. In M. Rizzo & D. Tranel (Eds.), *Head injury and postconcussive syndrome* (pp. 1–18). New York: Churchill Livingstone.

Rorschach, H. (1942). *Psychodiagnostics: A diagnostic test based on perception.* (Translated by P. Lemkau & B. Kronenburg). Berne: Huber.

Rosen, G.M. (1996). Posttraumatic stress disorder, pulp fiction, and the press. *Bulletin of the American Academy of Psychiatry and the Law, 24,* 267–275.

Rotgers, F. & Barrett, D. (1996). *Daubert v. Merrell Dow* and expert testimony by clinical psychologists: Implications and recommendations for practice. *Professional Psychology: Research and Practice, 27,* 467–474.

Rotter, J.B. (1950). *Incomplete Sentences Blank.* San Antonio, TX: The Psychological Corporation.

Rowland, L.P., & Sciarra, D. (1989). Head injury. In L.P. Rowland (Ed.), *Merritt's textbook of neurology.* Philadelphia: Lea & Febiger.

Schnyder, U. (1996). Prevention and therapy of post-traumatic disorders from a biopsychosocial viewpoint. *Schweiszerische Rundschau fur Medizin Praxis, 85,* 1603–1608.

Schwarzwald, J., Solomon, Z., Weisenberg, M., & Mikulince, M. (1987). Validation of the impact of event scale for psychological sequelae of combat. *Journal of Consulting and Clinical Psychology, 55,* 251–256.

Shaffer, J.W., Nussbaum, I.K., & Little, J.M. (1972). MMPI profiles of disability insurance claimants. *American Journal of Psychiatry, 129* (4), 63–67.

Simon, R.I. (1995). *Posttraumatic stress disorder in litigation. Guidelines for forensic assessment.* Washington, DC: American Psychiatric Press.

Slovenko, R. (1994). Legal aspects of post-traumatic stress disorder. *Psychiatric Clinics of North America, 17,* 439–446.

Sparr, L.F. (1995). Post-traumatic stress disorder. Does it exist? *Neurologic Clinics, 13,* 413–429.

Sparr, L.F., & Boehnlein, J.K. (1990). Posttraumatic stress disorder in tort actions: Forensic minefield. *Bulletin of the American Academy of Psychiatry and the Law, 18,* 283–302.

Starkstein, S.E. et al. (1990). Mania after brain injury: Neuroradiological and metabolic findings. *Annals of Neurology, 27,* 652–659.

Swanson, G.S., Blount, J., & Bruno, R. (1990). Comprehensive system Rorschach data on Vietnam combat veterans. *Journal of Personality Assessment, 54,* 160–169.

Sweet, J.J., & Westergaard, C. (1997). Psychopathology and neuropsychological assessment. In G. Goldstein & T.M. Incagnoli (Eds.), *Contemporary approaches to neuropsychological assessment* (pp. 325–358). New York: Plenum.

Sweetland, R.C., & Keyser, D.J. (Eds.). *Tests* (3rd edition). Austin, TX: Pro-Ed.

Tuokko, H., Vernon-Wilkinson, R., & Robinson, E. (1991). The use of the MCMI in the personality assessment of head-injured adults. *Brain Injury, 5,* 287–293.

Waitzkin, H. & Magana, H. (1997). The black box in somatization: unexplained physical symptoms, culture, and narratives of trauma. *Social Science and Medicine*, *45*, 811–825.

Warner, J.S., & Fenichel, G.M. (1996). Chronic post-traumatic headache often a myth? *Neurology*, *46*, 915–916.

Wetter, M.W., & Deitsch, S.E. (1996). Faking specific disorders and temporal response consistency on the MMPI-2. *Psychological Assessment*, *8*, 39–47.

Wilson, J.P. (1994). The historical evolution of PTSD diagnostic criteria: From Freud to DSM-IV. *Journal of Traumatic Stress*, *7*, 681–698.

Yates, W.R. (1996). Psychiatric conditions. In M. Rizzo & D. Tranel (Eds.), *Head injury and postconcussive syndrome*. New York: Churchill Livingston.

Youngjohn, J.R., Davis, D., & Wolf, I. (1997). Head injury and the MMPI-2: Paradoxical severity effects and the influence of litigation. *Psychological Assessment*, *9*, 177–184.

Zilberg, N.J., Weiss, D.S., & Horowitz, M.J. (1982). Impact of Event Scale: A cross-validation study and some empirical evidence supporting a conceptual model of stress response syndromes. *Journal of Consulting and Clinical Psychology*, *50*, 407–414.

Zung, W.W.K. (1965). A Self-rating Depression Scale. *Archives of General Psychiatry*, *12*, 63–70.

Chapter 6

INTERPRETING APPARENT NEURO-PSYCHOLOGICAL DEFICITS:
WHAT IS REALLY WRONG?

Thomas Kay
New York University School of Medicine, New York City

The forensic situation

Unique factors operating

Neuropsychological diagnosis is literally "seeing through nerves and soul" – a process that "starts with an observation of a certain level of behavioral functioning and attempts to explain that level of functioning as a complex interaction between brain and psyche" (Kay, 1992, p. 109). From this perspective, the process of neuropsychological evaluation is more complex than determining whether brain damage exists, and, if so, which parts of the brain may be involved. It is an undertaking that attempts to integrate information regarding cognitive capacity and multiple factors that may impact on mental functioning. The presence of cognitive deficits may or may not indicate brain damage; an adequate evaluation must be of sufficient scope and depth to consider all the possible causes of apparent neuropsychological deficits.

In clinical settings, this process of exploration usually takes place cooperatively among patient, examiner, and those who are seeking the information (e.g., family, other doctors, therapists). The goals are to determine the nature of the impairment and the need for treatment. The patient's self-presentation, and relationship to the examiner, is driven by a need to perform as adequately and honestly as possible for someone who is perceived to have his or her best interests at heart. Similarly, within the routine clinical context those receiving the information are expected to use it without distortion or personal agendas.

The context of forensic neuropsychological evaluation is quite different from that of the typical clinical evaluation. Forces operate that can change the behavior of the patient, the process of evaluation, and the role of the neuropsychologist. At least in the American system, patient and examiner are caught up in an adversarial process in which the patient is trying inherently to demonstrate that he or she has a significant injury, while the opposing side has an interest in disproving that claim. The examiner is retained and being paid by someone with an agenda (to either maximize the gains or minimize the loss of his or her client); there is often a colleague who is being paid by someone with an opposing agenda; and the results of each side's evaluation will be scrutinized in a critical, sometimes hostile, manner by the opposing side and their neuropsychological expert.

Inevitably, the pressures exerted in this adversarial system can potentially affect the process of evaluation and the interpretation of apparent neuropsychological deficits. It is incumbent upon the examiner to maintain a standard of objectivity, to treat all data equally (not selectively attending to those data that fit with the agenda of the retaining side), and to consider all possible explanations for the neuropsychological performance of the patient before formulating an opinion.

Additional factors exist that set the forensic situation apart from the clinical. Patients often come into the evaluation motivated to prove they have real injuries. They may have had previous health care assessment or treatment experiences that they perceive as invalidating. They may believe that the goal of the defense neuropsychologist is to deny, cast doubt upon, or minimize the validity or significance of their injuries. Some may feel their life is being opened unfairly to scrutiny, in an attempt to find some fatal flaw that will undermine their case. They may have been denied services or reimbursement by insurers as a result of skeptical "insurance doctors". The result is that when the patient appears for a *plaintiff* neuropsychological evaluation, they may be desperate to demonstrate the reality and significance of their injury; when a patient appears for a *defense* evaluation, he or she may, in addition, be primed with distrust and hostility that can affect both performance and the perceptions of the examiner. In both cases, the challenge is heightened for the neuropsychologist to accurately "see through" the complicating factors, as clearly as possible.

When the original injury is legitimate and was traumatic in nature, and the trial is occurring three or four years after the accident, the intense process of evaluations and depositions may reopen emotional wounds that have been healing over time. In such instances, patients may experience an increase of anxiety and depression during the pre-trial process, which may in turn affect test performance.

In addition, there are pressures, expectations and biases inherent in the forensic situation, with which the neuropsychologist must deal. Lawyers vary in how blatantly they communicate what they would like their expert to find. The best lawyers present an attitude of "just be honest and tell me what's going on", while others make it clear in the first phone call that they need an expert opinion that either, depending upon which side they represent, establishes or refutes the existence of neuropsychological damage. Neuropsychologists are subject to the same laws of (financial) reinforcement, needs to please, and needs to avoid anger and disapproval, as the rest of humanity. Consciously or unconsciously, the knowledge of what will bolster the retaining attorney's case is present and active in the psyche of the forensic neuropsychologist, as is true for all expert witnesses regardless of discipline. The temptation to selectively attend to and focus on those aspects of performance that are consonant with the agenda of the retaining side, while dismissing or giving short shrift to what contradicts it, can be strong.

It is dangerous for a neuropsychologist to enter a case feeling an obligation to further the retaining attorney's position; this will increase the possibility of biased perceptions or selective attention. It is not the duty of the plaintiff's neuropsychologist to validate the patient's injury, nor is it the duty of the defense's neuropsychologist to pick holes in a case of neuropsychological impairment. Both of these approaches serve only to decrease respect for the opinion of neuropsychologists in the forensic setting. Rather, it is the duty of the neuropsychologist, regardless of retaining side, to make the most honest determination possible of the plaintiff's neuropsychological status.

Countertransference issues are also often present. Many neuropsychologists engaged by plaintiff attorneys are clinically active in the treatment of neurologically impaired patients. Their empathy for patients may inappropriately turn them into advocates, rather than neutral observers within the forensic context. Conversely, neuropsychologists regularly retained by defense attorneys are often well-heeled in evaluating that subset of patients who do not respond to treatment and become chronic patients, and these experts may be inherently skeptical of persons complaining of injuries that are not apparent.

In summary, unique factors are operating within the forensic setting that complicate the process of neuropsychological assessment compared to that encountered routinely in clinical activities in private practice or academic medical settings. The neuropsychologist must be sensitive to issues of patient presentation and motivation, as well as affective states (especially depression and anxiety) that might affect performance. In addition, the neuropsychologist must remain vigilant for any tendency on his or her part to arrive at perceptions and interpretations because they meet the needs of the retaining attorney (cf. Guilmette & Hagan, 1997; van Gorp & McMullen, 1997). As noted in the Foreword, throughout this text, and in the Conclusion, the field of neuropsychology is placing strong emphasis on objectivity at present.

Issues of interpretation

Legitimate data and the need for a narrative explanation

Issues of data interpretation operate across all neuropsychological assessments. These are highlighted in the forensic setting when two neuropsychologists evaluate the same person or review the same records and reach quite different conclusions. Above and beyond the possible pressures to reach an opinion in support of the retaining side, differences of neuropsychological opinion can be due to a number of sources, first and foremost of which is the question of what are considered legitimate data.

There is no way to evaluate apparent neuropsychological deficits fully, if the data to be interpreted are limited to test scores. In order to interpret the meaning of test scores, we necessarily rely on the quality of the patient's presentation and test-taking behavior; a history of the symptomatology; an understanding of developmental history, premorbid functioning, and possible damage to the central nervous system; a formulation of the person's personality structure and style of coping with adversity and stress; and the situational variables in the person's life that may contribute to the expression of symptoms and affect functioning. Ideally, these data are gathered through observation, review of records, and interviews with persons who knew the patient well before and after the event in question. These are all legitimate sources of neuropsychological data. In the absence of such data, it is not possible to determine the complete meaning of test scores, and therefore to interpret apparent neuropsychological deficits.

There is a controversy in the field regarding whether it is necessary for a neuropsychologist to actually see and examine a patient in order to render an opinion, or whether he or she can rely on the data made available by others (i.e., render an opinion based on a record review). The author is extremely reluctant to testify in a case unless he has met and evaluated the patient. The reason for this is simple: if I rely on the data collected by others to formulate a case, I am limited to the data they considered relevant. While commenting on the records in a limited way (e.g., do the data as they stand unequivocally support the contention of brain injury?) is possible and legitimate, it is the author's opinion that the true value of a neuropsychological opinion is in the *formulation* of a case (i.e., an integrative explanation of the patient's behavior, complaints, and functioning). For this level of explanation, it is usually not sufficient to rely on the data collected by others. (One should also keep in mind that significant issues in a person's life may be at stake; would we want critical decisions made about our lives "within a high degree of neuropsychological certainty" by someone who never met us?)

Finally, the formulation in a forensic neuropsychological evaluation should go beyond the determination of whether brain damage exists or not. It should offer a plausible explanation of the behavior of the client, and describe how a combination of factors, only one of which is brain functioning, is operating to determine the patient's level of neuropsychological functioning. This requires a narrative: a story that integrates all the diverse neuropsychological data and makes compelling sense. This is true regardless of which side of the case the neuropsychologist is on.

Sensitivity and specificity

One of the most common errors made in forensic neuropsychological evaluations is the automatic interpretation of test scores in the deficit range as evidence for brain damage. Those tests most sensitive to any type of brain damage – tests of attention (especially complex, or divided attention), concentration, and short term memory, are also those least specific to brain damage. That is, multiple other conditions – among them pain, psychological distress, anxiety, depression, personality disorders, obsessive preoccupations, emotional disruption due to activation of old trauma, fatigue and medications – can affect these areas of cognitive functioning. Failure to consider critically all possible sources of observed neuropsychological deficit can lead to a mistaken conclusion of brain damage, as is mentioned throughout this text.

This being said, a second and very different common error in forensic neuropsychological evaluations is the dismissal of the significance of neuropsychological deficits in the presence of strong emotional overlay, with the presumptive conclusion that therefore brain damage cannot exist. Brain injury is not inconsistent with emotional reactivity; nor is it inconsistent with personality disorder. It is as erroneous to dismiss brain damage in the presence of a strong emotional component as it is to interpret every poor performance as evidence for brain damage. The art lies in sorting one out from the other. In some cases this is quite easy to do, and one can reach an opinion within reasonable medical probability; in some cases, one simply cannot, because the waters are too muddied.

Alternative explanations and diagnoses

It is part of the burden of the forensic neuropsychological evaluation to demonstrate that the opinion offered fits the data better than any other explanation. Ruling brain damage in or out requires a systematic consideration of alternative formal diagnoses (e.g., malingering, conversion disorders, factitious disorders, somatization disorders, adjustment reactions, depression, anxiety disorders, etc.) and dynamic explanations for the observed behaviors. Ruling in or out such alternative diagnoses requires analysis of information that goes beyond neuropsychological test scores.

Focus of this chapter

The most legitimate use of the neuropsychological evaluation in the forensic process is to clarify the nature of the claimant's complaints and possible injuries, and therefore facilitate the settlement of the case; or in cases where settlement is not possible, to provide expert testimony that will clarify the nature of the complaints and possible injury for a judge and/or jury. The process of formulating a neuropsychological opinion ideally reflects an objective process of evaluating all available data.

This chapter is intended for neuropsychologists retained on either side of a case. The interpretation of historical, behavioral, and cognitive data is an extremely complex undertaking that is made even more complex by the forces operating within the forensic setting. Clear cut cases of obvious neurological damage are the easiest, in that what is at issue is not the existence, but the extent and implications of neuropsychological damage. Of greatest complexity are those cases in which there is questionable neurological damage in the presence of strong emotional reactivity and a history that suggests a possible dynamic for post-traumatic emotional dysfunction or insufficient effort. This chapter is written most specifically with these difficult cases in mind. How does a competent and ethical neuropsychologist – regardless of which side has retained them – go about the process of interpreting what appear on testing to be neuropsychological deficits and reach an opinion that will stand up to scrutiny within a court of law?

Conceptual issues in the interpretation of neuropsychological deficits

A neuropsychological model of functional disability

Introduction
The author has developed a model that describes the interaction of factors that contribute to functional disability with apparent neuropsychological deficits (Kay, 1993; Kay, Newman, Cavallo, Ezrachi, & Resnick, 1992). While initially developed to clarify issues in mild traumatic brain injury, the model is applicable to any event or process that leads to possible brain damage. The impetus for developing such a model was the tendency both clinically and forensically to view patients with subjective neuropsychological complaints after mild head injury as either brain damaged (with all their problems emanating from neurological injury) or psychologically disturbed (with all of their problems emanating from dysfunctional emotional responses to trauma). The goal was to avoid such dichotomous thinking, and integrate both brain damage and psychological response (to the damage and/or trauma) into a unified model. While other authors have discussed the idea that multiple factors contribute to functional outcome after possible mild brain injury (e.g., Jacobson, 1995; Nemeth, 1996), the model presented here is unique, as far as the author knows, in terms of both specifying the interactive, modifying influence of factors upon each other, and introducing the idea of objective vs. subjective cognitive factors within an interactive model. The result is schematically presented in Figure 6.

The central thrust of the model is that, after some event that potentially causes brain damage, neurological, physical, and psychological factors interact to determine a person's level of functioning. The event could be head injury, stroke, anoxia, meningitis, encephalitis, toxic encephalopathy, or other causes. We will use head trauma as the illustrative example because it is the most common issue

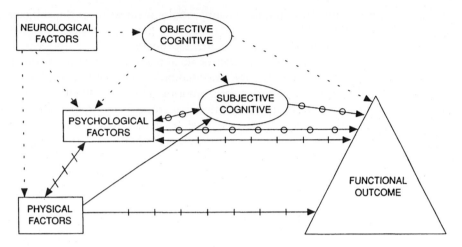

Fig. 6.   A neuropsychological model of functional outcome after mild traumatic brain injury.

for forensic neuropsychologists. The primary neurological, physical, and psychological factors are represented as boxes on the left of Figure 6. This is a dynamic model, in which forces move from place to place among the components, and ultimately impact functional outcomes in multiple ways, represented by the large triangle on the right. The arrows within the model are uni- or bi-directional, indicating the potential flow of forces among the components.

Objective and subjective cognitive deficits

A central concept within this model, and the one most relevant to this chapter, is a differentiation between *objective* and *subjective* cognitive components, as illustrated at the top of the model. The objective cognitive component is the true cognitive potential of the person as defined by neurological factors only – the brain's ability to function. True brain injury often results in objective cognitive deficit. It is this we would like to be measuring on neuropsychological testing. However, there is also the subjective cognitive component, which represents the person's experience of their own cognitive capacities at any given point in time; this is represented by a separate oval in the model. While objective cognitive functioning has only a single source of input (the arrow from neurological factors), the subjective cognitive component has multiple inputs. The person's experience of their capacity to process information is determined not only by their neurologically determined cognitive status, but by psychological and physical factors as well. For example, psychological factors, such as anxiety, depression, and emotional disruption independent of or due to reactivated prior emotional trauma, as well as physical factors, such as fatigue, pain, and medication, can impact on cognitive functioning above and beyond damage to the brain.

What we are really sampling in neuropsychological testing is the subjective cognitive component in this model. In relatively straightforward cases (i.e., when there is little non-neurological impact on mental functioning), the objective and subjective components are virtually identical; in a fluid representation of the model, they would then be nearly superimposed on each other. In complex cases of questionable brain damage, with strong emotional factors, a history that puts the person at risk for catastrophic reactions to trauma or disease, and when physical injury is also present (especially with a chronic pain component), the subjective cognitive factor is wedged increasingly farther away from the objective cognitive component. In such cases, the presence of apparent neuropsychological deficits may be relatively more reflective of non-neurological factors.

It is in these complex situations that the interpretation of observed neuropsychological deficits may become problematic, and open to differing opinion. In forensic cases, it is incumbent upon the neuropsychologist, regardless of retaining side, to honestly and responsibly develop a diagnostic opinion that accounts for the functional capacity of the person using all the factors in this model, not simply to look at the "neurological" to "objective cognitive" arrow in isolation (the "is there brain damage?" question).

Lawyers often limit the question posed to the neuropsychologist as to whether or not brain damage exists and is responsible for the patient's problems. When posed in this manner, the question assumes the outdated dichotomy ("organic" vs. "functional") that violates the complex dynamic represented in the above model. It is helpful in these situations for the neuropsychologist to broaden the referral question to include the possibility of psychological damage resulting from the trauma. This broadening of perspective is of potential interest to attorneys on both sides of the dispute. For example, it is often important for plaintiff lawyers to understand that genuine trauma, even with no or minimal brain damage, can trigger disabling psychological reactions. In these instances arguing that brain damage is the source of all problems is not accurate. Conversely, defense lawyers need to understand that: (1) they need not deny the existence of minimal brain injury in order to limit their client's liability when functional disability is exaggerated by non-injury related factors, and (2) even in the absence of brain dysfunction, psychological injury can occur as a result of a trauma or disease. It is not the neuropsychologist's role to determine courtroom strategy, but to give the lawyer a valid and clear explanation of the patient's behavior so that an appropriate strategy can be determined by the attorney. In an ideal world, the neuropsychological evaluation would take place early enough in the settlement process that the legal strategy could flow from it, rather than just prior to trial.

Primary factors affecting outcome
Returning to the model, another central concept is that of relative "vulnerability." Along each of the primary dimensions of the model – neurological, physical, and psychological – the person can be thought of as having relative strength or vulnerability, which helps determine the impact of trauma (or disease) on that person's ability to function. These vulnerabilities must be taken into account

when considering the impact of each factor on the ability to function. Each factor will be discussed below, with selective issues being discussed in more detail in later sections of this chapter.

Neurological factors. Neurological damage can occur after trauma, stroke, anoxia, or various types of encephalopathy. The amount of actual cell loss, the brain areas in which they occur, the extent to which they disrupt functional neural networks, and the non-structural impact on biochemical systems (e.g., neurotransmitters) will determine the amount of neuropsychological deficit. The effects of neurological damage may be cognitive (e.g., attentional, memory, or executive deficits), affective (e.g., depression, anxiety, emotional lability), or behavioral (e.g., disinhibition, lethargy). These neurobehavioral deficits require observation above and beyond cognitive testing; behavioral observation and collateral reports are an essential part of the data base.

Any individual may be more or less vulnerable to the effects of trauma or disease on the central nervous system. Age is a primary vulnerability factor; older persons, with more extensive pre-morbid cell loss, may be more vulnerable to the effects of trauma or disease, take longer to recover, and have less complete recovery.

Individual differences in brain integrity and neural complexity may affect sensitivity to damage later in life. Pre-existing learning difficulties must be identified and factored out when interpreting apparent deficits in attention and new learning. Low test scores should be interpreted more cautiously with persons with estimated pre-morbid low IQ levels. (In the forensic setting, obtaining all available academic records, including standardized test scores, is standard professional practice.) Conversely, persons with expected high pre-morbid IQ's who score in the average range may be judged to have acquired deficits relative to their baseline performance, if the domain sampled is known to correlate with IQ.

Pre-existing neurological damage is another dimension of vulnerability. This may be the result of previous trauma or illness, or self-inflicted damage in the form of excessive ingestion of drugs or alcohol. There is some evidence, for example, that a previous concussion, despite apparently complete clinical recovery, can make a person more vulnerable to the effects of a second concussion (Ewing, McCarthy, Gronwall, & Wrightson, 1980; Gronwall & Wrightson, 1975). Clinical experience suggests that persons with a long history of drug and alcohol abuse may be more vulnerable to effects of mild insult to the brain. The existence and possible effects of pre-morbid central nervous system damage must always be explored in the forensic context.

Physical factors. Separate from injury to the brain, other physical injuries may indirectly affect cognitive function and mimic brain damage. This is especially true in trauma, where the likelihood of multiple injuries is great. Chronic pain is particularly common after trauma and may influence cognitive functioning through psychological distress or distractibility. Both central (e.g., Risey & Briner, 1990) and peripheral (e.g., Grimm, Hemenway, Lebray & Black, 1991) vestibular dysfunction can cause cognitive interference at the interface of visual, somatosensory, and vestibular feedback to the brain, disrupting efficient processing of information. Fatigue, regardless of cause, can interfere with cognitive performance. Medications and sleep disturbance are two common causes of fatigue in persons with suspected brain injury.

Vulnerabilities also exist within this physical dimension. For example, there is great variability in the experience and tolerance of pain, which can be a variable affecting psychological distress, and therefore cognitive functioning. Physical vulnerabilities also exist that may make pain more intense and therefore more cognitively disruptive after trauma. For example, persons with chronic degenerative spinal cord disease may be more debilitated after a cervical sprain coincident with concussion. The resulting pain may exacerbate the cognitive changes resulting from the concussion.

Psychological factors. This is a broad area that includes prior psychological history, personality style and coping mechanisms, one's emotional response to the injury, and the psychological realities of the person's work and home environment, including interpersonal relationships and family situation. It in this arena in which an extensive clinical interview, that carefully explores pre-morbid functioning and personality dynamics, can be critical in formulating an explanation of post-traumatic reaction. Persons with pre-existing psychiatric or personality disorders have poorer outcomes after central nervous system insult. However, certain individuals who appeared to be functioning within the normal range pre-morbidly may nevertheless carry the psychological ingredients for vulnerability to the effects of trauma, physical injury and pain, or loss of cognitive function.

For example, the person's prior history with illness or trauma has a formative impact on how they deal with the trauma or illness at hand. How childhood injuries, illnesses or illnesses of other family members were responded (or not responded) to may provide valuable information regarding the patient's current dynamic of post-injury or post-illness behavior. Persons who suffered overt sexual, physical, or emotional abuse, especially if they have not been able to integrate it well into their ego structure (i.e., elements are still dissociated), may have old emotional responses to trauma reactivated by trauma in the present. The resulting emotional turmoil may be cognitively disruptive and mimic brain dysfunction. (This topic will be explored in more depth in the section on "Post-traumatic stress disorder".)

The unique psychodynamics of a person's source of self-esteem may also be relevant to functional disability and subjective neuropsychological complaints after acquired brain injury. For example, persons whose primary source of self esteem derives from their ability to perform consistently at a very high level may be particularly vulnerable to the effects of mild brain injury and the loss of their cognitive "edge". Such persons often lack a history of strong emotional support, are quite intelligent, have a tendency toward obsessive-compulsiveness, and are over-achievers in their work. Persons with these characteristics often have catastrophic emotional reactions to the inability to function at their previous high level, because it attacks the core of their sense of self worth.

Other "vulnerable" personality styles may include persons with strong dependency needs (who may respond to mild brain injury by becoming immobilized with increased anxiety, insecurity, and dependence on others); persons with tendencies toward grandiosity (who suffer narcissistic wounds from their cognitive deficits and are unable to acknowledge and therefore accommodate to their injury); and persons with borderline tendencies (who may become angry, hostile, unconsciously exaggerate their inability to function, and in some cases collapse into a dramatic personality regression).

These are clinically derived examples of vulnerable personality styles. A careful exploration of the individual's personality dynamics, using clinical interviewing, collateral reports, personality and (at times) projective testing, is often extremely useful in constructing a compelling story – that may or may not include elements of brain injury – to convincingly explain a person's behavior and level of functioning after possible brain injury.

Another element that comes into play in the psychological domain is the person's environmental demands at the time of the injury or illness, and to what extent physically or neurologically-based deficits interfere with effective functioning. Manual laborers may handle mild memory loss better than secretaries or lawyers, while back pain may be more disruptive and cause more emotional distress to the former than the latter.

The response of the family system is a final important psychological variable. Families often respond quite differently to mild than severe brain injury. After more severe brain injury the family is more likely to rally around the injured family member in a supportive way. After milder injuries, especially if disability persists in the face of apparent physical normalcy, there may be a tendency for animosity and alienation to develop within the family, especially the spousal, system. Pre-existing "cracks" in the marital relationship may be widened, and the resulting conflict incorporated into the symptom complex of the patient. If this situation persists untreated, complaints of pain, poor attention and concentration, memory loss, and disorganization worsen, and may reflect the emotional distress in the person's life, with or without the presence of underlying neurological damage.

Dysfunctional feedback loops

Within this model, two "dysfunctional feedback loops" are relevant to the interpretation of apparent neuropsychological deficits. The first is a chronic pain loop, indicated by the lines with hashmarks at the bottom of the model, linking physical factors, functional outcome, and psychological factors. In this scenario a real injury occurs, which causes pain, and the pain interferes with the ability to function. Depending on the personality dynamics of the individual, the inability to function may increase the anxiety or emotional distress of the individual, which would then feed back into the physically based pain by intensifying it. This will further decrease functional outcome, increasing psychological distress, intensifying the pain, etc. Eventually this feedback loop takes on a life of its own, and becomes functionally detached from the original source of the pain. Deactivating such self-perpetuating dysfunction is a primary goal of chronic pain programs.

This model suggests a parallel dysfunctional feedback loop in the cognitive realm (indicated in Figure 1 by lines with circles). An objective cognitive deficit (e.g., in the area of attention) may occur after a concussion and directly impact the person's ability to function. Depending on environmental demands and personality style, this decrease in functional ability may have psychological consequences, such as when anticipatory anxiety arises from repeated situations in which attentional deficit results in functional breakdown. Subsequently, when anticipatory anxiety is again triggered situationally, the anxiety itself interferes with attention. Because of the neurologically-based 'weak link' in attention, the anxiety worsens the subjective cognitive experience above and beyond the objective cognitive deficit. Functioning is correspondingly worsened, and the anticipatory anxiety consequently heightened. The result is that functional capacity is further reduced, anxiety increased, cognition worsened, etc. Parallel to chronic pain, this dysfunctional cognitive feedback loop may take on a life of its own, becoming detached from and operating independently of its neurological source. These clinical observations are consistent with the findings of Machulda, Bergquist, Ito, and Chew (1998) who found that postconcussive symptoms were associated with *perceived* level of stress.

For the vast majority of people, spontaneous neurological recovery after concussion results in normalized cognitive function and resumption of normal functional activity (Alves, Macciocchi, & Barth, 1993; Binder, Rohling, & Larrabee, 1997; Kay et al., 1992; Levin et al., 1987). It also happens in some cases that despite recovery of the brain, chronic pain and/or chronic subjective cognitive dysfunction remains. This can happen when psychological factors propel one or both of the dysfunctional feedback loops into self-sustaining activity. (The dotted lines in the model reflect the potential for such detachment from the primary factors.) These are the patients who, because of personality and/or environmental variables, may remain or even become more symptomatic, without having a permanent brain injury. However, in a small subset of persons, a threshold of neurological damage may be crossed (depending on the vulnerability of the brain and the severity of the injury), and permanent objective cognitive dysfunction may occur (Binder, 1997).

The neuropsychologist in the forensic situation must remain alert for the existence of such dysfunctional loops, especially in persons whose disability has not resolved after years. While not forgetting that real neurological injury may underlie and continue to fuel cognitive dysfunction, the neuropsychologist must also be alert for the possibility that there are no 'objective' cognitive deficits and that cognitive dysfunction now consists of, and is being maintained by, psychological factors. The worsening of cognitive deficits over time is often the result of emotional factors.

Issues of destabilization
A genuinely traumatic event of the type described in DSM-IV has the potential to destabilize a person's psychological balance. This can occur with or without damage to the brain, and is largely dependent on the personal history and

coping capacity of the individual. Individuals who become psychologically destabilized can manifest subjective cognitive impairments that mimic brain injury: impaired attention (especially in the face of competing stimuli or distraction), poor ability to absorb new information (often attributable to impaired attention), and apparent executive deficits, which take the form of disorganization and failure to follow through with tasks (although lack of energy, feeling overwhelmed, and not knowing where to start often appear to be the underlying causes upon inquiry). Rather than reflecting underlying brain damage, such symptoms may be the result of a psychological destabilization with elements of depression, anxiety, or dissociation. Severe psychological destabilization can be associated with altered brain capacity for processing information; this brain *dysfunction* (a term implying abnormal brain function not necessarily associated with gross structural change) is not the same as brain *damage* (a term referring to permanent structural changes within the brain). When this cluster of impairments appears, it is incumbent upon the neuropsychologist in the forensic setting to determine whether it is the result of brain damage, brain dysfunction, or psychological destabilization that may or may not be associated with transient or persistent brain dysfunction. Making this discrimination requires an in-depth knowledge of the patient's history and personality dynamics, as well as a careful consideration of the nature of the event in question and the possibility that brain injury may have occurred.

As noted above, some persons are more vulnerable than others to psychological destabilization after an injury involving external trauma. A person who has experienced significant sexual, physical, or emotional trauma in the past, may achieve emotional stability by dissociating that experience. That is, a sense of personal integrity is possible only through the coping mechanisms of denial, suppression or repression – defense mechanisms that serve to 'split off', or distance, the overwhelming emotional experience from the sense of self.

Even without brain injury, a physical trauma as an adult may serve to destabilize psychological balance by overwhelming coping (defense) mechanisms. The experience of physical pain and being out of control, when coupled with the perception of being treated in a dismissive or inattentive fashion by persons designated as caretakers, may so activate old emotional experiences that the ability to keep them split off from consciousness may be overwhelmed, and these old emotional forces may destabilize the person psychologically. Under such conditions, especially when depression, anxiety or dissociation are present, the cognitive capacity for attention, new learning, and organization may break down in a way that mimics brain injury. Hallmarks of these situations include: the presence of extreme subjective complaints that are not borne out, or appear inconsistently, on neuropsychological testing; a history consistent with emotional destabilization; and a strongly emotional and often regressive presentation on the part of the patient, especially when recounting the injury. It is not uncommon for the diagnosis of brain injury to be embraced by this type of person, because it can serve an explanatory and psychologically integrative function.

Emotional destabilization may also be precipitated and intensified when brain dysfunction is present. Altered brain functioning can only complicate the emotional components of trauma. More speculatively, the sudden disruption of cognitive efficiency may compromise the neurological underpinnings of coping. A coordinated balance of neural activation and suppression may underlie the means of managing emotions and memories that we refer to as "defense mechanisms". A concussion may temporarily disrupt the organization, independent functioning, or executive management of such neural networks. The result, in persons with a history of unresolved emotional trauma, may be a breakdown in defense mechanisms and a flooding in of previously dissociated material, which serves to further overwhelm the functional capacity of the concussed brain.

Destabilization may also occur in persons vulnerable to psychiatric disorders because of family history. It may be that neurological and biochemical integrity is necessary to suppress a latent mental illness or keep it inactive. For example, the occurrence of head trauma in persons with family histories of mental illness, who themselves showed little or no evidence of psychiatric disturbance prior to their injury, may trigger thought, mood, or anxiety disorders in these individuals. Destabilization at the neurological or biochemical level may be occurring in these cases. From a forensic point of view, it is important to clarify the combined role of pre-existing and traumatic factors. It may be neither true that the head trauma alone caused all the problems, nor that the problems are simply the result of a pre-existing condition. The reality is that, even if infrequent, head trauma can be responsible for triggering a psychiatric condition in a vulnerable individual.

Defining deficits

Certain principles of neuropsychological test interpretation are worth reviewing because of their critical importance in the forensic setting. Keeping a number of principles in mind will help the ethical forensic neuropsychologist stay on an objective track and avoid allowing the agenda of the retaining side to distort his or her interpretations. First, the normative base of the test being used, and the pre-morbid baseline of the patient, must be kept clearly in mind. Automatic claims of deficits whenever scores are low, and automatic claims of lack of deficit whenever scores are in the unimpaired range, are equally invalid. Second, single test scores should rarely be used to claim that a deficit exists. Significant intra-individual variability can occur in normal test performance; the claim that a deficit exists should be based on a consistent pattern of performance. It is wise to do redundant testing in legal cases. Consistent performance across a number of memory tests is a much more convincing indicator of presence or absence of deficit, than impaired or adequate performance on a single test. Third, the pattern of deficit must fit known patterns of neurological disorder. The neuropsychologist must have adequate knowledge both of the nature of the neurological injury and the cognitive, behavioral or personality impairments expected.

What data are necessary to interpret apparent deficits?

Obtaining academic records, pre-morbid medical records, and work evaluations is helpful in both establishing pre-morbid levels of functioning, and determining if pre-existing deficits or disease processes may be relevant to current performance. Obtaining medical records regarding the event responsible for the suspected neurological deficits is critical. When trauma is involved, especially when it is a motor vehicle accident, obtaining police, emergency medical service (EMS), and emergency room (ER) records is recommended. Hospital admission and discharge summaries are often inadequate sources of information; obtaining complete medical records from all hospitalizations and medical treatments requires time consuming work but pays off in the richness of detail. In particular, nursing notes, and office notes of therapists and treating doctors, often provide "behind the scenes" information often lost in formal summaries.

In addition to acute records, all subsequent medical and rehabilitation records should be obtained and reviewed in detail, if possible. Particularly relevant to the neuropsychologist are previous neuropsychological evaluations. In all cases, it is worth the effort to obtain raw data in the form of reported test scores, and whenever possible the actual protocols. Direct comparison of patient responses, not just total scores, can be helpful in establishing credibility in test taking over time. Inconsistencies in neuropsychological test performance need to be identified and discussed. In those cases in which malingering is suspected, inconsistent performance over time is an important piece of evidence. Conversely, patients who are legitimately having difficulty with test performance will often show consistency over time.

Collateral reports by persons who know the patient well are also valid data in the evaluation process – as long as these data are considered as critically as any other data. Of particular value are reports of disinterested persons who knew the patient both before and after the incident involved. Whenever possible, the neuropsychologist should interview the provider of information directly.

In the repertoire of neuropsychological procedures, none is more useful than the clinical interview. Rather than a brief screening device, the clinical interview should be the centerpiece of the evaluation process, an encounter between patient and doctor that establishes the working relationship, generates hypotheses, and sets the stage for formal testing. The author does not begin the clinical interview with questions about the trauma or event that resulted in the law suit. Rather, he asks the person to tell him his or her life story prior to the accident or illness, explaining why he is doing so. Inquiring first about the person, their family, childhood, growing up, schooling, relationships, value systems, sources of self esteem, life decisions, choices of spouse, children and parenting, serves three potential purposes. First, it sends the clear message that the examiner is truly interested in the person sitting with them, which helps engender trust and thus sets the stage for a valid neuropsychological evaluation. Second, the examiner is gathering the background information that will ultimately enable him or her to construct a meaningful and convincing narrative about how the event in question impacted on this particular person's life. Third, knowing personal background material often leads to hypotheses about reactions to trauma that can be explored when that point is reached in the interview. In the author's experience it is of particular value to inquire about the person's early perceptions of and relationship to their parents. While psychodynamic issues are often far from the mind of classically trained neuropsychologists, the reality is that these are the formative relationships in each of our lives, and the response of a person to trauma or neurological disease or accident may not fully be appreciated in some instances without reference to these influences.

Regardless of its place in the interview, the injury or illness in question should be the focus of careful clinical inquiry. When the event is a trauma with questionable loss of consciousness, it is critically important to inquire in a way that will elicit a detailed, moment by moment narrative of the event from the patient, rather than simply asking a question, such as "did you lose consciousness?" In suspected head trauma, the alteration or loss of consciousness, as well as the presence of retrograde or anterograde amnesia, are best estimated by gaps in the patient's narrative recall of events prior to, during and subsequent to the injury. It is also critical to inquire about symptoms experienced at the time of injury.

Of parallel interest to the content of the information is the affective tone in which accident information is presented. This may range from detached to emotionally florid. How the patient has dealt with the affect surrounding the trauma can provide important clues as to the influence of psychological factors on test performance and subjective symptomatology. Normal accounts of trauma will include intermittent appropriate flashes of feeling, which become less intense the greater the time since the injury. Emotionally florid or fresh descriptions of trauma may signal continued high post traumatic distress, which may interfere with cognition. Extremely detached descriptions of the trauma may suggest repressed affect that could be manifested as a somatization process.

The author prefers to conclude the interview with a detailed inquiry into the patient's current level of symptomatology and functioning. When considering symptoms, it is wise to initially inquire about changes in an open-ended, nondirective manner, to observe what the patient spontaneously reports. This can be followed up by directed inquiries into physical, cognitive, emotional, and behavioral domains. It is also useful to inquire about the future: expectations, goals, fears, plans. This can be extremely helpful in formulating a sense of the patient's response to the injury.

It is the author's opinion that formal neuropsychological testing should include both cognitive and personality/emotional assessment. The approach of conducting two separate psychological evaluations, one purely cognitive and the other purely psychodynamic, is much less likely to yield an integrated, comprehensive, valid portrait of the person. Formal cognitive testing should follow the guidelines for comprehensive neuropsychological testing (see Lezak, 1995, pp. 110–143), and be informed by what is known about the neurological nature of the injury. For example, hemiplegic patients should be evaluated in particular depth on tests sensitive to damage in the affected hemisphere. Patients with suspected diffuse, non-focal injuries should have particular attention paid to tests of speed and efficiency of information processing, attention, concentration, and memory. Test selection should consider the appropriateness of normative sample for the person being tested. Commonly accepted, well-normed tests with good validity and reliability should be used as often as possible. More experimental tests that are sensitive to elusive deficits (e.g., complex attention) can also be useful, as long as the normative database matches the patient to assure valid interpretation of results.

While there is considerable controversy in this area, it is the author's conviction that including raw scores and relevant normative data for interpretation in an appendix to the report is appropriate and should be the accepted professional standard. Particularly in the forensic setting, accountability is crucial; pure interpretation in the absence of foundational data is unacceptable. The scores should be available for scrutiny as part of the record.

The decision to go beyond cognitive testing and do personality testing (objective and/or projective) should be made on a case by case basis. In the case of a severe and obvious head trauma, for example, documenting the nature and extent of cognitive impairment may be the primary issue, and subtleties of personality style and emotional functioning may add little information of use to either side. On the other hand, in more complex cases in which there is a legitimate question of whether a neurological injury has occurred, and at issue is the etiology of apparent neuropsychological deficits, personality testing may be extremely helpful. However, interpretation of personality tests developed with psychiatric populations should be made cautiously, taking into account the fact that many patients with neurological injuries have legitimate medical concerns. The psychiatric interpretations of response sets appropriate in the absence of neurological impairment may not be valid for similar response sets in the presence of genuine neurological impairment.

A standard part of the neuropsychological evaluation involves reporting on the behavior of the patient during the testing situation. There are a number of ways in which a neuropsychologist can elicit additional non-testing data from the testing situation, which can be useful in the interpretation of apparent deficits.

First, create an opportunity to observe the patient in an unstructured, decision-making situation (e.g., going out to lunch). Observing the person when they must enter an unfamiliar setting, seek out options, make decisions, and relate to strangers in an appropriate way, can yield valuable neurobehavioral information, which may be consistent or at odds with subjective reports or test data. Second, request that the patient report to you any experiences during testing that are similar to what he or she is reporting as a cognitive deficit in their day to day life. This may help establish the ecological validity of the testing, and may also provide valuable insights as to the interpretation of apparent deficits. Third, test the limits, when feasible. Following standardized administration of tests, go back especially to failures and redo items, either having the patient talk through their thinking process, or providing alternate means of solving the problem. This may yield critical insights into the nature of cognitive breakdowns. For example, having the person "talk through" failed items on a test like

Wechsler's Picture Arrangement, normally taken silently, can give valuable insights as to whether failure was due to inattention to visual detail, breakdowns in inferential thinking, or sequencing problems. Similarly, testing the limits after poor performance on the Wisconsin Card Sort test may give valuable cues as to whether memory problems, executive dysfunction, or thought disorder underlie failure. As long as they are gathered in a way that does not influence ongoing performance and does not preclude valid retesting, these data can be valuable in the interpretation of apparent deficits. Carelessness or lack of effort may also come to light under such inquiry.

Factors influencing cognitive performance

In this section we will consider multiple conditions, above and beyond neurological damage, that may affect cognitive performance. It is beyond the scope of this chapter to review the effects of neurological damage on neuropsychological test performance. Excellent texts are available to guide the neuropsychologist in this regard (e.g., Heilman & Valenstein, 1993; Lezak, 1995; Walsh, 1985, 1994). It is assumed that the forensic neuropsychologist will be well versed in recognizing neurologically based cognitive deficits when they occur. This section will focus on those conditions other than acquired brain damage which may contribute to the appearance of cognitive impairment.

Miscellaneous contributing factors

Estimating premorbid intelligence is critical in interpreting apparent neuropsychological deficits, and is discussed thoroughly in Chapter 3. Even though many neuropsychological tests correlate with intelligence it is erroneous to assume that cognitive test scores that fall above or below a statistically derived "cutoff" for brain injury must reflect an acquired deficit or lack thereof. Estimates of premorbid intelligence, based on academic records and history, work success, collateral reports, and results of tests least affected by acquired brain injury, should be a standard part of the neuropsychological evaluation process.

Certain non-psychotic psychiatric disorders are associated with documented neuropsychological deficits and suspected neurological involvement. Examiners evaluating persons with severe pre-morbid personality disorders, such as obsessive-compulsive (Cox, 1997; Tallis, 1997) and borderline (Judd & Ruff, 1993; Swirsky-Sachetti et al., 1993) conditions, should take into account the possibility that deficits found on testing may be related to the pre-existing condition. Historical inquiry into level and changes of functioning is critical in determining to what extent observed deficits may be acquired. Of course, such persons may also sustain acquired neurological damage and submit to neuropsychological evaluations in a legal case. The resulting deficits may be a combination of pre-existing and acquired deficits that is difficult to disentangle.

A history should also be taken for past and current drug and alcohol use and abuse. Long term and extensive use of either may lead to cognitive changes.

There is evidence that even after a drug such as cocaine is no longer being used, neuropsychological dysfunction can persist (Berry et al., 1993; Strickland et al., 1993). When either drugs or alcohol are being used immediately prior to or during the course of evaluation, performance can be affected. While chronic use of drugs or alcohol can lead to cognitive deficits, concurrent use of either can exaggerate or create the impression of neurological deficits during the course of evaluation. The neuropsychologist must be aware of past and present drug and alcohol use and factor this knowledge into the interpretation of apparent deficits.

Another possible cause of interference with cognitive performance comes from prescription drugs the patient may be taking at the time of the evaluation. Interpreting the effects of drugs on test performance is complex. On the one hand, many medications used at therapeutic levels (e.g., antidepressants and mood stabilizers) serve to improve mental functioning and thus remove artificial cognitive depressors. Yet, side effects of being "in a fog" or "spacey" are not uncommon with such medications, and may represent a mild counteractive effect that interferes with concentration. Other drugs (e.g., anxiolytics, especially the benzodiazepines) are known to have negative effects on attention and memory; yet, the anxiety they relieve may have even more debilitating effects on cognitive performance. Still other classes of drugs (e.g., seizure medications) may prevent disruptive neurological events, but have varying effects on cognition (depending on the drug and the patient response, which can be quite variable). The 'bottom line' in the forensic setting is that it is incumbent upon the examiner to be aware of which drugs the patient is using, the potential impact of the drugs on cognition, and the impact of the medication on test performance. Regardless of the conclusion, it is best to discuss it in the report.

A potential problem that is common and obvious, yet often overlooked, is patient fatigue. This can take the form of the patient coming into the testing situation tired, or becoming tired over the course of the testing. A fatigued patient will perform suboptimally, which invalidates the ability to document the nature and extent of deficits. (Of course the tendency toward fatigue may itself be a neurobehavioral deficit that has profound functional effects and should be documented.) With patients subject to fatigue over time, numerous breaks should be given during the session and the patient should be monitored for deteriorating performance. Multiple short sessions may be needed to complete valid testing.

A patient arriving fatigued for a testing session presents a separate problem. During scheduling, inquiry should be made as to the patient's usual time of waking and travel time, to assure that testing is not beginning at a time that makes fatigue more likely. Inquiring at the beginning of the session as to the patient's sleep the night before and current level of fatigue should be routine. If sleep disruption is a chronic problem, this should be noted in the report and taken into account when interpreting apparent deficits.

As in all neuropsychological evaluations, forensic examiners should take special note of any other medical problems that might affect cognitive and sensory-motor testing. Peripheral hearing loss or visual impairment may interfere with comprehension or visual perception of problems, respectively. Cervical involvement or impingement of peripheral nerves in the neck, shoulder and arm area may affect manual motor performance. Any medical problem causing the patient distress during testing may interfere with attention and concentration. Especially after trauma to the head, peripheral vestibular damage may cause interference in visual tracking of information because of the interface between visual and vestibular systems, and symptoms after even peripheral vestibular disorders (e.g., perilymph fistulas) may be remarkably similar to the symptoms of post-concussion syndrome (Grimm et al., 1991).

Finally, it occasionally happens that an emotionally disruptive event unrelated to the legal case will occur that potentially influences test performance. An example would be the recent death of a significant other. The associated possible depression must be evaluated and considered in the interpretation of apparent deficits.

Pain

Pain is a common component of many litigated injuries and can directly or indirectly influence cognitive test performance. Many traumatically injured patients will enter the testing situation with pain, or develop it over the course of testing. Headache is particularly common. Patients with chronic pain can show neuropsychological deficits even in the clear absence of central nervous system involvement.

Taylor, Cox and Mailis (1996) gave two measures of complex attention, the Paced Auditory Serial Addition Test (PASAT) and the Consonant Trigram Test (CTT) to three groups of subjects: (1) whiplash patients with chronic symptoms who were stunned or dazed in an accident, but had no loss of consciousness; (2) patients with moderate to severe traumatic brain injury (TBI), with extended loss of consciousness, but no focal neurological damage; and (3) patients with chronic pain syndrome (CPS) without central nervous system trauma. All three groups were matched on age, education, and IQ, and were given self-report measures of pain and depression in addition to the two cognitive tests.

The whiplash and CPS groups did not differ on pain or depression; the TBI group reported less pain and depression than the other two groups. There were *no* significant differences among any groups on either of the cognitive tests. Pain duration was negatively related to PASAT performance within the CPS group, and to CTT performance within the whiplash and TBI groups. Depression was unrelated to either cognitive test.

It is striking that the chronic pain group (with no history of central nervous system involvement) was as cognitively impaired on tests of complex attention as the group with moderate to severe brain injury. Clearly, chronic pain can be associated with apparent neuropsychological deficit. The fact that duration of pain was correlated with impairment is of interest, and raises the question of what it is about pain or pain behavior that impacts on cognitive test performance.

To address this issue, Kewman et al. (1991) evaluated 73 musculoskeletal pain outpatients on self-report measures. They excluded patients with previously diagnosed cognitive dysfunction, degenerative disease, or any patients who had taken analgesic medications that day. The McGill Pain Questionnaire, a mental status exam, and four self-ratings of psychological distress were administered. One third of the patients reached criteria for impairment in at least one domain of the mental status examination, the most common being memory, verbal similarities, and calculations. Cognitive impairment was statistically associated with self reports of pain, education, and psychological distress. When the data were statistically controlled for education, cognitive impairment was still correlated with self reports of pain. However, when the data were controlled for psychological distress, cognitive impairment was no longer correlated with pain. The authors concluded that psychological distress appeared to be mediating cognitive impairment in the presence of pain.

These results fit with clinical experience. Pain may interfere with cognitive test performance in two ways. First, acute pain during the testing situation may interfere with performance, if it causes distraction. Whether it does so may depend on the source of the pain and how successful the patient is in dissociating from it. Second, chronic pain may induce a state of psychological distress that itself interferes with cognitive performance.

In the forensic setting, the neuropsychologist must be on the alert for both conditions. The patient's level of pain during testing must be monitored, and testing discontinued if it appears that the distraction from the pain is interfering with testing. If the pain is intractable and testing must go on despite it, the effects of distraction due to pain must be considered when interpreting apparent deficits. When there is a chronic pain situation, anxiety and dysphoric mood must be taken into account when interpreting test performance. The forensic neuropsychological evaluation is not complete in any patient with acute or chronic pain unless the possible effects of pain on test performance are discussed.

Anxiety

Sufficient anxiety in any individual will interfere with the processing of information. Individuals with acquired cognitive deficits may develop a conditioned anxiety response to situations in which they have encountered cognitive breakdown. Weakened cognitive processes are even more susceptible to interference from anxiety. This situation often carries over into the forensic neuropsychological evaluation. During testing, many patients become anxious when they begin to fail, or when they encounter tests on which they expect to do poorly.

It is critical that anxiety be carefully monitored, and its effect on testing noted, reported, and taken into consideration in reporting any suspected deficits. Interpretation of deficits will be made easier to the extent anxiety during testing can be reduced. First, ambient anxiety – the general level of anxiety the patient is carrying into the testing situation – must be noted and minimized. Asking the patient what they have been told about the testing situation, and how they are feeling about it, usually opens the door to talking about and defusing anxiety. Explaining what the testing situation entails, and assuring the patient that the long process will include plenty of breaks, also helps. Defense neuropsychologists have the added challenge of establishing positive rapport with a patient who may be suspicious, distrustful, and anxious about being misrepresented. It is important for examiners for the defense to explain to the patient that their job is simply to portray a picture as accurate as possible of the person's neuropsychological status for the attorney they represent, and all they ask is that the person be honest with them and try their best. The simplest and most effective way of reducing ambient anxiety regardless of side is to treat the person professionally and kindly, listen carefully, and make them feel understood – regardless of one's ultimate formulation of their problems.

The second, and more difficult, aspect is anxiety evoked during testing that threatens to interfere with test performance. An important discrimination that needs to be made is whether anxiety is causing poor test performance, or whether poor performance is causing anxiety. Typically, this can be accomplished by careful observation and inquiry about non-verbal cues. For example, when the Paced Auditory Serial Addition Test is described, many patients appear to tense up. Even more have an anxiety response when they start to fall behind in the first series. After noting the tendency toward anxiety interfering with cognition, it is imperative that the examiner stop the testing, do what is possible to reduce the anxiety, and then resume the testing to observe cognitive capacity under the condition of minimal anxiety. Techniques that are helpful include allowing the patient to express the anxiety, reassurance, and practice (if it does not violate the standardization of test administration).

In some situations, it may be apparent that anxiety does not become operative until the patient experiences cognitive breakdown. In other situations, the anxiety will precede the performance and interfere with it. In the latter situation, anxiety reduction, if possible, will either significantly improve performance, or it will become clear that even with anxiety under control the patient is unable to do the task adequately. Whatever scenario unfolds, active monitoring and intervention of the examiner is required. When it is judged that anxiety was interfering with test performance, this must be considered when interpreting apparent deficits, and taken into account when formulating an opinion on the nature and severity of neurologically based deficits.

Depression

Perhaps no single confounding variable is more clinically relevant to the interpretation of apparent neuropsychological variables than depression. Persons who suffer damage to the brain are at risk for neurobiologically-based depression, and reactive depressions are common after genuine physical trauma, with or without brain damage. Furthermore, as noted above, the forensic neuropsychological evaluation may take place at a time when patient's healing process is interrupted by a series of evaluations and depositions, and a reactive period of depression is not uncommon during this process.

Research findings
It is beyond the scope of this paper to review the extensive literature on depression and neuropsychological deficits. The reader is referred to a number of excellent recent reviews (e.g., Sweet et al., 1992). Despite the richness of the literature, it is far from easy to generalize about the effects of depression on neuropsychological test performance. Sweet et al. wrote:

> For now, we can conclude only that depression has real potential for negatively influencing neuropsychological functioning and, therefore, the results of neuropsychological measures used in clinical assessments. We are currently unable to predict which individuals will show impaired cognitive functioning concomitant with depression or in what specific fashion these effects will be manifested. (pp. 38–39)

With this in mind, we will review a few recent illustrative studies, and draw some inferences for the forensic setting. What complicates the situation is that the unpredictable effects of depression are superimposed upon questionable deficits arising from possible neurological damage in a situation within which large financial awards may be at stake *if* the patient is shown to be sufficiently injured.

A number of studies have found that neuropsychological impairments are most pronounced during acute depression, and improve as the depression is treated and symptoms remit. Trichard et al. (1995) compared 23 severely depressed inpatients with 15 normal controls matched for age, sex, education, and WAIS-R Vocabulary. Patients were assessed on hospital admission and discharge on the Stroop, Animal and Verbal Fluency, and a Figural Cancellation task. At admission, the patients were impaired on all three tests relative to controls; at discharge the patients were compararble to normals on the two fluency tests, but continued to be relatively impaired on the Stroop and Cancellation tests. The authors suggested that the results were consistent with the theory that acute depression suppresses frontal functions, but that periodically depressed persons suffer from a chronic attentional deficit even between depressive episodes. In studies such as these, conclusions are, of necessity, limited by the choice of tests administered.

The theme of inter-episode chronic impairment has been addressed in a number of other studies. For example, Marcos et al. (1994) compared 28 symptom free patients with a history of depression (none of whom had received ECT) to 19 controls matched on age, sex and education, with no neurological or psychiatric history. The depressed subjects performed worse than controls on all three memory tests (Wechsler Memory Scale Immediate and Delayed Visual Memory, Delayed Logical Memory, and Paired Associate Learning) and WAIS Block Design; the groups did not differ on WAIS Vocabulary, Similarities, Digit Span, or Coding, or on Trailmaking (A or B) or finger tapping. Here inter-episode deficits again appear, but now in the domain of memory, rather than attention. Again, choice of tests becomes critical to the conclusions; the Trichard et al. study used no memory tasks, and their attentional tasks were more complex than those used by Marcos et al. It may be that problems in complex attention underlie apparent memory deficits in chronically depressed persons between episodes.

An additional problem in the interpretation of the effects of depression is that severity of mood disturbance may not correlate well with neuropsychological deficits. Brown, Scott, Bench and Dolan (1994) compared 29 outpatients with major depressive disorder to 20 age and education matched controls. The depressed subjects were divided into three groups depending on their performance on an expanded mental status exam: 10 who were globally cognitively unimpaired, 10 who were globally borderline cognitively impaired, and 9 who were globally cognitively impaired. Even those depressed patients who did not show global cognitive impairment performed worse than controls on selected measures of attention, language and memory. However, the gradient of cognitive impairment across the three groups was unrelated to the severity of depression.

Recent studies have attempted to isolate specific areas of brain dysfunction that appear to underlie depression, and link these areas to neuropsychological deficits. Dolan et al. (1992) selected 10 subjects with widespread cognitive impairments and 10 with no cognitive impairments (matched demographically and for medications) from 33 outpatients with major depressive disorders, and compared their PET scans to each other and to a group of matched controls. Compared to the cognitively unimpaired depressed group, depressed patients with widespread cognitive impairment showed decreased rCBF in the left medial prefrontal gyrus (as well as increased rCBF in cerebellar vermis). When the cognitively impaired depressed group was compared to normal controls, they showed the above PET differences, but in addition decreased rCBF in left and right dorsolateral prefrontal cortex, left anterior cingulate, and right insula. They compared their findings to previous work that found left frontal and cingulate rCBF decrements in depressed subjects.

In a follow up study, Dolan, Bench, Brown, Scott and Frackowiak (1994) linked this approach more specifically to neuropsychological deficits. From their previous studies, they selected those neuropsychological tests that best discriminated controls from depressed patients, and levels of cognitive impairment within depressed patients. They subjected these variables to a principal components analysis, and derived a two factor solution that accounted for 50% of the variance. The two factors appeared most consistent with Memory and Attention. They then assigned factor scores to a group of 29 outpatients with major depressive disorder, and 16 matched controls. All received steady state PET scans. The primary finding was that both neuropsychological factors correlated with decreased rCBF in the anterior medial cortex (including the frontal pole). Posterior rCBF decrements were specific to the two factors.

Finally, related to the documentation of depression in the forensic setting, a number of studies have shown that different self-report measures of depression correlate differentially with neuropsychological deficits. Bornstein, Baker and Douglass (1991) found that among 23 non-medicated patients with major depression, poor memory performance on the Wechsler Memory Scale was associated with increased depression on the Hamilton Rating Scale of Depression, but not the MMPI Depression scale.

Such differences may be related to biological components of depression that are unequally tapped across self-report and examiner rating measures. Palmer et al. (1996) divided 36 persons meeting DSM-III-R criteria for major depressive disorder into two groups using the Hamilton Rating Scale of Depression. Twenty-two had primarily vegetative symptoms, and 14 had primarily psychological symptoms. They were compared to 40 normal controls matched on age and education. All were given a comprehensive neuropsychological battery.

The vegetative depressed group performed worse than controls on WAIS-R Full Scale and Performance IQ's, WMS-R Immediate Visual Reproductions, Rey Complex Figure Recall, Facial Recognition, and the Wisconsin Card Sort (both categories and perseverations). However, the psychologically depressed group did not differ from controls on any measures. Depression may be more likely to cause neuropsychological deficits if vegetative signs are present; the vegetative signs may be linked to frontal medial brain dysfunction.

Forensic applications

In the forensic setting, it is common for cases to come to trial a number of years following the trauma or event in question. Often there have been a number of neuropsychological evaluations over that period of time (especially if the patient has been in rehabilitation). This gives the neuropsychologist the opportunity to evaluate the effects of depression both historically and currently.

Historically, it is important to look carefully for deterioration in functional performance, increased subjective symptomatology over time, or decreasing test scores on neuropsychological assessments. All may be evidence that a depressive process was interfering with recovery and affecting performance. Clinically, the examiner must determine if the patient is still in that depressive period, or has emerged from it.

Concurrently, the neuropsychological evaluation must include an assessment of depression. This is made through a combination of clinical interviewing, behavioral observation, collateral reports, and administration of self-report measures. Convergence of self report measures may be more valuable than use of a single measure.

Given the lack of certainty in the research literature on which neuropsychological domains are affected predictably by depression, it is impossible to define strict decision making rules for deciding when depression is affecting performance. Certainly some common sense conclusions can be drawn, however. Depression can affect performance either through effects on effort, or through a neurologically based mechanism that affects underlying capacity. The latter represents a state of brain dysfunction which may be related to or separate from the cause of possible brain damage being litigated. It is certainly coincidental that frontal and limbic areas of the brain – which are particularly vulnerable to traumatic injury – are also areas which some recent literature finds to be a common substrate for depression associated with cognitive deficit. The work of

Dolan, Brown and colleagues suggests that attention and memory – again the two hallmarks of deficit after traumatic brain injury – are two consistent dimensions of deficit, and that the presence of both dimensions correlates with dysfunction at the interface between frontal cortical and limbic systems.

All of this should make the forensic neuropsychologist cautious in inferring structural brain damage in mildly injured and depressed persons who show nonspecific deficits in attention and memory. Additionally, subjective complaints of poor organization and inability to follow through on tasks, especially when they are not associated with executive deficits on testing, may be more reflective of brain-based inertia associated with medial frontal dysfunction and manifested as depression. Changes in affective expression and irritability may also be related either to fronto-limbic dysfunction due to traumatically based structural damage or a biological depression affecting similar areas.

Convergence of information about the nature and severity of the brain injury, the historical evolution of subjective complaints and dysfunction, and the current assessment, should be used in combination to differentiate as much as possible the effects of depression from brain injury. This is not always possible. At times, a suppressing effect of depression may be superimposed on a more specific pattern of deficits reflective of brain injury. The situation is most difficult to sort out in cases of suspected mild diffuse brain injury. Often the profile of deficits does not stand out in sharp relief until the depression is resolved, and often this does not happen until litigation is concluded. Finally, Sweet et al. (1992) offer a number of specific suggestions for test administration and interpretation in the presence of depression.

Post-Traumatic Stress Disorder

The concept
The underlying concept of Post-traumatic Stress Disorder (PTSD) is that a catastrophic event can be so overwhelming to the human psyche that it cannot be integrated into the sense of one's self and conscious experience, and therefore elements of the experience are split off, or dissociated, only to reappear periodically as intrusive thoughts, flashbacks, or traumatic memories, alternating with periods of emotional withdrawal and avoidance.

PTSD is relevant to the forensic neuropsychologist in three ways. First, PTSD may be superimposed on a legitimate TBI and result in deficits above and beyond the brain injury. Second, PTSD may masquerade as TBI, confusing the diagnosis. Third, trauma may reactivate unresolved emotional trauma from the past, again creating a symptom picture that mimics, but is not, TBI.

PTSD and neuropsychological deficits
Persons in acute stages of PTSD show cognitive deficits. Difficulty concentrating is one of the diagnostic criteria. Persons in acute PTSD are hyperaroused, emotionally hyperreactive, and in significant psychological distress, with corre-

sponding interference on their ability to attend, concentrate, remember, and organize themselves. Conversely, the numbing/avoidant aspect of PTSD is often accompanied by significant depression and somatization.

There is also research evidence that persons with chronic PTSD show persistent complaints of poor concentration (McFarlane, 1988; Sutker, Winstead, Galina, & Allain, 1991), impairment in selected intellectual functions and memory (Sutker, Galina, West, & Allain, 1990; Sutker et al., 1991; Sutker, Allain, & Johnson, 1993), or memory impairments alone (Bremner et al., 1993). In a study suggesting that chronic PTSD may cause brain changes associated with cognitive decline, Bremner et al. (1995) compared Vietnam veterans with PTSD to matched controls without PTSD. The PTSD group performed worse on memory tests and had decreased hippocampal (but not caudate or temporal lobe) volume. Within the PTSD group, poorer memory was correlated with decreased volume of the hippocampus, a brain structure implicated in new learning and memory. These findings were thought to be related to stress induced chronic gluticorticoid exposure increasing hippocampal neuron vulnerability to multiple factors, such as endogenously released excitatory amino acids, and ultimately leading to hippocampal degeneration and memory deficits.

PTSD and forensic neuropsychology
PTSD with TBI. Many patients who suffer possible brain injuries in trauma show elements of post-traumatic stress, but the incidence of diagnosable PTSD after head trauma is relatively rare (Sbordone & Liter, 1995). Nevertheless, it does occur, and the forensic neuropsychologist should be alert for the co-existence. In the author's experience, TBI and PTSD are most likely to occur together when the brain injury is suffered in a personal attack with no retrograde amnesia to protect the psyche, or when death of another person occurs in the accident.

Clinical examples include the following two situations. An executive was cornered in a rest room and beaten with a pipe until he was unconscious; he had full recall of the beating until he lost consciousness, and suffered both a serious brain injury and a full blown post-traumatic stress disorder.

A sixteen year old high school dropout with a disruptive family history suffered a minor head injury when the car in which she was a passenger crashed, killing her friend who was driving. She developed a severe post-traumatic stress disorder revolving around guilt over his death. She subsequently developed seizure-like episodes involving bad smells, fear, and shaking. MRI's and standard EEG's were normal, and medication did not control the events. Although suspected of either malingering or having psychogenic seizures, subsequent inpatient monitoring off medication captured numerous events with corresponding right temporal lobe spikes on EEG. The checkered history of the patient and the accompanying PTSD initially served as distractors to a neurologically-based seizure disorder. The presence of PTSD after TBI should not in and of itself prejudice the neuropsychologist away from a careful consideration of real brain injury. Further, its presence adds a level of injury above and beyond the neurological.

PTSD masquerading as TBI. The forensic neuropsychologist should be on the alert for cases in which a minor head injury has occurred, but in which PTSD is the real cause of dysfunction. A 21 year old music student suffered a concussion with about 20 minutes loss of consciousness when the jet in which he was travelling from Columbia to the United States crashed. He experienced an 'out of body' experience on the triage field, and hospital records indicated depression, sleep disturbance, and revulsion at his self image (he had multiple facial fractures), but no behaviors specific to brain injury. He lost the possibility of a professional music career because of an injury to his hand. He underwent a personality change from being gregarious to irritable and isolated; he lost interest in the pleasures and challenges of life. He was unable to maintain friendships or sexual relationships; he was let go from a number of jobs. Self-cutting became common when he felt desperate. He had nightmares about the accident. He felt he would die young. He also believed his body was grotesque and misaligned, and that he would need surgeries to control a deteriorating spinal condition. Neuropsychological testing by a plaintiff-retained neuropsychologist revealed severe cognitive deficits that she attributed to brain injury, and recommended intensive cognitive remediation. Defense examination included a fake-bad MMPI profile and a diagnosis of malingering. Retesting by the author after induction procedures to insure maximum effort revealed normal neuropsychological functioning, but suggested the diagnosis of PTSD with a histrionic personality style. A subsequent *defense* expert confirmed this diagnosis.

The lessons of this case for forensic neuropsychologists are that: (1) PTSD can occur coincidentally with recovered head injury. (2) Neuropsychological deficits need to be critically evaluated in terms of the record and the nature of the injury. In this case, despite the concussion, there was no medical evidence of sustained brain damage, and the severity of deficits was out of proportion to the neurological injury. (3) Alternate diagnoses need to be considered when interpreting neuropsychological deficits. In this case, the diagnosis of PTSD was overlooked. (4) Exaggerated responses on the part of the client do not eliminate the possibility of a legitimate psychological injury. In this case, probably due to a core hysterical style and cultural factors, the client communicated his distress through responses that exaggerated deficit. In fact, no significant cognitive deficit existed. Yet, there was a significant psychological injury (PTSD) that occurred as a result of the trauma.

Reactivation of old trauma. To this point we have considered the DSM-IV diagnosable PTSD that may occur at the time of injury, with or without accompanying brain damage. More elusive, and as yet undescribed in the literature, is the phenomenon of a lesser emotional trauma in the present reactivating old, unresolved emotional trauma, the sequelae of which mimic traumatic brain injury. The original emotional trauma may or may not have met DSM-IV criteria for PTSD at that time. In the author's experience, a significant number of instances of misdiagnosed TBIs fall into this category.

A person who suffered overwhelming emotional, physical, or sexual abuse as a child, may have split off certain aspects of the experience from conscious recollection and sense of self. A trauma in adulthood, with or without brain injury, may recreate the conditions that reactivate the dissociated experiences of the original trauma. In the author's experience, these include (1) overwhelming fear, (2) physical pain, (3) perceived lack of control in the situation, and (4) perceptions of inadequate caretaking on the part of those who are supposed to protect and help. All four of these elements are present in childhood abuse, and all four are not uncommon in adult accidents. When there is no insulation via retrograde or anterograde amnesia, there is more likely to be an intense emotional reaction and therefore reactivation of old trauma.

This reactivation phenomenon usually does not reach DSM-IV criteria for PTSD, but can influence cognitive functioning, and therefore affect the interpretation of apparent neuropsychological deficits. The dissociated emotional experience of old trauma flows into the person's experience of the current trauma, resulting in a level of emotional distress and cognitive disorganization that destabilizes psychological equilibrium, and may mimic traumatic brain injury. Especially common are disruptions of attention, concentration, memory, organization, initiation and mood.

A chiropractor in her mid 50's was referred by her physiatrist with a diagnosis of mild traumatic brain injury six months after a motor vehicle accident. She complained of fatigue, poor attention and concentration, decreased memory, disorganization, and an inability to follow through with any projects. Neuropsychological evaluation revealed above average cognitive performance in all areas. Personal history included emotional deprivation, physical abuse, a rigid authoritarian household, and a past 20 year marriage to a physically abusive alcoholic. History of the accident yielded an account infused with anger and bitterness that included elements of fear, pain, being out of control, and being treated badly. The conclusion was that the woman had not suffered a traumatic brain injury, but that the nature of the trauma triggered the re-emergence of old emotions from her past abuse, with a strong element of depression. Short term treatment aimed at her psychological symptoms resolved the subjective cognitive complaints.

The author has encountered numerous similar cases in forensic evaluation and suspects that the above scenario is more common than currently appreciated. Surprisingly large numbers of patients, especially women, upon inquiry report histories of abuse that are not typically explored in neuropsychological evaluations. Forensic neuropsychologists should be alert to this possibility.

Malingering and issues of effort

Within this chapter, discussion of psychological and neuropsychological interpretation has presumed that sufficient effort was present on the part of the plaintiff when completing formal tests. Before addressing issues of malingering and effort, it should be emphasized that the active attempt to elicit optimal performance is part of the professional responsibility of the neuropsychologist. Obtaining maximum effort is of particular importance in the forensic setting, in which unique factors may incline some persons, consciously or unconsciously, toward less than optimal effort.

The issues of malingering and lack of effort in testing are dealt with in depth in chapters 9 and 10, and will be addressed here only briefly. Malingering is the conscious attempt to do poorly for some external advantage – in forensic situations, usually financial. However, rather than thinking of malingering as an all or nothing situation, it may be helpful to think of two continua: one of effort and one of consciousness. The former describes to what extent the patient is bringing sufficient effort to the testing situation, while the latter describes how aware he or she is of it.

Patients vary considerably in how much effort they bring to the testing situation. Beyond the will of the patient at the moment, effort may be compromised by fatigue, depression, or drugs. In this situation, the patient may be using all the effort they have available at the moment, but the effort available may be sapped by one of the factors mentioned.

Patients may give less than optimal effort for other reasons. They may feel a need to show that they are injured, and withhold effort to dramatize their inability. They may be resentful and hostile toward the response of others to their injury, and perform without effort in a passive-aggressive manner. They may have learned early in life that helplessness enables dependency needs to be met, and unconsciously act this out in the testing situation. They may have fallen into the sick and disabled role as a way of stabilizing a shaky identity, and play out that role – consciously or unconsciously – in the testing situation. Alternatively, they may consciously be trying to deceive the examiner.

Patients may or may not be aware of decreased efforts during test taking. True malingering is the conscious attempt to perform badly and is deliberate. Unfortunately, the term is also used to include a wider variety of behaviors that are more impulsive or less conscious. The client who is angry and resentful and carries out all activities of living in a resistant, provocative manner – from how they clean their house to the work they do to how they take tests – is clearly withholding effort and has some awareness of it, but not in the same way as the sociopathic malingerer who is scheming for gain. Patients whose insufficient test-taking effort is driven by dysfunctional psychological forces may be only vaguely aware of their motives and actions.

The continuum of consciousness is most clearly reflected in the range of formal DSM-IV diagnoses available and most relevant for the neuropsychologist in forensic settings. Malingering and factitious disorders are conscious distortions of performance on the part of the patient, for purposes (respectively) of external or psychological gain. Somatoform disorders, on the other hand, including the somatization disorders, are unconscious manifestations of psychological conflicts or needs through somatic symptoms. Conversion disorders are specific sensory or motor symptoms that have no neurological basis and are also unconsciously driven.

Developing a formulation

Approaching the evaluation

The first step toward developing an objective formulation takes place during the first phone conversation with the retaining attorney. Regardless of side, establishing the proper relationship with the lawyer is critical to being able to interpret apparent deficits objectively. The discussion that follows applies only to requests leading to evaluation for expert testimony, not to a lawyer's referral to become clinically involved as a treating doctor, with future testimony possible in that capacity. In the latter situation, the case should be considered on its clinical merits.

It is important to establish early on your intent to evaluate the case as you see it, and that your role is to give the attorney neuropsychological information that is as accurate as possible. Rather than agreeing over the phone to conduct an evaluation, a series of steps of increasing involvement is the best means of avoiding the situation of feeling pressured to interpret findings in a certain direction.

The first step is an extended phone conversation in which the facts of the case are discussed. An important part of this process is to determine if the attorney has made up his or her mind about the reality of a possible brain injury. Plaintiff attorneys who have decided prematurely that this is a brain injury case with significant damages and are looking for a neuropsychologist to document that fact, or defense attorneys who have decided that no brain injury exists and are searching for a neuropsychologist willing to take that stance, are exerting pressure on the expert. Such positioning by the attorney may cause problems for the expert witness faced with a complex case in which neurological damage is uncertain.

Once the case has been discussed over the phone, it is next wise to request all the medical records before agreeing to take on the case. Do not ask for only psychological reports; the entire situation should be surveyed, if possible. After reviewing the records carefully, the neuropsychologist should discuss the case with the attorney, offering initial impressions, questions to be answered, and a set of hypotheses that the evaluation would test.

At this point the information provided by the neuropsychologist and the issues raised may help the lawyer decide the action he or she will take. Formal evaluation may be unnecessary. For example, a plaintiff lawyer may decide there is no basis for a traumatic brain injury and pursue the case with regard to other injuries. Or a defense lawyer may hear that there is convincing evidence for brain injury from the record as it exists, and take a different tack in the case. When these situations occur, the neuropsychologist has brought clarity to the case and increased the chances of settlement, despite the fact that no evaluation was performed.

Once an evaluation is completed, some lawyers will request a phone discussion prior to the report, others will simply request the report. In either case, the gradual decision to become involved in a case, once the facts are known and guidelines established, will go a long way toward freeing the neuropsychologist to interpret the presence or absence of deficits ethically and honestly, and develop an objective formulation free from pressure.

Framing the evaluation

It is extremely important that the neuropsychologist not enter the evaluation process with his or her mind made up about the nature and severity of the neuropsychological injury, especially when the etiology of the dysfunction is uncertain. Some cases may be clear examples of the presence or absence of brain injury, but these seldom pose problems of interpretation for the neuropsychologist. Ideally, the neuropsychologist enters the evaluation as part detective and part scientist, searching for clues and evidence, and keeping an open mind as he or she considers various possibilities. Especially when the etiology of dysfunction is unclear, it is helpful to utilize a hypothesis-testing approach. As data become available (from the record, from observation or interview, or from testing), the clinician can formulate diagnostic questions, and steer the evaluation to answer them. This may include returning to clinical interviewing part way through the testing, or changing the selection of tests. Flexibility of approach can be critical to being in a position to interpret apparent deficits and develop a convincing formulation of the case, and does not need to detract from comprehensiveness.

When writing the report, it should be written as a narrative of intellectual inquiry, not of advocacy or dismissal. The conclusions should emerge by power of logical persuasion based on unfolding evidence, and should carry the reader along with their logic. Contradictory or confusing data (whether in the medical records, the behavior of the client, or results of testing) should be discussed as such, and not ignored or dismissed as irrelevant. Doing so will enhance the credibility of the report, for whichever side it is written. What does not work well in ambiguous cases is taking a stand on the presence or absence of brain injury early on, and then advancing evidence for that position throughout the report.

When the report is written as an unfolding narrative, the results section – where the test data are presented – can be written much as the results section of a research paper. The data can be presented descriptively with little discussion or inference. The editorial comment should be on the implications for functioning in that domain, not about etiology or the premature formulation of an opinion. That process should come only after all the records have been reviewed, the content of clinical interviews presented, behavioral observations made, and the results of all tests described.

After summarizing the test results, the neuropsychologist is ready to offer an integration of all the relevant data in a consistent manner. This formulation section becomes like the discussion section of a research paper, in which the implications of the summarized data are discussed, and alternate ways of understanding the data are considered (differential diagnoses and dynamic formulations explored). It should end with the formulation that best fits the data and makes sense in terms of that real person and their life. The reader should experience the formulation as an inevitable conclusion based on all the available data, told as a story that makes sense. Then, in a final conclusion section, a terse summary of the opinion within reasonable medical certainty can be stated. In both the formulation and the concluding statement, the issue of brain damage is raised and answered, but the formulation and opinion should provide a broader and richer context of interpretation. If appropriate to the case, the report should conclude with a statement of implications for future function and recommendations for and prognosis of treatment.

Some lawyers may request limitations on issues addressed in this final section (e.g., the need for treatment or prognosis on return to work). Experts should not feel constrained as to what issues they address to answer the referral question. It is wise at the time of initial referral to clarify, with the attorney, the specific nature and scope of the referral question(s) and the issues to be addressed in the report.

Guidelines for interpreting apparent deficits

It is impossible to write simple guidelines for the interpretation of apparent neuropsychological deficits. The reality is that all the factors presented above need to be taken into account when interpreting test scores. What is provided below is the description of a process for the interpretation of findings, phrased as a series of questions.

Do apparent deficits exist?

This is the first question to ask when looking at the test scores. The question can be broken down into three subquestions. Are there any *absolute* deficits – that is, test performances that are so clearly impaired that they would be considered deficits for any functioning person? What *relative* deficits exist – performances that may not fall in the absolute deficit range, but are quite certain to be relative impairments for that particular person given what is known about academic and work history, and can be inferred about premorbid intelligence? What possible or *questionable* deficits exist, which may or may not be relative deficits? Note that at this point in the analysis, the neuropsychologist is not yet worried about explaining, interpreting, or even speculating on the etiology of the apparent deficits. Test scores reflective of impairment are simply being noted and categorized.

Do apparent deficits form a consistent pattern?

Deficits should never be inferred from a single test performance, especially in the forensic setting. As noted above, there can be a great deal of "noise" or individual variability even within non-injured, functional individuals. It is a mistake to interpret every test score in the deficit range as representing a neuropsychological deficit. Consistent patterns of deficits across a number of tests are much better indicators of potentially real neuropsychological deficits.

Is the pattern of apparent deficits consistent with the injury?

This is a two part assessment. If there is a consistent pattern of deficits, do they fit the (1) nature and (2) severity of the injury? Implicit in this pair of questions is the issue of whether the pattern of deficits matches an established neurological condition. Fronto-temporal traumatic lesions, mild concussions, and a bullet wound to the right posterior hemisphere all have quite different neurological, and therefore neuropsychological consequences. The neuropsychologist must look carefully at what is known about the nature of the injury, and ask whether it matches the apparent deficits. The other variable is severity of the injury. Concussions can cause problems with attention, concentration, and memory, but if a person with a transient loss of consciousness has severe impairments in all these areas and has become totally dysfunctional years later, there is a mismatch between injury and neuropsychological deficit severity that needs to be explained.

Does the pattern of apparent deficits match the subjective complaints?

The neuropsychologist must also consider the consistency between the patient's subject cognitive complaints and observed test results, and be prepared to explain any discrepancies. In the most straightforward cases, the patient has specific and circumscribed complaints that are specifically confirmed on neuropsychological testing, and these findings fit what is known about the injury or causative event. This rarely occurs, however, in contested forensic cases. Often there is a match between severe and widespread subjective complaints and severe and widespread apparent neuropsychological deficits. To this extent there is a test-to-complaint match, but the nature and severity of the apparent deficits may not match the nature and severity of the injury.

Not uncommonly there is a mismatch between complaints and test results. On occasion, denial of subjective complaints will conflict with patterns of apparent deficit on testing. This usually happens in a forensic setting when a person is attempting to protect a certain right or privilege and is denying cognitive deterioration (e.g., a doctor in the early stages of dementia who is in danger of losing his license because of patient complaints). It can also happen in cases of more severe traumatic brain injury in which brain-based lack of awareness prevents the patient from being able to recognize how impaired he or she has become. With neurologically milder injuries, certain personality styles (e.g., those involving grandiose and narcissistic tendencies) may interfere with acknowledging cognitive inability. Some patients are simply too frightened by the possibility of cognitive decline to admit they are having problems.

Also common is a mismatch in the other direction: persons with cognitive complaints who show intact, or relatively intact, neuropsychological test performance. There are at least three possible causes of this phenomenon. First, there is a possibility that the tests administered have failed to tap a true deficit, or more accurately, that it is impossible to confidently interpret a deficit because departures from high premorbid levels leave performance too high to confidently interpret as a relative deficit. This may be often the case in extremely intelligent individuals who use highly refined and efficient skills to do very high level work, in which they are juggling a large number of options or pieces of information simultaneously. In these persons, even a relatively subtle change in the speed and efficiency of information processing can make the difference between being able to do their job or not. What they report as memory problems, poor concentration, or inability to think clearly may be true at the level at which they used to perform, but may be very difficult to capture on neuropsychological tests.

A second possible cause of test performance being better than subjective reports is that a person may have a subtle brain injury which allows them to perform adequately in a single focus, structured situation, but prevents them from functioning consistently across time in more complex and challenging real world situations. That is, the person may really be forgetful and not carry out responsibilities in overwhelming situations at home and work, but show good memory and organization on formal testing under protected conditions.

The third possible reason subjective complaints are not reflected in neuropsychological testing is that rather than subtle brain injury, emotional factors are operating that cause the person to become overwhelmed in complex and demanding situations, but not in the projected testing situation. This can happen as a result of anxiety, depression, or the resurrection of old, unresolved emotional trauma. A case in point is the chiropractor with a history of abuse described above; despite her real subjective complaints of poor concentration, memory, and organization, and despite the fact that she was struggling to function in her environment, her neuropsychological testing appeared completely normal.

What can explain the mismatches?

As has already been suggested, the neuropsychologist must now explain any mismatches that occur *within patterns of test results* (malingering? anxiety? inconsistent attention? ebbs and flow of effort? medication taking effect?), *between patterns of test results and nature or severity of injury* (depression? other medical conditions? pain? regression into the sick role?), and *between test results and subjective complaints* (insensitivity of testing? PTSD?). The parenthetical examples are meant to be illustrative, rather than comprehensive. All of the possibly confounding variables discussed above must be taken into account when attempting to explain the mismatches. Out of the process of explaining mismatches emerges the objective formulation of the case.

What are the etiologies of the deficits?

Having addressed and hopefully resolved the issues of the mismatches, the neuropsychologist should now be in a position to succinctly state his or her opinion about the etiology of the apparent deficits. For the reader, this should emerge as a logical conclusion that emerges from the narrative, rather than an *ex cathedra* statement that is based simply on perceived expertise and one's opinion.

The etiological explanation is not necessarily simple and straightforward – although this is ideal. In complex and ambiguous cases, it often involves balancing a number of causal factors that may be interacting, including possible brain injury, personality style, affective response, and the effect of other medical or environmental conditions. Information about prognosis or possible treatment may flow from the etiology.

What dynamic explanation best explains the level of functioning?
Finally, even after the apparent neuropsychological deficits have been interpreted, it is incumbent upon the neuropsychologist in the forensic setting, regardless of side, to provide a convincing narrative that explains the functional level of the patient – whether that is malingering, simple brain damage, or a more complex story taking into account personality, family history, destabilized coping mechanisms, or the reawakening of old trauma. To the extent that forensic neuropsychologists take this task seriously, neuropsychological evaluations will move the system toward resolution of the dispute, rather than providing additional ammunition for irreconcilable positions.

Differential diagnoses and alternate formulations
Part of presenting a convincing argument for any diagnosis or formulation, regardless of side, requires a serious consideration of alternate diagnoses and other ways of conceptualizing the case. From a purely strategic point of view, it is also better to anticipate likely objections and alternative formulations, and deal with them thoughtfully in a report, rather than reacting defensively to them on the stand. If the neuropsychologist's formulation is, in fact, the one that follows best from the data, it will not be problematic to discuss alternate diagnoses or formulations, and emerge with one's opinion convincingly intact. To avoid doing this not only leaves the neuropsychologist vulnerable to arguments he or she has not considered, but suggests the possibility that it was easier to avoid conflicting data or interpretations than consider and incorporate or explain them.

A parting comment on ethics and competence
A recent issue of *The Clinical Neuropsychologist* (August 1997) contains a series of opinions and responses entitled "Ethical Considerations in Forensic Neuropsychological Consultation". A comment by Adams (1997) addresses the unethical nature of neuropsychologists who, without seeing the client or testifying, and therefore without having their opinions or use of information come under scrutiny, provide the defense team with selective ammunition to inappropriately discredit the plaintiff or the opposing neuropsychologist. He then addresses, however, what he considers a more basic ethical issue: "Sniper critics . . . are bad enough, but 'experts' completely unperturbed by, or even aware of, the poverty of their knowledge in neuropsychology are a far bigger problem in the forensic arena." (p. 295).

Justice to the plaintiff, the forensic process, and the reputation of neuropsychology are not served by well meaning, but naive, neuropsychologists appearing in court. Unfortunately, as Adams points out, board certification (or its absence) under the current system is no guarantee of competence (or incompetence). And lawyers are often in no position to judge the competence of their "experts".

This chapter has addressed certain aspects of interpreting apparent neuropsychological deficits. What is implicit is that doing so requires intelligence, experience, and the blended knowledge from two separable (some would argue) arenas: brain functioning and clinical psychology. In the courtroom we are asked to do something more complex and elusive than DNA analysis: interpret and explain multidetermined human behavior, with the future of a human being at stake. This should be a daunting and humbling task. It is our ethical responsibility to be aware of what we need to know to take on this task, to take this task deadly seriously, and to stay out of situations in which we do not belong.

References

Adams, K.M. (1997). Comment on ethical considerations in forensic neuropsychological consultation. *The Clinical Neuropsychologist, 11*, 294–295.

Alves, W., Macciocchi, S.N., & Barth, J.T. (1993). Postconcussive symptoms after uncomplicated mild head injury. *Journal of Head Trauma Rehabilitation, 8*(3), 48–59.

Berry, J., van Gorp, W.G., Herzberg, D.S., Hinkin, C., Boone, K., Steinman, L., & Wilkins. J.N. (1993). Neuropsychological deficits in abstinent cocaine abusers: Preliminary findings after two weeks of abstinence. *Drug and Alcohol Dependence, 32*, 231–237.

Binder, L.M., Rohling, M.L., & Larrabee, G.J. (1997). A review of mild head trauma Part I: Meta-analytic review of neuropsychological studies. *Journal of Clinical and Experimental Neuropsychology, 19*, 421–431.

Binder, L.M. (1997). A review of mild head trauma. Part II: Clinical implications. *Journal of Clinical and Experimental Neuropsychology, 19*, 432–457.

Bornstein, R.A., Baker, G.B., & Douglass, A.B. (1991). Depression and memory in major depressive disorders. *Journal of Neuropsychiatry and Clinical Neurosciences, 3*(1), 78–80.

Bremner, J.D., Scott, T.M., Delaney, R.C., Southwick, S.M., Mason, J.W., Johnson, D.R., Innis, R.B., McCarthy, G., & Charney, D.S. (1993). Deficits in short-term memory in posttraumatic stress disorder. *American Journal of Psychiatry, 150*, 1015–1019.

Bremner, J.D., Randall, P., Scott, T.M., Bronen, R.A., Seibyl, J.P., Southwick, S.M., Delaney, R.C., McCarthy, G., Charney, D.S., & Innis, R.B. (1995). MRI-based measurement of hippocampal volume in patients with combat-related posttraumatic stress disorder. *American Journal of Psychiatry, 152*, 973–981.

Brown, R.G., Scott, L.C., Bench, C.J., & Dolan, R.J. (1994). Cognitive function in depression: Its relationship to the presence and severity of intellectual decline. *Psychological Medicine, 24*, 829–847.

Cox, C.S. (1997). Neuropsychological abnormalities in obsessive-compulsive disorder and their assessments. *International Review of Psychiatry, 9*, 45–60.

Dolan, R.J., Bench, C.J., Friston, K.J., Brown, R., Scott, I, & Frackowiak, R.S. (1992). Regional cerebral blood flow abnormalities in depressed patients with cognitive impairments. *Journal of Neurology, Neurosurgery and Psychiatry, 55*, 768–773.

Dolan, R.J., Bench, C.J., Brown, R.G., Scott, L.C., & Frackowiak, R.S. (1994). Neuropsychological dysfunction in depression: The relationship to regional cerebral blood flow. *Psychological Medicine, 24* 849–857.

Ewing, R., McCarthy, D., Gronwall, D., & Wrightson, P. (1980). Persisting effects of minor head injury observable during hypoxic stress. *Journal of Clinical Neuropsychology, 2*, 147–155.

Grimm, R.J., Hemenway, W.G., Lebray, P.R., & Black, F.O. (1989). The perilymph fistula syndrome defined in mild head trauma. *Acta Oto-Laryngologica, 464*, (Suppl.), 1–40.

Gronwall, D., & Wrightson, P. (1975). Cumulative effect of concussion. *The Lancet, 2,* 995–997.

Guilmette, T., & Hagan, L. (1997). Ethical considerations in forensic neuropsychological consideration. *The Clinical Neuropsychologist, 11,* 287–290.

Heilman, K.M., & Valenstein, E. (1993). *Clinical neuropsychology* (3rd ed.). New York: Oxford University Press.

Jacobson, R.R. (1995). The post-concussional syndrome: Physiogenesis, psychogenesis and malingering. An integrative model. *Journal of Psychosomatic Research, 39,* 675–693.

Judd, P.H., & Ruff, R.M. (1993). Neuropsychological dysfunction in borderline personality disorder. *Journal of Personality Disorders, 7,* 275–284.

Kay, T. (1992). Neuropsychological diagnosis: Disentangling the multiple determinants of functional disability after mild traumatic brain injury. In L.J. Horn & N.D. Zasler (Eds.), *Physical medicine and rehabilitation: State of the art reviews: Vol.4, No.1. Rehabilitation of post-concussive disorders* (pp. 109–127). Philadelphia: Hanley & Belfus.

Kay, T, Newman, B., Cavallo, M., Ezrachi, O., & Resnick, M. (1992). Toward a neuropsychological model of functional disability after mild traumatic brain injury. *Neuropsychology, 6,* 371–384.

Kay, T. (1993). Neuropsychological treatment of mild traumatic brain injury. *Journal of Head Trauma Rehabilitation, 8*(3), 74–85.

Kewman, D.G., Vaishampayan, N., Zald, D., & Han, B. (1991). Cognitive impairment in musculoskeletal pain patients. *International Journal of Psychiatry in Medicine, 21,* 253–262.

Levin, H., Mattis, S., Ruff, R.M., Eisenberg, H.M., Marshall, L.F., Tabaddor, K., High, W.M., & Frankowski, R.F. (1987). Neurobehavioral outcome following minor head injury: A three center study. *Journal of Neurosurgery, 66,* 234–243.

Lezak, M.D. (1995). *Neuropsychological assessment* (3rd ed.). New York: Oxford University Press.

Machulda, M., Bergquist, T., Ito, V., Chew, S. (1998). Relationship between stess, coping, and postconcussion symptoms in a healthy adult population. *Archives of Clinical Neuropsychology, 13,* 415–424.

Marcos, T., Salamero, M., Gutierrez, F., Catalan, R., Gasto, C., & Lazaro, L. (1994). Cognitive dysfunctions in recovered melancholic patients. *Journal of Affective Disorders, 32,* 133–137.

McFarlane, A.C. (1988). The phenomenology of posttraumatic stress disorders following a natural disaster. *The Journal of Nervous and Mental Disease, 176*(1), 22–29.

Nemev, A.J. (1996). Behavior-descriptive data on cognitive, personality, and somatic residua after relatively mild brain trauma: Studying the syndrome as a whole. *Archives of Clinical Neuropsychology, 11,* 677–695.

Palmer, B.W., Boone, K.B., Wohl, M.A., Berman, N., & Miller, B.L. (1996). Neuropsychological deficits among older depressed patients with predominantly psychological or vegetative symptoms. *Journal of Affective Disorders, 41,* 17–24.

Risey, J., & Briner, W. (1990). Dyscalculia in patients with vertigo. *Journal of Vestibular Research, 1,* 31–37.

Sbordone, R.J., & Liter, J.C. (1995). Mild traumatic brain injury does not produce posttraumatic stress disorder. *Brain Injury, 9,* 405–412.

Strickland, T.L., Mena, I., Villanueva-Meyer, J., Miller, B.L., Cummings, J., Mehringer, C.M., Satz, P., & Myers, H. (1993). Cerebral perfusion and neuropsychological consequences of chronic cocaine use. *Journal of Neuropsychiatry and Clinical Neurosciences, 5,* 419–427.

Sutker, P.B., Galina, Z.H., West, J.A., & Allain, A.N. (1990). Trauma-induced weight loss and cognitive deficits among former prisoners of war. *Journal of Consulting and Clinical Psychology, 58,* 323–328.

Sutker, P.B., Winstead, D.K., Galina, Z.H., & Allain, A.N. (1991). Cognitive deficits and psychopathology among former prisoners of war and combat veterans of the korean conflict. *American Journal of Psychiatry, 148,* 67–72.

Sutker, P.B., Allain, A.N., & Johnson, J.L. (1993). Clinical assessment of long-term cognitive and emotional sequelae to World War II prisoner-of-war confinement: Comparison of pilot twins. *Psychological Assessment, 5*(1), 3–10.

Sweet, J.J., Newman, P., Bell, B. (1992). Significance of depression in clinical neuropsychological assessment. *Clinical Psychology Review, 12,* 21–45.

Swirsky-Sachetti, T., Gorton, G., Samuel, S., Sobel, R., Genetta-Wadley, A., & Burleigh, B. (1993). Neuropsychological function in borderline personality disorder. *Journal of Clinical Psychology, 49,* 385–396.

Tallis, F. (1997). The neuropsychology of obsessive-compulsive disorder: A review and consideration of clinical implications. *British Journal of Clinical Psychology, 36,* 3–20.

Taylor, A.E., Cox, C.A., & Mailis, A. (1996). Persistent neuropsychological deficits following whiplash: Evidence for chronic mild traumatic brain injury? *Archives of Physical Medicine and Rehabilitation, 77,* 529–535.

Trichard, C., Martinot, J.L., Alagille, M., Masure, M.C., Hardy, P., Ginestet, D., & Feline, A. (1995). Time course of prefrontal lobe dysfunction in severely depressed in-patients: A longitudinal neuropsychological study. *Psychological Medicine, 25,* 79–85.

van Gorp, W., & McMullen, W. (1997). Potential sources of bias in forensic neuropsychology evaluations. *The Clinical Neuropsychologist, 11,* 180–187.

Walsh, K. (1994). *Neuropsychology: A clinical approach* (3rd ed.). Edinburgh: Churchill Livingstone.

Walsh, K.W. (1985). *Understanding brain damage: A primer of neuropsychological evaluation.* Edinburgh: Churchill Livingstone.

Chapter 7

COMPLEXITIES IN THE EVALUATION OF EXECUTIVE FUNCTIONS

David C. Osmon
University of Wisconsin-Milwaukee, Milwaukee, Wisconsin

Introduction

Forensic evaluations typically require a neuropsychologist to address a claimant's overall adaptability. It is generally held that executive functions (EF) are among the principal brain abilities underlying real-world adaptability. Further, it is assumed that EF are able to be selectively measured and that these measurements allow inferences about the person's real-world functioning. Such inferences occur in various forms, including questions about ability to cope with major adult life tasks; competency to advocate for self in legal matters; ability to live independently; capability of participating in treatment interventions; need for financial, social, and physical assistance; among others. However, assumptions about EF measurement are largely untested and research has only recently begun to address these issues by examining: (a) the theoretical structure of EF and its components, (b) the extant clinical and experimental measures of EF, and (c) the ecological validity of EF and their role in adaptive behavior.

This chapter reviews the current empirical underpinnings of these three issues, summarizing the capability of neuropsychology to address a person's adaptability. The first section (Theoretic Considerations) addresses fundamental conceptual controversies, laying the groundwork for a definition of EF and a delineation of essential components of EF that need to be assessed. A strong theoretical model of component abilities of EF is crucial at this early stage in our understanding of frontal lobe functioning. Such a model facilitates a process analysis of EF necessitated by the lack of frontally-specific measures of EF. The second section (Psychometric Considerations) looks at psychometric factors related to the clinical assessment of EF. The third section (Forensic Considerations) examines the importance and prevalence of EF in everyday life, noting the role of assessing EF for predicting real-world adaptability.

Theoretic considerations

The theoretical structure of EF is still largely undetermined, and the very existence of a central executive is a point of serious contention that has yet to be resolved. While its existence is nearly axiomatic in most circles of neuropsychology, and presumed to be associated with the frontal lobe, from purely rational grounds there are those who argue that such a concept requires a "ghost in the machine" and, therefore, cannot be scientific. More empirical arguments against a central executive are based upon various research findings. Two such findings are: (a) diffuse pathologies obscure the relationship between frontal lobe functioning and EF performance (Ott, Lafleche, & Whelihan, 1996), suggesting that EF is a product of the operation of the brain as a whole and not one region, and (b) correlations between EF test performance are larger with metabolic activity in non-frontal structures than frontal regions (e.g., the thalamus: Goldenberg et al., 1992).

Another unresolved fundamental issue is the lack of a specific measure of EF and the implication of this for diagnosis in neuropsychology. It is disquieting to be unable to associate performance/impairment on a specific test with the integrity of, or damage to, one area of the brain, as has been the case with the Wisconsin Card Sorting Test (Mountain & Snow, 1993). However, results from the functional imaging literature have largely supported Luria's (1966) "functional systems" hypothesis. That is, the literature has shown the involvement of anterior and posterior cortical tertiary regions along with the appropriate perceptual secondary cortex in most cognitive tasks studied to date (see Roland, 1993). With both anterior and posterior cortical involvement in most tasks, it is no surprise that specificity in testing is difficult to demonstrate. Evidence of "functional systems", however, does not argue against the involvement of discrete cortical regions in specific cognitive abilities. Nor does such evidence argue against the existence of a central executive that regulates cognitive activity. In fact, a regulatory role for the frontal lobes would tend to auger for its involvement in any major cognitive activity. In order to sort through the morass of ideas about the components of EF, the following section reviews theoretical formulations of frontal functioning.

Theories of executive function
Since theories of EF are too numerous and complex to review completely in this forum, only selected ideas and theorists are included to make specific points. Separate components of EF are highlighted that have some demonstrated accessibility to measurement and should be included in any complete assessment of adaptive functioning. However, it is first necessary to discuss a definition of EF.

Definition of EF
Since no accepted definition of EF yet exists, it is not clear whether to conceive of "frontal lobe" function as the manifestation of one underlying, pervasive ability, such as working memory or mental set, or to assume the existence of multiple abilities. Since the frontal lobes make up more than one-quarter of the neocortical surface and other cognitive abilities (e.g., language, attention, memory) have all been found to fractionate into multiple elemental abilities, it can be assumed that EF is also componential. It is also heuristically convenient to encapsulate the components of EF under one rubric. The concept of mental set is a rubric comprehensive enough to serve this purpose. Explanation of this reasoning follows.

The concept of mental set is akin to the concept of learning set, which is the theoretical and experimental basis of modern day EF theory (Harlow, 1949; Jacobsen, 1935). This basis is rooted in the delayed response and object alternation studies which provide for an understanding of abstract thinking and how it moves beyond the immediate perceptual characteristics of the stimuli. Moving beyond concrete external stimuli requires an internal mental representation that can bridge action over time and over the concrete aspects of the environment. Holding that internal representation "on-line" is generally assumed to be the task of the prefrontal cortex, as evidenced by the "delay neurons" (Fuster, 1993). The internal representation has been referred to as working memory (Goldman-Rakic, 1987). However, Fuster describes two types of cells, the "looking forward" type tied to the response, which increases activity during the delay and prepares for a response; and the "looking back" type tied to the original stimulus, which decreases activity during the delay and represents the working memory. Since mental set (Osmon & Suchy, 1996) can encompass both working memory and preparatory set aspects of the internal representation, it seems the preferred rubric.

Mental set also captures the regulatory (Luria, 1966) role of the frontal lobes. Such regulation acts (Baddeley et al., 1997) to deploy memory, perceptual, attentional, language, and other cognitive abilities. It is this regulation that has led to the use of the term *executive* in reference to frontal lobe functioning. The mental set forms in the frontal lobe and all other cognitive activity is brought under the guiding influence of this set by virtue of recurrent backward connections directly with posterior, perceptual cortex (Goldman-Rakic, 1987) and indirectly through the reticular nucleus of the thalamus (Yingling & Skinner, 1975). Such influence is evident in the effects of different frontal lobe lesions on the Necker cube illusion (Cohen, 1959; Ricci & Blundo, 1990). Bilateral frontal lobe lesions disinhibit the reversing illusion of the Necker cube, while unilateral lesions inhibit the perceptual reversing. While the reason for this differential effect of the two lesions is unclear, it would appear that mental set can exert a 'top-down' influence upon perceptual and other posterior cortical processing.

In addition to forming the set, the frontal lobe acts to maintain or switch the current set in accordance with interoceptive and exteroceptive feedback that is integrated in the frontal regions. Nauta (1971) coined the term "interoceptive agnosia" to characterize this role of the fronto-subcortical limbic structures (see *Theoretical components of Executive function* and *Stroop-effect tasks* sections for further discussion of switching and maintaining set). Thus, the following working definition of EF is adopted for this chapter: *Executive functioning is the process of forming, maintaining, and switching mental set that regulates cognitive and socio-emotional function to bring it under control of both interoceptive and exteroceptive processing for the purpose of behaving in an adaptive fashion.*

With that definition in place, it is appropriate to selectively review various current theories in order to identify important dimensions to consider in the forensic evaluation of EF.

Theoretical components of EF

While no consensus has been reached on the fundamental components of EF, numerous candidates have been set forth. Several have been selected for review here, including working memory, interference, inhibition, conditional action, planning, and socio-emotional functioning. It must be kept in mind, however, that no one component has been demonstrated to be selectively assessed by any one instrument. Therefore, the ability to identify fractionated impairment of EF components is, at this point, a process of clinical judgment in need of greater actuarial basis.

Working memory is one of the most fundamental and best studied components of EF. Its empirical foundation in the animal literature on the delayed response paradigm has established this component as a function of the dorso-lateral prefrontal cortex (Brodmann areas 9, 46). This localization has been demonstrated by single unit studies (see Fuster, 1993) along with lesion (Goldman-Rakic, 1987) and functional imaging studies (see Roland, 1993) in both humans and animals with numerous working memory tasks. Anatomically, this region forms a polymodal integration site of higher order association areas of the posterior cortex, making it the ideal place for perceptual information of the moment to become integrated into a temporary memory representation. It is, therefore, viewed as the site at which the mental set forms (i.e., the goal or intention to behave).

Interference is a separate component of EF that underlies two different aspects of frontal function, namely attention and response selection. Attentional tasks, including Stroop-effect tasks (Bench et al., 1993; Pardo et al., 1990) and Go-No-Go tasks (Casey et al., 1997) have been consistently related to the anterior cingulate gyrus (A-CG). Response selection is an ability that has been attributed to the function of the supplementary motor area (SMA). Rostral aspects of SMA have been shown to activate prior to self-cued movement, indicating a role in selecting the response (Humberstone et al., 1996). Caudal SMA activation occurs at the onset of movement and continues throughout difficult motor actions, indicating a role in maintaining the motor/mental set. Interestingly, when the need to prepare a motor action is removed, the medial frontal lobe does not activate, indicating that the medial frontal aspect of response selection necessitates a motor/mental set component (Brett, Jenkins, Stein, & Brooks, 1996). The A-CG and SMA have a close relationship centered around the facility with which distracting, irrelevant information can be gated out in order to maintain the current mental set (Rubia et al., 1997). This attentional aspect of the medial frontal lobe is seen in numerous studies that have shown the A-CG to activate during specific attentional tasks (Posner & Petersen, 1989). Thus, maintaining mental set is seen as a process where motor planning and motivated attention co-mingle in the medial frontal area in order to deal with interference between competing stimuli or responses. Inhibition, described below, is a closely allied process.

Inhibition has recently been distinguished as a separate construct that, along with interference has the status of basic biological processes associated with frontal lobe functioning (Dempster & Brainerd, 1995). The two constructs appear to be semi-independent as evidenced by studies that show: (a) inhibition is not the only mechanism for resistance to interference (Bower, Thompson-Schill, & Tulving, 1994), (b) decreased inhibition does not always result in increased interference (Kane, Hasher, Stoltzfus, Zack, & Connelly, 1994), and (c) interference and inhibition are unrelated in some studies (see Ncill, Valdes, & Terry, 1995). Neuropsychological evidence suggests a separate anatomical system for the two cognitive abilities (interference and inhibition).

Inhibition relates to orbitofrontal processes that allow mental set to be switched because of sensitivity to reinforcement contingencies. Thus, commission errors on Go-No-Go tasks (e.g., Continuous Performance Test: Rosvold, Mirsky, Sarason, Bransome, & Beck, 1956) and impulsivity and disinhibition associated with the pseudopsychopathic syndrome are associated with damage to orbitofrontal areas (Blumer & Benson, 1975). These findings support the view that interference and inhibition are seen as the workings of opponent anatomical processors.

In an opponent processor view, interference is the function of a processor, the role of which is to gate out distracting stimuli. Gating attention in this manner allows the current mental set to be maintained and the appropriate response to be selected that keeps behavior moving forward to the next action of the current mental set. As mentioned above, this process is a function of the medial frontal lobe (A-CG and SMA). On the other hand, inhibition is seen as a process that complements interference by stopping ongoing behavior or switching mental set. Stopping behavior requires recognizing a novel stimulus that signals the need to switch mental sets rather than continuing with the next action that maintains set. Switching is accomplished by the sensitivity to external contingencies that is the purview of the orbitofrontal cortex, by virtue of its rich connections to the amygdala and basal forebrain limbic regions. Both of these limbic regions (amygdala areas coding aversive consequences and basal forebrain regions coding positive consequences) have strong connections to orbitofrontal cortex with little direct connection to the dorsolateral frontal cortex (see Passingham, 1993, pp. 154–171). It is disruption to these component processes that leads to various clinical presentations, including reduplicative paramnesia, inert perseverative responding, and stereotyped motor responding (Osmon, 1996).

Conditional action (or conditional association, as Petrides [1991] calls it) is a separate EF component that has been studied both with animals and with humans. This component refers specifically to the ability to make a conditional association based upon specific sensory cues. For example, in a visuo-spatial conditional task, the participant learns to point to one of six locations, depending upon which of six lights turns on. Feedback on each trial is used to make the conditional association. Petrides has demonstrated over a series of well thought out studies (using both lesioned animals and epilepsy surgery human participants) that the failure to select specific actions is contingent upon the inability to make conditional associations. It has been demonstrated that failure on conditional learning tasks is not a difficulty remembering the relevant information, motorically coordinating the actions, or discriminating the sensory information. Studies have also been done with tactile and abstract information beyond the spatial tasks mentioned above. The results consistently demonstrate the relationship between conditional action and the premotor cortex (Brodmann areas 6 & 8), making this ability a separate component of EF that goes beyond simple movement control.

Planning is a separate, although multifactorial, component of EF that gets at a facet of adaptive ability not addressed in more "basic" aspects of executive control. Specifically, planning refers to the ability to envision both the ideas contained in a strategy and the steps involved in a tactical approach to realizing the strategy. The ability requires the restraint to completely survey the problem space, noting and keeping track of all relevant details. It also involves being able to "see in the mind's eye" (or hear/feel in the mind's ear/body-schema) the consequences of a sequence of actions. Such envisioning occurs on the ideational apraxia item from the Reitan-Indiana Aphasia Screening Examination, when one knows that it is impossible to touch the right elbow with the right hand without actually having to attempt the action. The restraint and envisioning aspects of planning are well represented in tests such as Porteus Mazes (qualitative score: Porteus, 1973) and the Tower of London (see next section: Shallice, 1982). From the standpoint of mental set, planning tasks involve both forming and maintaining processes. This supposition is supported by functional imaging studies of the Tower of London, where both dorsolateral and medial frontal activation is seen (Rezai et al., 1993).

Socio-emotional aspects of EF are a separate component that has been little recognized and hard to operationalize until recently. While selective disruption of personality in the absence of significant cognitive difficulties does occur (e.g., Ackerly & Benton, 1948; Damasio & Anderson, 1993), it is rare, consequently little theoretical progress has been made until recently. Oder et al., (1992) have brought forth the importance of this aspect of EF and have noted that traditional frontal lobe measures do not access this area of functioning. The inability to measure this function with traditional tests may explain why this aspect of EF has been neglected and difficult to incorporate within theoretical models of frontal lobe functioning. Its importance has been evident in numerous lines of research, however. For example, lesion, imaging, and developmental research have all begun to demonstrate the role of the frontal lobe, especially the orbital regions, in socio-emotional functioning. Building upon prior work (Isaacson, 1982; Luria, 1966; Nauta, 1971; Papez, 1935), Damasio (1991) has posited an elaborated concept of the importance of somatic clues from the limbic system in the guidance of cognitive processes. Heller (1993) outlines the way in which frontal and right hemisphere systems act to code autonomic information in order to perceive and produce emotional experience. These ideas have received support from electrophysiological (Tranel & Damasio, 1994), lesion (Cicerone & Tanenbaum, 1997), and functional imaging studies (Johanson, et al., 1992). In addition, a body of literature has arisen using EEG to demonstrate the relationship between frontal physiology and personality functioning (Davidson, 1992). Finally, a truly exceptional compilation of developmental literature by Schore (1994) makes the point that the right orbitofrontal region, being stimulated by interpersonal-visual transactions with caregivers, begins the process of encoding the infant's earliest experiences and, therefore, matures earliest. Schore marshalls further evidence that the neocortex in general, and the prefrontal cortex in particular, is both "experience sensitive" and "experience dependent", concerning dendritic proliferation and synaptogenesis. He reasons that the relationship with the caregiver, being the single most influencing experience of the infant, is the crucible on which socio-emotional functioning is forged.

While these ideas are only formative, they serve to buttress the supposition that the frontal lobes are involved in socio-emotional functioning and personality. Furthermore, they signal the need to be mindful of assessing for socio-emotional deficits in frontal lobe damage.

Psychometric considerations

Given the above theoretical considerations, it is important to examine putative measures of EF for their componential composition. Measures of EF are typically multi-factorial as demonstrated by factor analysis (see Table 1). Specifically, *variable complexity*, a product of factor analysis, is defined as the number of factors on which a particular variable loads. The variable complexity associated with a battery of EF measures given to 63 college students was 1.7, indicating that many tests loaded onto more than one of the five factors found in this unpublished study. These results suggest that it is important to identify factor loadings for any given measure in order to determine which EF components are present on that measure. For example, as shown in the table, Trail Making-Part B (TMT-B) loads moderately on two factors (one & three) and weakly on two other factors (four & five). Two important aspects of TMT-B are: (a) the ability to understand the concept of alternating between numbers and letters, which probably accounts for loading on the first factor with tasks of concept formation and working memory, and (b) the ability to avoid distraction in order to think effectively and complete the task quickly, which likely accounts for loading with the maintenance of set factor of the Milwaukee Card Sorting Test (MCST) and fluid intelligence component of the Shipley. While the results of this factor analysis are only for purposes of demonstration, they do underscore the importance of understanding the componential composition of EF measures for test interpretation.

This section reviews several EF measures in order to identify important constituent factors of the tests. Knowing these factors and how they overlap across tests provides clues to interpreting underlying EF deficits across a battery of frontal tests.

Relationships among EF tests

Clinical interpretation of neuropsychological tests requires the combined knowledge of the neuroanatomical system activated by the task and the fractionated cognitive structure of the task. While it is traditional to divorce these two knowledge bases and sometimes fashionable to disavow any relationship between the two, the two are inseparable. Full knowledge of either is ultimately dependent upon knowledge of the other. Thus, this section reviews several EF tests from the standpoint of neuroimaging literature pertaining to neuroanatomical systems and the psychometric literature concerning the fractionated cognitive structure of relevant tasks.

Combining neuroanatomic and cognitive structure knowledge about a test allows a process analysis of performance in order to understand mental ability strengths and weaknesses of a person. As a general statement, it is import-

Table 1. Factor Analysis of a Battery of Frontal Lobe Tests in 63 College Students.

Variable	Factor 1	Factor 2	Factor 3	Factor 4	Factor 5
Information	.79	.06	−.19	.08	−.01
Concept	.65	−.21	.17	.02	.18
Brown-Peterson	−.60	.09	−.38	.25	.04
Necker	.49	.33	−.11	.33	.29
Design Fluency	.10	−.86	.00	-.00	−.06
Shipley CQ	.00	−.17	.78	.13	.11
MCST-F2	−.08	.24	.69	-.43	.08
TMT-B	.41	.07	.55	.26	−.26
MCST-F1	.07	.11	.00	−.77	−.20
Category Fluency	.22	.43	-.03	.63	−.22
Stroop CW	.11	.05	.04	.05	.89

Note. Information = WAIS-R subtest; Concept = composite of WAIS-R Similarities + Stanford Binet Verbal Relations; Brown-Peterson = composite of CVC + design versions; Necker = composite of spontaneous + forced reversals; Design Fluency = Gotman & Milner's version; Shipley CQ = Hartford Shipley Conceptual Quotient; MCST-F1/2 = factor scores 1/2 of Milwaukee Card Sorting Test; TMT-B = Trail Making Test-Part B; Category Fluency = composite of plants, animals, appliances; Stroop CW = Golden's third page.

ant to recognize that broad inferences from one level (anatomy) to the other (cognition) are to be understood only with a full appreciation of the constituent cognitive elements of a task. For example, it would be patently absurd to claim that since EF is a frontal lobe process, the Wisconsin Card Sorting Test (WCST) exclusively involves the frontal lobe. However, recognizing that constituent elements of the WCST (e.g., working memory) are accomplished by the frontal lobe is absolutely accurate. In addition, the recognition that other elements of the WCST (e.g, visual perception) are accomplished by posterior perceptual cortical regions is accurate and makes the task less than specific for detecting frontal lobe damage. Ultimately, our ability to interpret a battery of tests rests with the combined understanding of the fractionated elements of cognitive abilities that constitute the tests and where those fractionated elements of cognition are localized, to the extent they are. By examining the pattern of strengths and weaknesses among the fractionated cognitive abilities in a battery of tests overlapping on these constituent abilities, the cognitive impairments and areas of damage can be determined.

As a corollary to the above reasoning, scores that represent the constituent elements of EF allow the identification of the impaired cognitive element, accounting for performance decrements on the battery. While it would be simpler to include tests that are unifactorial, such pure measures do not exist for the wide range of types, locations, and courses of brain damage seen in practice. Thus, if two EF measures that have strong interference and working memory constituents (e.g., Stroop and WCST) are impaired and a third measure that is strongly affected by working memory and less so by interference (e.g., Brown-Peterson task) is unimpaired, then, through the process of elimination, interference is the impaired cognitive element. In the end, the ability to discern dissociated performance between elemental abilities is a necessary part of making accurate diagnostic, prognostic, and treatment statements.

A few of the better known EF tasks that have both neuroanatomic and cognitive structure literature are worth noting. Tasks are deemed accepted measures of frontal and executive processes when satisfying all three of the following criteria: (a) activating frontal lobe regions in functional neuroimaging studies, (b) being associated with impairment when frontal lobe lesions are present, and (c) being associated with one of the six fractionated aspects of EF reviewed above according to concurrent and/or construct validity studies. Negative studies using only one of the three criteria are not viewed as having ruled out the task from consideration, since one methodology alone is not a sufficient or necessary condition for declaring a task's EF nature. For example, in a study of frontal versus non-frontal lesioned brain injury patients, no group differences on the Category or WCST tests were found (Anderson, Bigler, & Blatter, 1995). While it might be concluded that neither test is specific for frontal damage, it is inaccurate to conclude that frontal functioning is unimportant in doing the tasks or that the tasks would not be helpful in interpreting frontal problems on a battery of tests. The usefulness of both tasks in assessing EF is demonstrated by Positron Emission Tomography (PET) studies showing differing regional activations in the frontal lobe during performance of the two tests (Adams et al., 1995). Combining knowledge of the differing regional activations with knowledge of the differing cognitive elements across the two tasks can tell much about a patient's cognitive functioning. Likewise, overinterpretations of lesion studies that seem to argue against the EF nature of certain tasks must be guarded against. For example, Reitan and Wolfson (1995) found no significant differences between frontal and non-frontal lesioned groups on the Category Test and Trail Making Test-Part B. However, the functional imaging data of Adams and associates clearly demonstrates the role of the frontal lobe in the Category Test. The lack of group differences may be attributable to posterior cortical damage in the non-frontal groups causing impaired functioning on non-EF components of the tasks, while EF components are impaired in the frontal group. Thus, a battery of EF tests in which frontal and non-frontal cognitive abilities overlap is necessary. Such a battery allows the responsible impairment to be teased apart from irrelevant cognitive elements on a given task. Table 2 summarizes putative elements of EF involved in full criteria (functional imaging, lesion, psychometric validity) and partial criteria EF tests.

The importance of fractionating the executive components within an EF task can be demonstrated in studies incorporating process analysis of test performance using fractionated scores. For example, Beatty (1993) compared performance on the WCST and the California Card Sorting Test (CCST). The more differentiated scores of the CCST allowed an important fractionation of cognitive processes not allowed by the highly correlated Categories Achieved and Perseverative Responses scores of the WCST. Such a fractionation demonstrated that elderly healthy participants on the WCST had difficulty forming the appropriate mental set, *not* difficulty switching mental set or a problem with perseveration. Similar process interpretations are possible with another card sorting task that allows fractionation of EF (Milwaukee Card Sorting Test: Osmon & Suchy, 1996). These fractionations of card sorting abilities are examples of how the cognitive structure of a task informs our understanding of its functional neuroanatomy. In the case of the CCST mentioned above, the 'forming set' difficulty reflects the gradual deterioration in dorsolateral prefrontal cellular integrity with age (Huttenlocher, 1979).

Full criteria tasks
Wisconsin Card Sorting Test (WCST). The functional neuroanatomy of the WCST has been well established by both Positron Emission Tomography (PET) and Functional Magnetic Resonance Imaging (fMRI) studies. The general neuroanatomical system for this task includes extrastriate visual cortex (perceptually processing the colored objects), inferior parietal lobule (symbolic and conceptual meaning of the three sorting principles), and various frontal regions (largely forming and maintaining mental set). A large contribution of the dorsolateral prefrontal regions (Brodmann areas [BA] 44–47, 9) are typically found (Weinberger, Berman, & Zec, 1986). The importance of medial frontal regions (especially BA 24, 32) are also identified in several

Table 2. Putative Elemental Abilities in Full and Partial Criteria Executive Function Tests.

Test	Working memory	Inhibition	Interference	Conditional action	Planning	Socio-emotional
			Criteria tests			
W/M/CST	++	+	+	+	−	−
Category	+	+	+	+	−	−
Fluency-L	+	+	+	−	−	−
Stroop	+	++	++	−	−	−
S-O-P	++	+	+	++	+	−
CPT	+	++	+	−	−	−
TOL	+	−	−	−	++	−
Dual Task	++	++	+	−	−	−
			Partial criteria tests			
Fluency-S	−	−	+	−	−	−
Fluency-D	+	+	+	−	−	−
TOH	+	+	−	−	−	−
Fluid IQ	+	+	+	−	+	−
B-P	++	+	−	−	−	−
RG	+	+	++	−	−	−
Olfactory	−	−	−	−	−	+a
TMT-B	+	+	+	−	+	−
BDS	−	+	+	++	+	−
Outcome	−	−	−	−	−	+
Shallice	−	−	−	−	+	−

Note. W/M/CST = Wisconsin/Milwaukee/California Card Sorting Test; Fluency-L/S/D = Letter/Semantic/Design fluency; S-O-P = Self-Ordered Pointing; CPT = Continuous Performance Test; TOL/H = Tower of London/Hanoi; Fluid IQ = Woodcock-Johnson-Revised factor score; B-P = Brown-Peterson; RG = Random Generation; TMT-B = Trail Making-Part B; BDS = Behavioral Dyscontrol Scale; Shallice = real-life planning tasks (Shallice & Burgess, 1991); Outcome = novel task measuring sensitivity to future outcomes (Bechara et. al., 1996).
aOlfactory test is associated with socio-emotional functioning based upon correlation rather than direct measurement of this ability.

studies (Adams et al., 1995). Premotor activation is commonly found (Osmon et al., 1996).

The cognitive structure of the WCST has been viewed as a rather global EF measure, including working memory, interference, inhibition, and conditional action components. Dorsolateral prefrontal activation implicates the working memory component used to keep in mind the necessary conceptual information to select the appropriate match on each trial. Medial frontal activation is associated with the interference component of the WCST. Interference occurs to a mild extent for those patients who are impaired in selecting the important attribute (color, shape, number) to attend to on each trial. Inhibition can also be important when a strong response habit to one principle must be overridden in order to sort to a principle that has less of a reinforcement history, as occurs when switching mental set between the color and shape principle. While switching mental set is clearly needed when the set reinforced for ten consecutive trials is no longer appropriate, this process accounts for a minor amount of overall mental work during the test. Because it accounts for a small amount of work, metabolic changes are minor and functional imaging results show little orbitofrontal activation. However, a given patient may be greatly affected by a switching difficulty in which the feedback can not be utilized to reconfigure set, as sometimes happens with damage from an anterior communicating artery aneurysm (ACoA: Malloy et al., 1993; Osmon, 1996) and associated inert perseverative responding. Premotor activation implies the conditional action that allows differentiated motor responses to similar appearing stimuli.

Work remains to be done in fractionating the processes on the WCST. Factor analysis of the MCST (Osmon & Suchy, 1996) has replicated the two "frontal" factors found on the WCST, perseveration and loss of set (Sullivan et al., 1993). However, a third factor has also been found, reflecting the verbal regulation of behavior and dissociation between knowledge and action (Verbal-Behavior Discord: see Figure 1). Four factors have been found on the California Card Sorting Test (Greve et al., 1995), suggesting possibly even greater fractionation that may have clinical relevance.

Category Test (CAT). Adams et al. (1995) found that performance on subtest VII of the CAT correlated with several prefrontal and non-frontal regions. This result is consistent with the general assumption that the CAT is a globally sensitive measure of brain integrity (Reitan & Wolfson, 1993). In fact, the CAT is often used as the sole indicator of cognitive functioning (Braun & Richer, 1993). The CAT does not differentiate between brain injury patients with and without imaging evidence of frontal lobe damage (Anderson, Bigler, & Blatter, 1995). Taken together, these studies suggest a widespread brain network involved in the task, including multiple frontal and multiple non-frontal regions.

Having such a widespread brain network makes the test generally sensitive to brain dysfunction of all kinds and locations. However, as a result, the test is also sensitive to many other factors, relating strongly to IQ and age (Boyle, Ward, & Steindl, 1994), emotional status (MacNiven & Finlayson, 1993), and cultural issues (Arnold et al., 1994; Cuevas & Osterich, 1990). These relationships detract from the specificity of the test for brain dysfunction. Furthermore, item analysis shows much noise in the test. It includes many items that do not relate to the test as a whole, are too easy or too difficult, and relate more to response bias and item stimulus characteristics instead of learning (Bertram, Abeles, & Snyder, 1990; Laatsch & Choca, 1991). Despite these limitations, the concept formation and reasoning aspects of the test are strong enough to maintain its popularity, making it a frequently administered neuropsychological test (Sellers & Nadler, 1992).

Part of the CAT's popularity certainly stems from the fact that its unique rule learning aspect assesses concept formation and problem-solving (Perrine, 1993). Whereas the WCST has rather obvious principles that are easily conceptualized, the CAT principles, except for the first and second subtests, are more difficult to

Fractionated Scores of the Milwaukee Card Sorting Test

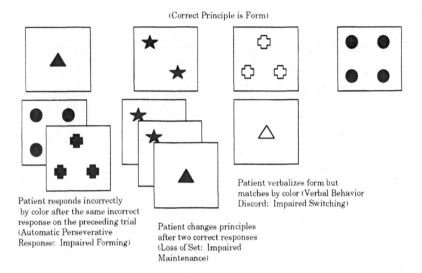

(Correct Principle is Form)

Patient responds incorrectly by color after the same incorrect response on the preceeding trial (Automatic Perseverative Response: Impaired Forming)

Patient changes principles after two correct responses (Loss of Set: Impaired Maintenance)

Patient verbalizes form but matches by color (Verbal Behavior Discord: Impaired Switching)

Fig. 1.   Under each of the first three key cards in the top portion of figure are demonstrated the scores which load highest on the three factors of the Milwaukee Card Sorting Test. The first factor reflects difficulty forming a mental set, in this case the second principle of the test (form). The patient perseverates in sorting to color (the first principle) in an inert and stereotyped fashion. The second factor reflects a difficulty maintaining mental set with an incorrect change in sorting principle after two consecutive correct responses. The third factor reflects a difficulty switching mental set because of discordance between what the patient verbalizes as their intended sort (form) and their actual sort (color). This type of error typically manifests as a flexible change in verbalization to the correct principle but a perseverated actual sort to the principle correct in the prior stage of the test.

discern. Thus, whereas the main component of the conceptually easy WCST is simply to identify the relevant attribute and maintain that mental set, the CAT includes the additional requirement to learn the rule used to guide behavior. However, the CAT's conceptual and abstraction components are not so different from other EF tests. For example, in a factor analysis the CAT loaded with the WCST, while the PASAT, Visual Search & Attention Test, and Trail Making-Part B loaded on a separate factor. In addition, CAT loads on one of the four factors of the California Card Sorting Test (Greve et al., 1995). Also, the CAT contains three factors itself that relate to the different principles that govern response correctness (Kelly, Kundert, & Dean, 1992). While correlating with many EF and non-EF neuropsychological measures, the CAT shows little relationship to the WCST (Perrine, 1993). Similar correlations have been found for both the short booklet and the original Category Test versions with other neuropsychological measures (including the WCST), suggesting that both versions measure similar EF processes (Gelowitz & Paniak, 1992).

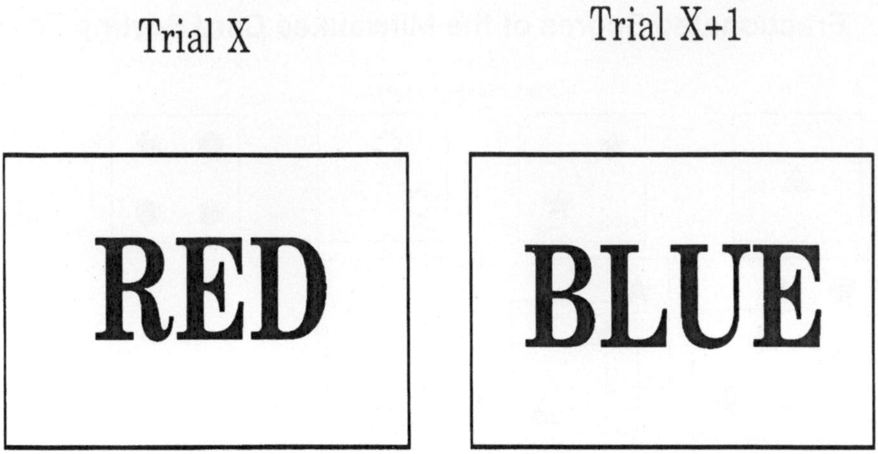

Trial X   Trial X+1

Fig. 2.   An example of negative priming on the Stroop task and its relation to the inhibition aspect of executive function. The first trial requires the person to inhibit
saying the word (red) in favor of saying the color of the ink (blue). On the next
trial the stimulus that was just inhibited (red) is now the target response (red ink
color). However, since it was just inhibited the reaction time to say this ink color
(red) is longer on average than a trial in which the target response is not inhibited on the preceding trial. Having a computerized version of the Stroop task
would allow a score that differentiates inhibition from interference.

The clinical use of the test as one part of an EF battery is verified by cluster
analysis (Goldstein, 1990). The CAT combines with the WCST, Trail Making-
Part B, and the Tactual Performance Test to create 5 clusters of schizophrenic
patients who differ in level and pattern of performance across the EF measures.
In addition, the way is yet open to create fractionated scores that might yield
more specific information about elemental aspects of EF. As the test currently
stands, the only useful information appears to be the total error score, and the
CAT therefore remains only a global indicator of brain integrity and, in conjunction with other EF tests, can function as a general indicator of frontal lobe
integrity. However, the significant psychometric limitations of the test, mentioned above, need to be kept in mind when using the test clinically.

Stroop-effect Tasks. The Stroop effect is one of the most robust cognitive
phenomena available for assessment (MacLeod, 1992). The most common clinical instantiation of the Stroop effect is the Stroop Color-Word Test (Golden,
1978), although another more recent version is available (Trenerry, Crosson,
DeBoe, & Leber, 1989). Numerous other versions of the effect are available
experimentally, but not yet developed for clinical use. Future development of
these experimental versions should yield many valuable interpretative techniques, including scores for negative priming (Fox, 1995), Simon effect (Simon
& Baker, 1995) and others. Current clinical versions of the Stroop effect are useful for evaluating several aspects of EF, including attention and the interference
and inhibition constructs.

Functional neuroimaging studies have demonstrated extensive frontal lobe activation during performance of Stroop effect tasks, suggesting that regional frontal activation accounts for different elements of EF. The strongest evidence for regional specialization of EF exists for the response selection aspect inherent in the Stroop effect (Bench et al., 1993; Pardo et al., 1990). When subtracting out all other aspects of the Stroop except the response selection component, the medial frontal area, especially in the right hemisphere, is the region activated. Without subtraction, dorsolateral frontal activation is also commonly seen, again more in the right than left hemisphere (Sengstock, Osmon, & Suchy, 1997). Lesion studies support the right hemisphere laterality of the Stroop effect (Vendrell et al., 1995), despite early evidence for left hemisphere involvement (Perret, 1973). In addition, functional imaging studies (Bench et al., 1993) and lesion studies implicate the orbitofrontal regions as a crucial area in the performance of the inhibition aspects of the Stroop task, as described below.

Kixmiller et al. (1995) found that ACoA patients had memory intrusion errors, suggesting that inhibition problems are related to orbitomesial dysfunction. In another study from a different laboratory, ACoA patients showed unusual sensitivity to proactive interference because of an inability to suppress irrelevant information at retrieval (Van der Linden et al., 1993). This suppression failure fits the earlier definition of inhibition problems (orbitofrontal process) as a difficulty withholding prepotent responses. This deficit is separate from interference difficulties having to do with gating out distracting information (medial frontal process). Such a distinction between inhibition and interference has clinical merit, as evidenced in a study of impulsive individuals who were functional or dysfunctional (Brunas-Wagstaff, Bergquist, & Wagstaff, 1994). Dysfunctional impulsives tended to produce errors on the Stroop task, relating to an inability to inhibit a more prepotent response. Functional impulsives produced fewer such errors but had slower processing speed because of difficulties 'gating out' irrelevant stimuli. Such evidence lends support to the neuroanatomical basis of fractionations of EF. Further evaluation of the cognitive components inherent in the Stroop task sheds light on the neuroanatomical system involved.

Inhibition, interference, and working memory are the crucial EF components of the Stroop task. The ability to inhibit a more prepotent response in favor of a less automatic response is evident in the need to ignore the word in favor of naming the color of ink in which the word is printed. Difficulties in the process of inhibition are manifested as errors in which the word is read instead of appropriately naming the ink color, and probably represent the most pathognomonic indicator on the Stroop task. Such errors would most likely be attributed to orbitofrontal dysfunction, as is suggested by the difference in intrusion errors in ACoA patients (orbitobasal damage) compared to Korsakov patients (whose dysfunction is more in the superior medial frontal region: Adams et al., 1993). Support for this interpretation is also found in the correlation between the Stroop effect and inert perseverative counting errors on the Random Number Generation task (Brugger et al., 1995). Stroop-like inhibition errors are indicated when the patient issues counting sequences (e.g., 1–2–3. . .) during random number generation. Such a deficit represents an inert perseveration, resulting from the overlearned, automatic nature of a counting response. Studies of impulsivity, a sign often associated with orbitofrontal dysfunction (Damasio, Tranel, & Damasio, 1990), also demonstrate inhibition processes on the Stroop task. Visser, Das-Smaal, and Kwakman (1996) found that negative priming on the Stroop task was associated with social impulsivity. Negative priming (see Fig. 2) was defined as slowed naming of an ink color that was the distractor on the preceding trial. As a distractor on the preceding trial, that stimulus was inhibited, requiring increased ability to overcome inhibition on the next trial. Therefore, better function of the orbitofrontal region is required, a characteristic not present in impulsive individuals. Interestingly, the Stroop interference score did not relate to social impulsivity, indicating that negative priming (i.e., inhibition) and interference are separate processes. The interference aspects of the Stroop effect is demonstrated by a separate set of studies outlined below.

While the inhibition aspect is pathognomonic, the Stroop effect is perhaps best known as a measure of attention by virtue of its known relationship to the construct of "resistance to interference." Attentional aspects appear to be separate from inhibition aspects, as they are associated with different types of errors and a different anatomical system. Also, Stroop tasks are better known as clinical measures of attention because of their use with populations having attentional difficulties.

Resistance to interference implies the ability to focus attention on task-relevant dimensions of a stimulus (e.g., ink color, *not* word; and current stimulus, *not* prior or spatially contiguous stimuli). This aspect of the Stroop task is separate from the inhibition aspect described above because it implies maintaining the current mental set by 'gating out' distracting information, rather than switching mental set by focusing on current reinforcement contingencies to override responses with strong habit strength. Thus, medial frontal structures are involved in the attentional aspects, not the orbitofrontal structures involved in the inhibition/switching aspects of the task. The involvement of medial frontal cortex in the interference aspects of the Stroop is verified by a study in which medial frontal lobectomy patients were compared with temporal lobectomy patients and normal controls (Richer et al., 1993). Medial frontal patients were slower on both the Stroop and an interference task in which distractors interfered with performance. In addition, several studies have used the Stroop task to evaluate "attentional" difficulties in Attention Deficit Disorder (ADD) and traumatic brain injury patients. Stroop deficits are a stable feature of ADD from 6–11 years of age compared to normal controls, despite normal developmental advances in other cognitive skills (Grodzinsky & Diamond, 1992). A major review of ADD found poor Stroop interference performance, especially in ADD-without-hyperactivity individuals (Barkley, Grodzinsky, & DuPaul, 1992). The occurrence of omission errors on a Go-No-Go task (Continuous Performance Test) and the lack of commission errors may implicate interference more than inhibition as the underlying difficulty. Also, the Stroop effect may have some

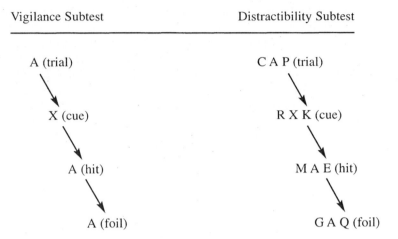

Fig. 3.   Two versions of Go-No-Go tasks (commonly known as the Continuous
Performance Test) can be used to help distinguish inhibition from interference components. The first version (e.g., Vigilance subtest on the Gordon Diagnostic System)
is shown here where a cue (X) heralds the potential occurrence of the imperative.
On the Distractibility subtest, the interference component of the task is enhanced
by including irrelevant stimuli (letters on each side of the relevant stimulus).

selectivity for ADD, as suggested by a study finding that Stroop deficits occurred
in ADD, but not Conduct Disorder or Normal Controls (Lavoie & Charlebois,
1994). Stroop deficits are also a common finding in head injury, a population
often associated with attentional difficulties (Levin, Grafman, & Eisenberg,
1987). However, emotional difficulties are also associated both with head injury
and Stroop performance, confounding the measurement of attention. It would
appear, however, that attentional difficulties occur beyond emotional issues and
help explain Stroop deficits in this population. Batchelor, Harvey, and Bryant
(1995) found that Stroop deficits could be only partially accounted for by emotional indices. And, Stroop effect deficits distinguished post-concussional syndrome patients who recovered within three months from those that did not, the
latter having continued impairment on the Stroop (Bohnen, Twijnstra, & Jolles,
1992). However, it should be noted that the head injury studies have not sorted out which aspect of the Stroop (inhibition or interference) is involved in the
deficit. A newer study, however, showed "supervisory strategy" effects accounting for Stroop deficits in head injury patients even when statistically controlling
for slowed processing, suggesting an orbitofrontal inhibition deficit (Azouvi et
al., 1996).

Working memory is also an important part of the Stroop task, as suggested
by the necessity of keeping in mind the rule to ignore the word and say the color
of the ink. The working memory component of the task is suggested by developmental studies in which young children are unable to perform the Day-Night
Stroop task (say "day" to a black card with moon and stars; say "night" to a

white card with a sun). Diamond and Taylor (1996) found that the most common error of children between the ages of 3 and 6 was to comply with only one rule, suggesting that working memory capacity was ineffective in holding both rules in mind at once. However, other issues are also at work in young children's inadequacy in this task, since a similar study (Gerstadt, Hong, & Diamond, 1994) found that even three-year-olds could perform at a high level on a two rule task without the interference component (say "day" to one abstract design and "night" to another design). Because the rules were verbal in nature, working memory may have been a negligible factor in the Stroop effect deficit, since Stroop effects are more right hemisphere in nature. In support of this notion is the finding that Stroop performance is associated with a spatial working memory task in frontally-dysfunctional schizophrenics (Rizzo et al., 1996). Also, Stroop measures account for as much variance in working memory measures associated with aged subjects as measures of speed of processing, the accepted crux of cognitive dysfunction associated with age (Salthouse & Meinz, 1995). Finally, the reverse Stroop effect occurs when the interference effect operates not at the perceptual encoding stage but on the working memory representation of the stimulus. This effect is found to be as strong or even stronger than the standard Stroop effect, implicating post-encoding and working memory processes in the Stroop task (Simon & Baker, 1995).

In summary, the strength of the Stroop effect as an EF measure is clear. However, the current versions of the Stroop test are limited in their ability to elicit the necessary information about fractionated aspects of EF. Inhibition and interference are confounded in current clinically available versions of Stroop tasks. Computerized versions of the Stroop effect are needed to allow scores that represent the different facets of information contained in the Stroop effect, including the Simon effect and negative priming.

Continuous Performance Test (CPT). CPT measures tax the maintenance of mental set by testing sustained attention and distractibility. As such, these measures include a Go-No-Go component in which many trials are foils (No-Go), acting to distract attention away from the current mental set. The definition of Go-No-Go procedures has been varied and contradictory in the literature, resulting in a quagmire of confusion about the theory of its neuroanatomy and cognitive structure. One clinical version of a Go-No-Go task has been the CPT (Rosvold et al., 1956), consisting of infrequent Go trials in which the subject presses a button when a particular letter or symbol is seen (both auditory stimuli and/or responses are used in some versions). The task is made more complex by complicating the imperative stimulus to include a working memory component (e.g., press when an A occurs but only after an X). Relative infrequency of target stimuli necessitate many No Go trials, requiring the subject to control the impulsive tendency to respond (inhibition, as described above). In one version of the task (Gordon Diagnostic System), a distraction subtest includes irrelevant stimuli in proximity to the imperative stimulus (see Fig. 3). This version enhances the attentional or interference component (separate from the inhibition component) of the task by requiring the irrelevant stimulus to be ignored and mental set to be maintained. Both components are present in the Go-No-Go task, as suggested by factor analytic studies showing inattention and response inhibition factors (Healey et al., 1993). It is these two separate components of Go-No-Go tasks that have led to confusion in the literature. Attempts to separate these multiple components using different scores have been only partly successful. Commission errors (i.e., responding to a No Go stimulus) have been used to represent the inhibition aspects related to orbitofrontal areas. This practice is supported by the association of commission errors with impulsivity: college students differing on the Impulsivity Scale (Reed & Derryberry, 1995), hyperactive children (Lassiter et al., 1994), aggression in ADD children (Nigg, Hinshaw, & Halperin, 1996), disruptive behavior (Halperin et al., 1995), and hyperactive children but not hyperactive children with anxiety (Pliszka, 1992). However, the commission error score is not a unitary entity and some aspects of commission errors are related to impulsivity, while others are not (Halperin et al., 1991). Thus, selective measures of inhibition and interference are not yet obtainable from the CPT test.

The functional imaging literature has demonstrated significant frontal lobe activation during the CPT, especially in the medial frontal regions (Rezai et al., 1993). Pathologic populations also show a relationship between the medial frontal regions and the CPT. Siegel et al. (1995) showed such a relationship in both schizophrenic and infantile autistic patients. The large literature on EEG and event-related potentials associated with Go-No-Go tasks also implicates the medial frontal region, although deep structures and the orbitofrontal areas are not able to be well localized with scalp electrodes. When surface electrodes across the convexity of the frontal lobes are used in monkey, dorsolateral activation is seen in Go-No-Go tasks (Sasaki et al., 1993). Taken together this literature suggests that the interference aspect is the most prominent part of the Go-No-Go tasks.

However, other regions of the frontal lobe have been implicated in functional imaging studies of Go-No-Go tasks. For example, Schroeder et al. (1994) found general hypofrontality for schizophrenics compared with healthy controls while performing the CPT. Further, autonomic non-responding schizophrenics had lower absolute lateral frontal activation compared to autonomically responsive schizophrenics (Hazlett et al., 1993), suggesting that concurrent limbic activation may be important for activating frontal lobes. In addition, disruption of CPT performance caused by sleep deprivation was associated with reduced PET activation in the thalamus, basal ganglia, and limbic areas (Wu et al., 1991). Lesion studies also support the involvement of other frontal regions in the Go-No-Go tasks. For example, Rueckert and Grafman (1996) found special significance for right frontal damage in performance on several sustained attention tasks, one of which was a CPT. Interestingly, errors implicating inattention/interference were found, including: (a) longer reaction time, (b) omission errors, (c) increasing performance decrement in later stages of the test. Thus, most of the evidence implicates the medial and dorsolateral aspects of the frontal lobe, and the presumed orbitofrontal component of Go-No-Go tasks has been difficult to verify.

The presumption of orbitofrontal cortex involvement in inhibition aspects of Go-No-Go tasks has been based mostly upon clinical and indirect evidence. For example, it is well known that traumatic brain injury is associated with contusing and lacerating damage to the basal portions of the frontal lobe where the inner table of the skull is ridgy and uneven. This damage is assumed to play a role in the disinhibition of behavior and decompensation of social decorum that accompanies moderate to severe head injury (Jennet & Teasdale, 1981). Likewise, the "frontal lobe syndrome" of pseudopsychopathy is associated with orbitofrontal damage (Blumer & Benson, 1975). Older animal studies have demonstrated a release of emotional control that occurs following orbitofrontal lesioning (Butter & Snyder, 1972). More recent integrations of this line of research suggests that the orbitofrontal region provides autonomic signals about the affective valence of a stimulus (Bechara et al., 1996). Such information about stimulus significance allows inhibition of responses and switching of mental set. When these inhibitory controls are dysfunctional, then commission errors occur and social decorum is lost because important affective information is not used to override the current mental set or the prepotent response to the situation. This interpretation is supported by research with psychopathic individuals who show difficulties both on orbitofrontal tasks that require inhibition, such as Go-No-Go and the Porteus Maze Qualitative score, and tasks of smell identification (LaPierre, Braun, & Hodgins, 1995). Further fractionations of orbitofrontal processes involved in inhibition and switching mental set are suggested by results of animal lesion studies. For example, Iversen and Mishkin (1970) found inferior convexity orbitofrontal lesions to disrupt Go-No-Go tasks, while more ventral and medial orbitofrontal lesions disrupted object reversal and spatial alternation tasks. However, further research is needed to clarify how these two areas relate to disinhibition and inability to switch mental set on Go-No-Go tasks.

Dual Task (DT). A growing literature supports Baddeley's (1986) contention that DT methods invoke frontal processes because of bottlenecks in response selection and other cognitive abilities that occur in the central executive. Lesion studies, functional neuroimaging methods, and cognitive paradigms provide converging evidence that DT methods are effective measures of EF that reflect frontal lobe functioning. A functional magnetic resonance imaging study has provided preliminary evidence that single tasks that do not individually activate the frontal lobes do cause frontal activation when performed simultaneously in the DT fashion (D'Esposito et al., 1995). DT impairment also occurs in various populations associated with frontal lobe dysfunction, including schizophrenia (Granholm, Asarnow, & Marder, 1996), closed head injury (Azouri et al., 1996), Parkinson's disease (Robertson, Hagelwood, & Rawson, 1996), multiple sclerosis (D'Esposito et al., 1996), and aging (Greenwood & Parasuraman, 1991). Cognitive studies have shown that concurrent finger tapping affects frontal memory and letter fluency abilities more than temporal lobe/hippocampal memory and semantic fluency functions (Moscovitch, 1994). Conversely, DT performance does not interfere with a perceptual feature integration phenomenon dependent upon posterior cortical functioning, illusory conjunction errors (Prinzmetal, Henderson, & Ivry, 1995). Furthermore, cognitive studies using psychological refractory period designs have shown that shorter stimulus onset asynchronies between the tasks increases reaction time. Increased reaction time argues that a central executive bottleneck in processing is responsible for the interference in concurrent task performance (Pashler, 1994). Furthermore, performance decrements on the primary task depend upon instructions on which task to attend to. Such instruction-dependent performance argues for a volitional EF component to the DT method (Foos, 1995). Finally, DT performance correlates with effortful and speeded tasks that have a preparatory attentional component but do not relate to posterior processes such as vigilance and selective attention (Rosenbaun & Taylor, 1996). In addition, DT performance is useful in ecologically valid prognostications about such real-world functions as driving. In fact, Crook, West, and Larrabee (1993) recommend including an ecologically valid DT in test batteries used with aged clients.

Tower of London (TOL). Planning is difficult to evaluate in the clinical situation using traditional testing methods, and richer information may be gleaned through ecological, qualitative approaches. For example, wayfinding in unfamiliar places was evaluated, finding that Alzheimer's patients had poorly structured decision plans. Those plans lead to many errors in trying to reach a destination and return to the starting point compared to healthy age-matched controls (Passini et al., 1995). Despite such severe difficulties, the patients performed well in problem situations which were well-defined and required development of subroutines where all relevant information was available in the problem-space. Such situations typify most clinical tests, making them less sensitive to planning deficits. One exception is the Tower of London test (TOL). Although not in frequent use, this test is well suited for clinical use because it lacks the drawback of approaches such as arithmetic story problems and other problem-solving tasks that attempt to measure decision-making and planning with subroutines. That drawback is dependence upon achievement (arithmetic knowledge) and other cognitive abilities (e.g., processing speed), such that planning skills become a minor component of the measured variance. Instead, the TOL hones in on the planning skill by using a task which does not depend upon prior learning and is a novel problem for most individuals. Planning skills, which require forward thinking and the impulse control to think through the problem to the end before initiating a response, are the crux of the TOL task (see Fig. 4). However, care must be taken in the design of a tower task and not all versions (e.g., Tower of Hanoi/Toronto) are equivalent in their ability to tap planning functions. For example, in a carefully analyzed study Goel and Grafman (1995) were able to demonstrate that the Hanoi version had little to do with forward thinking skills, requiring instead the recognition that a counter-intuitive "backward" move must be made (see Fig. 5). While this cognitive skill involved in the Hanoi version may be frontal (the authors suggest an analogy to the Stroop requirement of inhibiting a prepotent response), it is not the forward thinking skill involved in TOL.

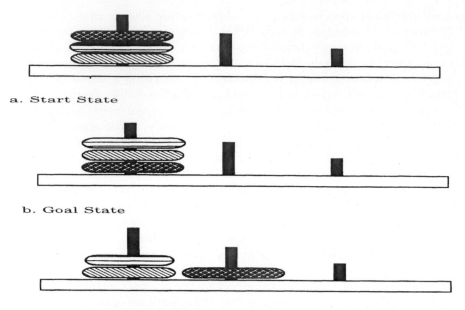

a. Start State

b. Goal State

c. Incorrect First Move

Fig. 4.   The Tower of London requires few moves, allowing an administration proce-
           dure that emphasizes "forward thinking." Thus, planning skill is most import-
           ant in this task rather than other frontal abilities. Lack of planning results in first
           move errors as shown above (correct sequence: Disk 1 to right peg; Disk 2 to
           middle peg; Disk 3 to middle peg; Disk 1 to left peg; Disk 3 to left peg; and Disk
           2 to left peg).

Planning skill on the TOL appears to be related to medial and superior dorsolateral prefrontal regions according to functional imaging studies (Baker et
al., 1996; Rezai et al., 1993). In addition, a well-designed series of studies by
Owen and Robbins and associates (e.g., Owen et al., 1995) have demonstrated the relationship between the TOL and frontal/frontostriatal dysfunction in
focal frontal lesion subjects, Korsakov's and Parkinson's patients. Results of
these efforts suggest that both working memory and motor planning are
important and unique components of TOL. As Owen et. al. (1995) showed,
working memory is important in order to hold in mind the series of moves
involved in the plan and becomes more important as the number of moves
required to achieve the end position increases (up to 5 moves in the TOL).
Owen and associates also showed that motor planning (as measured by correlation between "thinking time" on initial moves and overall TOL accuracy)
was the crucial component for Parkinson's patients, while working memory
was more important for focal frontal lesion patients. Besides working memory and motor planning, inhibition of stereotyped or impulsive actions may also
be an important component of TOL, especially in patient populations.
Correlations between TOL and Porteus Mazes have been found and support
the role of inhibition on TOL (Krikorian, Bartok, & Hay, 1994). It appears

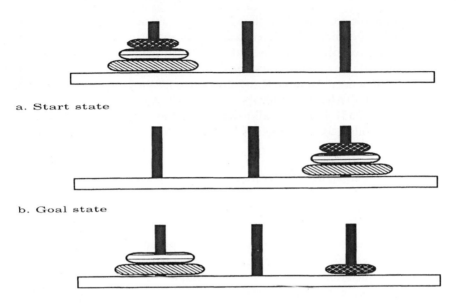

a. Start state

b. Goal state

c. Counterintuitive backward move

Fig. 5.   Counterintuitive backward move required on some "Tower" tasks. As Goel and
          Grafman (1995) note, some "Tower" tasks (e.g., Hanoi but not London ver-
          sions) require the recognition that the global goal (getting entire stack on right-
          most peg) must be postponed while subgoals are accomplished. Thus, the first
          disk must be placed on the target peg (counterintuitively) in order to place the
          intermediate disk off the target peg. Then, the third move clears the target peg
          in order to place the largest disk on the target peg on the fourth move. From
          here, the problem can be solved in the fewest possible moves (i.e., Disk 1 to right
          peg; disk 2 to middle peg; disk 1 to middle peg; disk 3 to right peg; disk 1 to left
          peg; disk 2 to right peg; disk 1 to right peg). Note also the subgoal of stacking
          disk 1 on disk 2 (third move) in order to clear the right peg to place disk 3.
          While this example requires few moves and can be done "in the head", the true
          Hanoi version includes five disks and several items with moves greater than the
          immediate memory span. Because of the number of moves required and the
          necessity for keeping track of subgoals, it is difficult to do mentally and a "per-
          ceptual strategy" is resorted to by most. As Goel and Grafman note, such a
          strategy takes this test out of the realm of planning functions.

that the Porteus qualitative scores, but not quantitative scores, tap into the
impulsivity dimension (LaPierre, Braun, & Hodgkins, 1995), and further
research into the relationship with TOL is needed.

   As measured by TOL, planning skill is an important area of evaluation in a
battery of EF tests. It would appear to have important application to evaluating
real-world functioning as long as certain aspects of administration and scoring
that ensure adequate measurement of planning are attended to. Specifically,
administration procedures must be designed to require subjects to mentally solve
the problem without actually making motor actions (in order to avoid the tendency of subjects to begin without first thinking through the movements ahead

of time). In addition, "thinking time" (latency to first move) is an important fractionated score under the correct administration procedure. It is also useful to have qualitative information available on the individual movements actually made on each trial in order to track the path to solution. Computerized versions of the task allow the most flexible scoring options. As a result, the modified version of the CANTAB TOL appears to be the best instantiation at this time (Owen et. al., 1995). However, all tower tasks must be evaluated carefully for test-retest reliability, as this issue has been problematic in the past.

Partial criteria tests

Several tests are often used as EF measures that do not meet full criteria established above but deserve some mention. Specifically, fluency measures, fluid intelligence, random generation, Trail Making Test-Part B and Behavioral Dyscontrol Scale tasks are briefly discussed.

Fluency. Fluency measures have proven to be popular clinical measures of EF for at least two reasons. First, they provide easy to use measures of important EF processes, including forming and switching mental set in order to generate numerous responses. Second, specific and laterality-sensitive measures of frontal lobe functioning are possible when comparing phonemic (generating words beginning with a particular letter) with semantic (generating words from a particular category, like animals) and verbal (generating words) with design (generating designs) fluency formats. While both phonemic and semantic fluency seem to include both frontal and posterior brain processes (Martin et al., 1994), dual-task methodology, aging, and lesion studies with various populations suggest that phonemic fluency is more related to frontal regions, while semantic fluency is more related to posterior brain regions (Moscovitch, 1994). Likewise, phonemic fluency is more left-frontal-related, while design fluency is more right-frontal-related (Jones-Gotman & Milner, 1977). A recently developed design fluency measure appears to provide a better clinical instrument than the original research version because it standardizes the administration and scoring of the task in a more replicable fashion (Ruff & Light, 1987). In addition, a "category" design fluency measure has been reported with some preliminary evidence that it relates more to posterior right hemisphere functioning (Mickanin et al., 1994), although more research is needed before clinical use is warranted. Although the existence of these various 'frontal' and 'posterior' versions of fluency measures adds to neuropsychological diagnostic capability, the crucial executive component of fluency measures is still unclear. Is the frontal nature of fluency tasks related strictly to phonemic-expressive Broca's area cognitive abilities? The frontal nature of design fluency measures and the lesser relationship of category versions of fluency measures to frontal functioning suggests that the generative aspects of the tasks are crucial and separate components contributing to the frontal-executive nature of the tasks. However, more research is needed to resolve this question.

Fluid intelligence (Gf). Gf is a concept that has been infrequently addressed and accepted within the neuropsychology of EF, but is becoming a powerful explanatory construct within the traditional psychometric literature. It is also useful for measuring an aspect of frontal lobe functioning that has long been recognized as important (Luria & Tsvetkova, 1990) but has been relatively neglected within American neuropsychology, namely *problem-solving* skill (Heidrich & Denney, 1994). While the CAT with its large rule-learning component is argued to be a problem-solving measure, its relationship to Gf has been insufficiently demonstrated. In addition, it has been shown to correlate greatly with the WAIS-R, a rather poor measure of Gf (Duncan, Burgess, & Emslie, 1995; Kaufman, Ishikuma, & Kaufman, 1994), but a good measure of crystallized intelligence (Gc: Logue & Allen, 1971). Other neuropsychological measures might be used to assess Gf (e.g., Verbal Concept Attainment Test: Bornstein, 1983; WAIS-R PIQ: Boone, 1995); however, recent psychometric developments require consideration of measures developed using the Cattell-Horn theory of Gc and Gf (e.g., Culture Fair Intelligence Test: Cattell, 1973; Woodcock-Johnson-Revised: McGrew, 1994). The importance of such measures is amply demonstrated in a study of frontally lesioned patients whose Wechsler FSIQs were superior, but their Culture Fair IQs were 22–38 points lower and 20–60 points lower than matched controls who had similar FSIQs, while posterior-lesioned patients had similar IQs across the two measures (Duncan, Burgess, & Emslie, 1995). Also, accepted measures of Gf have been shown to correlate better than Gc measures with: (a) brain size and size of dorsolateral prefrontal cortex (Raz et al., 1993), (b) cognitive losses with age that are not completely explained by mental speed or visual-spatial factors (Wang & Kaufman, 1993), and (c) dependence in elderly individuals (Poon et al., 1992). Importantly, Gc measures correlate better than Gf measures with auditory event-related potentials recorded over temporal and vertex sites, indicating the more posterior brain localization of Gc processes (Widaman et al., 1993). Considered together, the above literature argues for assessment of the construct of fluid intelligence because it addresses the conceptually satisfying ability of problem-solving skill. In addition, its clinical utility as a counterpoint to crystallized intelligence measures is certain, and it is likely to have benefit in predicting and maintaining everyday functioning (Diehl, Willis & Schaie, 1995; Hayslip, Maloy & Kohl, 1995; Poon, Messner, Martin, Noble, 1992).

Random Generation (RG). RG is a creative approach to measuring EF that is based upon asking the subject to generate a random string of numbers or letters (Baddeley, 1966) and is reviewed here for its intellectual appeal and uniqueness in EF measurement.

Normal individuals tend to avoid repetitions based upon lay misconceptions of randomness; however, the task also elicits prominent perseverative behavior in the form of stereotyped series (1–2–3, etc.) in patients. Working memory and right frontal abilities appear to be important skills in the task, based upon good performance being correlated with a relative superiority in design fluency compared with letter fluency (Brugger, Monsch, & Johnson, 1996). A PET study demonstrates the task's predilection for activating superior dorsolateral frontal lobe regions, implicating internal response selection functions as being important in the task (Itagaki et al., 1995). Cognitive studies demonstrate the relationship between RG and other effortful EF tasks, such as mental arithmetic (Logie, Gilhooly, & Wynn, 1994) and syllogistic reasoning (Gilhooly et al., 1993).

Until better developed clinical norms are available, the task should serve only a supporting role within a battery of EF measures. However, careful qualitative use of the task may be helpful in situations where use of more complicated EF measures is impossible (e.g., moderate to severely impaired Alzheimer's patients). Three strategies in generating random strings have been found and may prove useful in developing scoring systems. Ginsburg and Wiegersma (1991) found that individuals engage in repetition avoidance (not giving the same response consecutively), series responding (stereotyped responses such as 1-2-3 or a-b-c), and cycling (strategies which loop through cycles of responses: 1-3-6-1-3-6). A useful clinical comparison task is suggested by a methodology used to generate a mathematical model of RG (Rabinowitz et al., 1989). In this study a RG task was compared to a free association task in which subjects were asked to associate the RG responses (in this case the numbers 1 through 6) to nonsense syllables in order to approximate random generation. Using this comparison task, a discrepancy score could be computed that might be a more powerful measure by virtue of it being individually tailored to the patient's idiosyncratic performance. Since the comparison task is an EF measure in its own right (internal response selection), the discrepancy score would likely reflect selective aspects of EF (e.g., working memory).

Trail Making Test-Part B (TMT-B). TMT-B has enjoyed an on-again-off-again designation as an EF test (Reitan & Wolfson, 1993). It has intuitive appeal as an EF measure because of the necessity to retain in working memory one sequence (e.g., numbers) while doing mental work on the other sequence (alphabet). In addition, the need to control mental processes while switching between the two sequences, as well as the volitional, guided visuomotor requirements to search the random array of stimuli on the page both seem to tax frontal lobe mechanisms. However, its EF nature has been questioned because the test is generally sensitive to brain integrity, being a measure heavily dependent upon speed of processing (Reitan & Wolfson, 1995). Despite this controversy it is imperative to remember that lack of *specificity* for frontal lobe functioning does not equate with lack of *sensitivity* to frontal lobe functioning. In fact, preliminary functional imaging results demonstrate that TMT-B correlates with frontal lobe changes on SPECT (Curran et al., 1993). These findings support numerous studies that find both TMT-B deficits in frontally involved populations and concurrent validity between TMT-B and other EF measures. Specifically, Ricker, Axelrod, and Houtler (1996) found that both oral and written versions of TMT-B are more impaired in anterior than posterior lesion patients, with the oral version correlating with established EF measures, yet being unrelated to expressive language tasks. In addition, along with other EF measures (WCST, fluency, Stroop) TMT-B correlates with habituation of N100 indices from evoked responses that relate to frontal lobe volume (Soinenen et al., 1995). The TMT-B has also been shown to be impaired in populations typically associated with EF deficits, including Alzheimer's and schizophrenic patients. Lafleche and Albert (1995) found mild AD patients to be impaired on four EF measures, one being TMT-B, while numerous other tasks that did not require "concurrent manipulation of information" remained intact compared with aged-matched healthy controls. The TMT-B has also been shown to be related to the full spectrum of schizoid disorders. Gass and Daniel (1990) found that psychotic indicators on the MMPI were associated with TMT-B impairment. EF measures (WCST, TMT-B, fluency) were impaired in schizophrenic and depressed patients relative to controls (Franke et al., 1993a). In addition, siblings of schizophrenics were impaired on TMT-B, WCST, and fluency relative to controls and had higher scores on measures of physical anhedonia (Franke et al., 1993b). Finally, schizotypal personality disorder subjects are impaired on TMT-B and WCST while performing equivalently on non-EF and even some EF measures relative to other personality disorder and normal control subjects (Trestman et al., 1995). While the above results await confirmation from quantitative functional imaging studies, the balance of the evidence strongly suggests that TMT-B has EF components. Because it is a generally sensitive measure of brain integrity, TMT-B results need to be interpreted within a battery of tests that overlap with its EF and non-EF components. Such a strategy is useful when needing to evaluate interference, inhibition, and working memory aspects of EF.

Further research regarding TMT-B component abilities is needed, since the common belief that it is a measure of attention (as distinct from processing speed) appears to be a misconception. For example, on one factor analysis TMT-B loaded more strongly with tests of visual-nonverbal intelligence than measures of attention (Larrabee & Curtiss, 1995). Likewise, WAIS-R Block Design scores related to TMT-B in healthy very old subjects while other indices were unrelated, including education, MMSE, and memory scores (Wahlin et al., 1996). In support of these findings, TMT-B along with other visuo-motor measures (Target Test, Tactual Performance Test, and Grooved Pegboard) were the best predictors among a total of 15 measures in discriminating non-verbal learning disability children from reading and spelling disability and normal control children (Harnadek & Rourke, 1994). Along a different line of study, TMT-B has been found to be more visuo-motor demanding than TMT-A because the trace path is 56 cm longer and the visual environment is busier (Gaudino, Geisler, & Squires, 1995). Also, Fossum, Holmberg and Reinvang (1992) concluded that TMT-B was harder than TMT-A because it was both more symbolically complex and more difficult in spatial arrangement. Ratios of TMT-B to TMT-A scores have been found useful in some circumstances (Lamberty et al., 1994) and further study of their use in EF measurement seems warranted. In summary, the ease of use makes the TMT-B an attractive test to use in a comprehensive battery with several EF measures. However, interpretation is limited regarding specific EF until further research is available to elucidate relevant neuroanatomy and cognitive structure.

Behavioral Dyscontrol Scale (BDS). Motor planning and formulation of an intention to behave in a self-directed, multi-action manner have long been implicated as frontal-executive functions (Luria, 1966). However, systematic evaluation of these functions has not been a routine part of most test batteries because of the lack of easily quantifiable measures of such abilities. The recent development of the Behavioral Dyscontrol Scale (BDS) may represent an important step toward rectifying this gap in EF tools (Grigsby & Kaye, 1992). This brief test (approximately 10 minutes administration and scoring time) provides a factorially rich measure of abilities not captured by traditional EF tests. The test is centered around Lurian psychomotor tasks, including Go-No-Go, conflictual instruction motor imitation, and seriated motor programming tasks. It has also been shown to relate to adaptive living abilities better than non-EF measures in a variety of populations (e. g., Suchy, Blint, & Osmon, 1997). Further research is needed and is expected to support both the frontal neuroanatomy and EF cognitive structure of this test. Its current clinical use, as a measure of frontal psychomotor functions that goes a step beyond the simple tapping speed and pegboard assembly measures, appears warranted, although caution is advised until more research from other laboratories is available.

Forensic considerations

Luria (1966) argued for a hierarchical view of cognitive abilities with EF representing the highest level of functioning. Judging from the ubiquitous activation of the frontal lobe in imaging studies (Roland, 1993) and the role of EF in all manner of cognitive activities from memory (Tulving et al., 1994) to attention (Posner & Petersen, 1989) to perception (Zatorre et al., 1994) to language (Petersen et al., 1988) as well as all manner of neuropsychiatric disorders (Cummings, 1985) it would appear that this view is warranted. Despite the importance of EF, it can be difficult to translate how paper-and-pencil EF impairment relates to real-world functioning. In addition, it is not clear that all important fundamental components of EF have been identified. As a result, the ability to recognize EF impairment in all its manifestations in real-world behavior is problematic. Until EF components can be isolated and independently measured it will be difficult to compose a battery of frontal tests that will serve to detect EF impairment in a wide variety of real-world situations. Consequently, prognostic predictions and treatment recommendations about ability to adapt to daily living will be limited, even when EF impairment is detected. A simple example serves to illustrate the complexities of EF assessment.

Something as simple as scheduling and prioritizing several activities to be completed in a day's work is an executively complex activity (Shallice & Burgess, 1991). It is necessary to both hold in mind (working memory) the necessary activities and retrieve the order of actions scheduled for the day (episodic memory). Is it possible to tease apart working memory ability and contention scheduling (i.e., the Shallice and Burgess mechanism for resolving conflicts for control of the cognitive aparatus that would help schedule and prioritize activities, in this example) aspects separately by laboratory-like paper and pencil tests and also be certain that these abilities, as they relate to real-world activities, have been properly evaluated?

In addition to the memory aspects of this task, executive aspects of attention and mental set become important in this daily task. For example, it is important to remain vigilant to changing contingencies which require "on-line" revision of the order of activities in the plan. Thus, if one is unable to attend to how the road construction along the route to the shopping mall will preclude timely completion of the remaining tasks on the schedule, then ineffective adaptive functioning ensues. It is exactly this type of executive control that is reflected in Miller, Galanter, and Pribram's (1960) theory of frontal lobe functioning and Shallice and Burgess' (1991) model of the Supervisory Attentional System and their assessment technique. Will assessing all six of the above-reviewed components of EF provide the necessary survey of abilities that will explain a person's difficulties in such a real-world activity or predict their ability to return to such activities post-morbidly? Theoretically, if all basic elemental components of EF are known and able to be selectively measured in a quantifiable fashion, then such a survey could be accomplished. However, confidence in the ability of traditional assessment instruments to do so has not been high, hence, the interest in more ecologically-valid techniques. For example, Shallice & Burgess describe a real-life scheduling task that assesses how effectively a person schedules activities for which too little time is available to complete everything. In this task, one must recognize that the better pay off is to spend a little time on a number of tasks. Frontally damaged patients were found unable to demonstrate the flexibility of action required to perform well in this task, a process that mimics the vagaries of real life. Thus, with further research such ecologically sound measures may help to assess the complex interplay of EF components that make up real-world activities. Two other examples that involve this complex interplay are instructive and are reviewed below - communication and mental envisioning.

Frontal lobe aspects of communication are also relevant to adaptive functioning, including dynamic, generative narrative (Luria, 1970) and the understanding of metaphorical and/or conceptual-semantic linguistic information (Binder et al., 1997; Bottini et al., 1994; Gardner et al., 1983). For example, much of daily life involves making one's meaning known to others while being involved in novel and unusual situations (generative narrative). Daily living also means understanding other's meanings, especially in complex conceptual and metaphorical ways. Failing these abilities, it is difficult to navigate the interpersonal environment to do even simple tasks, such as communicating with a bank clerk about the nuances of different accounts or engaging in the enjoyment of social interactions using humor, innuendo, and other numerous ways of communicating through indirect reference.

Envisioning possibilities is also an important part of cognating in novel real-life situations and is a cognitive process that has been shown to require frontal lobe activation (Pribram, 1991; Zatorre et al., 1994). Such abilities are likely to be related to the planning functions of the frontal lobe where the day's events must be arranged into a schedule. Again using the above example, if driving to the shopping mall, mailing letters, buying groceries, and picking up shirts at the cleaners must all be accomplished before returning home at an appointed time, then possible routes to take must be envisioned in order to select the most efficient path. Difficulty juxtaposing various paths in the "mind's eye" would be a likely result of some types of frontal damage, according to inferences from the above studies.

Various combinations of dysfunction affecting the elementary EF can produce a host of clinical presentations. Presentations of EF impairment can manifest as isolated cognitive difficulties affecting real-world activities, as in the above scenario or even as neuropsychiatric disorder. With the wide ranging involvement of EF in so many areas of life, the paramount importance of EF in adaptive living should come as no surprise. Various studies have demonstrated the superiority of EF compared to other cognitive abilities in predicting real-life functioning. Some of this literature is reviewed in the following section.

Adaptability and executive functions

The nature of the major adult life activities (work, love, play) poses a special challenge to the neuropsychologist trying to specify and prognosticate about the practical, functional effects of brain dysfunction. Considering a patient's adaptability in such complex activities, composed of many different mental abilities and personality factors, is a risky proposition. As shown by Green (1996) different abilities predict different aspects of functioning. Memory was found to relate generally to all functions examined, while attention best predicted social problem-solving and EF best predicted community functioning. Furthermore, prediction is made all the more difficult by the degrees of freedom open for compensation and therapeutic "work-arounds" in such a multifaceted situation as real life. In fact, retrospective chart reviews of the vocational predictions of cerebral palsy victims after 20 years showed a tendency to predict poorer outcomes than were eventually attained (O'Grady, Crain, & Kohn, 1995). This tendency is generally thought to be representative of many functional outcome predictions. However, prognosis is not entirely beyond neuropsychology's grasp. In this same study, 90–100% accuracy was attained when looking at predictions of those individuals who were expected to become competitively employed and those expected to be unemployed. It was only those individuals falling in-between those two extremes (predicted to require sheltered workshop employment) that were poorly predicted. Therefore, it seems that conservative prognostic statements can be made safely and that knowing predictive limitations is most important when prognosticating.

As the hypothetical scheduling example above serves to illustrate, ecologically valid measures of EF are needed that relate to and capture the rich manner in which EF manifest in daily living skills. Creative approaches using new techniques with a broader scope of mental ability are proving useful in the quest to predict the complexity of daily functioning. For example, Godbout and Doyon (1995) used scripting theory (i.e., the notion that the frontal lobe generates and stores the serial behavioral actions of routine activities of daily life) and others have used novel scoring techniques of EF measures to demonstrate relationships between EF measures and daily functioning. While conservativism is necessary in clinical work, creative approaches to the problem of predicting daily functioning needs to be encouraged in this research.

Several studies have shown that EF are important, and usually better than other cognitive measures (e.g., Dementia Rating Scale: Nadler et al., 1993), in making ecologically valid predictions about real-world functioning. Also, fractionated aspects of EF relate differentially to various aspects of daily living. For example, executive motor and cognitive functions are differentially successful in predicting different aspects of adaptive ability. Kaye, Grigsby, Robbins, and Korzun (1990) found that the Behavioral Dyscontrol Scale (BDS) outperformed the Mini Mental Status Examination in predicting autonomous functioning, impulsivity, and apathy in geriatric rehabilitation patients. Furthermore, Suchy, Blint, & Osmon (1997) found that the motor aspects of the BDS predicted eventual release from a nursing home placement to independent living in former geriatric, acute rehabilitation stroke patients. On the other hand, having better executive functioning, irrespective of motor functioning, on the BDS predicted immediate release from the acute rehabilitation unit to independent living, suggesting that cognitive health overrides motor functioning when making initial determinations about independent living status. The related ability, "frontal apraxia," has also been investigated in studies of independent functioning (Schwartz et al., 1995).

In addition, methods other than traditional paper and pencil tests are useful in assessing EF and their role in daily functioning. For example, Martzke, Swan, & Varney (1991) found that collateral interview and the neighborhood signs of anosmia were important in identifying disabled head injury patients. It was also suggested that these patients may have been missed using traditional neuropsychological measures. Also, personality assessment has been important when cognitive deficits remediate with time, as is often the case in frontal lobe injury. For example, Yarnell and Rossie (1990) found good educational and vocational outcomes in neurodiagnostically-verified bifrontally injured patients who had no to minimally detectable cognitive deficits yet had clinically evident personality change. In addition, questionnaire approaches have also enjoyed some initial successes (Frontal Personality Scale [FLOPS]: Grace, Malloy, & Stout, 1997, and Brook Adaptive Functioning Questionnaire: Dywan, Roden, & Murphy, 1995).

In general, these studies use EF tests or rating scales in comparison with measures of other cognitive abilities to show how well test scores relate to such skills as maintaining an independent lifestyle, activities of daily living, driving a car, and social and community functioning, among other aspects of daily living. Some of the lessons learned from this literature are that the traditional assessment package may need to be broadened beyond the standard laboratory-like paper-and-pencil instruments of neuropsychology and that different instruments designed for specific aspects of EF (e.g., motor, cognitive, emotional, etc.) have utility in different situations. Clearly, one EF instrument will not serve all purposes. Different components of EF are suited to different predictions and sometimes personality variables rather than cognitive abilities are more important in understanding prognostically complex situations. Likewise, neighborhood signs are sometimes very important predictive variables, especially when few good cognitive measures are available to assess a given brain system (e.g., orbitofrontal area).

Conclusion

Complexities in evaluating EF stem from the yet undifferentiated development of theories about frontal lobe functioning. Nowhere is this better demonstrated than in the recent developments in the measurement of intelligence (Horn & Cattell, 1966). Recently, the technology of assessment is catching up with theoretical developments to encompass the measurement of both crystallized and fluid aspects of intelligence (McGrew, 1994). Lagging technology sidetracked progress for a time and resulted in an overemphasis on measuring the crystallized component of intelligence. Part of the reason for this detour is the lack of understanding of the elemental components of EF. Those components are the crux of fluid intelligence.

While theoretical progress in understanding EF continues, neuropsychology still lacks even an agreed upon definition of EF. As a result, clinical assessment of EF remains more art than science. Until such time as a solid definition of EF and its components is available and until those components can be measured with sufficient sensitivity and specificity, the evaluation of EF will be a process of clinical judgment. Good judgments, in the meantime, will depend upon surveying EF with a variety of measures, including: several traditional tasks that measure a wide range of component EF; more qualitative, ecologically-sensitive techniques that mimic the complexity of real-world activities; questionnaires that address behavioral and personality characteristics from both the perspective of the patient and a caretaker; and structured interview techniques that reveal idiosyncratic manifestations of EF impairment. While traditional measures need to be supplemented with the more qualitative techniques, care must be used in interpreting measures that are not strictly objective and quantitative. Their use should be limited to secondary roles after hypotheses have been tested with quantitative, well-normed techniques. Qualitative measures serve the purpose of substantiating conclusions derived from quantiative techniques, broadening understanding of those conclusions, and suggesting further hypotheses that can then be tested with quantitative techniques. Such a multi-pronged approach has the best chance of capturing the subtleties and complexities of assessing EF in the clinic.

As theoretical progress in understanding EF continues, a reconciliation of the psychometric approach to intelligence and the field of neuropsychology will be possible. Combining the psychometric strengths of measuring crystallized and fluid abilities with the neuropsychological explanatory power of posterior and anterior cortical functions promises to infuse new life into the science of individual differences.

References

Ackerly, S. S., & Benton, A. L. (1948). Report of a case of bilateral frontal lobe defect. *Research Publications of the Association for Research in Nervous and Mental Disease, 27,* 479–504.

Adams, K. M., Gilman, S, Koeppe, R. A., Kluin, K. J. et al. (1993). Neuropsychological deficits are correlated with frontal hypometabolism in positron emission tomography studies of older alcoholic patients. *Alcoholism Clinical and Experimental Research, 17,* 205–210.

Adams, K. M., Gilman, S., Koeppe, R., Kluin, K. et al. (1995). Correlation of neuropsychological function with cerebral metabolic rate in subdivisions of frontal lobes of older alcoholic patients measured with (-sup-1-sup-8F) fluorodeoxyglucose and positron emission tomography. *Neuropsychology, 9,* 275–280.

Anderson, C. V., Bigler, E. D., & Blatter, D. D. (1995). Frontal lobe lesions, diffuse, damage, and neuropsychological functioning in traumatic brain-injured patients. *Journal of Clinical and Experimental Neuropsychology, 17,* 900–908.

Anderson, S. W., Damasio, H., Jones, R. D., & Tranel, D. (1991). Wisconsin Card Sorting performance as a measure of frontal lobe damage. *Journal of Clinical and Experimental Neuropsychology, 13,* 909–922.

Arnold, B. R., Montgomery, G. T., Castaneda, I., & Longoria, R. (1994). Acculturation and performance of Hispanics on selected Halstead-Reitan neuropsychological tests. *Assessment, 1,* 239–248.

Azouvi, P., Jokic, C., Van der Linden, M., Marlier, N. et al. (1996). Working memory and supervisory control after severe closed-head injury: A study of dual task performance and random generation. *Journal of Clinical and Experimental Neuropsychology, 18,* 317–337.

Baddeley, A. D. (1966). The capacity for generating information by randomization. *Quarterly Journal of Experimental Psychology, 18,* 119–129.

Baddeley, A. D. (1986). *Working memory.* New York: Oxford University.

Baddeley, A. D., Della Sala, S., Papagno, C., & Spinnler, H. (1997). Dual-task performance in dysexecutive and nondysexecutive patients with a frontal lesion. *Neuropsychology, 11,* 187–194.

Baker, S. C., Rogers, R. D., Owen, A. M., Frith, C. D. et al. (1996). Neural systems engaged by planning: A PET study of the Tower of London task. *Neuropsychologia, 34,* 515–526.

Barkley, R. A., Grodzinsky, G., & DuPaul, G. J. (1992). Frontal lobe functions in attention deficit disorder with and without hyperactivity: A review and research report. *Journal of Abnormal Child Psychology, 20,* 163–188.

Batchelor, J., Harvey, A. G., & Bryant, R. A. (1995). Stroop Colour Word Test as a measure of attentional deficit following mild head injury. *The Clinical Neuropsychologist, 9,* 180–186.

Beatty, W. W. (1993). Age differences on the California Card Sorting Test: Implications for the assessment of problem solving by the elderly. *Bulletin of the Psychonomic Society, 31,* 511–514.

Bechara, A., Tranel, D., Damasio, H., & Damasio, A. R. (1996). Failure to respond autonomically to anticipated future outcomes following damage to prefrontal cortex. *Cerebral Cortex, 6,* 215–225.

Bench, C., Frith, C., Grasby, P., Friston, L., Paulesu, E., Frackowiak, R., & Dolan, R. (1993). Investigations of the functional anatomy of attention using the Stroop test. *Neuropsychologia, 31,* 907–922.

Bertram, K. W., Abeles, N., & Synder, P. J. (1990). The role of learning in performance on Halstead's Category Test. *The Clinical Neuropsychologist, 4,* 244–252.

Binder, J. R., Springer, J. A., Bellgowan, P. S. F., Swanson, S. J., Frost, J. A., & Hammeke, T. A. (1997). A comparison of brain activation patterns produced by auditory and visual lexical-semantic language tasks. *NeuroImage, 5,* 558.

Blumer, D., & Benson, D. F. (1975). *Psychiatric aspects of neurologic disease.* New York: Grune & Stratton.

Bohnen, N., Twijnstra, A., & Jolles, J. (1992). Performance in the Stroop color word test in relationship to the persistence of symptoms following mild head injury. *Acta Neurologica Scandinavica, 85,* 116–121.

Boone, D. E. (1995). A cross-sectional analysis of WAIS-R aging patterns with psychiatric inpatients: Support for Horn's hypothesiss that fluid cognitive abilities decline. *Perceptual and Motor Skills, 81,* 371–379.

Bornstein, R. A., (1983). Verbal Concept Attainment Test: Cross validation and validation of a booklet form. *Journal of Clinical Psychology, 39,* 743–745.

Bower, G. H., Thompson-Schill, S., & Tulving, E. (1994). Reducing retroactive interference: An interference analysis. *Journal of Experimental Psychology: Learning, Memory, and Cognition, 20,* 51–66.

Boyle, G. J., Ward, J., & Steindl, S. R. (1994). Psychometric properties of Russell's short form of the Booklet Category Test. *Perceptual and Motor Skills, 79,* 128–130.

Braun C. M., & Richer, M. (1993). A comparison of functional indexes, derived from screening tests, of chronic alcoholic neurotoxicity in the cerebral cortex, retina and peripheral nervous system. *Journal of Studies on Alcohol, 54,* 11–16.

Brett, M., Jenkins, I. H., Stein. J. F., & Brooks, D. J. (1996, May). *Movement selection without preparation does not activate the SMA.* Paper presented at the Second International Conference of Functional Mapping of the Human Brain, Copenhagen, Denmark.

Brugger, P., Monsch, A. U., & Johnson, S. A. (1996). Repetitive behavior and repetition avoidance: The role of the right hemisphere. *Journal of Psychiatry and Neuroscience, 21,* 53–56.

Brugger, P., Pietzsch, S., Weidmann, G., Biro, P. et al. (1995). Stroop-type interference in random-number generation. *Psychological Reports, 77,* 387–390.

Brunas-Wagstaff, J. Bergquist, A., & Wagstaff, G. F. (1994). Cognitive correlates of functional and dysfunctional impulsivity. *Personality and Individual Differences, 17,* 289–292.

Butter, C. M., & Snyder, D. R. (1972). Alterations in aversive and aggressive behaviors following orbital frontal lesions in rhesus monkeys. *Acta Neurobiologica Experimentalis, 32,* 525–566.

Casey, B. J., Orendi, J. L., Trainor, R. J., Schuber, A. B., & Noll, D. C. (1997, June). *Functional MRI study of sex differences in prefrontal activation during performance of a Go-No-Go task.* Paper presented at the Second International Conference of Functional Mapping of the Human Brain, Boston, MA.

Cattell, R. B. (1973). *Technical supplement for the Culture Fair Intelligence Tests Scales 2 and 3.* Champaign, IL: Institute for Personality and Ability Testing.

Cicerone, K. D., & Tanenbaum, L. N. (1997). Disturbance of social cognition after traumatic orbitofrontal brain injury. *Archives of Clinical Neuropsychology, 12,* 173–188.

Cohen, L. (1959). Perception of reversible figures after brain injury. *Archives of Neurology and Psychiatry, 81,* 765–775.

Crook, T. H., West, R. L., & Larrabee, G. J. (1993). The driving-reaction time test: Assessing age declines in dual-task performance. *Developmental Neuropsychology,* 9, 31–39.

Cummings, J. L. (1985). *Clinical neuropsychiatry.* New York: Grune & Stratton.

Curran, S. M., Murray, C. M., Van Beck, M., Dougall, N. et al. (1993). A single photon emission computerized tomography study of regional brain function in elderly patients with major depression and with Alzheimer-type dementia. *British Journal of Psychiatry, 163,* 155–165.

Damasio, A. R. (1991). Concluding comments. In H. S. Levin, H. M. Eisenberg, & A. L. Benton (Eds.), *Frontal lobe function and dysfunction.* New York: Oxford University Press.

Damasio, A. R., & Anderson, S. W. (1993). The frontal lobes. In K. M. Heilman & E. Valenstein (Eds.), *Clinical Neuropsychology* (3rd ed.). New York: Oxford University Press.

Damasio A. R., Tranel D., & Damasio H. (1990). Individuals with sociopathic behavior caused by frontal damage fail to respond autonomically to social stimuli. *Behavioral Brain Research, 41,* 81–94.

Davidson, R. J. (1992). Anterior cortical asymmetry and the nature of emotion. *Brain and Cognition, 20,* 125–151.

Dempster, F. N., & Brainerd, C. J. (1995). *Interference and inhibition in cognition.* New York: Academic Press.

D'Esposito, M., Detre, J. A., Alsop, D. C., & Shin, R. K. (1995). The neutral basis of the central executive system of working memory. *Nature, 378,* 279–281.

D'Esposito, M., Onishi, K., Thompson, H., Robinson, K. et al. (1996). Working memory impairments in multiple sclerosis: Evidence from a dual-task paradigm. *Neuropsychology, 10,* 51–56.

Diamond, A., & Taylor, C. (1996). Development of an aspect of executive control: Development of the abilities to remember what I said and to "Do as I say, not as I do". *Developmental Psychobiology, 29,* 315–334.

Diehl, M., Willis, S. L., & Schaie, K. W. (1995). Everyday problem solving in older adults: Observational assessment and cognitive correlates. *Psychology and Aging, 10,* 478–491.

Duncan, J., Burgess, P., & Emslie, H. (1995). Fluid intelligence after frontal lobe lesions. *Neuropsychologia, 33,* 261–268.

Dywan, J., Roden, R., & Murphy, T. (1995). Orbitofrontal symptoms are predicted by mild head injury among normal adolescents. *Journal of the International Neuropsychological Society, 1,* 121.

Foos, P. W. (1995). Working memory resource allocation by young, middle-aged, and old adults. *Experimental Aging Research, 21,* 239–250.

Fossum, B., Holmberg, H., & Reinvang, I. (1992). Spatial and symbolic factors in performance on the Trail Making Test. *Neuropsychology, 6,* 71–75.

Fox, E. (1995). Negative priming from ignored distractors in visual selection: A review. *Psychonomic Bulletin and Review, 2,* 145–173.

Franke, P., Maier, W., Hardt, J., Frieboes, R. et al. (1993a). Assessment of frontal lobe functioning in schizophrenia and unipolar major depression. *Psychopathology, 26,* 76–84.

Franke, P., Maier, W., Hardt, J., & Hain, C. (1993b). Cognitive functioning and anhedonia in subjects at risk for schizophrenia. *Schizophrenia Research, 10,* 77–84.

Fuster, J. M. (1993). Single-unit studies of the prefrontal cortex. In E. Perecman (Ed.), *The frontal lobes revisited.* New York: IRBN Press.

Gardner, H., Brownell, H. H., Wapner, W., & Nichelow, D. (1983). Missing the point: The role of the right hemisphere in processing of complex linguistic materials. In E. Perecman (Ed.), *Cognitive processing in the right hemisphere.* New York: Academic Press.

Gass, C. S., & Daniel, S. K. (1990). Emotional impact on Trail Making Test performance. *Psychological Reports, 67,* 435–438.

Gaudino, E. A., Geisler, M. W., & Squires, N. K. (1995). Construct validity in the Trail Making Test: What makes Part B harder? *Journal of Clinical and Experimental Neuropsychology, 17,* 529–535.

Gelowitz, D. L., & Paniak, C. E. (1992). Cross-validation of the Short Category Test-Booklet Format. *Neuropsychology, 6,* 287–292.

Gerstadt, C. L., Hong, Y. J., & Diamond, A. (1994). The relationship between cognition and action: Performance of children 3 1/2–7 years old on a Stroop-like day-night test. *Cognition, 53,* 129–153.

Gilhooly, K. J., Logie, R. H., Wetherick, & Wynn, V. (1993). Working memory and strategies in syllogistic-reasoning tasks. *Memory and Cognition, 21,* 115–124.

Ginsburg, N., & Wiegersma, S. (1991). Response bias and the generation of random sequences. *Perceptual and Motor Skills, 72,* 1332–1334.

Godbout, L., & Doyon, J. (1995). Mental representation of knowledge following frontal-lobe or postrolandic lesions. *Neuropsychologia, 33,* 1671–1696.

Goel, V., & Grafman, J. (1995). Are the frontal lobes implicated in "planning" functions? Interpreting data from the Tower of Hanoi. *Neuropsychologia, 33,* 623–642.

Golden, C. J. (1978). *The Stroop Color and Word Test: Manual.* Chicago: Stoelting.

Goldenberg, G., Oder, W., Spatt, J., & Podreka, I. (1992). Cerebral correlates of disturbed executive function and memory in survivors of severe closed head injury: A SPECT study. *Journal of Neurology, Neurosurgery, and Psychiatry, 55,* 362–368.

Goldman-Rakic, P. S. (1987). Circuitry of the prefrontal cortex and the regulation of behavior by representational knowledge. In F. Plum & V. Mountcastle (Eds.), *Handbook of Physiology* (Vol. 5; pp. 373–417). Bethesda, MD: American Physiological Society

Goldstein, G. (1990). Neuropsychological heterogeneity in schizophrenia: A consideration of abstraction and problem-solving abilities. *Archives of Clinical Neuropsychology, 5,* 251–264.

Grace, J., Malloy, P. F., & Stout, J. (1997, November). *Assessing frontal behavioral syndromes: Reliability and validity of the Frontal Lobe Personality Scale.* Paper presented at the National Academy of Neuropsychology meeting in New Orleans, LA.

Granholm, E. Asarnow, R. F., & Marder, S. R. (1996). Dual-task performance operating characteristics, resource limitations, and automatic processing in schizophrenia. *Neuropsychology, 10,* 11–21.

Green, M. F. (1996). What are the functional consequences of neurocognitive deficits in schizophrenia? *American Journal of Psychiatry, 153,* 321–330.

Greenwood, P., & Parasuraman, R. (1991). Effects of aging on the speed and attentional cost of cognitive operations. *Developmental Neuropsychology, 7,* 421–434.

Greve, K. W., Farrell, J. F., Besson, P. S., Crouch, J. A. (1995). A psychometric analysis of the California Card Sorting Test. *Archives of Clinical Neuropsychology, 10,* 265–278.

Grigsby, J., & Kaye, K. (1992). *The Behavioral Dyscontrol Scale: Manual.* Denver, CO: Authors.

Grodzinsky, G. M., & Diamond, R. (1992). Frontal lobe functioning in boys with attention-deficit hyperactivity disorder. *Developmental Neuropsychology, 8,* 427–445.

Halperin, J. M., Newcorn J. H., Matier, K., Bedi, G. et al. (1995). Impulsivity and the initiation of fights in children with disruptive behavior disorders. *Journal of Child Psychology and Psychiatry and Allied Disciplines, 36,* 119–1211.

Halperin, J. M., Wolf, L. E., Greenblatt, E. R., Young, G. et al. (1991). Subtype analysis of commission errors on the continuous performance test in children. *Developmental Neuropsychology, 7,* 207–217.

Harlow, H. F. (1949). The formation of learning sets. *Psychological Review, 56,* 51–65.

Harnadek, M. C. S., & Rourke, B. (1994). Principal identifying features of the syndrome of nonverbal learning disabilities in children. *Journal of Learning Disabilities, 27,* 144–154.

Hayslip, B., Maloy, R. M., & Kohl, R. (1995). Long-term efficacy of fluid ability interventions with older adults. *Journals of Gerontology Series B Psychological Sciences and Social Sciences, 50B,* 141–149.

Hazlett, E. A., Dawson, M. E., Buchsbaum, M. S., & Nuechterlein, K. H. (1993). Reduced regional brain glucose metabolism assessed by positron emission tomography in electrodermal nonresponder schizophrenics: A pilot study. *Journal of Abnormal Psychology, 102,* 39–46.

Healey, J. M., Newcorn, J. H., Halperin, J. M., Wolf, L. E. et al. (1993). The factor structure of ADHD items in DSM-III-R: Internal consistency and external validation. *Journal of Abnormal Child Psychology, 21,* 441–453.

Heidrich, S. M., & Denney, N. W. (1994). Does social problem solving differ from other types of problem solving during the adult years? *Experimental Aging Research, 20,* 105–126.

Heller, W. (1993). Neuropsychological mechanisms of individual differences in emotion, personality, and arousal. Special Section: Neuropsychological perspectives on components of emotional processing. *Neuropsychology, 7,* 476–489.

Horn, J. L., & Cattell, R. B. (1966). Refinement and test of the theory of fluid and crystallized intelligence. *Journal of Educational Psychology, 57,* 253–276.

Humberstone, M., Clare, S., Hykin, J., Morris, P. G., Sawle, G. V. (1996, May). *The roles of pre-SMA and SMA-proper in motor initiation and decision making: Studies using 'Event Related' fMRI.* Paper presented at the Second International Conference of Functional Mapping of the Human Brain, Copenhagen, Denmark.

Huttenlocher, P. R. (1979). Synaptic density in human frontal cortex-developmental changes and effects of aging. *Brain Research, 163,* 195–205.

Isaacson, R. L. (1982). *The limbic system* (2nd. ed.). New York: Plenum Press.

Itagaki, F., Niwa, S. I., Itoh, K., & Momose, T. (1995). Random number generation and the frontal cortex. *International Journal of Psychophysiology, 19,* 79–80.

Iversen, S., & Mishkin, M. (1970). Perseverative interference in monkeys following selective lesions of the inferior prefrontal convexity. *Experimental Brain Research, 11,* 376–386.

Jacobsen, C. F. (1935). Function of the frontal association area in primates. *Archives of Neurology and Psychiatry, 33,* 558–569.

Jennett, B., & Teasdale, G. (19981). *Management of head injuries.* Philadelphia: Davis.

Johanson, A., Smith, G., Risber, J., Silfverskiold, P. et al. (1992). Left orbital frontal activation in pathological anxiety. *Anxiety, Stress and Coping: An International Journal, 5,* 313–328.

Johnstone, B., Holland, D., & Hewett, J. E. (1997). The construct validity of the Category Test: Is it a measure of reasoning or intelligence. *Psychological Assessment, 9,* 28–33.

Jones-Gotman, M., & Milner, B. (1977). Design fluency: The invention of nonsense drawings after focal cortical lesions. *Neuropsychologia, 15,* 653–674.

Kane, M. J., Hasher, L., Stoltzfus, E. R., Zack, R. T., & Connelly, S. L. (1994). Inhibitory attentional mechanisms and aging. *Psychology and Aging, 9,* 103–112.

Kaufman, A. S., Ishikuma, T., & Kaufman, N. L. (1994). A Horn analysis of the factors measured by the WAIS-R, Kaufman Adolescent and Adult Intelligence Test (KAIT), and two new brief cognitive measures for normal adolescents and adults. *Assessment, 1,* 353–366.

Kaye, K., Grigsby, J., Robbins, L. J., & Korzun, B. (1990). Prediction of independent functioning and behavior problems in geriatric patients. *Journal of American Geriatrics Society, 38,* 1304–1310.

Kelly, M. D., Kundert, D. K., & Dean, R. S. (1992). Factor analysis and matrix invariance of the HRNB-C Category Test. *Archives of Clinical Neuropsychology, 7,* 415–418.

Kixmiller, J. S., Verfaellie, M., Chase, K. A., & Cermak, L. S. (1995). Comparison of figural intrusion errors in three amnesic subgroups. *Journal of the International Neuropsychological Society, 1,* 561–567.

Krikorian, R., Bartok, J., & Gay, N. (1994). Tower of London procedure: A standard method and developmental data. *Journal of Clinical and Experimental Neuropsychology, 16,* 840–850.

Laatsch, L., & Choca, J. (1994). Cluster-branching methodology for adaptive testing and the development of the Adaptive Category Test. *Psychological Assessment, 6,* 345–351.

Lafleche, G., & Albert, M. S. (1995). Executive function deficits in mild Alzheimer's disease. *Neuropsychology, 9,* 313–320.

Lamberty, G. J., Putnam, S. H., Chatel, D. M., Bieliauskas, L. A. et al. (1994). Derived Trail Making Test indices: A preliminary report. *Neuropsychiatry, Neuropsychology, and Behavioral Neurology, 7,* 230–234.

LaPierre, D., Braun, C. M. J., & Hodgins, S. (1995). Ventral frontal deficits in psychopathy: Neuropsychological test findings. *Neuropsychologia, 33,* 139–151.

Larrabee, G. J., & Curtiss, G. (1995). Construct validity of various verbal and visual memory tests. *Journal of Clinical and Experimental Neuropsychology, 17,* 536–547.

Lassiter, K. S., D'Amato, R. C., Raggio, D. J., Whitten, J. C. M. et al. (1994). The construct specificity of the Continuous Performance Test: Does inattention relate to behavior and achievement? *Developmental Neuropsychology, 10,* 179–188.

Lavoie, M. E., & Charleboie, P. (1994). The discriminant validity of the Stroop Color and Word Test: Toward a cost-effective strategy to distinguish subgroups of disruptive preadolescents. *Psychology-in-the-Schools, 31,* 98–107.

Levin, H. S., Grafman, J., & Eisenberg, H. M. (Eds.). (1987). *Neurobehavioral recovery from head injury.* New York: Oxford University Press.

Lezak, M. D. (1995). *Neuropsychological assessment* (3rd ed.). New York: Oxford University Press.

Logie, R. H., Gilhooly, K. J., & Wynn, V. (1994). Counting on working memory in arithmetic problem solving. *Memory and Cognition, 22,* 395–410.

Logue, P. E., & Allen, L, (1971). WAIS predicted Category Test score with the Halstead Neuropsychological Battery. *Perceptual and Motor Skills, 33,* 1095–1099.

Luria, A. R. (1966). *Higher cortical functions in man.* New York: Basic Books.

Luria, A. R. (1970). *Traumatic aphasia.* Mouton: The Hague.

Luria, A. R., & Tsvetkova, L. S. (1990). *The neuropsychological analysis of problem solving.* Orlando, FL: Paul M. Deutsch Press.

MacLeod, C. M. (1992). The Stroop task: The "gold standard" of attentional measures. *Journal of Experimental Psychology: General, 121,* 12–14.

MacNiven, E., & Finlayson, M. A. (1993). The interplay between emotional and cognitive recovery after closed head injury. *Brain Injury, 7,* 241–246.

Malloy, P., Bihrle, A., Duffy, J., & Cimino, C. (1993). The orbitomedial frontal syndrome. *Archives of Clinical Neuropsychology, 8,* 185–201.

Martin, A., Wiggs, C. L., Lalonde, F., & Mack, C. (1994). Word retrieval to letter and semantic cues: A double dissociation in normal subjects using interference tasks. *Neuropsychologia, 32,* 1487–1494.

Martzke, J. S., Swan, C. S., & Varney, N. R. (1991). Posttraumatic anosmia and orbital frontal damage: Neuropsychological and neuropsychiatric correlates. *Neuropsychology, 5,* 213–225.

McGrew, K. (1994). *Clinical interpretation of the Woodcock-Johnson Tests of Cognitive Ability Revised.* Boston: Allyn & Bacon.

Mickanin, J., Grossman, M., Onishi, K., Auriacombe, S. et al. (1994). Verbal and nonverbal fluency in patients with probable Alzheimer's disease. *Neuropsychology, 8,* 385–394.

Miller, G. A., Galanter, E. H., & Pribram, K. H. (1960). *Plans and the structure of behavior.* New York: Henry Holt & Co.

Mountain, M. A., & Snow, W. G. (1993). Wisconsin Card Sorting Test as a measure of frontal pathology: A review. *The Clinical Neuropsychologist, 7,* 108–118.

Moscovitch, M. (1994). Cognitive resources and dual-task interference effects at retrieval in normal people: The role of the frontal lobes and medial temporal cortex. *Neuropsychology, 8,* 524–534.

Nadler, J. D., Richardson, E. D., Malloy, P. F., Marran, M. D. et al. (1993). The ability of the Dementia Rating Scale to predict everyday functioning. *Archives of Clinical Neuropsychology, 8,* 449–460.

Nauta, W. J. H. (1971). The problem of the frontal lobe: A reinterpretation. *Journal of Psychiatry Research, 8,* 167–187.

Neill, W. T., Valdes, L. A., & Terry, K. M. (1995). Selective attention and the inhibitory control of cognition. In F. N. Dempster & C. J. Brainerd (Eds.), *Interference and inhibition in cognition.* New York: Academic Press.

Nigg, J. T., Hinshaw, S. P., & Halperin, J. M. (1996). Continuous Performance Test in boys with attention deficit disorder: Methylphenidate dose response and relations with observed behaviors. *Journal of Child Clinical Psychology, 25,* 330–340.

Oder, W., Goldenberg, G., Spatt, J., & Podreka, I. (1992). Behavioural and psychosocial sequelae of severe closed head injury and regional cerebral blood flow: A SPECT study. *Journal of Neurology, Neurosurgery, and Psychiatry, 55,* 475–480.

O'Grady, R. S., Crain, L. S., & Kohn, J. (1995). The prediction of long-term functional outcomes of children with cerebral palsy. *Developmental Medicine and Child Neurology, 37,* 997–1005.

Osmon, D. C. (1996). Understanding symptoms of medial frontal lobe disorder: A clinical case study. *Journal of Clinical Psychology in Medical Settings, 3,* 23–39.

Osmon, D. C., & Suchy, Y., (1996). Fractionating frontal lobe functions: Factors of the Milwaukee Card Sorting Test. *Archives of Clinical Neuropsychology, 11,* 541–552.

Osmon, D. C., Zigun, J. R., Suchy, Y., & Blint, A. (1996). Whole brain fMRI activation on Wisconsin-like card sorting measures: Clues to test specificity. *Brain and Cognition, 30,* 308–310.

Ott, B. R., Lafleche, G., & Whelihan, W. M. (1996). Impaired awareness of deficits in Alzheimer disease. *Alzheimer Disease and Associated Disorders, 10,* 68–76.

Owen, A. M., Sahakian, B. J., Hodges, J. R., Summers, B. A. et al. (1995). Dopamine-dependent frontostriatal planning deficits in early Parkinson's disease. *Neuropsychology, 9,* 126–140.

Papez, J. W. (1937). A proposed mechanism of emotion. *Archives of Neurology and Psychiatry, 38,* 725–744.

Pardo, J., Pardo, P., Janer, K., & Raichle, M. (1990). The anterior cingulate cortex mediates processing selection in the Stroop attentional conflict paradigm. *Proceedings of the National Academy of Sciences, 87,* 256–259.

Pashler, H. (1994). Dual-task interference in simple tasks: Data and theory. *Psychological Bulletin, 116,* 220–244.

Passingham, R.E. (1993). *The frontal lobes and voluntary action.* New York: Oxford University Press.

Passini, R., Rainville, C., Marchant, N., & Joanette, Y. (1995). Wayfinding in dementia of the Alzhemier type: Planning abilities. *Journal of Clinical and Experimental Neuropsychology, 17,* 820–832.

Perret, E. (1973). The left frontal lobe of man and the suppression of habitual responses in verbal categorical behavior. *Neuropsychologia, 12,* 323–330.

Perrine, K. (1993). Differential aspects of conceptual processing in the Category Test and Wisconsin Card Sorting Test. *Journal of Clinical and Experimental Neuropsychology, 15*, 461–473.

Petersen, S. E., Fox, P. T., Posner, M. I., Mintun, M., & Raichle, M. E. (1988). Positron emission tomographic studies of the cortical anatomy of single-word processing. *Nature, 331*, 585–589.

Petrides, (1991). Learning impairments following excisions of the primate frontal cortex. In H. S. Levin, H. M. Eisenberg, & A. L. Benton (Eds.), *Frontal lobe function and dysfunction*. New York: Oxford University Press.

Pliszka, S. R. (1992). Comorbidity of attention-deficit hyperactivity disorder and overanxious disorder. *Journal of the American Academy of Child and Adolescent Psychiatry, 31*, 197–203.

Poon, L. W., Messner, S., Martin, P., Noble, C. A. et al. (1992). The influences of cognitive resources on adaptation and old age. *International Journal of Aging and Human Development, 34*, 31–46.

Porteus, S. D. (1973). *Porteus Maze Test: Fifty years' application*. Palo Alto, CA: Pacific Books.

Posner, M. I., & Petersen S. E. (1990). The attention system of the human brain. *Annual Review of Neuroscience, 13*, 25–42.

Pribram, K. H. (1991). *Brain and perception: Holonomy and structure in figural processing*. Hillsdale, NJ: Lawrence Erlbaum.

Prinzmetal, W., Henderson, D., & Ivry, R. (1995). Loosening the constraints on illusory conjunctions: Assessing the roles of exposure duration and attention. *Journal of Experimental Psychology: Human Perception and Performance, 21*, 1362–1375.

Rabinowitz, F. M., Dunlap, W. P., Grant, M. J., & Campoine, J. C. (1989). The rules used by children and adults in attempting to generate random numbers. *Journal of Mathematical Psychology, 33*, 227–287.

Raz, N., Torres, I. J., Spencer, W. D., Milman, D. et al. (1993). Neuroanatomical correlates of age-sensitive and age-invariant cognitive abilities: An in vivo MRI investigation. *Intelligence, 17*, 407–422.

Reed, M. A., & Derryberry, D. (1995). Temperament and response processing: Facilitatory and inhibitory consequences of positive and negative motivational states. *Journal of Research in Personality, 29*, 59–84.

Reitan, R. M., & Wolfson, D. (1993). *The Halstead-Reitan Neuropsychological Test Battery: Theory and clinical interpretation*. (2nd ed.). Tucson, AZ: Neuropsychology Press.

Reitan, R. M., & Wolfson, D. (1995). Category Test and Trail Making Test as measures of frontal lobe functions. *The Clinical Neuropsychologist, 9*, 50–56.

Rezai, K, Andreasen, N. C., Alliger, R., Cohen, G., Swayze II, V., & O'Leary, D. S. (1993). The neuropsychology of the prefrontal cortex. *Archives of Neurology, 50*, 636–642.

Ricci, C., & Blundo, C. (1990). Perception of ambiguous figures after focal brain lesions. *Neuropsychologia, 28*, 1163–1173.

Richer, F., Decary, A., Lapierre, M. F., Rouleau, I. et al. (1993). Target detection deficits in frontal lobectomy. *Brain and Cognition, 21*, 203–211.

Ricker, J. H., Axelrod, B. N., & Houtler, B. D. (1996). Clinical validation of the oral trail making test. *Neuropsychiatry, Neuropsychology, and Behavioral Neurology, 9*, 50–53.

Rizzo, L., Danion, J-M., Van der Linden, M., Grange, D. et al. (1996). Impairment of memory for spatial context in schizophrenia. *Neuropsychology, 10*, 376–384.

Robertson, C., Hazlewood, R., & Rawson, M. D. (1996). The effects of Parkinson's disease on the capacity to generate information randomly. *Neuropsychologia, 34*, 1069–1078.

Roland, P. (1993). *Brain activation*. New York: John Wiley & Sons.

Rosenbaun, G., & Taylor, M. J. (1996). Attentional processing in schizophrenia: Experimental induction of the crossover effect. *Cognitive Therapy and Research, 20*, 195–208.

Rosvold, H. E., Mirsky, A. F., Sarason, I., Bransome, E. D., Beck, L. H. (1956). A continuous performance test of brain damage. *Journal of Consulting Psychology, 20*, 343–350.

Rubia, K., Overmeyer, S., Taylor, E., Bullmore, E., Brammer, M., Williams, S., Simmons, A., & Andrew, Ch. (1997, June). *Response Selection and "Temporal Bridging" in a delay task*. Paper presented at the Second International Conference of Functional Mapping of the Human Brain, Boston, MA.

Rueckert, L., & Grafman, J. (1996). Sustained attention deficits in patients with right frontal lesions. *Neuropsychologia, 34*, 953–963.

Ruff, R., & Light, R. (1987). The Ruff figural fluency test: A normative study with adults. *Developmental Neuropsychology, 3*, 37–51.

Salthouse, T. A., & Meinz, E J. (1995). Aging, inhibition, working memory, and speed. *Journal of Gerontology: Series B Psychological Sciences and Social Sciences, 50B*, 297–306.

Saski, K., Gemba, H., Nambu, A., & Matsuzaki, R. (1993). No-go activity in the frontal association cortex of human subjects. *Neuroscience Research, 18*, 249–252.

Schore, A. N. (1994). *Affect regulation and the origin of the self: The neurobiology of emotional development*. Hillsdale, NJ: Lawrence Erlbaum.

Schroeder, J., Buchsbaum, M. S., Siegel, B. V., Geider, F. J. et al. (1994). Patterns of cortical activity in schizophrenia. *Psychological Medicine, 24*, 947–955.

Schwartz, M. F., Montgomery, M. W., Fitzpatrick-DeSalme, E. J., Ochipa, C. et al. (1995). Analysis of a disorder of everyday action. *Cognitive Neuropsychology, 12*, 863–892.

Sellers, A. H., & Nadler, J. D. (1992). A survey of current neuropsychological assessment procedures used for different age groups. *Psychotherapy in Private Practice, 11*, 47–57.

Shallice, T., & Burgess, P. (1991). Higher-order cognitive impairments and frontal lobe lesions in man. In H. S. Levin, H. M. Eisenberg, & A. L. Benton (Eds.), *Frontal lobe function and dysfunction*. New York: Oxford University Press.

Sengstock, S., Osmon, D. C., & Suchy, Y. (1997). Investigation into the Stroop paradigm: A Functional Magnetic Resonance Imaging study. Unpublished manuscript.

Shallice, T. (1982). Specific impairments of planning. *Philosophical Transactions of the Royal Society London, 298*, 199–209.

Siegel, B. V., Nuechterlein, K. H., Abel, L., & Wu, J. C. (1995). Glucose metabolic correlates of continuous performance test performance in adults with a history of infantile autism, schizophrenics, and controls. *Schizophrenia Research, 17*, 85–94.

Simon, J. R., & Baker, K. L. (1995). Effect of irrelevant information on the time to enter and retrieve relevant information in a Stroop-type task. *Journal of Experimental Psychology: Human Perception and Performance, 21*, 1028–1043.

Soininen, H. S., Karhu, J., Partanen, J., Paakkonen, A. et al. (1995). Habituation of auditory N100 correlates with amygdaloid volumes and frontal functions in age-associated memory impairment. *Physiology and Behavior, 57*, 927–935.

Suchy, Y., Blint, A., & Osmon, D. C. (1997). Behavioral Dyscontrol Scale: Criterion and predictive validity in an inpatient rehabilitation unit population. *The Clinical Neuropsychologist, 11*, 258–265.

Sullivan, E. V., Mathalon, D. H., Zipursky, R. B., Kersteen-Tucker, Z., Knight, R. T., & Pfefferbaum, A. (1993). Factors of the Wisconsin Card Sorting Test as measures of frontal-lobe function in schizophrenia and in chronic alcoholism. *Psychiatry Research, 46*, 175–199.

Tranel, D., & Damasio, H. (1994). Neuroanatomical correlates of electrodermal skin conductance responses. *Psychophysiology, 31*, 427–438.

Trenerry, M. R., Crosson, B., DeBoe, J., & Leber, W. R. (1989). *Stroop Neuropsychological Screening Test: Manual*. Odessa, Florida: PAR.

Trestman, R. L., Keefe, R. S. E., Mitropoulou, V., Harvey, P. D. et al. (1995). Cognitive function and biological correlates of cognitive performance in schizotypal personality disorder. *Psychiatry Research, 59*, 127–136.

Tulving, E., Kapur, S., Craik, F. I. M., Moscovitch, M., & Houle, S. (1994). Hemispheric encoding/retrieval asymmetry in episodic memory: Positron emission tomography findings. *Proceedings of the National Academy of Sciences, 91*, 2016–2020.

Van der Linden, M., Bruyer, R., Roland, J., Schils, J. P. (1993). Proactive interference in patients with amnesia resulting from anterior communicating artery aneurysm. *Journal of Clinical and Experimental Neuropsychology, 15*, 525–536.

Vendrell, P. Junque, C., Pujol, J., Jurado, M. (1995). The role of prefrontal regions in the Stroop task. *Neuropsychologia, 33*, 341–352.

Visser, M., Das-Smaal, E., & Kwakman, H. (1996). Impulsivity and negative priming: Evidence for diminished cognitive inhibition in impulsive children. *British Journal of Psychology, 87*, 131–140.

Wang, J. J., & Kaufman, A. S. (1993). Changes in fluid and crystallized intelligence across the 20- to 90-year age range on the K-BIT. *Journal of Psychoeducational Assessment, 11*, 29–37.

Wahlin, T. B. R., Backman, L., Wahlin, A., & Winblad, B. (1996). Trail Making Test performance in a community-based sample of healthy very old adults: Effects of age on completion time, but not on accuracy. *Archives of Gerontology and Geriatrics, 22*, 87–102.

Weinberger, D. R., Berman, K. F., & Zec, R. F. (1986). Physiologic dysfunction of dorsolateral prefrontal cortex in schizophrenia: I. Regional cerebral blood flow evidence. *Archives of General Psychiatry, 43*, 114–125.

Widaman, K. F., Carlson, J. S., Saetermoe, C. L., & Galbraith, G. C. (1993). The relationship of auditory evoked potentials to fluid and crystallized intelligence. *Personality and Individual Differences, 15*, 205–217.

Wu, J. C., Gillin, J. C., Buchsbaum M. S., Hershey, T. et al. (1991). The effect of sleep deprivation on cerebral glucose metabolic rate in normal humans assessed with positron emission tomography. *Sleep, 14*, 155–162.

Yarnell, P. R., & Rossie, G. V. (1990). Bifrontal brain trauma and "good outcome" personality changes: Phineas P. Gage syndrome updated. *Journal of Neurologic Rehabilitation, 4*, 9–16.

Yingling, C. D., & Skinner, J. E. (1975). Regulation of unit activity in nucleus reticularis thalami by the mesencephalic reticular formation and the frontal granular cortex. *Electoencephalography and Clinical Neurophysiology, 39*, 635–642.

Zatorre, R. J., Evans, A. C., & Meyer, E. (1994). Neural mechanisms underlying melodic perception and memory for pitch. *Journal of Neuroscience, 14*, 1908–1919.

Chapter 8

ECOLOGICAL VALIDITY:
PREDICTION OF EVERYDAY AND VOCATIONAL FUNCTIONING FROM NEUROPSYCHOLOGICAL TEST DATA

Robert J. Sbordone
Private Practice, Irvine, California

Thomas J. Guilmette
Southern New England Rehabilitation Center, St. Joseph's Hospital, Providence, Rhode Island and Eleanor Slater Hospital, Cranston, Rhode Island

Introduction

With the occurrence of acquired brain damage (e.g., traumatic brain injuries, carbon monoxide poisoning, toxic exposure, etc.) caused by the fault of another party, the forensic neuropsychologist is expected to determine, based on the patient's test data, not only whether the patient has sustained a brain injury, but also predict how the injury will affect this patient's ability to live independently, return to work or school, or maintain competitive employment, even though the neuropsychological tests utilized were not designed to make such predictions (Sbordone, 1996). For example, when a neuropsychologist testifies about such issues in the courtroom, based on his or her test data, the neuropsychologist's opinions typically have little or no experimental support and are often based upon a number of faulty assumptions (Bach, 1993). As a result of the involvement of neuropsychologists in the forensic arena, the issues of test reliability and validity must be broadened to include the issue of whether our tests have adequate ecological validity. Thus, neuropsychologists' claims of high test reliability and validity may be overshadowed by the fact that a given test may lack ecological validity, in the sense that a particular test score may or may not have bearing on the patient's ability to function in his or her environment. This chapter will examine the issue of ecological validity and critically evaluate the effectiveness of neuropsychological tests to predict the everyday functioning of brain-injured patients and their ability to function in the workplace or school.

Ecological validity

Definition

Ecological validity can be defined as the functional and predictive relationship between the patient's performance on a set of neuropsychological tests and the patient's behavior in a variety of real world settings (e.g., home, work, school, community, etc.; Sbordone, 1996).

Implicit within this definition is the assumption that the neuropsychologist's choice of tests to be administered will assess cognitive and behavioral functions that are germane to the settings the patient currently functions within, is considering returning to, or plans to function within in the future. In order to accomplish this, the neuropsychologist should gather as much information as possible about these settings from the patient, the patient's family, friends, co-workers and significant others, as well as specialists in the field of vocational and occupational therapy to determine their demand characteristics. Such information should provide the neuropsychologist with an appreciation of the demands each of these settings place on the patient's cognitive strengths and deficits, goals, objectives, premorbid skills, and biological systems (e.g., medical conditions, health, disabilities, etc.).

The demands of a particular setting may vary considerably as a consequence of its nature, purpose, and objectives. For example, the demand characteristics within the patient's workplace may fluctuate considerably as a function of such factors as: economics, absenteeism, sickness, organizational or administrative changes, competition from other companies, changes in the marketplace, changes in equipment, marketing goals and/or strategies, etc. Depending upon the interface between the patient's cognitive strengths and deficits, and the demand characteristics of the particular setting that the patient finds himself within, the setting may either compensate for or exacerbate the patient's cognitive and/or behavioral impairments. When the former occurs, the patient may develop a secondary psychological disorder (e.g., generalized anxiety reaction, adjustment reaction with depressed mood, or major depression). When the latter occurs, the cognitive functioning of the brain-injured patient may significantly diminish according to the principle of conditional neurological lesion (Sbordone, 1987, 1988, 1991).

Principle of a conditional neurological lesion

This principle argues that the behavioral manifestations of a neurological insult, such as a brain injury, are a function of the degree to which the individual is:

stressed, fatigued, emotionally distressed or confronted by excessive metabolic demands. Thus, the cognitive and behavioral functioning of the brain-injured patient would be expected to diminish if the person becomes anxious or depressed, or if exposed to stress or failure, or is terminated from his or her job. For example, a brain-injured patient who is mildly to moderately impaired with respect to his/her organizational and planning skills may function fairly well in an environment that is familiar, highly structured, and contains supportive and caring individuals, who have accepted the patient's cognitive deficits and are willing to provide timely cues and prompts. If, on the other hand, this patient is placed in a rather disorganized and unfamiliar environment that contains individuals who are critical and offer little in the way of cues, prompts or support, this patient will most likely develop secondary psychological problems that will cause his or her cognitive functioning to significantly diminish. The net effect may be the inability of this patient to function effectively in this environment.

Test environment

Brain-injured patients have traditionally been tested in quiet environments that are relatively free of extraneous or distracting stimuli to optimize their test performance. This type of environment has been a standard in the field, for the expressed purpose of reducing extraneous negative influences within the evaluation process (Cronbach, 1984). In this manner, assessments across time and between different psychologists are made more comparable. However, a key issue is whether we can legitimately generalize neuropsychological test data obtained under such artificial conditions to the patient's everyday and vocational functioning. This is especially important if everyday environments relevant to the patient are noisy, disorganized, chaotic, contain numerous extraneous and distracting stimuli, or in other respects are highly dissimilar to the conditions under which the test data were obtained.

Modifications of test protocol

Neuropsychologists may consciously or unconsciously modify standard test instructions or procedures when they evaluate brain injured-patients, particularly when the patient has severe cognitive or behavioral impairments. For example, the examiner may repeat the test instructions several times, or permit the patient to spend several minutes warming up in order to take the test, as well as provide frequent cues and prompts during testing to facilitate performance. Unfortunately, these modifications may be recorded only infrequently by neuropsychologists or examiners, since considerable emphasis has traditionally been placed on the patient's quantitative test performance. Thus, while the patient's quantitative performance under such conditions may have placed him or her within the mild to moderately impaired range with respect to test norms, the patient otherwise may have performed in the severely impaired range or not have been able to take the test, if the neuropsychologist or examiner had not modified the test protocol to compensate for the patient's deficits (e.g., attention, initiation, etc.).

More importantly, the patient's quantitative test performance under such conditions may significantly underestimate the patient's cognitive and behavioral impairments and result in inaccurate predictions about the patient's ability to return to work or school, live independently, and function effectively within the community. In order to increase the generalizability of our test results from the office to real world settings, it is incumbent upon the clinician to record variations in the instructions that were given to the patient, or the test procedures utilized, and not focus exclusively on the test scores.

Conditionality

Definition
Conditionality refers to specific modifications of the testing protocol and/or compensatory interventions utilized by the examiner during testing that permit the patient to ignore extraneous or distracting stimuli, attend to the examiner and task, comprehend and recall the test instructions, perform according to test requirements, compensate for problems of initiation and organization, develop positive emotional and/or psychological states and attitudes, and perform at an optimal level throughout the entire test or battery (Sbordone, 1996).

Masking of the patient's cognitive and behavioral impairments
The cognitive and behavioral impairments of brain-injured patients can be masked if the patient is tested in a quiet and highly structured setting, particularly if the examiner exercises considerable patience and treats the patient in a gentle and calm manner. As a consequence, a brain-injured patient may not manifest behavioral impairments (e.g., poor frustration tolerance, use of crude or coarse language, irritability, and/or aggressive outbursts) in this setting, even though such behaviors are apparent and problematic in the patient's home and workplace.

Faulty recommendations
As a result of conditionality and the quiet and highly structured test setting, a neuropsychologist may find that a patient's quantitative test scores on a standardized neuropsychological battery fall within the normal range. Based upon these test findings and the examiner's behavioral observations, neuropsychologists may erroneously conclude that the patient is able to function independently in the community and/or return to work, particularly if they are unaware that the patient's major handicap is an inability to regulate his or her emotions and behavior in relatively non-structured or stressful environments. Unfortunately, in the instance of a behaviorally impaired individual

who performs well on psychometric measures of ability, inappropriate neuropsychological recommendations may also result in an emotionally traumatic experience for the patient and his or her family that may, in turn, exacerbate the patient's cognitive and behavioral difficulties. The complex task of evaluating such executive dysfunction is further discussed in Chapter 7.

Hazards of blind interpretation of neuropsychological test data

Many health and rehabilitation professionals, not to mention attorneys, have naively assumed that a brain-injured patient's cognitive and behavioral impairments can be determined simply by comparing the patient's neuropsychological test scores to normative standards. Unfortunately, the performance of patients during neuropsychological testing is frequently confounded by a variety of complex factors, which often go unrecognized. The following case example illustrates this.

Case example

A 47-year-old right-handed white male was referred by his attorney for neuropsychological testing. He arrived for his scheduled appointment at the designated time and stated that he had been rendered unconscious for several hours as a result of a motor vehicle accident that had occurred three years ago. He complained of numerous cognitive and somatic problems and denied a prior history of similar symptoms or head trauma. He indicated that he had graduated from high school with A's and B's and had been earning in excess of $100,000 a year as a construction superintendent prior to his accident. Neuropsychological testing revealed widespread cognitive impairments ranging from severe to profound. Based on the test data and the information provided by the patient, the neuropsychologist concluded that these impairments were caused by the severe brain injury the patient had sustained in the motor vehicle accident and that the severity of cognitive impairments precluded any hope of ever living independently and being competitively employed.

Approximately three months later, the patient was seen by another neuropsychologist, who carefully reviewed the patient's medical records, which revealed that the patient had never been rendered unconscious and was alert and oriented at the scene of the accident and in the emergency room. These records also revealed that the patient had a pre-existing history of severe cluster headaches, chronic low back pain, orthopedic problems, head trauma, psychiatric problems, and alcohol and drug abuse. A review of the patient's academic records revealed that he had only completed nine years of formal education and had been diagnosed as having an attention deficit-hyperactivity disorder and had been placed in special education classes for severe learning problems. These records also revealed that the patient was frequently absent or tardy and was frequently described by his teachers as irresponsible, manipulative, impulsive, and aggressive. Additional records revealed that this patient had been incarcerated in the past for insurance fraud, burglary, selling drugs, driving under the influence, and use of heroin and cocaine.

The second neuropsychologist also learned that two days prior to testing, the patient had fallen down a flight of stairs while intoxicated and had injured his lower back, which resulted in his drinking a quart of gin the night before testing. In addition, the patient admitted that he had taken several pain medications just prior to testing, which caused him to feel confused and disoriented. He also admitted that he had developed an intense dislike toward the examiner, had felt very "stressed" during testing, and had frequently entertained thoughts of physically assaulting the examiner.

The above example illustrates some of the numerous confounding factors that alone or in combination could have negatively impacted this patient's neuropsychological test performance, regardless of whether or not he had sustained a brain injury. The first neuropsychologist's opinion was based entirely upon patient self report and test data. This type of clinical decision-making process can be considered 'narrow' in that it fails to consider numerous confounding factors described above and hence can render interpretation of etiology and prediction of everyday and vocational functioning invalid.

Confounding factors

Sbordone and Purisch (1996) have identified a number of confounding factors that can produce poor performance during neuropsychological testing, regardless of whether or not the patient is brain-injured. These factors are: prior brain injury or insult, congenital or pre-existing neurological conditions, absences (seizures), acute pain, symptoms and impairments secondary to physical injuries, peripheral sensory impairment, peripheral motor impairment, current and chronic medical illness, sleep deprivation, excessive fatigue, alcohol/drug abuse, medication use, psychiatric illness, recent psychosocial stressors, suboptimal motivation and malingering, negative patient/examiner interaction, cultural/linguistic discrepancies, vocational and avocational background, test sophistication, and practice effects. The forensic neuropsychologist must determine the relative contribution that these factors may have on a patient's test performance so that predictions regarding the patient's everyday and vocational functioning will be ecologically valid.

Use of a vector analysis approach to evaluate the ecological validity of neuropsychological test data

Sbordone and Purisch (1996) have recommended that neuropsychologists utilize a vector analysis approach to evaluate the ecological validity of neuropsychological test data. Specifically, they argued that if the information about the patient's behavior obtained from *different* sources (e.g., significant others, medical, academic, vocational records, etc.) lines up *consistently* (i.e., from mathematics, conceptually produces a common line or vector) with the patient's neuropsychological test data, then there is a high probability that the test data is ecologically valid. If on the other hand, the test data are generally inconsistent with this information (e.g., the test data indicate that the person is severely impaired in the face of medical records that indicating no loss of consciousness or head trauma and appreciable work and educational attainments subsequent to the accident), the ecological validity of the test data appears low. Thus, the neuropsychologist should explore alternative explanations of the test data and avoid arriving at premature diagnoses, opinions, or predictions about the patient's ability to function in everyday and vocational settings.

Prediction of everyday functioning

The literature pertaining to the prediction of everyday functioning from neuropsychological test data mirrors the difficulties inherent in predicting vocational functioning which are discussed later in this chapter. That is, everyday skills can encompass such a vast domain and diverse abilities (e.g., dressing, grooming, bathing, handling finances, driving, household management, cooking, etc.) that the prediction of competency makes it difficult to compare studies, particularly when examined across different populations, and arrive at any definitive conclusions. However, within a forensic context, clinicians need to be aware of the existing literature in this domain, particularly with regard to the limitations of the confidence of their clinical judgement.

Over 10 years ago, McSweeny and colleagues (McSweeny, Grant, Heaton, Prigatano, & Adams, 1985) reported that neuropsychological measures can be used to predict everyday-life functioning in impaired persons (i.e., 303 patients with chronic obstructive pulmonary disease and concomitant neuropsychological impairment), but not in normals. They found that the best overall predictor of the quality of life was Trails B. However, more specific neuropsychological tests served as better predictors for specific areas of functioning (i.e., motor/psychomotor tests related better to physical mobility, self-care, home management, and socialization; while aphasia symptoms influenced communication). Although these authors acknowledged that the correlations between neuropsychological measures and everyday functioning were modest, their results supported some previous research of even 20 years earlier (Lorenze & Cancro, 1962) which described the value of neuropsychological tests in predicting outcome following stroke. Within the stroke population, the assessment of left neglect and aphasia test results can provide strong evidence for impaired dressing and communication skills, respectively. For example, research has demonstrated a significant relationship between constructional impairment and left neglect in right hemisphere strokes and dressing apraxia (Warren, 1981; Williams, 1967). Results from the Behavioral Inattention Test (Wilson, Cockburn & Halligan, 1987), a standardized functional and behavioral measure of neglect and inattention, have been found to be significantly related to problems with activities of daily living (ADL) tasks (Wilson, 1996). With aphasia, there is an obvious relationship between the deficits noted with language skills during aphasia testing and pragmatic communication impairments.

The nature of the aphasia exam itself lends itself to greater ecological validity than many "laboratory" measures of cognitive functioning. That is, the examiner is directly assessing the skills that the patient will need to demonstrate outside of the testing environment (e.g., comprehension, fluency, naming, etc.) to communicate effectively. As with vocational functioning, the better a clinician can test skills that approximate a specific "real world" ability, the greater the likelihood of accurate prediction.

In a review of the available literature, Williams (1996) indicated that the overall relationship between neuropsychological measures and everyday skills is low to moderate, with Pearson correlations generally ranging from .2 to .5. With regard to the prediction of everyday abilities from neuropsychological data alone, Williams (1996) further opined:

> There is no study which indicated that neuropsychological tests were strongly predictive of any everyday functional skill. Certainly patients who do very poorly, or are untestable, will have great problems at home and may not be able to function in any job. However, we do not have the clinical or experimental evidence to support predictive statements for patients who score higher. For patients in the borderline to normal ranges on our tests, available research suggests our predictions are essentially guesses (Williams, 1996, p.141).

The generally modest ability of neuropsychological tests to predict everyday functioning suggests that although test measures can provide some beneficial information under some circumstances, they cannot be used in isolation and clinicians will need to acknowledge the limitations of their measures in this domain. In addition, it is also important to bear in mind the distinction between a test being used as a diagnostic measure of some clinical syndrome or disorder versus a test being used to predict everyday skills. In the former, neuropsychological measures are used routinely and are usually quite helpful at identifying differences between clinical groups or between clinical and normal subjects. Predicting performance outside of a neuropsychology laboratory, however, is a different matter. Thus, a test can be diagnostically useful, but still not correlate significantly with everyday skills.

In order to provide more specific information about selected areas of everyday functioning and neuropsychological measures, the following sections will cover the prediction of driving, memory, and executive functioning.

Prediction of driving abilities

After a neurologic insult, such as a traumatic brain injury or stroke, the issue of driving safely may be a significant concern, particularly for younger adults who wish to resume a normal lifestyle as quickly as possible. Driving can play a critical role in patients' independence, socialization, and participation in activities within the community (including rehabilitative treatment and employment, the latter having significant financial consequences). However, given the danger of unsafe driving for the patient and others, the assessment of driving ability has become an important area of inquiry within rehabilitation neuropsychology.

Cubic and Gouvier (1996) reported that with few exceptions "most studies have found positive results in predicting driving skills from psychological assessment techniques" (pp. 211). For example, Sivac, Olson, Kewman, Won, and Henson (1981) found that neuropsychological measures assessing visual perception, reasoning, and attention (specifically, stereoscopic depth perception, Picture Completion, Picture Arrangement) were more related to in-traffic driving than to a functional assessment of driving skills completed on a closed-course in a parking lot for a group of 23 people with various types of brain damage. Gouvier et al. (1981) also demonstrated that driving skills could be predicted from psychological tests and driving simulators.

In their review of this issue, Hopewell and Van Zomeren (1990) found that some neuropsychological indices could be helpful in providing evidence to limit or prohibit driving. However, they also reported that a person's previous driving/accident history, pattern and extent of substance abuse, and psychiatric factors contributed to the prediction of accident risk more than performance on psychomotor tasks. They suggested that the patient's driving history and a number of psychosocial factors also need to be incorporated into decisions about driving safety.

Prediction of memory functioning

The preponderance of research in the area of memory functioning and neuropsychological assessment has focused on the ability of memory tests to differentiate between clinical and normal subjects. That is, most research has attempted to address the fundamental questions of how sensitive memory tests are to brain damage and the utility of memory measures in diagnosing or identifying organic brain syndromes. The issue of how well memory tests predict everyday memory functioning has not, until recently, received much attention from researchers. Parenthetically, this has been true in the field of neuropsychology as a whole, which may explain, in part, why the relationship between test scores and everyday functioning is so modest.

One formal, but elementary, method for assessing patients' everyday memory skills is simply to ask them. There are several self-report measures of everyday memory that are available such as the Memory Functioning Questionnaire (Gilewski, Zelinski, & Schaie, 1990) and the Memory Assessment Clinics-Self-Rated Everyday Memory Scale (Crook & Larrabee, 1990). However, self-report scales generally have a weak association between a patient's memory report and their actual test performance, in part, due to the detrimental effect of a brain injury on a patient's insight and awareness of his or her deficits. Self-report of memory is also very susceptible to depression. Thus, self-rating scales are useful in that they provide information about a patient's belief about his or her memory, but should not be used as a sole measure of memory functioning.

As an adjunct to self-rated memory reports, ratings of the patient's everyday memory from family members or significant others can also provide helpful information. The Memory Functioning Questionnaire noted above can also be completed by the patient's family. The Memory Assessment Clinics rating scale has a parallel version which is intended to be used specifically with the patient's family member or significant other. A comparison of the patient's and family member's ratings can provide some information about the patient's level of insight, particularly if the family member's report is significantly worse than the patient's, and the consistency of reporting within the family. However, with all self-reported measures administered for forensic purposes, the issue of possible exaggeration of the patient's deficits by family members or significant others needs to be carefully addressed. Furthermore, a plaintiff's neuropsychological expert should expect to be vigorously challenged by opposing counsel on any opinions derived from self-report measures alone.

As indicated above, traditional objective memory measures were not designed with the intent of predicting everyday functional memory skills and, as such, are somewhat limited in their ability to do so. A measure created specifically to assess everyday memory, however, is the Rivermead Behavioral Memory Test (RBMT; Wilson, Cockburn, & Baddeley, 1985) in which the patient is asked to perform functional tasks, such as remembering an appointment, a newspaper article, faces, and to deliver a message. The RBMT has been shown to be significantly correlated with rehabilitation staff ratings of memory lapses exhibited by patients with subjective and relative-ratings of memory functioning (Wilson, Cockburn, Baddeley, & Horns, 1989). The RBMT may be less sensitive to subtle memory deficits than other more traditional memory tests, such as the Wechsler Memory Scale-Revised (WMS-R), Wechsler Memory Scale-III (WMS-III) or verbal learning tests (e.g., Rey or California), although Wilson (1991) reported that the RBMT, unlike the WMS-R, was able to discriminate between severe TBI patients (5–10 years post injury), who were living independently within the community, from those who were not.

In contrast, Kaitaro, Koskinen and Kaipio (1995) found that verbal tests of memory (e.g., Logical Memory and Verbal Paired Associates from the Wechsler Memory Scale-Revised) correlated significantly with subjects' or their relatives' estimates of the occurrence of everyday memory problems five years after severe TBI, although measures of visual memory from the WMS-R did not. These authors reported that their findings supported earlier evidence of the ecological validity of verbal memory tasks with memory problems with everyday life (Sunderland, Harris, & Baddeley, 1983) and with limited social contacts following TBI (Oddy & Humphrey, 1980). Additionally, Millis, Rosenthal and Lourie (1994) found that the total scores from the Rey Auditory Verbal Learning Test (and Trails B) were significantly related to psychosocial outcome at one year post injury in a group of 23 TBI patients.

Some memory tests may have more face validity with regard to predicting everyday memory problems by creating memory challenges that seem more analogous to daily memory tasks, although their actual ability to do so has not yet been demonstrated. For example, the California Verbal Learning Test (Delis, Kramer, Kaplan, & Ober, 1987), is a list learning task (i.e., five learning trials with immediate and delayed recall conditions) that uses 16 shopping items, rather than a list of unrelated words, to more closely resemble the types of memory demands that people may encounter in their daily lives. The Memory Assessment Scales (Williams, 1991) contains a Names-Faces subtest, which appears more similar to a memory task we all encounter in our daily lives, although there is no evidence yet that it is more ecologically valid than recalling a list of words or short stories read aloud. Of note, presumably in an effort to make memory testing more similar to everyday functioning, the third edition of the newly revised Wechsler Memory Scale (WMS-III) includes a faces memory task as well as a requirement for the patient to recall hypothetical family members seen previously in pictures of several family-oriented events. Nonetheless, these memory instruments require more research to determine their relationship with everyday memory problems.

Ecological assessment of executive functions and problem solving

As with vocational skills, executive abilities can have a dramatic effect upon a patient's adaptive functioning. The ability to organize oneself, plan and coordinate a plan of action, follow-through on tasks and activities, and modify one's approach based upon changing conditions in the environment (all aspects of executive functioning) can have far-reaching implications for everyday functioning, particularly for more complex activities such as home management and handling finances. Problems with impulsivity and disinhibition can also adversely affect everyday skills. However, in spite of a wide variety of tests that neuropsychologists may use to assess executive and frontal functions, there is little systematic research that has established the ecological validity of these measures in predicting "real world" abilities. The difficulties associated with assessing executive functions in the laboratory or office and relating them to everyday competence are the same as those found with predicting vocational abilities. In short, the demands placed on the brain-injured patient within the highly structured test environment are less than adequate and too restrictive to capture many of the difficulties the patient may have with these abilities in his or her daily life.

Acker (1990) described several additional factors that interfere with the assessment of executive functions, including: The complexity of frontal lobe structures and functions; the heterogeneity of frontal lobe problems; the push and assistance from the examiner during testing for maximum performance; and a focus on outcome scores and a neglect of process. Consequently, it is important to supplement formal test procedures with observations and descriptions of the patient during testing and from others who observe the patient under more varied and frequent circumstances. It is also important to obtain a careful history of the patient's planning, organizational skills, and initiative prior to his or her injury in order to accurately gauge the effects, if any, of an injury upon these abilities.

In spite of the problems noted above, some formal executive function procedures may have some promise in predicting everyday skills. For example, the Behavioral Dyscontrol Scale (BDS; Grigsby & Kaye, 1992; Grigsby, Kaye, & Robbins, 1990) is a relatively new instrument created specifically for the purpose of assessing the individual's ability to engage in ADL's. The BDS is a 10-minute measure based primarily on simple cognitive and motor tasks used by Luria in his studies of frontal lobe functioning. The BDS has been shown to predict executive aspects of independent functioning, particularly apathy and disinhibition, in a geriatric population (Kaye, Grigsby, Robbins, & Korzum, 1990). It was also found to be superior to the Folstein Mini-Mental State Exam in predicting an independent versus an assisted living situation and functional independence three months after discharge in a rehabilitation population (Suchy, Blint and Osmon, 1997).

The Dementia Rating Scale (DRS; Mattis, 1988) has been found to significantly predict an elderly patient's performance of everyday functions in a hospital setting (Nadler, Richardson, Malloy, Marran and Brinson, 1993) and accounted for 27% to 49% of the variance. In particular, the Initiation/Perseveration subscale was weighed most heavily in each analysis reflecting "the significant contribution of Initiation/Perseveration (executive functioning) in predicting everyday functions" which the authors reported "is consistent with theoretical expectations, which suggest that these 'executive' (or frontal lobe) functions are necessary for the planning, organizing, and initiation of functional activities" (Nadler et al., 1993).

Dunn and colleagues (Dunn et al., 1990) also found a significant relationship between neuropsychological measures (Halstead-Reitan Neuropsychological Test Battery) and functional daily living skills in geriatric patients. However, in contrast to the findings described above with the DRS, Dunn et al. (1990) did not find a particularly strong relationship between executive functions and everyday abilities. Contrary to their expectations, they did not find a positive relationship between the Category Test (a measure believed to reflect executive skills) and a patient's ability to handle emergencies that presumably require novel problem solving, logical analysis and complex decision making. The authors hypothesized that their results reflected more the importance of temporal lobe functioning given the consistent and significant association between their measure of community competence and Speech Sounds Perception, Seashore Rhythm, and Tactual Performance-Memory.

An evolving test that was designed specifically to predict everyday problems arising from executive dysfunction is the Behavioral Assessment of The Dysexecutive Syndrome (BADS; Alderman, Evans, Burgess, & Wilson, 1993). The BADS is composed of practical problem solving tasks, paper and pencil problem solving activities, route following, and measures requiring demanding attention and ability to follow directions. Although research is ongoing on this instrument, a preliminary study has found the BADS a better predictor of everyday problems than the Wisconsin Card Sorting Test (Wilson, 1993).

As with the RBMT and the BADS, tests that present activities that more closely resemble tasks found in everyday functioning appear to have promise for future research in this area. Consistent with this theme, Diehl, Willis and Schaie (1995) reported on the Observed Tasks of Daily Living (OTDL), which they described as a new behavioral measure of everyday problem solving for older adults, focusing specifically on food preparation, medication intake, and telephone use. The OTDL was designed to simulate everyday problem solving situations in the three domains listed above by using real-life materials, such as ordinary food packages, and questions that elicit a sequence of observable behaviors that can be scored objectively using a standardized observation protocol. Although further research is needed, the authors concluded that performance-based assessment instruments measuring domain specific skills may be particularly important in determining the functional competencies of cognitively challenged older adults.

Prediction of everyday functioning: Summary

Research currently suggests that there is no one specific neuropsychological measure that can accurately predict all everyday skills for all persons. Wilson (1993) suggests that "we need a complete range of neuropsychological assessments in order to provide the clearest picture of a brain-injured patient's present neuropsychological condition and possible future progress after returning to everyday life in the community" and that "the tester should be willing to combine information gathered from neuropsychological tests with information gathered from behavioral observations, interviews, rating scales, self-report measures, and other assessment procedures" (Wilson, 1993, p. 214). In short, it is not enough to rely solely on the patient's neuropsychological test performance in predicting everyday competence. The behavioral domains are too complex and our understanding of everyday skills too elementary. It is important that we seek convergent and consistent sources of information in order to increase our understanding and ability to predict an individual patient's everyday competencies. This is particularly true within a forensic context when all assumptions are subject to challenge under cross examination.

Although our current neuropsychological measures can provide some assistance in this area, their contribution is generally modest, particularly for scores that fall in the average to low average range. Severely deficient scores are likely to be associated with impaired functional skills, although the prediction of everyday abilities from an intact neuropsychological profile is more difficult to make. Research also suggests that assessment can likely be more predictive if the tasks used during testing closely match or simulate everyday abilities. Although there are currently few assessment measures available that can provide this opportunity, this approach appears to be an important trend for the future.

Prediction of vocational functioning from neuropsychological test data

Identification of the "perfect" neuropsychological tests to assess and predict the ability to work remains elusive. Although research has demonstrated that neuropsychological measures can be helpful in identifying brain-injured patients who may or may not be successful in the workplace, the predictive relationship is far from perfect. However, within a forensic context and in their role as "experts", neuropsychologists may feel increased pressure to ascribe predictive relationships between test data and work potential in ways that have not been empirically demonstrated. Thus, as forensic consultants, neuropsychologists should become familiar with how well and under what conditions their test data may be best able to provide meaningful information about a patient's work potential. In addition, there may be other information, which will be described later, that can be used to supplement the patient's test data to better understand his/her vocational strengths and weaknesses.

Research on the prediction of work performance following traumatic brain injuries (TBI), where return to work rates have reportedly ranged from 12.5% to 71% (Greenspan et al., 1996), is difficult to synthesize as a result of the methodological differences across these studies. For example, these studies frequently utilized different predictor and outcome variables (e.g., having returned to work part-time, returned to work full time, or returned to academic enrollment full or parttime). Whether the patient was working at the time of his or her injury was not taken into account in some of these studies. In addition, the participation of the patient in a vocational rehabilitation program is another factor that needs to be considered. Finally, failure to examine systematically the effects of neuropathology (e.g., contusion, hydrocephalus, etc.), injury severity, age, premorbid abilities, family support systems, and type of job the brain-injured patient is returning to, etc., has obscured research findings and limits the generalizability of these studies. In spite of these methodological issues, there appear to be some measures that can assist the neuropsychologist in directly addressing the question of the brain-injured patient's vocational potential.

Some specific tests or test batteries have been correlated with the patient's ability to return to work. For example, Newman, Heaton, and Lehman (1978) found that the Russell Average Impairment Rating (based on the Halstead-Reitan Battery) correlated .47 with chronic unemployment. Additional measures, including the Category and Trail Making Tests, and perceptual errors, also correlated significantly with unemployment (.36, .32, and .41, respectively).

Ip, Dornan, and Schentag (1995) reported that the performance IQ from the WAIS-R, the WMS-R, and Visual Memory Index, Parts A and B of the Trail Making Test, the Halstead-Reitan Test Battery, and Grooved Pegboard Test (non-dominant hand) were all significantly related to the ability of TBI patients to return to work or school. Of these variables, the TBI patient's Performance IQ served as the best predictor for return to school or work status. Fraser, Dikmen, McLean, Miller, and Temkin (1988) also reported that at one year post injury, the Wechsler Performance IQ was the most significant discriminator between employed and unemployed TBI patients, based on their follow-up of 102 TBI patients. However, Fabiano and Crewe (1995) found that in addition to Performance IQ, Verbal IQ and Full Scale IQ, and four additional WAIS-R subtests (Picture Arrangement, Block Design, Similarities and Digit Symbol) were also highly related to employment outcomes in a group of 94 subjects who had sustained severe head injuries, when they were evaluated nearly six years post injury. Since these authors, however, did not use any other measures of cognitive or neuropsychological functioning, their ability to predict employability relative to WAIS-R data cannot be determined.

Ruff et al. (1993) examined neuropsychological test performance at baseline or six months post injury and employability at six and 12 months following TBI. Their findings revealed that speed of mental processing, motor speed (finger tapping), spatial integration (WAIS-R Block Design), and intact vocabulary (WAIS-R Vocabulary) were all significantly related to returning to work or school. The most powerful predictors from among their test data, however, were speed of information processing and WAIS-R Vocabulary. Girard et al. (1996) also found that speed of mental processing was significantly related to productivity, including returning to work following TBI. They reported that the following tests were most helpful in predicting positive outcomes: WAIS-R Block Design, Digit Symbol, Picture Completion, Arithmetic, and Full Scale IQ; Symbol Digit Modalities Test, Trails B, and WRAT-R Arithmetic. However, these test results, even when combined with demographic variables, accounted for less than 30% of the total variance in predicting outcome.

Memory may also play a role in return to work. In a large 15 year follow-up of Vietnam veterans who sustained penetrating head injuries, neuropsychological measures of verbal and visual memory loss (along with five other non-cognitive measures) were found to contribute significantly to unemployability (Schwab et al., 1993).

A combination of neuropsychological and other non-cognitive factors have been found to contribute significantly to employment status. For example, in a study that predicted the occupational status of 483 TBI ambulatory workers, the Performance IQ, Digit Span, Picture Completion, Block Design, Digit Symbol, Trial Making Part B, and the Wechsler Memory Scale score were found to be the seven best predictors (Bowman, 1996). Other factors, such as demographic and emotional variables, further contributed to accurate prediction of being employed or unemployed, although the author stated that cognitive performance "is more important in the specific question of return to work" (Bowman, 1996, p. 391).

In a discussion of the ecological validity of neuropsychological data to predict vocational functioning, Guilmette and Kastner (1996) concluded, in general, that global severity of cognitive dysfunction is most strongly associated with unemployability (i.e., all things being equal, the more severe the brain damage, the less likely the patient will be competitively employed). In addition, assessment of work capabilities is generally more valid for specific jobs than for the world of work in general. Thus, it is generally preferable to attempt to predict success at a specific job for which the job requirements and skills necessary for success are known, so that the test battery can be tailored accordingly. For example, the skills and abilities necessary to be successful as a secretary are most likely different from those necessary to drive a forklift. Consequently, these factors should be considered when the neuropsychologist is assembling his or her test battery in an effort to measure those skills and functions that are presumed to be important for success in a specific occupation. In selected instances, specific tests have already been developed that assess some of the skills necessary for specific jobs. For example, tests such as the Minnesota Clerical Test and the Clerical Aptitude Test have been developed to assess clerical aptitude, while the Minnesota Rate of Manipulation, Purdue Pegboard, and the Crawford Small Parts Dexterity tests have occupational norms to assess motor speed and dexterity. If working on a case with a vocational rehabilitation counselor, these data may be readily available.

Psychosocial and neuropsychological variables

Psychosocial and behavioral dysfunction are common sequelae of severe TBI. Irritability, immaturity, egocentricity, mood swings, or aggression can have significant effects in the workplace, particularly social interactions with co-workers and supervisors. Thus, even patients with intact cognitive functioning following TBI may be seriously handicapped for competitive employment if they have problems of behavioral dyscontrol. For example, Heaton, Chelune, and Lehman (1978) found that including MMPI data along with neuropsychological variables significantly increased their ability to correctly classify patients as employed or unemployed. Brooks et al. (1987) found that based on interviews with relatives of brain injury survivors, 30% of employment outcomes could be explained by self-care and emotional control variables.

Psychological problems and violent behavior were two of the seven most important variables in predicting vocational disabilities in a large group of Vietnam veterans with penetrating head injuries (Schwab et al., 1993). These authors found that these two variables in combination with the two most important cognitive factors (verbal and visual memory loss) and three neurological variables (paresis, seizures and visual field loss) had a cumulative and nearly equipotent effect on work failure. Similarly, Bowman (1996) reported that although neuropsychological variables accounted for more of the variance than emotional factors in predicting employment following TBI (21% vs. 7%, respectively), MMPI raw K, and scales 4 (Psychopathic Deviant), 8 (Schizophrenia) and 0 (Social Introversion) were also helpful in the prediction process.

Crepeau and Scherzer (1993) reported that emotional disturbance was one of four factors that correlated most highly and reliably with unemployment after they completed a meta-analysis of a series of studies that evaluated back-to-work variables following TBI. The three other factors included executive dysfunction, deficits with activities of daily living, and less vocational rehabilitation services.

The role of executive functions and behavioral assessments

Lezak (1987) concluded, based on a five year follow-up of 42 adults who had sustained severe TBI, that the most significant residual impairments were those functions "required for pursuing a course of study or getting or keeping a job, filling in unscheduled or unstructured times satisfactorily, and making and maintaining close social relationships – the executive functions" (p. 64).

Similarly, Crepeau and Scherzer (1993) have found that executive dysfunction was one of four of the most reliable correlates of unemployment. They emphasized that deficits in planning, organization, cognitive flexibility, unawareness of deficits, and loss of initiation needed to be taken into account when assessing the TBI patient's ability to return to work. In addition, Varney (1988) found that 92% of TBI patients who had developed total anosmia (which is a common indication of orbital frontal damage) were still unemployed two years after they had been medically cleared to return to work.

Deficits of executive functions as a result of damage to frontal and prefrontal structures following TBI have been described previously in detail (Malloy et al., 1993; Prigatano, 1992; Stuss & Gow, 1992; Stuss, Gow, & Hetherington, 1992) and are discussed in detail within Chapter 7 of this text. The formal and objective assessment of executive functions may be difficult due to the nature of the testing process itself. For example, patients who exhibit deficits of executive functions in their daily lives may not exhibit such deficits or the severity of their deficits when they are tested in a highly structured setting. This obviously underscores the need to obtain collateral information from family members or significant others who may know the patient well enough to provide a more accurate assessment of his or her organizational skills, problems of initiation, problem solving, judgment, awareness of deficits, ability to modify his or her behavior, etc. The need in forensic cases to consider the accuracy of self-report by the family, significant others, and plaintiff will be discussed later in this chapter.

Lezak (1983) developed the Tinker Toy Test to assess the patient's executive functions in terms of the patient's ability to spontaneously formulate goals, engage in planning, and carry out such plans. This test was administered to 50 TBI patients and 50 normal controls (Bayless, Varney, & Roberts, 1989). These authors found that although a score of 6 or greater on the Tinker Toy Test was not necessarily predictive of employment, a score of less than 6 was strongly associated with unemployment. While not in widespread clinical usage, this measure serves as an example of the relevance of some behavioral measures to employment.

The Behavioral Assessment of Vocational Skills (BAVS) was developed by Butler et al. (1989) to assess the patient's executive functioning in a more functional manner and shows some promise in predicting employment. For example,

Butler et al. (1983) administered a series of neuropsychological measures to a group of 20 patients with brain injuries, who were participating in an outpatient rehabilitation clinic. These patients were also given the task of assembling a wooden wheelbarrow from printed instructions. However, these patients were also intentionally interrupted, asked to work for several minutes on another task, and given negative criticism about their performance while trained observers rated their skills in such areas as following directions, organization, attention, frustration tolerance, problem solving, and judgment. While neuropsychological data (from the WAIS-R, Trails A and B, Logical Memory and Visual Reproduction from the WMS, and the Wisconsin Card Sorting Test) were related to some measures of vocational functioning during a three month trial of volunteer work, a multiple regression analysis revealed that the patient's score on the BAVS was the only significant predictor of vocational performance.

Another behavioral approach, The Functional Assessment Inventory (FAI; Crewe & Athelstan, 1981), has been used to describe patient potential for vocational rehabilitation. The FAI uses rating scales to assess strengths and weaknesses based on interviews with patients by rehabilitation counselors on such dimensions as cognitive and motor functioning, personality and behavior, vision, hearing, and economic disincentives. Rao and Kilgore (1992) found that the use of three global functional scales (Disability Rating Scale, Patient Evaluation and Conference System, and Levels of Cognitive Functioning Scale) predicted return to work with 73.5% to 84.4% accuracy in a group of 57 TBI patients 26 months post injury. Their results suggest that the greater the level of functional disability, the less likely a TBI patient is capable of returning to work. A similar finding was also reported by Greenspan et al. (1996). These latter authors used the Functional Independence Measure (FIM), a rating scale consisting of 18 separate areas of functioning, including grooming, bathing, ambulation and cognition. They concluded that "It is not the severity of the injury per se, but its relationship to functional independence that influences return to work" (Greenspan et al., 1996, p. 215).

Motivational factors

Within the forensic arena, the patient's ability or inability to work may contribute significantly to the monetary compensation that is awarded to the patient, both in terms of potential wage losses, as well as the psychological damage of being unemployable and unproductive. However, in forensic evaluations, when the plaintiff has a financial stake in the outcome, neuropsychologists can expect to be confronted with the issue of motivational factors (e.g., malingering and/or exaggeration of symptomatology). If a plaintiff performs so poorly on neuropsychological tests that the neuropsychologist believes that the plaintiff is unable to return to work, the plaintiff is often entitled to greater damages than other plaintiffs whose vocational impairments are less profound (Guilmette & Kastner, 1996). Thus, poor neuropsychological test scores may be used to justify a plaintiff's claim for financial compensation or disability (Binder, 1990). As a consequence, neuropsychological testing in forensic settings may be particularly vulnerable to malingering (Rogers, et al, 1993). Questions concerning validity of test performances are common. A comprehensive analysis and review of malingering and related diagnostic issues can be found in Chapters 9 and 10.

Reliability of the plaintiff's subjective complaints

The self-reported complaints of patients who have sustained traumatic brain injuries (e.g., Binder, 1996; Gouvier et al., 1992) have been frequently utilized by neuropsychologists to corroborate their test data. Unfortunately, the base rates of such complaints demonstrate that these symptoms are often routinely reported by individuals who have never sustained a traumatic brain injury, or who have no history of documented neurological impairment (e.g., Gouvier et al., 1988). Furthermore, Lees-Haley and Brown (1993) have reported that individuals who are involved in personal injury litigation with no history of brain injury, toxic exposure, or documented neuropsychological impairment, are likely to complain of symptoms that are commonly associated with traumatically brain-injured patients. Finally, Wong et al. (1994) have reported that normal subjects had no difficulty simulating symptoms of mild traumatic brain injuries on a self-report questionnaire. If neuropsychologists rely solely on an individual's subjective complaints and neuropsychological test scores, detection of malingering becomes even more difficult (Gouvier et al., 1992; Lees-Haley & Brown, 1993). In the same vein, reports of other interested parties (e.g., spouse) may not be objective, suggesting the need to seek observations of disinterested parties regarding the plaintiff's real world functioning.

Potential sources of bias

While neuropsychologists have focused considerable attention on the detection of malingering, little attention has been given to how attorneys can bias neuropsychologists and reduce their ability to render objective and honest opinions. Van Gorp and McMullen (1997) have recently described how neuropsychologists can become biased by the financial arrangements made with the retaining attorney (e.g., agreeing to accept a lien attached to the final settlement judgment reached in the case), during the initial contact with the attorney (e.g., the attorney may try to influence the neuropsychologist by strongly and confidently presenting his or her views of their client or the plaintiff prior to a review of any medical/school records and the neuropsychological evaluation), and the record review process (e.g., the attorney may send specific records that may prejudice the neuropsychologist's opinion of the plaintiff prior to the evaluation). Such information may bias the neuropsychologist's interview of the plaintiff, the tests

selected to be administered, the specific normative data utilized, and interpretation of the test data.

Attorneys may "coach" their clients to provide physicians and neuropsychologists with symptoms and clinical histories that are consistent with a brain
injury, even though the plaintiff's medical records prior to obtaining the services
of an attorney revealed no evidence of brain injury.

Case example

A 29-year-old male slipped and fell in the parking lot of a hardware store that
was covered with snow and ice and landed on his buttocks. He immediately got
up and finished loading his supplies into a car and drove approximately 12 miles
home. The next day, he went to see his family physician and complained of stiff-
ness and soreness in his lower back. He remained out of work for a week and
returned to work without incident. Several months later, he became very
depressed as a result of tragic personal losses. He then consulted with an attorney, who suggested to him that he may have sustained a brain injury during his
slip and fall accident. The attorney then referred him to a neurologist. When he
saw the neurologist, he complained of symptoms consistent with a closed head
injury that were remarkably similar to the information he had received from his
attorney about closed head injuries. He also informed the neurologist that he
had been rendered unconscious for several minutes as a result of striking his
head on the ground while falling in the parking lot. He stated that since that
time he had experienced severe problems with his recent memory, attention, and
problem solving skills, even though he did not report any alteration of consciousness or head trauma when he was examined the day following his accident by his family physician. He was then referred for neuropsychological
testing and was found to have moderate to severe neuropsychological impairments.

Based on these test scores, the neuropsychologist recommended that the
patient quit his job. He was then encouraged by his attorney to quit his job since
it would result in a larger financial award. Unfortunately, quitting his job created serious financial difficulties which resulted in frequent arguments with his
wife, conflicts with his children, and feelings of worthlessness as a result of his
perceived inability to contribute to his family's welfare. These problems exacerbated his depressed mood and resulted in his being placed in a psychiatric unit
for several weeks. All of these symptoms and problems were then attributed by
his "treating doctors" to the "brain injury" he had sustained as a result of his
slip and fall accident.

Coaching a client to simulate brain damage

Unscrupulous attorneys may obtain a list of the neuropsychological tests that
are scheduled to be performed by the neuropsychologist who has been retained
by the defense prior to agreeing to send their client in for neuropsychological

testing. Given advance knowledge of what specific tests are going to be administered to their client, attorneys can easily familiarize themselves with such tests by consulting a number of different textbooks or an undisclosed neuropsychological expert, and thus "coach" their client to simulate test performances consistent with brain dysfunction. In addition, these attorneys can also alert their client to specific tests and procedures that have been designed to detect malingering. Thus, the plaintiff may be instructed to perform well on these specific tests to "prove" that they are not malingering. When the neuropsychologist is later confronted during deposition or trial about whether or not he or she administered any specific tests to detect malingering, the neuropsychologist might be forced to conclude that the plaintiff showed no evidence of malingering or motivational difficulties on the specific tests that were designed to detect such problems. It might then only be possible to detect malingering if the plaintiff deliberately performed poorly on some of the remaining neuropsychological tests and the expert was familiar with the latest research in this area (see Chapter 9)!

Need for additional information

As we have seen, research conducted to date suggests that neuropsychological, psychological, or emotional, behavioral and executive function data can contribute significantly to the prediction of vocational success. However, when interpreted in isolation, the contribution of these variables overall is modest. Research has consistently shown that some factors, including demographics, family, availability of rehabilitation services, substance abuse, injury parameters, and economic circumstances, also provide important and meaningful information about the likelihood of employment following TBI (Bowman, 1996; Crepeau and Scherzer, 1993; Greenspan et al., 1996; Paniak et al., 1992; Ruff et al., 1993; Sander et al., 1996). Although a definitive actuarial formula or equation has not yet been developed to provide specific guidelines as to how these above variables should be integrated, the consistency with which they are reported to affect employability underscores the need to consider their influence with respect to a brain-injured patient's ability to work. In a forensic context, the need for objective collateral and background information is even more critical, given the economic factors that may be influencing a plaintiff.

For the neuropsychologist who is asked to provide an independent, expert evaluation of a litigant, the necessity of obtaining background information cannot be over-emphasized. As has been reviewed elsewhere in this chapter and in this text, every effort should be made to obtain the academic, military service (if appropriate) and employment records. Medical records, such as emergency room reports, CT/MRI results, discharge summaries, etc. are important to review in order to determine the extent of the initial injury. In addition, medical records for any previous injury or illness which could adversely affect the patient's performance on neuropsychological measures or ability to work (e.g., previous history of head injury, cerebrovascular accident, etc.) should also be reviewed (Sbordone, 1991).

Most, but not all studies (e.g., Bowman, 1996; Ip et al., 1995) have revealed that injury severity is positively related to unemployment (Crepeau & Scherzer, 1993; Dikmen & Machamer, 1995; Sander et al., 1996). Generally, poor work potential following TBI is related to longer periods of coma (Dikmen & Machamer, 1995; Fabiano & Crewe, 1995; Fraser et al., 1988; Ruff et al., 1993), lower admitting Glasgow Coma Scale scores (Dikmen & Machamer, 1995), and longer acute care hospital stays (Sander et al., 1996). For example, Ruff et al. (1993) found that patients who had been in coma for less than 20 days were 2.7 times more likely to return to work during the first six months post injury than those who were in coma for more than 20 days. Post-traumatic amnesia (PTA) has also been associated with injury severity and thus employment potential (Crepeau & Scherzer, 1993). Paniak et al. (1992) found that 33 days of PTA was the optimal cut-off for predicting whether a TBI patient could successfully return to work.

Although not empirically demonstrated, the location of brain damage as evidenced from neuroimaging procedures (e.g., CT or MRI) may also play a role in predicting work potential, given its relationship to neurobehavioral outcome (Levin et al., 1987). For example, massive bifrontal lobe damage may result in severe behavioral or executive function disorders that could seriously affect the patient's work behavior. In this respect, Martzke, Swan and Varney (1991), and Varney (1988) have reported that the presence of anosmia (inability to smell/taste) is strongly related to orbital frontal damage and unemployment, presumably due to the proximity of the olfactory nerve to orbital frontal lobe structures. In addition, damage to dominant hemisphere temporal lobe structures resulting in language impairment could be devastating to a writer, attorney, teacher, or salesman; while left spatial neglect secondary to damage to the right parietal lobe could produce deleterious vocational consequences in many occupations (e.g., carpentry, architecture, bricklaying, drafting, commercial artist, etc.).

The patient's chronological age appears to be related to poor return-to-work rates (Crepeau & Scherzer, 1993; Dikmen & Machamer, 1995; Ip, Dornan & Schentag, 1995; Ruff et al., 1993). For example, individuals who were older than 60 years of age (and perhaps even persons older than 45 years of age) are less likely to return to work than younger persons. Premorbid educational level and work history also need to be considered in approximating the likelihood of a patient returning to work (Dikmen & Machamer, 1995). For example, Greenspan et al. (1996) reported that individuals who failed to complete high school were less likely to return to work at one year post injury than those who were high school graduates. The patient's pre-existing history of dependability on the job, absenteeism, ability to manage frustration at work, and length of longest employment, are most likely important factors in predicting future work success, particularly following a mild TBI. A related issue is economic incentive to return to work.

Thus, if a patient has little or no economic incentive, either real or perceived, to return to work following a relatively mild TBI, he or she may not even

attempt to enter the work force and all efforts directed toward the patient's rehabilitation may not significantly alter the patient's work potential. Thus, the
forensic neuropsychologist needs to be aware of the patient's desire and incentive to work as part of his or her assessment.

Other issues that appear to relate to a patient's return to work potential
include social supports, substance abuse, and the availability of rehabilitation
services. In general, the greater the level of family support, including marriage,
increases vocational potential (Bowman, 1996; Greenspan, et al; Rao & Kilgore,
1992). Not surprisingly, ongoing substance abuse problems tend to decrease the
patient's ability for gainful employment (Girard et al., 1996; Ip, Dornan &
Schentag, 1995). This issue should be addressed by the neuropsychologist during a clinical or forensic evaluation. The availability and duration of rehabilitation services also appear to be important indicators with respect to the patient's
ability to return to work (Crepeau & Scherzer, 1993). Thus, a recommendation
for further rehabilitation, particularly vocational rehabilitation, may be appropriate *if* the patient has not yet been enrolled in such a program. Obtaining a
formal vocational rehabilitation evaluation, if available, may also enhance the
results of the neuropsychological assessment. Integrating the results of these two
assessments may provide convergent data about the likelihood of employment.

To augment data obtained from testing and prior records, the forensic neuropsychologist also needs to consider the type of work and work environment
that the patient is expected to return to. For example, the neuropsychologist
should be aware of the following variables: specific job skills necessary for success, routine versus unpredictability of the patient's work day, availability of
supervision, amount of interaction with co-workers and/or the public, degree of
match between job and the patient's strengths and minimizes his or her weaknesses, level of distraction in the work environment, and availability of part-
time work or likelihood of integrating the patient slowly into the work
environment. A formal work site evaluation by a vocational specialist or occupational therapist may also be helpful in some cases. Furthermore, interviewing
family members to assess the patient's organizational skills, initiation, motivation, behavioral controls, emotional lability, social skills, and fatiguability can
also provide important information that may not be available through formal
testing procedures, since the presence of these behavioral characteristics in the
patient premorbidly is crucial in assessing whether or not there has been a
change in a patient's behavior since his or her TBI.

Obtaining a comprehensive picture of the patient for forensic purposes can
be quite labor intensive since it requires that the patient's history, injury characteristics, current cognitive status, emotional/behavioral functioning, and family circumstances be considered in predicting vocational potential. Since there is
no single variable that can account for all vocational outcomes, neuropsychologists must consider as many factors as possible. For example, as Dikmen and
Machamer (1995) described, the patient with a Glasgow Coma Scale score of
13–15 had about an 80% chance of returning to work at one year post injury
(a rate consistent with a trauma control group); however, if the individual had

a pre-existing unstable work history, low educational level, was over the age of 50 and sustained severe injuries to other parts of his or her body, the probability of returning to work dropped to approximately 15%.

A thoughtful and methodical approach that considers the variables described in this chapter will hopefully assist the forensic neuropsychologist in the assessment of vocational potential of patients who have sustained brain injuries. It seems abundantly clear that further research is necessary to clarify the relationships among the variables described above.

References

Acker, M.B. (1990). A review of the ecological validity of neuropsychological tests. In D.E. Tupper & K.D. Cicerone (Eds.). *The neuropsychology of everyday life: Assessment and basic competencies.* Boston: Kluwer Academic Publishers.

Alderman, N., Evans, J., Burgess, P., & Wilson, B.A. (1993). Behavioral assessment of the dysexecutive syndrome. *Journal of Clinical and Experimental Neuropsychology, 15,* 69–70 (Abstract).

Bach, P.J. (1993). Demonstrating relationships between natural history, assessment results, and functional loss in civil proceedings. In H.V. Hall & R.J. Sbordone, (Eds.), *Disorders of executive functions: Civil and criminal law applications* (pp. 135–159). Delray Beach, FL: St. Lucie Press.

Bayless, J.D., Varney, N.R., & Robers, R.J. (1989). Tinker Toy Test Performance and vocational outcome in patients with closed head injuries. *Journal of Clinical and Experimental Neuropsychology, 11,* 913–917.

Binder, L.M. (1990). Malingering following minor head trauma. *The Clinical Neuropsychologist, 4,* 25–26.

Binder, L.M. (1993). Assessment of malingering after mild head trauma with the Portland Digit Recognition Test. *Journal of Clinical and Experimental Neuropsychology, 15,* 170–183.

Bowman, M.L. (1996). Ecological validity of neuropsychological and other predictors following head injury. *The Clinical Neuropsychologist, 10,* 382–396.

Brooks, N., McKinlay, W., Symington, C., Beattie, A., & Capsie, L. (1987). Return to work within the first seven years after severe head injury. *Brain Injury, 1,* 5–19.

Crepeau, F., & Scherzer, P. (1993). Predictors and indicators of work status after traumatic brain injury: A meta-analysis. *Neuropsychological Rehabilitation, 3,* 5–35.

Crewe, N.M., & Athelstan, G.T. (1981). Functional assessment in vocational rehabilitation: systematic approach to diagnosis and goal setting. *Archives of Physical Medicine and Rehabilitation, 62,* 299–305.

Cronbach, L.J. (1984). *Essentials of psychological testing* (4th edition). New York: Harper and Row.

Crook, T.H., & Larrabee, G.J. (1990). A self-rating scale for evaluating memory in everyday life. *Psychology and Aging, 5,* 48–57.

Cubic, B.A., & Gouvier, W.D. (1996). The ecological validity of perceptual tests. In R. Sbordone & C. Long (Eds.), *Ecological validity of neuropsychological testing.* Delray Beach, FL: GR Press/St. Lucie Press.

Cullum, C.M., Heaton, R., & Grant, I. (1991). Psychogenic factors influencing neuropsychological performance: Somatoform disorders, factitious disorder, malingering. In H.O. Doerr & A. Carlin (Eds.), *Forensic neuropsychology.* New York: Guilford Press, 141–174.

Delis, D.C., Kremer, J.H., Kaplan, E., & Ober, B.A. (1987). *California Verbal Learning Test. Research edition. Manual.* San Antonio: The Psychological Corporation Harcourt Brace Jovanovich.

Diehl, M., Willis, S.L., & Schaie, K.W. (1995). Everyday problem solving in older adults: Observational assessment and cognitive correlates. *Psychology and Aging, 10,* 478–491.

Dikmen, S., & Machamer, J.E. (1995). Neurobehavioral outcomes and their determinants. *Journal of Head Trauma Rehabilitation, 10.*

Fabiano, R.J., & Crewe, N. (1995). Variables associated with employment following severe traumatic brain injury. *Rehabilitation Psychology, 40,* 223–231.

Fraser, R., Dikmen, S., McLean, A., Miller, B., & Temkin, N. (1988). Employability of head injury survivors: First year post-injury. *Rehabilitation Counseling Bulletin, 31,* 276–288.

Gilewski, M.J., Zelinski, E.M, & Schaie, K.W. (1990). The memory functioning questionnaire for assessment of memory complaints in adulthood and old age. *Psychology and Aging, 5,* 482–490.

Girard, D., Brown, J., Burnett-Stolnack, M., Hashimoto, N., Hier-Wellmer, S., Perlman, O.Z., & Seigerman, C. (1996). The relationship of neuropsychological status and productive outcomes following traumatic brain injury. *Brain Injury, 10,* 663–676.

Goldstein, G. (1996). Functional considerations in neuropsychology. In R. Sbordone & C. Long (Eds.), *Ecological validity of neuropsychological testing.* Delray Beach, FL: GR Press/St. Lucie Press.

Gough, H.G. (1950). The f minus k dissimulation index for the MMPI. *Journal of Consulting Psychology, 14,* 408–413.

Gouvier, W.D., Presthold, P.H., & Warner, M.S. (1988). A survey of common misconceptions about head injury and recovery. *Archives of Clinical Neuropsychology, 3,* 331–343.

Gouvier, W.D., Maxfield, M.W., Schweitzer, J.R., Horton, C.R., Shipp, M., Neilson, K., & Hale, P.N. (1989). Psychometric prediction of driving performance among the disabled. *Archives of Physical Medicine and Rehabilitation, 70,* 745–750.

Gouvier, W.D., Cubic, B., Jones, G., Brantley, P., & Cutlip, Q. (1992). Post-concussion symptoms and daily stress in normal and head-injured college populations. *Archives of Clinical Neuropsychology, 2,* 193–211.

Greenspan, A.I., Wrigley, J.M., Kresnow, M., Branche-Dorsey, C.M., & Fine, P.R. (1996). Factors influencing failure to return to work due to traumatic brain injury. *Brain Injury, 10,* 207–218.

Greiffenstein, M.F., Gola, T., & Baker, W.J. (1995). MMPI-2 Validity versus domain specific measures in detection of factitious traumatic brain injury. *The Clinical Neuropsychologist, 9* (3), 230–240.

Grigsby, J., & Kaye, K. (1992). *The Behavioral Dyscontrol Scale: Manual.* Denver, CO: Authors.

Grigsby, J., Kaye, K., & Robbins, L.J. (1990). Frontal lobe disorder, behavioral disturbance, and independent functioning among the demented elderly. *Clinical Research, 38,* 81A.

Grigsby, J., Kaye, K., & Robbins, L.J. (1992). Reliabilities and factor structure of the behavioral dyscontrol scale. *Perceptual and Motor Skills, 74,* 883–892.

Guilmette, T.J., & Kastner, M.P. (1996). The Prediction of vocational functioning from neuropsychological data. In R.J. Sbordone & C.J. Long (Eds.). *Ecological validity of neuropsychological testing* (pp. 387–411). Delray Beach, FL: GR Press/St. Lucie Press.

Heaton, R.K., Chelune, G.J., & Lehman, A.W. (1978). Using neuropsychological and personality tests to assess the likelihood of patient employment. *Journal of Nervous and Mental Diseases, 166,* 408–416.

Heaton, R.K., Smith, H.H., Lehman, R.A., & Vogt, A.T. (1978). Prospects for faking believable deficits on neuropsychological testing. *Journal of Consulting and Clinical Neuropsychology, 46,* 892–900.

Hiscock, M., & Hiscock, C.A. (1989). Refining the forced-choice method for the detection of malingering. *Journal of Clinical and Experimental Neuropsychology, 11*, 967–974.

Hopewell, C.A., & Van Zomeren, A.H. (1990). Neuropsychological aspects of motor vehicle operation. In D.E. Tupper & K.D. Cicerone (Eds.), *The neuropsychology of everyday life: Assessment and basic competencies.* Boston: Kluwer Academic Publishers.

Ip, R.Y., Dornan, J., & Schentag, C. (1995). Traumatic brain injury: Factors predicting return to work or school. *Brain Injury, 9*, 517–532.

Iverson, G.L., & Franzen, M.D. (1996). Using multiple objective memory procedures to detect simulated malingering. *Journal of Clinical and Experimental Neuropsychology, 18*, 38–51.

Kaitaro, T., Koskinen, S., & Kaipio, M.L. (1995). Neuropsychological problems in everyday life. A 5-year follow-up study of young severely closed-head-injured patients. *Brain Injury, 9*, 713–727.

Kaye, K. Grigsby, J., Robbins, L.J., & Korzum, B. (1990). Prediction of independent functioning and behavior problems in geriatric patients. *Journal of American Geriatric Society, 38*, 1304–1310.

Kiernan, R., & Matthews, C. (1976). Impairment index versus t-score averaging in neuropsychological assessment. *Journal of Consulting and Clinical Psychology, 44*, 951–957.

Larrabee, G.J., & Crook, T.H. (1996). The ecological validity of memory testing procedures: developments in the assessment of everyday memory. In R. Sbordone & C. Long (Eds.), *Ecological validity of neuropsychological testing.* Delray Beach, FL: GR Press/St. Lucie Press.

Lees-Haley, P.R., & Brown, R.S. (1993). Neuropsychological complaint base rates of 170 personal injury claimants. *Archives of Clinical Neuropsychology, 8*, 203–209.

Levin, H.S., Amparo, E., Eisenberg, H.M., Williams, D.H., High, W.M., McArdle, C.B., & Weiner, R.L. (1987). Magnetic resonance imaging and computerized tomography in relation to neurobehavioral sequelae of mild to moderate head injuries. *Journal of Neurosurgery, 66*, 706–713.

Lezak, M.D. (1983). *Neuropsychological assessment* (2nd ed.) New York: Oxford University Press.

Lezak, M.D. (1987). Relationships between personality disorders, social disturbances and physical rehabilitation following traumatic brain injury. *Journal of Head Trauma Rehabilitation, 2*, 57–69.

Lorenze, E.J., & Cancro, R. (1962). Dysfunction in visual perception in hemiplegia: Its relation to activities of daily living. *Archives of Physical Medicine and Rehabilitation, 43*, 514–517.

Malloy, P., Bihrle, A., Duffy, J., & Cimino, C. (1993). The orbitomedial frontal syndrome. *Archives of Neuropsychology 8*, 185–201.

Martzke, J., Swan, C., & Varney, N. (1991). Post-traumatic anosmia and orbital frontal damage: neuropsychological and neuropsychiatric correlates. *Neuropsychology, 5*, 213–225.

Mattis, S. (1988). *Dementia Rating Scale.* Odessa, FL: Psychological Assessment Resources.

McSweeny, A.J., Grant, I., Heaton, R.K., Prigatano, G.P., & Adams, K.M. (1985). Relationship of neuropsychological status to everyday functioning in healthy and chronically ill persons. *Journal of Clinical and Experimental Neuropsychology, 7*, 281–291.

Millis, S.R., Rosenthal, M., & Lourie, I.F. (1994). Predicting community integration after traumatic brain injury with neuropsychological measures. *International Journal of Neuroscience, 79*, 165–167.

Mittenberg, W., Azrin, R., Millsaps, C., & Heilbrooner, R. (1993). Identification of malingered head injury and the Wechsler Memory Scale - Revised. *Psychological Assessment*, 5, 34–40.

Nadler, J.D., Richardson, E.D., Malloy, P.E., Marran, M.E., & Hostetler Brinson, M.E. (1993). The ability of the dementia rating scale to predict everyday functioning. *Archives of Clinical Neuropsychology*, 8, 449–460.

Newman, O.S., Heaton, R.K., & Lehman, R.A.W. (1978). Neuropsychological and MMPI correlates of patients' future employment characteristics. *Perceptual and Motor Skills*, 46, 635–642.

Nies, K.J., & Sweet, J. (1994). Neuropsychological assessment and malingering: A critical review of past and present strategies. *Archives of Clinical Neuropsychology*, 9, 501–552.

Oddy, M., & Humphrey, M. (1980). Social recovery during the first year following severe head injury. *Journal of Neurology, Neurosurgery and Psychiatry*, 43, 798–802.

Pankratz, L., & Binder, L.M. (1997). Malingering on intellectual and neuropsychological measures. In R. Rogers (Ed.), *Clinical assessment of malingering and deception* (2nd ed). New York: Guilford.

Paniak, C.E., Shore, D.L., Rourke, B.P., Finlayson, M.A., & Moustacalis, E. (1992). Long-term vocational functioning after severe closed head injury: a controlled study. *Archives of Clinical Neuropsychology*, 7, 529–540.

Prigatano, G.P. (1992). Personality disturbances associated with traumatic brain injury. *Journal of Consulting and Clinical Psychology*, 60, 360–368.

Rao, N., & Kilgore, K.M. (1992). Predicting return to work in traumatic brain injury using assessment scales. *Archives of Physical Medicine and Rehabilitation*, 73, 911–916.

Rogers, R., Harrell, E.H., & Liff, C.D. (1993). Feigning neuropsychological impairment: A critical review of methodological and clinical considerations. *Clinical Psychology Review*, 13, 255–274.

Ruff, R.M., Marshall, L.F., Crouch, J., et al (1993). Predictors of outcome following severe head trauma: Follow-up data from the traumatic coma data bank. *Brain Injury*, 7, 101–111.

Sander, A.M., Kreutzer, J.S., Rosenthal, M., Delmonico, R., & Young, M.E. (1996). A multicenter longitudinal investigation of return to work and community integration following traumatic brain injury. *Journal of Head Trauma Rehabilitation*, 11, 70–84.

Sbordone, R.J. (1987). A neuropsychological approach to cognitiverehabilitation within a private practice setting. In B. Caplan (Ed.), *Handbook of contemporary rehabilitation psychology* (pp. 323–342). New York: Charles Thomas.

Sbordone, R.J. (1988). Assessment and treatment of cognitive communicative impairments in the closed head injury patient: A neurobehavioral systems approach. *Journal of Head Trauma Rehabilitation*, 3, 55–62.

Sbordone, R.J. (1991). *Neuropsychology for the attorney*. Orlando: PMD Press.

Sbordone, R.J. (1996). Ecological validity: some critical issues for the neuropsychologist. In R.J. Sbordone & C.J. Long (Eds.), *The ecological validity of neuropsychological testing* (pp. 15–41). Orlando: GR/St. Lucie Press.

Sbordone, R.J, & Purisch, A.D. (1996). Hazards of blind analysis of neuropsychological test data in assessing cognitive disability: The role of psychological pain and other confounding factors. *Neurorehabilitation*, 7, 15–26.

Schretlen, D., Brandt, J., Krafft, L., & Van Gorp, W. (1991). Some caveats in using the Rey 15 Item Memory Test to detect malingered amnesia. *Psychological Assessment: A Journal of Consulting and Clinical Psychology*, 3, 667–672.

Schwab, K., Grafman, J., Salazar, A.M., & Kraft, J. (1993). Residual impairments and work status 15 years after penetrating head injury: Report from the Vietnam Head Injury Study. *Neurology*, 43, 95–103.

Sivak, M., Olson, P.L., Kewman, D.G., Won, H, & Henson, D.L. (1981). Driving and perceptual/cognitive skills: Behavioral consequences of brain damage. *Archives of Physical Medicine and Rehabilitation, 62*, 476–483.

Stuss, D.T., & Gow, C.A. (1992). "Frontal dysfunction" after traumatic brain injury. *Neuropsychiatry, Neuropsychology, & Behavioral Neurology, 5*, 272–282.

Stuss, D.T., Gow, C.A., & Hetherington, C.R. (1992). "No longer Gage": Frontal lobe dysfunction and emotional changes. *Journal of Consulting and Clinical Psychology, 60*, 349–359.

Suchy, Y., Blint, A., & Osmon, D.C. (1997). Behavioral Dyscontrol Scale: Criterion and predictive validity in an inpatient rehabilitation unit population. *The Clinical Neuropsychologist, 11*, 258–265.

Sunderland, A., Harris, J.E., & Baddeley, A. (1983). Do laboratory tests predict everyday memory? A neuropsychological study. *Journal of Learning and Verbal Behavior, 23*, 341–357.

Van Gorp, W.G., & McMullen, W.J. (1997). Potential sources of bias in forensic neuropsychological evaluations. *The Clinical Neuropsychologist, 11*, 180–187.

Varney, N. (1988). The prognostic significance of anosmia in patients with closed head trauma. *Journal of Clinical and Experimental Neuropsychology, 10*, 250–254.

Warren, M. (1981). Relationship of constructional apraxia and body scheme disorders to dressing performance in adult CVA. *The American Journal of Occupational Therapy, 35*, 431–437.

Williams, J.M. (1991). *Memory Assessment Scales.* Odessa, FL: Psychological Assessment Resources.

Williams, J.M. (1996). A practical model of everyday memory assessment. In R. Sbordone & C. Long (Eds.), *Ecological validity of neuropsychological testing.* Delray Beach, FL: GR Press/St. Lucie Press.

Williams, N. (1967). Correlation between copying ability and dressing activities in hemiplegia. *American Journal of Physical Medicine, 46*, 1332–1340.

Wilson, B.A., Cockburn, J., & Baddeley, A.D. (1985). *The Rivermead Behavioral Memory Test.* Flempton, Bury St. Edmunds, Suffolk, England: Thames Valley Test Company.

Wilson, B.A., Cockburn, J. , & Halligan, P. (1987). *The Behavioral Inattention Test.* Flempton, Bury St. Edmunds, Suffolk, England: Thames Valley Test Company.

Wilson, B.A., Cockburn, J., Baddeley, A.D., & Hiorns, R. (1989). The development and validation of a test battery for detecting and monitoring everyday memory problems. *Journal of Clinical and Experimental Neuropsychology, 11*, 855–870.

Wilson, B.A. (1993). Ecological validity of neuropsychological assessment: do neuropsychological indexes predict performance in everyday activities? *Applied and Preventive Psychology, 2*, 209–215.

Wilson, B.A. (1996). The ecological validity of neuro-psychological assessment after severe brain injury. In R. Sbordone & C. Long (Eds.), *Ecological validity of neuropsychological testing.* Delray Beach, FL: GR Press/St. Lucie Press.

Wong, J.L., Regennitter, R.P., & Barrios, F. (1994). Base rate and simulated symptoms of mild head injury among normals. *Archives of Clinical Neuropsychology, 9*, 411–425.

Chapter 9

MALINGERING:
DIFFERENTIAL DIAGNOSIS

Jerry J. Sweet

Evanston Hospital, Evanston, Illinois, and Northwestern University Medical School, Chicago, Illinois

Since the late 1980s, the topic of malingering has attracted a tremendous amount of attention from clinical neuropsychologists. From an initially controversial, albeit seemingly accurate, position that neuropsychologists were essentially ineffective in identifying malingerers (see Faust, Hart, & Guilmette, 1988; Faust, Hart, Guilmette, & Arkes, 1988; rejoinders by Bigler, 1990; McCaffrey & Lynch, 1992; and Matarazzo, 1991; and reviews by Franzen, Iverson, & McCracken, 1990; Rogers, Harrell, & Liff, 1993), a vast literature has produced test procedures and techniques that make it possible to detect malingering prospectively (see review and conclusions by Nies & Sweet, 1994).

Within this chapter, the latest clinical research will be reviewed and suggestions offered regarding the most effective current detection methods. There is little doubt that if the present rate of expansion continues, the malingering literature pertaining to clinical neuropsychology will produce new methods for detection on a regular basis. Despite this, there is a strong need for security regarding the actual decision rules and cutoff scores used to make individual diagnostic decisions. Therefore, in most cases, neuropsychologists who publish research in this area typically have stopped publishing the decision rules and cutoffs within journal articles; instead, requiring psychologists to make direct

The author gratefully acknowledges Drs. Scott R. Millis and Michael G. Mercury for their helpful critiques of the final manuscript version of this chapter.

contact with authors to obtain these essential data. Because it is critical to prevent coaching and protect the longevity of effective techniques, this chapter will exclude these data, in keeping with prevailing journal publication policy.

The following discussion of malingering will emphasize the utility of scientist-practitioner clinical research, which is to say the use of empirically-based procedures and decision rules. We will discuss consideration of moderator variables and basic concepts regarding differential diagnosis, and then review clinical strategies in a framework suggested by the recommendations of the review of Nies and Sweet (1994).

Importance of moderator variables

A moderator variable is any factor that meaningfully affects the predictive relationship between other variables (Anastasi, 1987). In other words, a moderator variable has differential predictive power among individuals, often represented as a third variable that affects a traditional test-criterion correlation (Rogers, 1995). To construct a simple example, assume that grip strength has x amount (i.e, as expressed by a test-criterion correlation size) of ability to predict group membership for normal and multiple sclerosis patients. Gender is known to be a moderator variable that differentially affects the predictive ability of grip strength. Therefore, to enhance clinical prediction of group membership using grip strength in this instance, gender should be considered. Consideration of gender can be accomplished through the use of appropriate norms before determination of group membership. In a sense, gender in this example may represent an alternative hypothesis to group membership in the multiple sclerosis group, that when addressed effectively can be eliminated as an explanation for the findings. The need to at least consider moderator variables in the analysis and interpretation of data is fundamental to all neuropsychological assessment activities, which is not to say that all measures are affected, all individuals are affected, or that the same moderator variables apply to all measures. Table 1 in Chapter 4 outlines some of the more common moderator variables in neuropsychological assessment.

In general, failure to consider moderator variables may result in a false positive diagnostic conclusion with regard to brain injury. That is, such factors as normal aging or low education may be confused with acquired brain dysfunction. Further, failure to adequately address moderator variables may preclude identification of important treatment recommendations. For example, if a clinically significant depression is not identified, and instead mistaken for brain injury, a potentially treatable emotional condition may remain unidentified (see review by Sweet, Newman, & Bell, 1992). Similar diagnostic and treatment error can also occur with other emotional and pain conditions, as previously discussed by Kay in Chapter 6.

With regard to the issue of malingering, failure to consider moderator variables could result in a patient appearing to be "too" impaired given the expectations associated with the neurological condition in question. This is particularly true when powerful variables, such as age and education, are ignored. For example, confusing the possible effects of brain injury with low education and advanced age (by not using appropriate norms) might render a given set of neuropsychological test results to be interpreted as so severely impaired as to be deemed not credible, whereas accounting for the effects of age may allow a more accurate appraisal of brain injury effects. Consideration of moderator variables may serve to reduce false positives with regard to malingering.

Basic assumptions underlying differential diagnosis of malingering

DSM-IV criteria insufficient

The current Diagnostic and Statistic Manual of Mental Disorders, Fourth Edition of the American Psychiatric Association, known as DSM-IV (American Psychiatric Association, 1994) indicates that malingering should be strongly suspected if any combination of four criteria are present. These criteria are: (1) medicolegal context of presentation, (2) marked discrepancy between the person's claimed stress or disability and objective findings, (3) lack of cooperation with diagnostic evaluation or treatment, and (4) presence of antisocial personality disorder. Basically, if one were to follow these criteria literally, many individuals would be viewed as malingerers who are in fact not malingering. All litigating individuals referred for evaluation (who automatically meet the first criteria) could be viewed as malingerers if: sufficient "objective" (usually interpreted as medical) findings were not found, as is the case with most mild traumatic brain injury and all somatoform cases, or lack of cooperation in treatment, as can be the case with patients suffering from serious frontal lobe pathology, or who have a pre-existing antisocial personality disorder, which in teenage and young adult males is often a risk factor for traumatic brain injury. On these bases alone, the DSM-IV criteria appear unsatisfactory. However, rather than ignore these criteria, it may be a better idea to view them as representing several of a larger number of factors that may increase the risk of malingering, but, in isolation, are insufficient evidence of the diagnosis.

Low incidence

By far, the vast majority of malingerers seen by clinical neuropsychologists will be those whose evaluations are part of a worker's compensation, personal injury, or disability (Social Security or long term disability) evaluation. A meta-analysis by Binder and Rohling (1996) of 18 studies consisting of 2,353 subjects reinforces the notion that neuropsychologists can expect to encounter the effects of financial incentives in these types of evaluations. Although not suggesting that all effects were due to malingering, these authors point out that less injured individuals were more likely to be pursuing financial incentives and were also more likely to display greater impairment. Results suggested that, ". . . absent financial compensation, some patients would have fewer problems and some patients' problems would be eliminated." (p. 9). However, according to published estimates, even within such a context, the incidence of malingering is well *under* a simple majority of cases. Estimates of the base rate of occurrence of neuropsychological malingering in litigating or 'benefit seeking' populations range from approximately 7.5–15% (Trueblood & Schmidt, 1993) to 8.5–14% and 10–25% (Frederick, Sarfaty, Johnston, & Powel, 1994) to 18–33% (Binder, 1993). Unless one has an unusual referral base, more frequent diagnosis of malingering in a forensic neuropsychological practice would not be expected. (Within referral samples of litigating patients, if one almost never or never detects malingering, or, conversely, too frequently concludes that malingering is present, either clinician bias (see discussions by Van Gorp & McMullen, 1997; Sweet & Moulthrop, 1999) or lack of current information on effective assessment procedures may be the cause.) Nevertheless, within the context of litigation, because of the potentially large amount of money at stake in a single case, the question of malingering will almost always need to be addressed because of the need to reassure the defense attorney or insurance company whose assets are at stake.

Not dichotomous
The diagnostic process is often characterized, some would say mistakenly, as a "present" (abnormal, diagnosis applied) or "absent" (normal, no diagnosis given) determination. Unfortunately, this approach, while understandable on one level, may not be accurate if viewed as 'all there is'. As one considers malingering as a diagnostic conclusion, it is important to note that there is no reason to believe that if present in one instance, then all the information obtained from the patient can be explained by malingering. Patients may perform at the level of their ability on some measures, while malingering on others. Cases have been reported in the literature (e.g., Palmer, Boone, Allman, & Castro, 1995) of patients with independently documented brain dysfunction who nevertheless perform in some manner consistent with malingering on some portion of their evaluation. Many experienced clinicians will have seen an occasional case such as this. Thus, valid performances on some measures do not rule out malingering, and, conversely, malingered performances on some measures do not rule out valid performances on others. In fact, in a minority of cases, both conditions may be present simultaneously.

Selective presentation
Related to malingering not necessarily being dichotomous in nature is the notion of selectivity of presentation. As one can imagine, without a great deal of accurate information pertaining to the manifestations of legitimate neurological conditions, individuals who wish to feign deficit must guess or reason their way through complex neuropsychological measures. Available research informs us that malingerers frequently are not good at constructing a credible neuropsychological profile. For example, Gouvier, Prestholdt, and Warner (1988) and

Willer, Johnson, Rempel, and Linn (1993) found that the general public has a number of misconceptions about head injury and associated deficits. Also, consider the incongruity of possible disorder (in this case traumatic brain injury) with nature and severity of test performances in the following studies. Wiggins and Brandt (1988) found that brain injury simulators frequently failed autobiographical items. Binder, Villanueva, Howieson, and Moore (1993) found that some mild head trauma malingerers demonstrated recognition memory deficits. Trueblood and Schmidt (1993) found that some mild head trauma malingerers demonstrated basic sensory-perceptual deficits. A review of the literature on legitimate mild traumatic brain injury cases (see Chapter 11 by Ruff within this text) reveals that such deficits are far from expected, indicating lack of accurate information on the part of the general public. Such lack of information works to the advantage of the clinician attempting to detect malingering.

Without an informed basis upon which to feign deficits, malingerers may choose many or just particular measures on which to display their deliberately poor effort based on their own idiosyncratic belief system regarding brain injury. Interestingly, when debriefing a large group of research simulators Goebel (1983) found that 36% attempted to fake on every test, while the remainder faked on only some of the tests of the Halstead-Reitan battery. On purely logical grounds, there is no way to predict *a priori* the particular measures that will be affected in a given malingerer. Stated differently, consecutive malingerers seen in the same office may appear similar or quite different from one another on given tests. For example, even individuals who are malingering memory impairment may choose different individual or sets of memory tests on which to feign deficit or choose to do poorly on all memory tests.

Degree of intention
Intentionality of behavior may represent what many psychologists really would prefer as their area of study. After all, how many psychologists chose the field initially to better understand "what was the person trying to do that brought about that behavior or, alternatively, what motivated the person to do that?" However, the sober lessons of dissecting human behavior in clinical research and in clinical decision-making often support the premise that, unless we are willing to violate basic rules of logic (e.g., as in the logical error of 'affirming the consequent', when we detect an apparent 'effect' and then in retrospect assume to know its 'cause', or similarly *post hoc ergo propter hoc*, when we believe that an event that occurs before an 'effect' must therefore be its 'cause'), we cannot determine intentionality with complete certainty.

Is it possible to intend only a 40% malingering effort or 80% malingering effort, as opposed to a 100% intentional malingered effort? At least hypothetically, the answer is yes. At present, there appears to be little research literature directly addressing this issue, if for no other reason than we can never be absolutely certain of an individual's intent. Whether we can determine an individual's true degree of intention to malinger is not readily apparent at present.

Degree of exaggeration

Degree of exaggeration is more accessible to empirical investigation, which has allowed development of a relevant literature. Numerous studies have document-ed that volunteer simulators can be trained to feign various levels of impaired performance. These studies can be differentiated according to (1) manipulating sophistication of knowledge about the malingered behavior, (2) directly asking subjects to perform at varying levels of accuracy, or by (3) examining the range of performances of subjects asked to malinger on a specific measure. With regard to the first type of study, it has become fairly well established that varying the sophistication of knowledge of the effects of head injury will lead malingering confederates to appear more or less obviously impaired (Ellwanger, Rosenfeld, Sweet & Bhatt, 1996; Martin, Bolter, Todd, Gouvier, & Niccols, 1993; Rosenfeld, Ellwanger, & Sweet, 1995). That is, given more information about the true nature of the disorder, subjects can more accurately feign deficit. The sec-ond type of research, directing confederates to perform at specific levels of accu-racy, has found that subjects can successfully produce errors on behavioral tasks at differential rates in accordance with instructions to do so (Ellwanger, Rosenfeld, Bermann, Nolan, & Sweet, 1996). Lastly, the range of performances of subjects asked to malinger on a given task, whether a specific test of malin-gering or a commonly used neuropsychological measure, is typically broad. For example, on simple forced-choice digit recognition measures (e.g., Binder, 1993; Hiscock & Hiscock, 1989), only a relatively small percentage of malingerers will perform significantly below chance, while the majority perform in the range of chance to well above chance levels (Nies & Sweet, 1994). This wide range of results may nevertheless be discriminable from performances of genuine brain injury performances, as can be seen with the "90% rule" of Guilmette, Hart, and Guiliano (1993) and Guilmette et al. (1996). Basically, this rule stems from research that has found genuine brain injury performances to be above a thresh-old of 90% correct on an easy digit recognition procedure, and has been found to classify the vast majority of brain-injured and simulating or malingering sub-jects correctly. Studies by other investigators have supported the use of a 90% rule for objectively easy forced-choice digit recognition procedures (e.g., Prigatano & Amin, 1993; Rosenfeld, Sweet, Chuang, Ellwanger, & Song, 1996). Of course, this rule does not apply to more difficult digit recognition procedures. Other examples of simulators and malingerers producing widely varying perfor-mance (i.e., larger standard deviations than clinical samples or normals) can be found in research on use of common neuropsychological measures. For example, two independent samples of simulators on the Category test were found to have standard deviations of 36 and 31 errors, respectively (Tenhula & Sweet, 1996). By comparison to any widely available norms presently in use, such large stan-dard deviations represent a wide distribution of feigned performance.

Variable deficits

Memory impairment is one of the most frequent complaints of neurological patients, and therefore, not surprisingly, also of malingerers. The type of memory deficit malingered may even include autobiographical information (Wiggins & Brandt, 1988; Rosenfeld, Ellwanger, & Sweet, 1995), as opposed to memory for new information, which is the subject of much malingering research (e.g., Griffenstein, Baker, & Gola, 1994; Millis & Putnam, 1994; Millis, Putnam, Adams, & Ricker, 1995; Mittenberg, Azrin, Millsaps, & Heilbronner, 1993; Sweet et al., in press). However, one cannot assume that memory deficits will be the type of deficit selected for feigning. Multiple investigators have found that malingerers and simulators will elect to feign a variety of sensory-perceptual, motor, and cognitive impairments (Binder & Willis, 1992; Frederick, Sarfaty, Johnston, & Powel, 1994; Heaton, Smith, Lehman, & Vogt, 1978; Mittenberg, Rotholc, Russell, & Heilbronner, 1996; Trueblood & Schmidt, 1993).

Multiple strategies

From a rational perspective, Beetar and Williams (1995) have suggested that malingerers may choose from a number of possible strategies when completing neuropsychological measures. These malingering strategies include: random responding, intentional wrong responses, delayed responding, and inattentiveness. Beetar and Williams interpreted their findings with neuropsychological and malingering measures to support the use of these four strategies by malingerers. However, rather than simply infer apparent strategies employed when faking an impairment, some investigators have chosen to debrief simulators participating in malingering studies. For example, Goebel (1983) used debriefing interviews to ask each of 141 simulating subjects to explain their approach to producing malingered performances on the Halstead-Reitan battery. In summary, 36% gave slowed performance or attempted to appear dull, confused, or too slow to comprehend, while 30% gave wrong answers, 14% displayed motor incoordination, and numerous other strategies were employed at lower frequencies.

Even within a single domain of ability, such as memory function, malingerers may choose different means by which to make their point regarding impairment. For example, Iverson (1995) interviewed 160 simulators to discern how each attempted to malinger memory impairment on a variety of measures. He found a number of different strategies, including: acting as though material presented was forgotten (going 'blank', asking examiner to repeat questions, confusing directions), showing emotional upset (i.e., frustration, aggravation), slowing response times, refusing to answer, giving bizarre or random responses, and forgetting items from earlier in the interview or testing.

Even on a single measure of ability, malingerers may act differently. Using debriefing questionnaires, Tenhula and Sweet (1996) found simulators selected a variety of different strategies. These authors reported that on the Booklet Category Test 31% showed a slow or variable pace of responding, 24% pretended memory loss for previous responses, and 20% acted disinterested, tired, or inattentive. Interestingly, only 16% attempted to produce errors exclusively on items perceived to be of greater difficulty, suggesting that the vast majority of their simulators did not consider item difficulty an important dimension in producing malingered performance.

Multiple dimensions of insufficient effort

Insufficient effort is test performance that is *not* caused by brain dysfunction, *not* attributable to moderator variables (e.g., age, depression, education, fatigue, etc.) and significantly *worse* than, or at least *different* from, performance standards known to reflect genuine neurological. Insufficient effort can be viewed as involving multiple dimensions consisting of different diagnostic categories, each conceived of as a continuum ranging from 0 to 100%, only one of which is distinguished by the necessary criteria of (1) deliberateness, and (2) the purpose of obtaining external reward. It is only this latter dimension of insufficient effort that represents true malingering. Other possible causes of insufficient effort that may be encountered include relatively rare cases of either true depressive pseudodementia (not just common effects of depression on neuropsychological measures) or atypical somatoform cases in which individuals appear to have adopted a powerless, helpless, incompetent demeanor or become so convinced of their deficiency(ies) that legitimate effort is not forthcoming. In these instances, the cause appears more psychological than conscious and deliberate. However, there is one other legitimate condition in which deliberate manufacture of symptoms is part of the actual disorder. Factitious disorder represents a manifestation of severe psychopathology in which symptoms are manufactured by the patient. In this disorder, however, the exclusive motivation is to receive attention and undergo treatment attempts from medical systems, toward the fulfillment of psychological needs. Factitious disorder *cannot* be diagnosed in conditions in which external reward is possible.

A further distinction among these various dimensions of insufficient effort can typically be found in the degree to which the individual will participate in and undergo serious and protracted medical treatment regimens; malingerers typically will not, while this is actually a goal for many individuals with somatoform and factitious disorders.

To err in diagnosing one of the non-malingering dimensions of insufficient effort as malingering is essentially to accuse an individual of a potentially criminal act (e.g., fraud, perjury), while possibly also denying needed clinical services (e.g., treatment of depression). With the heavy weight of misdiagnosis in mind, in most cases, decision rules suggested by clinical researchers are established at levels that decrease sensitivity (i.e., 'true' positives, in this case, correct identification of malingerers) in favor of increasing specificity (i.e., 'true' negatives, in this case, correct identification of non-malingerers).

Prospective assessment

Based on the now large, and still growing, literature on detection of neuropsychological malingering, there are a number of decision rules that can be applied to common neuropsychological instruments and a variety of tests of effort chosen that can serve to facilitate an accurate diagnosis. However, detection of malingering will be made more difficult, if this diagnostic possibility is not part of the rationale for construction of the particular battery of instruments chosen for administration to forensic referrals. Basically, since the issue will undoubtedly be raised in virtually all forensic cases (in the author's experience even with those patients who have an unquestionable history of acquired brain dysfunction), it is recommended that clinicians anticipate the need to address this issue empirically in *each* forensic case. The phrase "forewarned is forearmed" definitely applies here.

Clinical strategies

This section will outline and describe various approaches used by clinical neuropsychologists to identify malingering. The specific groupings of theses strategies are those recommended by Nies and Sweet (1994) after reviewing the literature. As has been stated numerous times in the literature, detection of malingering is not a task that can be assigned to a single instrument or single strategy. The author continues to recommend a multidimensional, multimethod approach to determining the presence or absence of malingering. Readers will note that the first three strategies are related specifically to particular test performances, while the second three involve a greater degree of clinical judgement, integration of overall test performances, and generalization to information regarding real world activities.

STRATEGY #1. Use of specific tests of insufficient effort

One of the major differences between the litigation assessments of clinical neuropsychologists at present and those of a decade ago is the common inclusion within test batteries of specific measures of insufficient errort and motivation. Table 1 lists literature citations of peer-reviewed publications in which these measures have been the subject of study.

Digit recognition procedures

The fact that the consciousness of the field has risen considerably regarding the need to proactively and purposefully address this diagnostic concern can be seen in the increasing use of digit recognition procedures in litigated assessments. Digit recognition procedures exist in various forms (e.g., Hiscock Forced-Choice Procedure: Hiscock & Hiscock,1989; Portland Digit Recognition Test (PDRT): Binder, 1993a; Multi-Digit Memory Test (MDMT): Martin, Bolter, Todd, Gouvier, & Niccolls, 1993) and are treated here as a single group of tasks. Basically, the commonality in these procedures rests on visual presentation of digit strings followed by a forced-choice recognition of the original digits. With varying degrees of face validity, often related to instructional set, interpolated tasks, or number of seconds delay in eliciting response, these procedures are deliberately constructed to have the appearance of challenging memory function, when in fact most are relatively easy. Given the forced-choice format, typically limited to no more than two choices, chance and below chance performances can be determined. While below chance performances will almost always discriminate insufficient effort from brain dysfunction, only a minority

Table 1. List of Measures Specifically Intended to Detect Insufficient Effort and Motivation with Relevant Research Citations of Studies Supporting their Effectiveness* or Limited Use in Identifying Malingering.

Forced-Choice Digit Recognition Procedures
(all methods known variously as Hiscock Forced-Choice Procedure, Multi-Digit Memory Test, Portland Digit Recognition Test, or Symptom Validity Test and all modifications of such in either manual or computer format with variable levels of difficulty and delays from stimulus to recognition, measured reaction times, abbreviated versions or event-related potential versions)

Back, Boone, Edwards, Parks, Burgoyne, and Silver (1996)*
Beetar and Williams (1995)
Binder (1992)
Binder (1993a)
Binder (1993b)
Binder and Kelly (1996)
Binder and Rohling (1996)
Binder, Villanueva, Howieson, and Moore (1993)
Binder and Willis (1991)
Ellwanger, Rosenfeld, Sweet, and Bhatt (1996)
Frederick, Sarfaty, Johnston, and Powel (1994)
Greiffenstein, Baker, and Gola (1994)
Guilmette, Hart, Giuliano, and Leininger (1994)
Guilmette, Whelihan, Hart, Sparadeo, and Buongiorno (1996)
Hiscock and Hiscock (1989)
Inman, Vickery, Berry, Lamb, Edwards, & Smith (1998)
Martin, Bolter, Todd, Gouvier, and Niccolls (1993)
Martin, Hayes, and Gouvier (1996)
Prigatano and Amin (1993)
Prigatano, Smason, Lamb, & Bortz (1997)
Pritchard and Moses (1992)
Rose, Hall, Szalda-Petree (1995)
Rose, Hall, Szalda-Petree, and Bach (1998)
Rosenfeld, Ellwanger, Nolan, Bermann, and Sweet (in press)
Rosenfeld, Sweet, Chuang, Ellwanger, and Song (1996)
Slick, Hopp, Strauss, Hunter, and Pinch (1994)
Slick, Hopp, Strauss, and Spellacy, F. (1996)
Trueblood and Binder (1997)

Forced Choice Sensory Detection
Binder (1992)
Haughton, Lewsley, Wilson, and Williams (1979)
Pankratz (1983)
Pankratz, Binder, and Wilcox, (1987)
Pankratz, Fausti, and Peed (1975)
Pritchard and Moses (1992)

Memory for 16 Item Test (modification of Rey MFIT)
Arnett and Franzen (1997)
Iverson and Franzen (1996)
Paul, Franzen, Cohen, and Fremouw (1992)

Rey Memory for 15 Item Test (Rey MFIT; using a variety of clinical cutoffs)
Arnett, Hammeke, and Schwartz (1995)*
Back, Boone, Edwards, Parks, Burgoyne, and Silver (1996)

Table 1 continues

Table 1 (continued).

Beetar and Williams (1995)*
Bernard (1990)*
Bernard, Houston, and Natoli (1993)*
Frederick, Sarfaty, Johnston, and Powel (1994)
Goldberg and Miller (1986)*
Greiffenstein, Baker, and Gola (1994)
Greiffenstein, Baker, and Gola (1996a)
Griffen, Normington, and Glassmire (1996)
Guilmette, Hart, Giuliano, Leininger (1994)*
Lee, Loring, and Martin (1992)
Millis and Kler (1995)*
Morgan (1991)*
Schretlen, Brandt, Krafft, and Van Gorp (1991)*
Simon (1994)*

Rey II Memory Test
Griffen, Glassmire, Henderson, and McCann (1997)

Rey Dot Counting
Arnett and Franzen (1997)
Back, Boone, Edwards, Parks, Burgoyne, and Silver (1996)
Beetar and Williams (1995)
Binks, Gouvier, and Waters (1997)
Frederick, Sarfaty, Johnston, and Powel (1994)*
Greiffenstein, Baker and Gola (1994)*
Martin, Hayes, and Gouvier (1996)*
Paul, Franzen, Cohen, and Fremouw (1992)
Rose, Hall, Szalda-Petree, and Bach (1998)*

Rey Word Recognition List (as described in Lezak, 1983)
Frederick, Sarfaty, Johnston, and Powel (1994)*
Greiffenstein, Baker, and Gola (1994)
Greiffenstein, Baker, and Gola (1996a)

Test of Memory Malingering (TOMM)
Tombaugh (1997)
Rees, Tombaugh, Gansler, & Moczynski (1998)

Other Recall and Recognition Tasks
Arnett and Franzen (1997)
Brandt, Rubinsky, and Lassen (1985)
Chouinard and Rouleau (1997)
Inman, Vickery, Berry, Lamb, Edwards, and Smith (1998)*
Iverson and Franzen (1996)
Iverson, Franzen, and McCracken (1991)
Iverson, Franzen, and McCracken (1994)
Frederick, Sarfaty, Johnston, and Powel (1994)*
Rose, Hall, Szalda-Petree, and Bach (1998)*
Wiggins and Brandt (1988)

---

*Studies with results suggesting limitations or cautions in use of at least one of the measures under study, sometimes due to low sensitivity or low specificity, or both, with some populations.

of malingerers actually score this poorly (Nies & Sweet, 1994). Fortunately, on the simplest of these procedures even levels of performance well above chance can provide good discrimination with some of these procedures. For example, in differentiating malingerers from patients with traumatic brain injury, the 90% correct rule appears to identify individuals on the MDMT with insufficient effort for whom brain dysfunction alone does not account for their performance (Guilmette et al., 1993; Guilmette et al., 1994). Importantly, placement of the digit recognition procedure early in the test battery may optimize the diagnostic utility of the procedure (Guilmette, Whelihan, Hart, Sparadeo, & Buongiorno, 1996). As detailed in Rosenfeld's chapter within this text (see Chapter 11), digit recognition procedures have even compared favorably to the use of event-related potential EEG recordings in identifying insufficient effort (e.g., Ellwanger, Rosenfeld, & Sweet, 1997; Rosenfeld, Ellwanger, & Sweet, 1995; Rosenfeld et al., 1996). Shortening the digit recognition procedure does not appear to alter its diagnostic accuracy (Binder, 1993b; Guilmette et al, 1994). Evaluation of response latency, in addition to number of responses correct, may improve detection of malingering using digit recognition procedures (Rose, Hall, & Szalda-Petree, 1995; Rose, Hall, Szalda-Petree, & Bach, 1998).

Rey Memory for 15 Item Test (Rey MFIT)
The accuracy with which one can reproduce simple visual stimuli from immediate memory has been shown to have relevance to the assessment of insufficient effort and malingering. Particularly when accompanied by instructions that highlight an apparently strong demand on memory capacity, individuals who are engaging in insufficient effort may have an inordinate amount of difficulty drawing the 15 simple, redundant, and familiar items that constitute the Rey Memory for 15 Item Test (known by variants of this name and pictured in Lezak, 1995, as well as Goldberg & Miller, 1986, and Lee, Loring, & Martin, 1992). Seemingly because of its brevity and ease of use, the Rey MFIT has become a popular measure within neuropsychological test batteries applied to litigated patients. Unlike the digit recognition procedures described in the previous section, the Rey MFIT has generated a research literature that is not universal in its support of the procedure.

As can be seen in Table 1, more than half of the studies investigating sensitivity and specificity of the Rey MFIT indicate concerns or limitations in application to some populations. For example, Arnett, Hammeke, and Schwartz (1995), in identifying neurological patients and simulators, found hit rates that ranged widely when applying a variety of previously recommended clinical cutoffs. Ultimately, their research found that only very conservative criteria allowed an acceptably low false positive rate, these same criteria resulting in relatively low true positives, ranging from 47% to 64%. Similar results comparing a variety of actual clinical patients with simulators and malingerers have also suggested unacceptably high false positives (resulting in low specifity) and/or unacceptably low true positives (resulting in low sensitivity) (e.g., Beetar & Williams, 1995; Bernard, 1990; Bernard, Houston, & Natoli, 1993; Schretlen,

Brandt, Krafft, & Van Gorp, 1991; Simon, 1994). Performance on the Rey MFIT has been shown to have a significant correlation with IQ (Goldberg & Miller, 1968; Schretlen et al., 1991), such that some patients with low IQ or severe brain dysfunction may score poorly on this measure (Morgan, 1991). Performance on the Rey MFIT has also been found to be influenced by educational level (Back et al., 1996). As with digit recognition procedures, placement at the beginning of the neuropsychological test battery may increase utility (Bernard, Houston, & Natoli, 1993). A direct comparison of diagnostic utility between a forced-choice digit recognition procedure and the Rey MFIT found the former superior to the latter in discriminating brain-damaged patients and depressed patients from simulators (Guilmette et al., 1994). Paul, Franzen, Cohen, and Fremouw (1992) have reported on a 16 item modification of the Rey MFIT, although a second study by Iverson and Franzen (1996) found low sensitivity to simulators.

While the above described studies make it clear that clinicians need to be cautious with the interpretation of the Rey MFIT in some populations, there is little reason to believe that the population with perhaps the greatest frequency of relevant application of the procedure (i.e., mild traumatic brain injury) will present the same diagnostic difficulties. Greiffenstein, Baker, and Gola (1994, 1996a) have found adequate specificity rates with clinical samples involving probable malingering, persistent postconcussive syndrome and traumatic brain injury patients. Patients with mild traumatic brain injury *are* able to complete the Rey MFIT above threshholds of insufficient effort. Overall, when using the Rey MFIT, clinicians need to consider: (1) the relevance of samples compared within individual studies to the specific differential diagnoses at hand, and (2) the evidence suggesting that the Rey MFIT will identify fewer malingerers than a forced-choice digit recognition procedure. The Rey MFIT should not be used in isolation as a means of detecting malingering.

Rey Dot Counting
Lezak (1983) seems to have been the primary source of information concerning this procedure and describes the Rey Dot Counting technique as involving a comparison of separate response times for counting grouped and ungrouped dots. When respondents do not perform counting at a faster pace with grouped dots, insufficient effort may be the cause. Although available to clinicians and researchers for the same period of time as the Rey MFIT, relatively few published studies on the Rey Dot Counting technique have appeared to date.

Greiffenstein, Baker, and Gola (1994) did not find grouped and ungrouped time scores to differ across three clinical groups, one of which consisted of probable malingerers. While Beetar and Williams (1995) report slower response times for simulators than for normal controls, sensitivity and specificity are not reported and the study did not evaluate individuals with brain dysfunction. Martin, Hayes, and Gouvier (1996) found that although cutoffs could be established for simulators, controls and mild head injured patients that resulted in high specificity (88–100%), Dot Counting scores had low sensitivity to simulated malingering (10–44%). Paul, Franzen, Cohen, and Fremouw (1992) did not find time differences between two groups of simulators and patients with brain dysfunction, but did find that simulators made more errors. These investigators reported sensitivity of 63% for simulated malingering and 92% specificity. However, Frederick et al. (1994) reported, based on comparison of numerous other measures of motivation on several samples of simulators and patients, that Dot Counting was a poor detector of response bias, with "...false-positive rates roughly equivalent to hit rates". (p. 121). Further, Rose et al., (1998) found high specificity, but low sensitivity (only 9%) when attempting to classify closed head injury and simulated performances of Dot Counting.

Finally, in a systematic analysis of possible test components, Binks, Gouvier, and Waters (1997) found that time differences were less discriminating than incorrect responses. Using discriminant function analysis of six Dot Counting scores, these authors report 89% of non-simulators and 81% of simulators correctly classified.

At present, the potential of the Rey Dot Counting technique to identify malingerers is not clear, but does not appear to reside in the original notion of its use (i.e., comparison of timed performances under different conditions). Clinicians will need to await further research before relying on this technique in the assessment of adversarial cases.

Rey Word Recognition List (WRL)
Lezak (1983) described yet another procedure by Rey, this one involving a comparison of recognition of a 15 word list to the first recall trial of the 15 item Rey Auditory Verbal Learning Test. Individuals with recall greater than recognition are thought to be demonstrating insufficient effort. As with Dot Counting, little research has been published on the WRL.

Frederick et al. (1994) used the WRL, in addition to other measures, and found it to be poor at discriminating naive and sophisticated simulated memory impairment from normal performance. These authors also found only approximately 24% detection rate of forensic malingerers using the WRL. Greiffenstein, Baker, and Gola (1994), using the recognition score of the 15 items without comparison to a recall list, found differences between traumatic brain-injured, persistent postconcussive, and probable malingering patients. Setting a cutoff at an acceptable level of specificity (93% when compared to postconcussive patients and 88% when compared to brain-injured patients) resulted in a sensitivity to malingering of 61% and 59%, respectively. In a subsequent study, Greiffenstein, Baker, and Gola (1996a) compared traumatic brain-injured patients and probable malingerers and found similar sensitivity and specificity for the WRL. Although appealing as a reversal of expectations from traditional verbal recall and recognition formats, the WRL is not yet proven as a useful measure of malingering.

Related recall and recognition tasks
In the same vein as the WRL, to detect verbal learning performances that do not make sense, Brandt, Rubinsky, and Lassen (1985) used a 20 item word list to

differentiate simulators from neurological patients based on forced-choice recognition of the original list, but not on recall. In a subsequent study, Wiggins and Brandt (1988) again found that recognition, but not recall, of the 20 word list was worse for the simulators than for actual memory-disordered patients. Following up on these findings, Iverson, Franzen, and McCracken (1991) used a 21 item word list with recall and forced-choice recognition formats to establish cutoffs to differentiate normals and memory impaired from simulators. The recognition data was the most effective, with high specificity and sensitivity. Subsequent validation studies by Iverson, Franzen, and McCracken (1994) and Iverson and Franzen also found recognition data to be more salient than recall data, and resulted in modification of the cutoff for the 21 item recognition task. In an independent study of Iverson's task, and other measures, Frederick et al. (1994) reported high specificity, but low sensitivity with simulators for the recognition portion of the 21 item task. In a second independent study, Rose et al., (1998) found an overall hit rate of only 51% in attempting to classify closed head injury and simulators using the Iverson's 21 item task. Inman, Vickery, Berry, Lamb, Edwards, and Smith (1998) also reported low sensitivity, high specificity, and an overall low hit rate with the Iverson 21 item task. In summary, although the 21 item recognition task of Iverson and colleagues has resulted in variable success thus far, the general approach of investigating recognition memory appears to warrant further investigation. As Inman et al. (1998) suggest, perhaps the brevity of the procedure and the high sensitivity support use as a screening tool, with other measures.

Other procedures
New specific methods of malingering continue to be created, but have not yet undergone sufficient validation research. For example, Strauss, Spellacy, Hunter, and Berry (1994) reported on the use of a simple reaction time to an auditory stimulus as a means of differentiating head injured patients and simulators. Findings such as this are in keeping with the often reported observations that the performance of malingerers is slowed, and also in keeping with other researchers who have included reaction time in assessment of simulators or malingerers (Beetar & Williams, 1995; Ray, Engum, Lambert, Bane, Nash, & Bracy, 1997; Rose et al., 1995).

Davis, King, Klebe, Bajszar, Bloodworth, and Wallick (1997) and Davis, King, Bloodworth, Spring, and Klebe (1997) describe studies of computerized versions of implicit memory tasks originally used in cognitive neuropsychology experiments. Both studies, involving priming with normal controls and simulators in the former and pattern classification with normal controls, simulators and amnesic patients in the latter, produced response patterns that differentiated the simulators. The priming task appeared to correctly classify a greater number of subjects, with the pattern classification task resulting in 20% false positive rate for controls and for amnesics. Further research on both tasks, particularly with the type of brain-injured patients likely to be in litigation, will be necessary to determine their value to neuropsychologists engaged in forensic assessments.

Inman et al. (1998) reported an impressive series of studies on a new forced-choice measure, known as the Letter Memory Test, that appears to have promise. Basically, the computer-adminstered, 45 item task manipulates face difficulty along 2 dimensions: number of letters to be remembered and number of recognition alternatives. Initial studies established hit rates comparable to the Multi-Digit Memory Test and better than the 21 item test.

STRATEGY #2. Evaluation of insufficient effort on common measures

Numerous commonly used neuropsychological measures lend themselves to the same analysis of chance level, below chance level, or unrealistically low (although perhaps above chance level) performances. For the most part, these are forced-choice procedures. Clearly, this area, looking at level of performance (and the section that follows looking at nonsensical or unique performance) among actual neuropsychological measures, has been one of the most fruitful with regard to research demonstrating clinically efficacious and practical means of identifying those exhibiting insufficient effort and malingering.

Given the size of the relevant literature on this topic, we will discuss only a few of these test findings. For a complete listing, see Table 2 for literature citations of peer-reviewed publications in which commonly used neuropsychological measures have been the subject of study with regard to insufficient effort and malingering. Again, the reader is reminded that no single indicator can be relied upon to detect or rule out malingering.

As a prototypical example of poorer than expected performance being indicative of insufficient effort, evaluation of Digit Span, whether related to intellectual or memory testing, has been prominent in the literature. Typically, most investigators have found simulators and malingerers to perform more poorly on Digit Span than patients with documented brain dysfunction (e.g., Martin, Hayes, & Gouvier, 1996; Mittenberg, Theroux-Fichera, Zielinski, & Heilbronner, 1995; Trueblood & Schmidt, 1993). In fact, Mittenberg et al. (1995) found that only on Digit Span did simulators perform significantly worse than brain-injured patients, while almost all other scores from the Wechsler Adult Intelligence Scale-Revised (WAIS-R), including IQ scores, were comparable.

Regarding other performances of simulators or malingerers that are below expectations (including occasionally chance or below chance levels, when applicable) set for clinical groups (usually head injured), the results of total errors on the Category test (e.g., Ellwanger, Tenhula, Sweet, & Rosenfeld, 1997; Tenhula & Sweet, 1996), total words recalled and recognized on measures such as the California Verbal Learning Test (e.g., Millis, Putnam, Adams, & Zych, 1995; Sweet et al., in press; Trueblood, 1993), errors on Seashore Rhythm Test (Mittenberg, Rotholc, Russell, & Heilbronner, 1996; Trueblood & Schmidt, 1993), categories completed on the Wisconsin Card Sorting Test (Bernard, McGrath, & Houston, 1996), and errors on portions of the Sensory-Perceptual Examination (Binder & Willis, 1991; Heaton, Smith, Lehman, & Vogt, 1978; Mittenberg et al., 1996; Trueblood & Schmidt, 1993) all exemplify this general finding. Only rarely in the literature have simulators or malingerers been

Table 2.   List of Neuropsychological Measures and Relevant Citations of Studies Supporting their Effectiveness in Identifying Malingering.

California Verbal Learning Test
Coleman, Rapport, Millis, Ricker, and Farchione (1998)
Millis, Putnam, Adams, and Ricker (1995)
Millis and Putnam (1997)
Sweet, Wolfe et al. (in press).
Trueblood and Schmidt (1993)
Trueblood (1994)

Category or Booklet Category Test
Bolter, Picano, and Zych (1985)
Ellwanger, Tenhula, Sweet, and Rosenfeld (1997)
Tenhula and Sweet (1996)

Digit Span
Beetar and Williams (1995)
Binder and Willis (1991)
Greiffenstein, Baker, and Gola (1994)
Heaton, Smith, Lehman, and Vogt (1978)
Iverson and Franzen (1994)
Iverson and Franzen (1996)
Martin, Hayes, and Gouvier (1996)
Meyers and Volbrecht (1998)
Mittenberg, Theroux-Fichera, Zielinski, and Heilbronner (1995)
Rawling and Brooks (1990)[a]
Suhr, Tranel, Wefel, and Barrash (1997)
Trueblood and Schmidt (1993)

Finger Tapping
Binder and Willis (1992)
Heaton, Smith, Lehman, and Vogt (1978)

Halstead-Reitan Neuropsychological Battery
Heaton, Smith, Lehman, and Vogt (1978)
Mittenberg, Rotholc, Russell, and Heilbronner (1996)
Reitan and Wolfson (1996)
Trueblood and Schmidt (1993)
Trueblood and Binder (1997)

Knox Cube Test
Iverson and Franzen (1994)

Luria-Nebraska Neuropsychological Battery
Mensch and Woods (1986)
McKinzey, Podd, Krehbiel, Mensch, and Conley Trombka (1997)

Memory Assessment Scales
Beetar and Williams (1995)

Rey Auditory Verbal Learning Test
Bernard (1990)
Bernard (1991)

Table 2 continues

Table 2 (continued).

Bernard, Houston, and Natoli (1993)
Binder, Villanueva, Howieson, and Moore (1993)
Chouinard and Rouleau (1997)
Cradock, Gfeller, and Falkenhain (1994)
Greiffenstein, Baker, and Gola (1994)
Greiffenstein, Baker, and Gola (1996a)
Suhr, Tranel, Wefel, and Barrash (1997)

Rey Complex Figure Test
Chouinard and Rouleau (1997)

Seashore Rhythm Test
Gfeller and Cradock (1998)
Trueblood and Schmidt (1993)

Sensory-Perceptual Exam (selected portions)
Binder and Willis (1992)
Heaton, Smith, Lehman, and Vogt (1978)
Trueblood and Schmidt (1993)

Warrington Recognition Memory Test
Cradock, Gfeller, and Falkenhain (1994)
Iverson and Franzen (1994)
Iverson and Franzen (1998)
Millis (1992)
Millis (1994)
Millis and Putnam (1994)

Wechsler Adult Intelligence Scale-Revised
Mittenberg, Theroux-Fichera, Zielinski, and Heilbronner (1995)
Trueblood (1994)
Rawling and Brooks (1990)[a]

Wechsler Memory Scale
Greiffenstein, Baker, and Gola (1994)
Iverson and Franzen (1996)
Rawling and Brooks (1990)[a]

Wechsler Memory Scale-Revised
Bernard (1990)
Bernard, Houston, and Natoli (1993)
Bernard, Houston, and Natoli (1993)
Denney (1999)
Greiffenstein, Baker, and Gola (1994)
Greiffenstein, Baker, and Gola (1996a)
Martin, Franzen, and Orey (1998)
Mittenberg, Azrin, Millsaps, and Heilbronner (1993)

Wisconsin Card Sorting Test
Bernard, McGrath, and Houston (1996)

---

[a]The malingering criteria suggested by Rawling and Brooks (1990) were found by Milanovich, Axelrod, and Millis (1996) to have poor specificity (i.e., numerous individuals from clinical populations were misclassified). Further research prior to clinical use has been suggested by these authors.

found to perform better than patients documented to have brain dysfunction. For neuropsychological measures within which chance levels are able to be determined, only an occasional investigator has found simulators or malingerers performing worse than chance. Importantly, with the exception of the Seashore Rhythm Test (Charter, 1994), most neuropsychological measures do *not* elicit performances from normal individuals that fall within the range of random or chance responding. In fact, only very severely brain impaired individuals can be expected, and even then only occasionally, to perform this poorly on most neuropsychological measures on a legitimate basis. Such individuals are not typically part of highly contested litigation proceedings, at least not when the issue of damages is at stake (as opposed to liability), since with very severely impaired individuals "damages" are often amply documented in medical records.

STRATEGY #3. Examination of nonsensical or unique malingering responses or response patterns on common measures
Another area in which neuropsychologists are now able to make determinations regarding malingering, which even a decade ago had not been empirically validated, concerns nonsensical or unique responses or response patterns within commonly administered neuropsychological measures. As previously, we will limit discussion to illustrative examples.

The findings of Wiggins and Brandt (1988) and others have suggested that the general public may have little understanding of human memory functions and memory disorders. Therein lies fertile ground for assessment of malingering; individuals pretending to be memory disordered may display nonsensical responses or patterns of responding that are not seen in genuine brain-injured patients. For example, Mittenberg, Azrin, Millsaps, and Heilbronner (1993) found that an effective means of identifying simulators was to compare two of the memory index scores (General Memory and Attention/Concentration) from the Wechsler Memory Scale-Revised (WMS-R) that would be in an opposite direction from actual head-injured patients. Similarly, Millis et al. (1995) found CVLT scores such as discriminability and long-delay cued recall to be effective in differentiating brain injury from malingering, as did Sweet et al. (in press) in a replication study using normals, brain-injured patients, simulators, and malingerers. In keeping with the basic notion that recognition memory can be expected to be relatively intact and better than recall in most head injured patients, a series of studies have used the Warrington Recognition Memory Test (RMT) to detect simulators and malingerers (Cradock, Gfeller, and Falkenhain, 1994; Iverson & Franzen, 1994, 1998; Millis, 1992, 1994; Millis & Putnam, 1994). These studies have consistently reported that malingerers perform worse on this recognition memory task than individuals with established brain dysfunction.

Unique performances have also been documented on cognitive measures. For example, following up on initial unpublished research on easy items of the Category Test by Bolter, Picano, and Zych (1985), Tenhula and Sweet (1996) documented and cross-validated the use of Bolter's easy items and five additional groups of unexpected item errors on the Booklet Category Test with normal controls, simulators, and traumatic brain injury patients. Of the six measures, five demonstrated high specificity (92–98%), adequate sensitivity (51–76%), and good overall hit rates (84–92%). A follow-up study by Ellwanger, Tenhula, Sweet, and Rosenfeld (1997) compared these same six types of Category Test items and found comparable results with simulators.

Interesting comparisons of other abilities have also been found to have merit in identifying simulators and malingerers. For example, Mittenberg et al. (1995) found a comparison of WAIS-R Vocabulary scale scores and Digit Span scale scores to identify simulators and head injured patients effectively. Application of this comparison by Mittenberg et al. to previously published data by other investigators also resulted in effective identification, as did an independent replication study (Millis & Ross, 1996). Mittenberg et al. (1996) have also used discriminant function coefficients with a variety of cognitive, sensory, and motor tasks from the Halstead-Reitan Battery to differentiate simulators and head injured patients. Again in this study, application of these coefficients to previously published independent data on malingerers and head injured patients was quite effective. An independent study of Mittenberg's discriminant function (McKinzey and Russell, 1997) looking only at specificity has suggested a slightly higher false positive rate (22.5% versus 16.2% found by Mittenberg) among non-litigating head-injured patients.

In general, the use of multivariate procedures has the advantage of being able to evaluate more complex patterns than are usually evaluated with simple decision rules. Sophisticated and highly complex patterns that identify malingerers may be more difficult for examinees to evade and more difficult for unscrupulous attorneys to coach.

Overall, the above described findings on common neuropsychological tests have tremendous relevance to detection of malingering, when it can be shown that normal and brain-injured patient groups do not present with such findings. Similarly, less well known neuropsychological measures, such as the computer-administered Cognitive Behavioral Driver's Inventory (Ray et al., 1997) may also be shown to have relevance to malingering, in the same manner. However, not all initial successful attempts to identify patterns of malingering responses on actual neuropsychological measures have been successful on replication. For example, the initially promising malingering criteria researched by Rawling and Brooks (1990), were subsequently found to produce an unacceptable number of false positives among a variety of patients with documented neurological and psychiatric disorders (Milanovich, Axelrod, & Millis, 1996).

As a final example of pattern analysis, while at present only found to differentiate documented traumatic brain injury from postconcussion syndrome, etiology-specific motor dysfunction profiles have been reported by Greiffenstein et al. (1996b). The postconcussion motor profile is basically one that does not make sense from a neurobehavioral standpoint and is divergent from that of traumatically brain-injured patients. Because unusual and unexpected motor findings can be present in malingering (e.g., Heaton et al., 1978), one wonders whether further comparisons of performance patterns on such tests as Grip Strength, Finger Tapping, and Grooved Pegboard will result in a particular pattern representing malingering. This type of research and the entire area of clinical research concerning unique and nonsensical responses undoubtedly will continue to be important in coming years.

STRATEGY #4. Examination of excessive inconsistency
In a story titled *Problem of Thor Bridge*, Sir Arthur Conan Doyle stated, through his most famous fictional character, Sherlock Holmes, "We must look for consistency. Where there is want of it, we must suspect deception". This insightful observation applies to our discussion of neuropsychological malingering.

A common theme in the malingering literature has been that malingerers exhibit behaviors on formal testing that are *inconsistent* in some respect. Typically, the type of inconsistency of relevance to malingering is that which is *excessive* and involves comparison of performances within and across *tests* from a single evaluation, as well as within and across multiple evaluation *sessions*.

Not all inconsistency is indicative of malingering. For example, some degree of apparent change in performance within a single session may be due to fatigue or situational variables, while apparent change in scores across time may be attributable to changes in emotional state, medications, test stability (viz., test-retest reliability), inexactness of the measurement process (viz., standard error of measure), 'practice effect' or other factors. By excessive, we mean that the amount of inconsistency present is beyond that expected and attributable to the normal and expected effects of known variables operating within or across testing sessions. Thus, inconsistency is not indicative of malingering simply because substantial changes in test scores occur across time, *if* the change is associated with a variable that itself has changed and is known to produce meaningful effects on the tests in question. For example, depression is known to affect certain neuropsychological tests more than others, and, when found to correspond with changes in test scores across time, may provide a reasonable explanation of apparent inconsistency (e.g., Sweet, 1983; Goulet Fisher, Sweet, & Pfaelzer-Smith, 1986; Newman & Sweet, 1986).

For the most part, determination of excessive inconsistency as a hallmark of malingering requires an intensive examination of the individual case at hand, and will not be as closely dependent on empirical research as are strategies #1, #2, and #3, described earlier.

However, a few relevant research studies are available. For example, Greiffenstein, Baker, and Gola (1994) describe a method of scoring Digit Span, called Reliable Digit Span, that considers the scores of only those trials of a given digit length that are *both* passed within a given administration. Their data suggests that a cutoff can be set for Reliable Digit Span that will discriminate between genuine head-injured patients and malingerers. The findings of Meyers and Volbrecht (1998) support this technique.

Also relevant to expectations of consistency are studies by Reitan and Wolfson (1996; 1997), who have established means, standard deviations, and composite consistency scores on the Halstead-Reitan Battery for small groups of litigating and non-litigating head-injured patients. While not directly applicable to the determination of malingering, since litigating subjects in the study had not been determined by independent criteria to be malingerers, this research nevertheless represents an empirical approach to establishing objective expectations of consistency across testing sessions for groups of head-injured patients. Future research on these composite scores with simulators and malingerers will be necessary to determine the usefulness of this approach in identifying individual malingerers.

STRATEGY #5. Comparison of test data to real-life behavior

Ultimately, malingering is a real world behavior, not just a behavior seen in the office; without a real world context, a determination of malingering cannot be made. That is, without the presence of a powerful external reward, as can occur within the context of litigation, malingering cannot be diagnosed. With regard to neuropsychological assessment in general, and determination of malingering in particular, a great deal of meaning can be derived from comparison of test data with the real life activities and behaviors of the patient. Especially in litigated cases, clinicians can benefit from information pertaining to the life of the patient outside the testing situation. While some of this information can come from the patient (e.g., driving and work status, recreational pursuits, etc.), independent sources of informaton can be invaluable. Comparison of independently obtained information with reported activities of the patient and with in office behavior and test performance can either reveal the self-serving discrepancy of a possible malingerer or assist in ruling out malingering as a consideration. While not possible for many cases that are seen as routine clinical referrals, given time constraints and limited resources in such instances, the context of litigaton often provides a time frame and the resources that allow additional information gathering. In litigated cases, there is seldom a justification for clinicians to focus exclusively on office behavior and test data, while ignoring the real world of the individual being evaluated. Although not a common occurrence, the classic irrefutable video portraying an individual repeatedly and comfortably engaging in an activity claimed to be impossible during a clinical interview can definitively answer the question of whether test data are valid. Conversely, lack of discrepancy between formal evaluation information and that obtained from the real world of the individual can serve to confirm or bolster the conclusion that malingering is not present.

STRATEGY #6. Determination of self-serving versus real-life losses

Malingerers are not known for protracted, meaningful self-sacrifice. Hence, we have the commonly held belief that malingerers rarely engage in meaningful, protracted, or invasive treatments. In fact, malingering behavior can be viewed as extremely *self-serving*. While not essential in ruling out or confirming malingering, some difficult to diagnose cases may be clarified by examining the real life consequences of apparent impairment. When verified through objective, independent sources or historical information that document significant and lasting real-life losses attributable to the apparent impairment, one can reasonably question whether malingering is present. Stated differently, when as a direct result of the claimed disorder it can be verified that the claimant has lost such items as house or property due to bankruptcy or a previously contented spouse due to divorce or a previously satisfying and secure vocation, there is less reason to suspect that malingering is present. While helping to rule out malingering, such information does not necessarily rule out psychological disorder, such as somatoform disorder or major depression, as an alternative to brain injury. Conversely, the individual who claims inability to drive a mail truck, while happily describing a recent motor home excursion to the Grand Canyon during which he was the driver, is revealing more than he may realize. The self-serving nature of such statements is obvious. To restate an earlier observation, determination of malingering involves real-life context. In litigated cases, self-serving versus real-life losses may help identify a malingerer versus a brain-injured patient (see Sweet & Kuhlman, 1995 for illustrative juxtaposition of brain injury versus malingering in two litigated cases).

Conclusions regarding status and limitations of detecting malingering

Trueblood and Binder (1997), by improving on both the methodological aspects of clinical survey research and also including up to date clinical data pertaining to malingering, found evidence of improved detection of malingerers by clinicians compared to earlier negative findings reported by Faust and colleagues (Faust, Guilmette, & Hart, 1988; Faust, Guilmette, Hart, & Arkes, 1988). Whether due to improved survey methodology, inclusion of better test information with which to make a distinction between traumatic brain injury and malingering, or to a heightened consciousness and greater knowledge base among clinicians, it is clear that neuropsychologists today are much better at detecting malingering than a decade ago. The progress that has been made in dagnostic accuracy is important, given the individual and societal costs associated with errors in diagnosis (see Price & Stevens, 1997).

Not all insufficient effort is malingering. There are multiple dimensions of insufficient effort; determination of insufficient effort as caused by malingering occurs within a real life context, not simply examination of test performances. Malingering can have variable degrees of intention, selectivity in presentation, variable degrees of exaggeration, and more than one strategy on the part of the malingerer or across malingerers. Moreover, malingering need not be an all or none concept; co-occurrence of malingering and brain dysfunction is possible, although not commonplace.

Malingering is a relatively low base rate behavior. Unless one has an atypical referral base, concluding that malingering is present significantly more or less frequently than the base rates reported in various litigation samples may suggest bias or a lack of information regarding modern and effective diagnostic procedures. Procedures that have been deemed to have clinical utility in detecting malingering are those that have been found to have high specificity and reasonable sensitivity, and have undergone replicated study and peer reviewed publication. In order to achieve high specificity, most measures sacrifice some degree of sensitivity. At this point, clinicians have a number of such procedures from which to choose.

The need to assess malingering in forensic cases is best dealt with prospectively. At present, it is recommended that clinicians consider the following test-related strategies in attempting to rule out or detect malingering: (1) administration of specific measures of insufficient effort, (2) evaluation of insufficient effort on common measures with forced-choice format, and (3) examination of nonsensical or unique malingering responses or response patterns on other common measures. Also, it is recommended that clinicians assess a broader integration of context and test data by considering the additional strategies: (4) examination of *excessive* inconsistency, (5) comparison of test data to real-life behavior, and (6) determination of self-serving versus real life losses.

With continuation of the tremendous interest in neuropsychological malingering that has developed in the last 10 years, the 'state of the art' is being constantly improved. In fact, so rapid is the production and publishing of new relevant research that clinicians may find it challenging to keep abreast of developments. For those interested in new studies, a handful of journals seem to contain the majority of malingering research. These journals include: *The Clinical Neuropsychologist, Archives of Clinical Neuropsychology, Journal of Clinical and Experimental Neuropsychology, Psychological Assessment,* and *Journal of the International Neuropsychological Society.*

We can no longer take comfort in the belief, now shown to be false, that we will simply know when malingering is evident, without addressing the issue deliberately and prospectively. We would also do well to rethink the 'throwaway' statement, often found at the end of the behavioral observations section of formal reports, that we were once taught by our mentors to make (i.e., "Present findings are a valid indication of the patient's functioning."). Such a statement should now be reconstructed to indicate the limitations of mere observation of overt behavior that appeared or did not appear to indicate that sufficient effort was invested on tasks presented to the patient. Gross observations of behavior as a means of determining the presence or absence of malingering are no longer acceptable as sole criteria for ruling out insufficient effort. Without prospective consideration of strategies to detect or rule out malingering, we are left to ponder the philosopher Teuber's admonition, "absence of evidence is not the same as evidence of absence".

References

American Psychiatric Association (1994). *Diagnostic and statistical manual of mental disorders* (4th ed.). Washington, DC: Author.
Anastasi, A. (1987). *Psychological testing* (6th ed.). Toronto: Macmillan.

Arnett, P., & Franzen, M. (1997). Performance of substance abusers with memory deficits on measures of malingering. *Archives of Clinical Neuropsychology, 12,* 513–518.

Arnett, P., Hammeke, T., & Schwartz, L. (1995). Quantitative and qualitative performance on Rey's 15-Item Test in neurological patients and dissimulators. *The Clinical Neuropsychologist, 9,* 17–26.

Back, C., Boone, K., Edwards, C., Parks, C., Burgoyne, K., & Silver, B. (1996). The performance of schizophrenics on three cognitive tests of malingering, Rey 15-Item Memory Test, Rey Dot Counting, and Hiscock Forced-Choice Method. *Assessment, 3,* 449–457.

Baker, G., Hanley, J., Jackson, H., Kimmance, S., & Slade, P. (1993). Detecting the faking of amnesia: Performance differences between simulators and patients with memory impairment. *Journal of Clinical and Experimental Neuropsychology, 15,* 668–684.

Beetar, J., & Williams, J. (1995). Malingering response styles on the Memory Assessment Scales and symptom validity tests. *Archives of Clinical Neuropsychology, 10,* 57–72.

Bernard, L. (1990). Prospects for faking believable memory deficits on neuropsychological tests and the use of incentives in simulation research. *Journal of Clinical and Experimental Neuropsychology, 5,* 715–728.

Bernard, L., & Fowler, W. (1990). Assessing the validity of memory complaints: Performance of brain-damaged and normal individuals on Rey's task to detect malingering. *Journal of Clinical Psychology, 46,* 432–436.

Bernard, L. (1991). The detection of faked deficits on the Rey Auditory Verbal Learning Test: The effect of serial position. *Archives of Clinical Neuropsychology, 6,* 81–88.

Bernard, L., Houston, W., & Natoli, L. (1993). Malingering on neuropsychological memory tests: Potential objective indicators. *Journal of Clinical Psychology, 49,* 45–53.

Bernard, L., McGrath, M., Houston, W. (1996). The differential effects of simulating malingering, closed head injury, and other CNS pathology on the Wisconsin Card Sorting Test: Support for the "pattern of performance" hypothesis. *Archives of Clinical Neuropsychology, 11,* 231–245.

Bernard, L., McGrath, M., & Houston, W. (1993). Discriminating between simulated malingering and closed head injury on the Wechsler Memory Scale-Revised. *Archives of Clinical Neuropsychology, 8,* 539–551.

Berry, D., Baer, R., & Harris, M. (1991). Detection of malingering on the MMPI: A meta-analysis. *Clinical Psychology Review, 11,* 585–598.

Berry, D., Wetter, M., Baer, R., Youngjohn, J., Gass, C., Lamb, D., Franzen, M., MacInnes, W., & Buchholz, D. (1995). Overreporting of closed-head injury symptoms on the MMPI-2. *Psychological Assessment, 7,* 517–523.

Bigler, E. (1990). Neuropsychology and malingering: Comment on Faust, Hart, and Guilmette (1988). *Journal of Consulting and Clinical Psychology, 58,* 244–247.

Binder, L. (1992). Malingering detected by forced choice testing of memory and tactile sensation: A case report. *Archives of Clinical Neuropsychology, 7,* 155–163.

Binder, L. (1993a). Assessment of malingering after mild head trauma with the Portland Digit Recognition Test. *Journal of Clinical and Experimental Neuropsychology, 15,* 170–182.

Binder, L. (1993b). An abbreviated form of the Portland Digit Recognition Test. *The Clinical Neuropsychologist, 7,* 104–107.

Binder, L., & Kelly, M. (1996). Portland Digit Recognition Test performance by brain dysfunction patients without financial incentives. *Assessment, 3,* 403–409.

Binder, L., & Rohling, M. (1996). Money matters: A meta-analytic review of the effects of financial incentives on recovery after closed-head injury. *American Journal of Psychiatry, 153,* 7–10.

Binder, L., Villanueva, M., Howieson, D., & Moore, R. (1993). The Rey AVLT recognition memory task measures motivational impairment after mild head trauma. *Archives of Clinical Neuropsychology, 8,* 137–147.

Binder, L., & Willis, S. (1992). Assessment of motivation after financially compensable minor head trauma. *Psychological Assessment: A Journal of Consulting and Clinical Psychology, 3*, 175–181.

Binks, P., Gouvier, W. D., & Waters, W. (1997). Malingering detection with the Dot Counting Test. *Archives of Clinical Neuropsychology, 12*, 41–46.

Brandt, J., Rubinsky, E., & Lassen, G. (1985). Uncovering malingered amnesia. *Annals of the New York Academy of Sciences, 44*, 502–503.

Charter, R. (1994). Determining random responding for the Category, Speech-Sounds Perception, and Seashore Rhythm Tests. *Journal of Clinical and Experimental Neuropsychology, 16*, 744–748.

Chouinard, M., & Rouleau, I. (1997). The 48-pictures Test: A two-alternative forced-choice recognition test for the detection of malingering. *Journal of the International Neuropsychological Society, 3*, 545–552.

Coleman, R., Rapport, L., Millis, S., Ricker, J., & Farchione, T. (1998). Effects of coaching on detection of malingering on the California Verbal Learning Test. *Journal of Clinical and Experimental Neuropsychology, 20*, xx-xx.

Cradock, M., Gfeller, J., & Falkenhain, M. (1994, November). *Detecting feigned memory deficits with the Rey Auditory Verbal Learning Test (RAVLT) and the Recognition Memory Test (RMT)*. Presented at the Annual Conference of the National Academy of Neuropsychology, Orlando, Florida.

Davis, H., King, J., Bloodworth, M., Spring, A., & Klebe, K. (1997). The detection of simulated malingering using a computerized category classification test. *Archives of Clinical Neuropsychology, 12*, 191–198.

Davis, H., King, J., Klebe, K., Bajszar, G., Bloodworth, M., & Wallick, S. (1997). The detection of simulated malingering using a computerized priming test. *Archives of Clinical Neuropsychology, 12*, 145–153.

Denney, R. (1999). A brief symptom validity testing procedure for logical memory of the Wechsler Memory Scale-Revised which can demonstrate verbal memory in the face of claimed disability. *Journal of Forensic Neuropsychology, 1*, 5–26.

Ellwanger, J., Rosenfeld, J.P., Bermann, R., Nolan, K., & Sweet, J. (1996). Revised combined oddball and matching-to-sample procedure for detection of simulated malingering of cognitive deficit with P300. *Psychophysiology, 33*, s35.

Ellwanger, J., Rosenfeld, J.P., & Sweet, J. (1997). P300 event-related brain potential as an index of recognition response to autobiographical and recently learned information in closed-head injury patients. *The Clinical Neuropsychologist, 11*, 428–432.

Ellwanger, J., Rosenfeld, J. P., Sweet, J., & Bhatt, M. (1996). Detecting simulated amnesia for autobiogphical and recently learned information using the P300 event-related potential. *International Journal of Psychophysiology, 23*, 9–23.

Ellwanger, J., Tenhula, W., Sweet, J., & Rosenfeld, J. P. (1997, February). *Identifying malingerers through the use of the Category Test and event-related potentials (ERPs)*. Presented at the Annual Meeting of the International Neuropsychological Society, Orlando, FL.

Faust, D., Hart, K., & Guilmette, T. (1988). Pediatric malingering: The capacity of children to fake believable deficits on neuropsychological testing. *Journal of Consulting and Clinical Psychology, 56*, 578–582.

Faust, D., Hart, K., Guilmette, T., & Arkes, H. (1988). Neuropsychologist's capacity to detect adolescent malingering. *Professional Psychology: Research and Practice, 19*, 508–515.

Fox, D., Lees-Haley, P., Earnest, K., & Dolezal-Wood, S. (1995). Post-concussive symptoms: Base rates and etiology in psychiatric patients. *The Clinical Neuropsychologist, 9*, 89–92.

Fox, D., Lees-Haley, P., Earnest, K., & Dolezal-Wood, S. (1995). Base rates of postconcussive symptoms in health maintenance organization patients and controls. *Neuropsychology, 9*, 606–611.

Franzen, M., Iverson, G., & McCracken, L. (1990). The detection of malingering in neuropsychological assessment. *Neuropsychology Review, 1*, 247–279.

Frederick, R., & Foster, H. Jr. (1991). Multiple measures of malingering on a forced-
choice test of cognitive ability. *Psychological Assessment: A Journal of Consulting
and Clinical Psychology, 3*, 596–602.

Frederick, R., Sarfaty, S., Johnston, J.D., & Powel, J. (1994). Validation of a detector of
response bias on a forced-choice test of nonverbal ability. *Neuropsychology, 8*,
118–125.

Gfeller, J., & Cradock, M. (1998). Detecting feigned neuropsychological impairment
with the Seashore Rhythm Test. *Journal of Clinical Psychology, 54*, 431–438.

Goebel, R. (1983). Detection of faking on the Halstead-Reitan Neuropsychological Test
Battery. *Journal of Clinical Psychology, 39*, 731–742.

Goldberg, J., & Miller, H. (1986). Performance of psychiatric inpatients and intellectually deficient individuals on a task that assesses the validity of memory complaints.
Journal of Clinical Psychology, 42, 792–795.

Goulet Fisher, D., Sweet, J., & Pfaelzer-Smith, E. (1986). Influence of depression on
repeated neuropsychological testing. *Clinical Neuropsychology, 8*, 14–18.

Gouvier, W.D., Prestholdt, P., & Warner, M. (1988). A survey of common misconceptions
about head injury and recovery. *Archives of Clinical Neuropsychology, 3*, 331–343.

Greiffenstein, M., Baker, W.J., & Gola, T. (1994). Validation of malingered amnesia measures with a large clinical sample. *Psychological Assessment, 6*, 218–224.

Greiffenstein, M., Baker, W.J., & Gola, T. (1996a). Comparison of multiple scoring methods for Rey's malingered amnesia measures. *Archives of Clinical Neuropsychology,
11*, 283–293.

Greiffenstein, M., Baker, W.J., & Gola, T. (1996b). Motor dysfunction profiles in traumatic brain injury and postconcussion syndrome. *Journal of the International
Neuropsychological Society, 2*, 477–485.

Greiffenstein, M., Gola, M., & Baker, W.J. (1995). MMPI-2 validity scales versus domain
specific measures in detection of factitious traumatic brain injury. *The Clinical
Neuropsychologist, 9*, 230–240.

Griffin, G.A., Glassmire, D., Henderson, E., & McCann, C. (1997). Rey II: Redesigning
the Rey Screening Test of malingering. *Journal of Clinical Psychology, 53*,
757–766.

Griffin, G.A., Nomrington, J., & Glassmire, D. (1996). Qualitative dimensions in scoring the Rey Visual Memory Test of malingering. *Psychological Assessment, 8*,
383–387.

Guilmette, T., Hart, K., & Giuliano, A. (1993). Malingering detection: The use of a
forced-choice method in identifying organic versus simulated memory inpairment.
The Clinical Neuropsychologist, 7, 59–69.

Guilmette, T., Hart, K., Giuliano, A., & Leininger, B. (1994). Detecting simulated memory impairment: Comparison of the Rey Fifteen-Item Test and the Hiscock forced-
choice procedure. *The Clinical Neuropsychologist, 8*, 283–294.

Guilmette, T., Whelihan, W., Hart, K., Sparadeo, F., & Buongiorno, G. (1996). Order
effects in the administration of a forced-choice procedure for detection of malingering in disability claimants' evaluations. *Perceptual and Motor Skills, 83*,
1007–1016.

Haughton, P. Lewsley, A., Wilson, M., & Williams, R. (1979). A forced-choice procedure
to detect feigned or exaggerated hearing loss. *British Journal of Audiology, 14*,
135–138.

Heaton, R., Smith, H., Lehman, R., & Vogt, A. (1978). Prospects for faking believable
deficits on neuropsychological testing. *Journal of Consulting and Clinical
Psychology, 46*, 892–900.

Hiscock, M., & Hiscock, C. (1989). Refining the forced-choice method for the detection
of malingering. *Journal of Clinical and Experimental Neuropsychology, 11*,
967–974.

Inman, T., Vickery, C., Berry, D. T., Lamb, D., Edwards, C., & Smith, G. (1998). Development and initial validation of a new procedure for evaluating adequacy of effort given during neuropsychological testing: The Letter Memory Test. *Psychological Assessment, 10,* 128–139.

Iverson, G. (1995). Qualitative aspects of malingered memory deficits. *Brain Injury, 9,* 35–40.

Iverson, G., & Franzen, M. (1994). The Recognition Memory Test, Digit Span, and Knox Cube Test as markers of malingered memory impairment. *Assessment, 1,* 323–334.

Iverson, G., & Franzen, M. (1996). Using multiple objective memory procedures to detect simulated malingering. *Journal of Clinical and Experimental Neuropsychology, 18,* 38–51.

Iverson, G., & Franzen, M. (1998). Detecting malingered memory deficits with the Recognition Memory Test. *Brain Injury, 12,* 275–282.

Iverson, G., Franzen, M., & McCracken, L. (1991). Evaluation of an objective assessment technique for the detection of malingered memory deficits. *Law and Human Behavior, 15,* 667–676.

Iverson, G., Franzen, M., & McCracken, L. (1994). Application of a forced-choice memory procedure designed to detect experimental malingering. *Archives of Clinical Neuropsychology, 9,* 437–450.

Lee, G., Loring, D., & Martin, R. (1992). Rey's 15-Item Visual Memory Test for the detection of malingering: Normative observations on patients with neurological disorders. *Psychological Assessment, 4,* 43–46.

Lees-Haley, P., & Brown, R. (1993). Neuropsychological complaint base rates of 170 personal injury claimants. *Archives of Clinical Neuropsychology, 8,* 203–209.

Lezak, M. (1983). *Neuropsychological assessment* (2nd ed). New York: Oxford University Press.

Lezak, M. (1985). *Neuropsychological assessment* (3rd ed). New York: Oxford University Press.

Martin, R., Bolter, J., Todd, M, Gouvier, W.D., & Niccolls, R. (1993). Effects of sophistication and motivation on the detection of malingered memory performance using a computerized forced-choice task. *Journal of Clinical and Experimental Neuropsychology, 15,* 867–880.

Martin, R., Franzen, M., & Orey, S. (1998). Magnitude of error as a strategy to detect feigned memory impairment. *The Clinical Neuropsychologist, 12,* 84–91.

Martin, R., Hayes, J., & Gouvier, W. D. (1996). Differential vulnerability between post-concussion self-report and objective malingering tests in identifying simulated mild head injury. *Journal of Clinical and Experimental Neuropsychology, 18,* 265–275.

Matarazzo, J. (1991). Psychological assessment is reliable and valid: Reply to Ziskin and Faust. *American Psychologist, 46,* 882–884.

McCaffrey, R., & Lynch, J. (1992). A methodological review of "method skeptics" reports. *Neuropsychology Review, 3,* 235–248.

McKinzey, R., & Russell, E. (1997). Detection of malingering on the Halstead-Reitan Battery: A cross-validation. *Archives of Clinical Neuropsychology, 12,* 585–589.

McKinzey, R., Podd, M., Krehbiel, M., Mensch, A., & Conley Trombka, C. (1997). Detection of malingering on the Luria-Nebraska Neuropsychological Battery: An initial and cross-validation. *Archives of Clinical Neuropsychology, 12,* 505–512.

Meyers, J., & Volbrecht, M. (1998). Validation of reliable digits for detection of malingering. *Assessment, 5,* 303–307.

Milanovich, J., Axelrod, B., & Millis, S. (1996). Validation of the Simulation Index-Revised with a mixed clinical population. *Archives of Clinical Neuropsychology, 11,* 53–59.

Millis, S. (1992). The Recognition Memory Test in the detection of malingered and exaggerated memory deficits. *The Clinical Neuropsychologist, 6,* 406–414.

Millis, S. (1994). Assessment of motivation and memory with the Recognition Memory

Test after financially compensable mild head injury. *Journal of Clinical Psychology,*
50, 601–605.

Millis, S., & Kler, S. (1995). Limitations of the Rey Fifteen Item Test in the detection of malingering. *The Clinical Neuropsychologist, 9,* 241–244.

Millis, S., & Putnam, S. (1994). The Recognition Memory Test in the assessment of memory impairment after financially compensable mild head injury: A replication. *Perceptual and Motor Skills, 79,* 384–386.

Millis, S., & Putnam, S. (1997). The California Verbal Learning Test in the assessment of financially compensable mild head injury: Further developments. *Journal of the International Neuropsychological Society, 3,* 225–226.

Millis, S., Putnam, S., Adams, K., & Ricker, J. (1995). The California Verbal Learning Test in the detection of incomplete effort in neuropsychological evaluation. *Psychological Assessment, 7,* 463–471.

Millis, S., & Ross, S. (1996). *Dissimulation indices of the Wechsler Adult Intelligence Scale-Revised: A replication and extension.* Presented at the European Meeting of the International Neuropsychological Society, Veldhoven, Netherlands.

Mittenberg, W., Rotholc, A., Russell, E., & Heilbronner, R. (1996). Identification of malingered head injury on the Halstead-Reitan Neuropsychological Battery. *Archives of Clinical Neuropsychology, 11,* 271–281.

Mittenberg, W., Theroux-Fichera, S., Zielinski, R., & Heilbronner, R. (1995). Identification of malingered head injury on the Wechsler Adult Intelligence Scale-Revised. *Professional Psychology: Research and Practice, 26,* 491–498.

Mittenberg, W., Azrin, R., Millsaps, C., & Heilbronner, R. (1993). Identification of malingered head injury on the Wechsler Memory Scale- Revised. *Psychological Assessment, 5,* 34–40.

Morgan, S. (1991). Effect of true memory inpairment on a test of memory complaint validity. *Archives of Clinical Neuropsychology, 6,* 327–334.

Nies, K., & Sweet, J. (1994). Neuropsychological assessment and malingering: A critical review of past and present strategies. *Archives of Clinical Neuropsychology, 9,* 501–552.

Newman, P., & Sweet, J. (1986). The effects of clinical depression on the Luria-Nebaska Neuropsychological Battery. *International Journal of Clinical Neuropsychology, 7,* 109–114.

Palmer, B., Boone, K.B., Allman, L., & Castro, D. (1995). Co-occurrence of brain lesions and cognitive deficit exaggeration. *The Clinical Neuropsychologist, 9,* 68–73.

Paul, D., Franzen, M., Cohen, S., & Fremouw, W. (1992). An investigation into the reliability and validity of two tests used in the detection of dissimulation. *International Journal of Clinical Neuropsychology, 14,* 1–9.

Pankratz, L. (1983). A new technique for the assessment and modification of feigned memory deficit. *Perceptual and Motor Skills, 57,* 367–372.

Pankratz, L., Binder, L., & Wilcox, L. (1987). Evaluation of an exaggerated somatosensory deficit with symptom validity testing. *Archivwes of Neurology, 44,* 798.

Pankratz, L., Fausti, S., & Peed, S. (1975). A forced choice technique to evaluate deafness in the hysterical or malingering patient. *Journal of Consulting and Clinical Psychology, 43,* 421–422.

Price, J. R., & Stevens, K. (1997). Psycholegal implications of malingered head trauma. *Applied Neuropsychology, 4,* 75–83.

Prigatano, G., & Amin, K. (1993). Digit Memory Test: Unequivocal cerebral dysfunction and suspected malingering. *Journal of Clinical and Experimental Neuropsychology, 15,* 537–546.

Pritgatano, G., Smason, I., Lamb, D., & Bortz, J. (1997). Suspected malingering and the digit memory test: A replication and extension. *Archives of Clinical Neuropsychology, 12,* 609–619.

Pritchard, D., & Moses, J. (1992). Tests of neuropsychological malingering. *Forensic Reports, 5,* 287–290.

Rawling, P., & Brooks, D. (1990). Simulation Index: A method for detecting factitious errors on the WAIS-R and WMS. *Neuropsychology, 4,* 223–238.

Ray, E., Engum, E., Lambert, E., Bane, G., Nash, M., & Bracy, O. (1997). Ability of the Cognitive Behavioral Driver's Inventory to distinguish malingerers from brain-damaged subjects. *Archives of Clinical Neuropsychology, 12,* 491–503.

Reitan, R., & Wolfson, D. (1996). The question of validity of neuropsychological test scores among head-injured litigants: Development of a dissimulation index. *Archives of Clinical Neuropsychology, 11,* 573–580.

Reitan, R., & Wolfson, D. (1997). Consistency of neuropsychological test scores of head-injured subjects involved in litigation compared with head-injured subjects not involved in litigation: Development of the Retest Consistency Index. *The Clinical Neuropsychologist, 11,* 69–76.

Rees, L., Tombaugh, T., Gansler, D., & Moczynski, N. (1998). Five validation experiments of the Test of Memory Malingering (TOMM). *Psychological Assessment, 10,* 10–20.

Rogers, R., Harrell, E., & Liff, C. (1993). Feigning neuropsychological impairment: A critical review of methodological and clinical considerations. *Clinical Psychology Review, 13,* 255–274.

Rogers, T. (1995). *The psychological testing enterprise: An introduction.* Pacific Grove, Cal: Brooks/Cole.

Rose, F., Hall, S., Szalda-Petree, A. (1995). Portland Digit Recognition Test-Computerized: Measuring response latency improves the detection of malingering. *The Clinical Neuropsychologist, 9,* 124–134.

Rose, F., Hall, S., Szalda-Petree, A., & Bach, P. (1998). A comparison of four tests of malingering and the effects of coaching. *Archives of Clinical Neuropsychology, 13,* 349–363.

Rosenfeld, J. P., Ellwanger, J., Nolan, K., Bermann, R., & Sweet, J. (in press). P300 scalp amplitude distribution as an index of deception in a simulated cognitive deficit model. *International Journal of Psychophysiology.*

Rosenfeld, J. P., Ellwanger, J., & Sweet, J. J. (1995). Detecting simulated amnesia with event-related brain potentials. *International Journal of Psychophysiology, 19,* 1–11.

Rosenfeld, J.P., Sweet, J.J., Chuang, J., Ellwanger, J., & Song, L. (1996). Detection of simulated malingering using forced choice recognition enhanced with event-related potential recording. *The Clinical Neuropsychologist, 10,* 163–179.

Schretlen, D., Brandt, J., Krafft, L., & Van Gorp, W. (1991). Some caveats in using the 15-Item Memory Test to detect malingered amnesia. *Psychological Assessment, 3,* 667–672.

Simon, M. (1994). The use of the Rey Memory Test to assess malingering in criminal defendants. *Journal of Clinical Psychology, 50,* 913–917.

Sinnett, E. R., & Holen, M. (1995). Chance performance in the absence of norms for neurpsychological tests-A clinical note. *Journal of Clinical Psychology, 51,* 400–402.

Slick, D., Hopp, G., Strauss, E., Hunter, M., & Pinch, D. (1994). Detecting dissumulation: Profiles of simulated malingerers, traumatic brain-injury patients, and normal controls on a revised version of Hiscock and Hiscock's forced-choice memory test. *Journal of Clinical and Experimental Neuroposychology, 16,* 472–481.

Slick, D., Hopp, G., Strauss, E., & Spellacy, F. (1996). Victoria Symptom Validity Test: Efficiency for detecting feigned memory impairment and relationship to neuropsychological tests and MMPI-2 validity scales. *Journal of Clinical and Experimental Neuropsychology, 18,* 911–922.

Strauss, E., Spellacy, F., Hunter, M., & Berry, T. (1994). Assessing believable deficits on measures of attention and information processing capacity. *Archives of Clinical Neuropsychology, 9,* 483–490.

Suhr, J., Tranel, D., Wefel, J., & Barrash, J. (1997). Memory performance after head injury: Contributions of malingering, litigation status, psychological factors, and medication use. *Journal of Clinical and Experimental Neuropsychology, 19,* 500–514.

Sweet, J. (1983). Confounding effects of depression on neuropsychological testing: Five illustrative cases. *Clinical Neuropsychology, 5,* 103–109.

Sweet, J., & Kuhlman, R. (1995). Evaluating malingering in brain injury claims: Genuine injury versus proven malingerer. *Trial Diplomacy Journal, 18,* 1–7.

Sweet, J., & Moulthrop, M. (1999). Self-examination questions as a means of identifying bias in adversarial assessments. *Journal of Forensic Neuropsychology, 1,* 73–88.

Sweet, J., Newman, P., & Bell, B. (1992). Significance of depression in clinical neuropsychological assessment. *Clinical Psychology Review, 12,* 21–45.

Sweet, J., Wolfe, P., Sattlberger, L., Numan, B., Rosenfeld, P., Clingerman, S., & Nies, K. (in press). Further investigation of traumatic brain injury versus insufficient effort on the California Verbal Learning Test. *Archives of Clinical Neuropsychology.*

Tenhula, W., & Sweet, J. (1996). Double cross-validation of the Booklet Category Test in detecting malingered traumatic brain injury. *The Clinical Neuropsychologist, 10,* 104–116.

Tombaugh, T. (1997). The Test of Memory Malingering (TOMM): Normative data from cognitively intact and cognitively impaired individuals. *Psychological Assessment, 9,* 260–268.

Trueblood, W., & Binder, L. (1997). Psychologists' accuracy in identifying neuropsychological test protocols of clinical malingerers. *Archives of Clinical Neuropsychology, 12,* 13–27.

Trueblood, W., & Schmidt, M. (1993). Malingering and other validity considerations in the neuropsychological evaluation of mild head injury. *Journal of Clinical and Experimental Neuropsychology, 15,* 578–590.

Tsushima, W., & Wong, J. (1992). Comparison of legal and medical referrals to neuropsychological examination following head injury. *Forensic Reports, 5,* 359–366.

Van Gorp, W., & McMullen, W. (1997). Potential sources of bias in forensic neuropsychological evaluations. *The Clinical Neuropsychologist, 11,* 180–187.

Wiggins, E., & Brandt, J. (1988). The detection of simulated amnesia. *Law and Human Behavior, 12,* 57–78.

Willer, B., Johnson, W., Rempel, R., & Linn, R. (1993). A note concerning misconceptions of the general public about brain injury. *Archives of Clinical Neuropsychology, 8,* 461–465.

Wong, J., Regennitter, R., & Barrios, F. (1994). Base rate and simulated symptoms of mild head injury among normals. *Archives of Clinical Neuropsychology, 9,* 411–425.

Youngjohn, J., Burrows, L., & Erdal, K. (1995). Brain damage or compensation neurosis? The controversial post-concussion syndrome. *The Clinical Neuropsychologist, 9,* 112–123.

Zielinski, J. (1994). Malingering and defensiveness in the neuropsychological assessment of mild traumatic brain injury. *Clinical Psychology: Science and Practice, 1,* 169–184.

Chapter 10

COGNITIVE PSYCHOPHYSIOLOGY IN DETECTION OF MALINGERED COGNITIVE DEFICIT

J. Peter Rosenfeld
Northwestern University, Evanston, Illinois

Joel W. Ellwanger
University of California, San Diego, VA Medical Center, San Diego, California

Introduction

In cases of head injury and other acquired disorders in which malingered cognitive deficit is suspected, neuropsychologists are increasingly involved as providers of expert testimony. Since development of a "malingering profile" based on existing standard psychological tests has not been perfected (Nies & Sweet, 1994), researchers and clinicians have turned to quasi-tests on which normals and non-litigating patients perform well, but on which suspected malingerers and experimental simulators perform poorly. The Hiscock forced choice procedure (HFCP), developed by Pankrantz (1983) and Hiscock and Hiscock (1989) is an example of such a quasi-test. In its most basic embodiment, the subject is briefly presented with a three-digit sample number. After a delay following the clearing of the sample from the screen, two three-digit numbers are presented at the left and right, respectively, of the screen. The subject must choose the matching number, known as "the match to sample." The match's position varies randomly over trials. Although it is generally the case that non-litigating patients with head injury perform at levels greater than 90% correct on this test, and may be mostly discriminated from experimental simulators (e.g., Guilmette, Hart, Giuliano, & Leininger, 1994), there is always the chance of a false positive error, which could have serious consequences if the result were the false accusation of malingering in an honest litigant. Indeed we have report-ed that in a sample of 14 non-litigating, closed-head-injury cases, although the mean score on the first two blocks of the HFCP was 93.5% correct (with 10 cases scoring 100% on the first block and seven cases scoring 100% on the sec-ond block), nevertheless, on the first block, there were two individual scores of 62.5% and 75%, and on the second block there were scores of 58%, 79%, and 83% in three of these non-litigating patients (Ellwanger, Rosenfeld, & Sweet, 1997). Thus, sole reliance on an arbitrary cutoff score (e.g., 90%) on a strictly behavioral test is likely to lead to a non-negligible false positive rate.

In the past several years, our research efforts have been directed at provid-ing physiological enhancements to the behavioral tests used by neuropsycholo-gists. In particular, we have utilized the P300 (sometimes referred to as P3) component of the Event-Related Potential (ERP). P300 is elicited in the electro-encephalogram (EEG) by rare, meaningful stimuli (Johnson, 1988). For exam-ple, if a subject attends to a video screen on which different names are presented, one at a time, every three seconds, his EEG will show the P300 ERP only in response to occasional presentation of his own name, given that the other names presented are not personally meaningful to him.

EEG contains many other ongoing amplitude variations (probably represent-ing other, endogenous psychological activities) on which the P300 may be super-imposed. Thus, to clearly see the P300, one must average many EEG epochs that are time-locked to key stimuli of one (i.e., rare and meaningful) type. Non-time-locked amplitude variations will then average out, and timelocked components, such as P300, emerge. Figure 1 shows two superimposed ERP averages at each of three scalp sites for an individual subject. The heavier curve is an average of 24 ERP responses to presentation of the subject's birthdate as a stimulus. The lighter curve is an average of 24 responses to another, personally irrelevant date. In the Figure, the arrow points to the downgoing P300 response (at Pz) evident in the average represented by the heavier curve, and largely absent from the lighter curve. P300 is a positive-going voltage (positive is shown downgoing traditionally) that is usually recorded with (minimally) electrodes at Fz, Cz, and Pz scalp sites. Typically, the size order of P300 amplitudes is $Fz < Cz < Pz$. The latency of the P300 peak can be anywhere from 300 to 1000 milliseconds post-stimulus, usual-ly depending on the complexity of the stimulus. The approximately 500 msec latency of the response in Figure 1 is typical for simple, meaningful, alphanumer-ic stimuli. It is noted that P300 is measured in two ways. The more typical way is to subtract the average amplitude of the maximum positive segment of EEG dur-ing the time window from 300 to 800 msec post-stimulus, from the average value

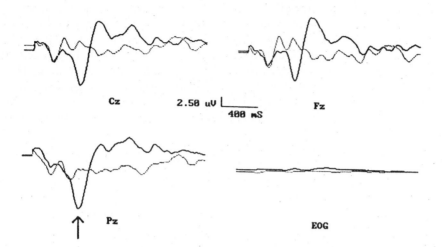

Fig. 1.   ERP responses at Fz, Cz, Pz, and EOG, to one subject's birthdate, presented with probability = .12, darker curve, and to other dates, presented with probability (p) = .88. Note (arrow in Pz waves, lower left) the prominent P300 response in the ERP in response to the rarer, more meaningful stimulus, but largely absent in the other waveform (Positive is down, negative is up in all ERP figures). EOG is electrooculogram, recording of eye movements whose flatness indicates freedom from eye movement artifacts.

of the EEG for 100–500 msec prior to the stimulus. This method is called the *baseline-to-peak* or *b-p* measure. We have found that in deception detection situations, another method is often preferable: instead of subtraction from the pre-stimulus baseline, the maximum positive segment in the 300–800 msec time window is subtracted from the immediately subsequent negative peak. This method is called the *peak-to-peak* or *p-p* method. In this chapter we will usually report p-p values, but occasionally present both b-p and p-p values.

The background of our use of P300 in detecting malingered cognitive deficit is in the field of detection of deception. Previously, we had shown that subjects would produce P300s uniquely in response to the names of items they had pretended to steal or in response to anti-social or illegal acts of which they were actually guilty (Johnson & Rosenfeld, 1992; Rosenfeld, 1992; Rosenfeld, Angell, Johnson, & Quian, 1991; Rosenfeld, Cantwell, Nasman, Wojdac, Ivanov, & Mazzeri, 1988; Rosenfeld, Nasman, Whalen, Cantwell, & Mazzeri, 1987). Others have published related demonstrations (e.g., Allen, Iacono, & Danielson, 1992; Farwell & Donchin 1991).

P300-based tests of recognition of autobiographical information

In the field of detecting malingered amnesia, Wiggins and Brandt (1988) found that on questions about birthdates, phone numbers, and mothers' maiden names, the percentages of experimentally instructed amnesia simulators giving incorrect or implausible answers were 42%, 42%, and 29%, respectively, whereas these percentages were zero in non-simulator controls, as well as in true neurologically-based amnesics. We thus designed an experiment (Rosenfeld, Ellwanger, & Sweet, 1995a) in which two groups of experimentally instructed simulators, naive and sophisticated (see Nies & Sweet, 1994), were first given recognition tests while their ERPs were recorded. There were three conditions. In the first condition, each subject saw his (her) mother's maiden name on a video screen, 12 times, interspersed with presentations of eight other (ethnically matched) names, each also presented 12 times. Thus the probability of the autobiographical item was about 11%, making it an *oddball* in the sense of rareness and meaningfulness. For the two other conditions, the autobiographical oddballs were the subjects' phone numbers and birthdates, interspersed with other (non-meaningful) phone numbers and birthdates, respectively. ERPs were averaged separately for oddball versus frequent other stimuli and it was found that the mean P300 oddball amplitude (at Pz) was significantly greater (about twice as great) in comparison with the frequent amplitude in all three conditions. On unexpected pencil and paper tests of recall and recognition given after the ERP recording session (during which subjects had only to repeat stimuli to assure us of their attention), the distinct majority of naive subjects "incorrectly" answered most items on both recall and recognition tests, whereas sophisticated subjects "recalled" 15% of the items, but "correctly" recognized about 50% of the items. While this pattern of behavioral simulation was consistent with the behavioral data of Wiggins and Brandt (1988), the ERP data suggested clear recognition.

A weakness of this first study was its failure to assess ERP data within individuals; (the significant effects described were results of statistical tests on group means). This weakness was remedied by a follow-up study (Ellwanger, Rosenfeld, Sweet, & Bhat, 1996) in which the basic methodology of the first study was replicated; however, there were two significant extensions: (1) statistical tests were done within individuals, and (2) the two other types of stimuli used in addition to birthdates were different than stimuli used by Rosenfeld et al. (1995a): In the second study, we used (a) newly-learned words (i.e., non-autobiographical information), and (b) the experimenter's name. In the former condition, we taught subjects a list of words to learn before the ERP recording and recognition session. In this latter session, we presented and separately averaged P300 responses to one set of 14 familiar (learned) words, and to eight sets of 14 completely novel words. Thus, again, the oddball probability = .11 (1/9). In the case of the session involving experimenter's name, during the ERP recording session, there were 12 repetitions of the experimenter's name (Joel), and 12 repetitions of each of the other eight male names. Prior to the session, when the experimenter (Joel Ellwanger) first contacted a subject to schedule a lab visit, he introduced himself by name. When the subject arrived at the lab, the first contact was reinforced by the experimenter's greeting ("Hi. I'm Joel Ellwanger"). We utilized the experimenter's name as an oddball because head-injured patients are often reported to forget initially the names of the therapists they come to see.

Table 1. Mean Behavioral Performance as a Function of Subject Type and Test Type.

Group	Oddball type	Correct response rate
Non-simulators	Birthdate	1.00
(truth-tellers)	Experimenter name	1.00
	Learned words	0.85
Naive Simulators	Birthdate	0.269
	Experimenter name	0.194
	Learned words	0.310
Sopisticated Simulators	Birthdate	0.395
	Experimenter name	0.208
	Learned words	0.434

Table 2. Sensitivities of Within-subject ERP Test.

Group	Oddball type	Proportion of correctly classified subjects
Truth-tellers	Birthdate	1.0
Naive simulators	Birthdate	0.9
Sophisticated simulators	Birthdate	0.86
Truth-tellers	Experimenter's name	0.80
Naive simulators	Experimenter's name	0.77
Sophisticated simulators	Experimenter's name	0.79
Truth-tellers	Learned words	0.43
Naive simulators	Learned words	0.50
Sophisticated simulators	Learned words	0.53

The group results were that for both groups used, amnesia simulators and truth-tellers, the P300 responses to all three types of oddball-memory items were significantly greater than the responses to frequent-other items. In *individual* tests, bootstrapping techniques (detailed in Ellwanger et al., 1996) were used to determine, within each subject, whether or not one could state with 95% confidence that P300 evoked by the oddball-memory item was greater than the response to other items, despite the fact that simulators did indeed simulate inability to recognize memory items (Table 1). The results of the bootstrapping analyses on P300s are shown in Table 2 where the proportions of correct diagnoses are given. Oddball-frequent P300 differences were clearly greater in truth-tellers (non-simulators) than in simulators; however, ERP test sensitivities in all subjects are best for birthday (autobiographical) stimuli ($> .9$), and worst for learned word stimuli (.5), with experimenter name stimuli producing medium sensitivities of about .8. However, we learned during de-briefing that three subjects actually did forget the experimenter's name. The average ERP at Pz in these forgetters is shown in Figure 2, in which the average response to the experimenter's name is superimposed on the average response to the other foil names.

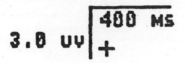

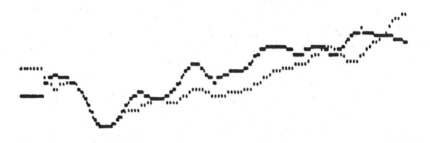

Fig. 2.   Averaged ERP responses to experimenter's name (darker curve) presented with
p = .12 and to other names (p .88) in 3 subjects who could not recall the experimenter's name. Note the lack of P300 in both waveforms from 400 to 800 msec
post-stimulus; (the stimulus appears 100 msec into the sweep and is indicated
by a discontinuity in the traces). The ERP to the experimenter name is *less* positive than that to the other names.

It is clear from the figure that there is no P300 in response to the experimenter's
name; indeed the response to other names is more positive-going than the
response to "Joel". In other words, in cases of non-recognition, which might
occur in some patients with major cerebral insults, there does not appear to be
a strong likelihood of a false positive. Indeed, the false positive issue was specifically considered in a further study:

In this follow-up experiment (Ellwanger et al., 1996), we explored the false
positive rate of our methods by comparing a new malingering group given the
birthdate stimulus, with a control group which was treated exactly like the
malingering group, except that in this group, the subject's actual birthdate was
never presented, although it was expected. Instead, the control group saw only
nine personally irrelevant dates, each repeated 12 times, although its members
were instructed to expect to see their own birthdates occasionally. The results
were that the same P300/bootstrapping methods which classified 9/13 malingerers (about 70%) correctly, also found zero of 11 controls to have recognized
any of the nine birthdates. (In the controls, we randomly chose one of the nine
dates and computed an average P300 to this date. This average was then compared with the average response to all other stimuli.) In other words, the false
positive rate was zero.

One question that could be posed about the field relevance of these studies concerns the fact that P300 is typically reported to be reduced in head-injured patients (Campbell, Houle, Lorrain, Deacon-Elliot, De Proulx, 1986; Deacon & Campbell, 1991; Rugg, Cowan, Nagy, Milner, Jacobson, & Brodes, 1988). Our prior studies used normal models of simulated malingering; however, the probable target population of these methods is the head-injured population. If this population has a reduced P300, regardless of true cognitive ability, it could be the case that our P300-based methods might fail in the very population for which they were developed. As a first step designed to get at this question, we ran 14 documented head-injured outpatients recruited from the Brain Injury Association of Illinois. Based on the duration of post-traumatic amnesia or loss of consciousness, the severity of head injuries seen varied from mild to very severe (Jennett, 1983) (other demographics can be found in Ellwanger, 1996; Ellwanger, Rosenfeld, & Sweet, 1997). These patients, instructed to do their best, were run in our autobiographical oddball paradigm in which their birthdates were used as oddballs in one run, and newly learned words (as described above in experiments with normal simulators) were used as oddballs in another run. In particular, we were interested in whether or not our individual diagnostic analyses, within subjects, could distinguish P300s evoked by oddball-memory items, versus frequent, other items. The results of bootstrap analyses were that for new word learning, only 6 of 11 (about 55%) patients' oddball and frequent P300s were found to be different. For the birthday stimuli, however, in 9 of 11 cases (82%), the P300 evoked by the oddball stimulus was found to be greater than that evoked by the frequent (other date) stimulus. These sensitivity values are similar to what we found in normal subjects (Table 2). While we too found that the P300s in patients were smaller than those in normals ($p < .01$), nevertheless the difference in P300 amplitude, between response to oddball versus response to frequent stimuli (our key dependent measure in the bootstrap tests), was enough to allow our methods to identify P300 responses to the birthday stimulus categories in most of our head-injured cases.

It seems clear from Ellwanger et al. (1996) and the other studies so far reviewed, that for suspected malingerers who are artless enough to claim an inability to recognize basic autobiographical information, the P300 recognition test has considerable potential for countering the false claims. This would be true even in patients with genuine head injury, who are exaggerating or inventing cognitive deficit. Moreover, the false positive rate in these situations appears to be close to zero.

Of course a *serious, remaining problem* is that many malingerers have the sophistication to realize that claims of inability to recognize basic autobiographical information have low credibility. They will thus not make such claims. It is for such individuals that quasi-tests, such as the HFCP match-to-sample tests (described above), were developed.

P300-enhanced match-to-sample tests

As already noted, purely behavioral tests such as the HFCP run the risk of false positive error. Thus, we have also developed methodology in which P300 recording during the HFCP is utilized so as to help (in)validate a subject's professed inability to discriminate matches and mismatches to a sample. In order to accomplish this goal, the standard, commercially available form of the HFCP (described above) had to be somewhat modified (Rosenfeld, 1997; Rosenfeld, Sweet, Chuang, Ellwanger, & Song, 1996). In particular, in the ERP-enhanced version of the test, a *sample* 3-digit number is presented on a display terminal for three seconds. Following a subsequent three-second delay, a second *test* number appears on the screen for about two seconds, and the ERP is recorded from 104 milliseconds prior to test presentation until 1.944 seconds thereafter when the test number is cleared. The test number is also a three-digit number, which either perfectly matches the sample (*match* trial) or which completely mismatches the sample in all three numerical positions (*mismatch* trial). Besides slight timing differences, the essential difference between the ERP-enhanced HFCP and the commercially available (purely behavioral) version is that in the latter, both match and mismatch stimuli appear simultaneously on one test screen trial, and the subject must choose one or the other as a match, whereas in the P300-enhanced procedure, matches and mismatches must appear separately and the subject must choose to say either "match" or "mismatch". This requirement is due to the necessity of having separate ERPs to the matches and mismatches. Also, since the P300 component of the ERP was our major interest, and since it is known to be larger as the probability of its eliciting stimuli declines (Johnson, 1986, 1988), we reduced the match trial probability to .33 in one condition, and to .17 in another. (This manipulation assessed the effect of match probability on our ability to detect simulators.)

In our first study with the P300-enhanced HFCP (Rosenfeld et al., 1996), we utilized a group of normals coached as "sophisticated" simulators. Following the ERP run, subjects took the commercial version of the HFCP (the Multidigit Memory Test or MDMT, copywritten and produced by Wang Neuropsychological Laboratory). Our first important result was that there were no significant *behavioral* differences between simulated malingering performance on the commercial-behavioral HFCP and our ERP-enhanced HFCP, despite the procedural differences. Indeed there was good correlation of simulation performance in both HFCP tests for both the .17 match probability group (r = .59, p < .02), and for the .33 match probability group (r = .56, p < .05). As far as hit rates were concerned, combining results from the .17 and .33 match probability groups, the average percent correct on the commercial HFCP was 54.45; on the ERP-enhanced HFCP, the percentage was 55.6. These very similar "hit" (i.e., simulation) rates are both close to the theoretically random 50% correct which one would see if a subject flipped a coin on each trial as a means of deciding whether to respond "match" or "mismatch". These results are important because they suggest that the same simulation behavior discernable with the

commercial HFCP is likewise picked up by the somewhat procedurally different ERP-enhanced HFCP. The implication is that if the ERP recording procedure does allow independent indexing of malingering behavior via the P300,
then its administration can replace administration of the standard HFCP.

How does the ERP recording enhance the behavioral data collection in the
case of the HFCP? When a subject has a 50% hit rate (or a rate not significantly
different than chance, as we saw in Rosenfeld et al., 1996), he is in effect claiming an inability to discriminate matches and mismatches. If this claim was actually true, then there should be no difference seen between P300s evoked by
matches and mismatches. Thus, if differences are seen between P300s to matches
and mismatches, verbal claims of inability to distinguish these two stimulus
types become harder to sustain.

Some group results of our first study of the P300-enhanced HFCP paradigm
are seen in Figure 3. The two independent variables were stimulus type (match
vs mismatch) and match probability ($p = .17$ vs. $p = .33$). We did indeed find
a highly significant effect of stimulus type ($p < .001$), but not of match probability. The reason for the latter negative finding is strongly suggested by the figure: As match (oddball) probability is reduced, the P300 to the match increases
and the P300 to the mismatch decreases; that is, there was an interaction effect

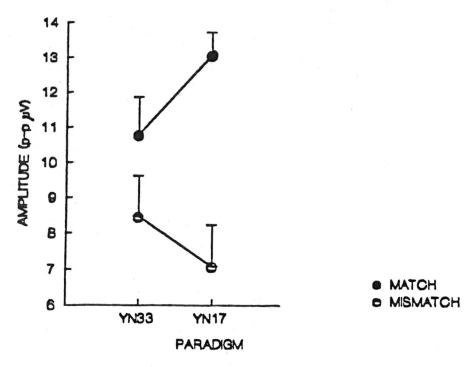

Fig. 3.   Computer calculated, p-p P300 Pz amplitudes in uv to matches and mismatches
          as a function of match probability = p = .33 ("YN33") vs. p = .17 ("YN17").

that closely approached significance, $p = .054$. Thus, the match-mismatch difference appears affected, as it should be, by oddball probability, and suggests use of low values of this variable in future applications.

For practical diagnosis, the question is: what proportion of simulators can be identified with the P300-enhanced HFCP? This reduces to the question of how many *individual* simulators will show significantly larger P300s to matches than to mismatches. In Rosenfeld et al. (1996), we utilized t-tests on the single trials data of individuals to examine whether the *individual's* average P300 to the match differed ($p < .05$) from his average P300 to the mismatch. We examined data from one scalp site only, (Pz) as in all previously reviewed studies. We detected 67% of the simulators in this way, a modest hit rate. However, it should be recalled that this was a first study, and that due to memory size limitations of the older computers utilized for this study (mid-1980s technology), the numbers of single sweeps we had available for analysis was limited. Moreover, even 67% detection may represent considerable savings in false claim payments. (As will be discussed below, more recent versions of our P300-enhanced HFCP have yielded higher detection rates).

Potential problems with P300 methods

At this point, it is important to consider some basic reservations that could be expressed about all the studies so far reviewed. One could first of all ask to what extent might P300 index non-conscious processing. That is, if it were the case that P300 can be elicited by meaningful, rare stimuli *outside a subject's conscious awareness*, then our methods would be of questionable value; a plaintiff's attorney could argue that the difference in amplitude between P300 to autobiographical versus non-autobiographical items of information (or between match and mismatch) in no way impugns a subject's conscious experience of non-recognition of meaningful information. Fortunately, (as reviewed in Rosenfeld et al., 1995a), the evidence supporting P300's indexing of only consciously perceived information is quite solid, and indeed follows from the leading theoretical views of the ERP's functional significance. For example, P300 *amplitude* has been related to stimulus categorization, surprise, expectancy, and context updating (Donchin & Coles, 1988), as well as to meaningfulness (Johnson, 1988). Such cognitive processes imply conscious processing of information. The *latency* of P300 has been associated with stimulus evaluation time (Johnson, 1988), another cognitive process that would seem (intuitively) more likely to happen consciously than automatically. The closest approximation of evidence we know of suggesting that P300 discriminates stimulus categories outside awareness come from our laboratory (Leiphart, Rosenfeld, & Gabrieli, 1993). Even in this study, however, P300 discriminated familiar (previously studied) vs. novel words better than honest behavioral responses only in a perceptual identification (i.e., *implicit* memory) task. In this same study, during an *explicit* memory task of recognition, such as the tasks utilized in research so far reviewed here, behavioral indices of categorization correlated highly with P300 amplitude. Paller and Kutas (1992), in a similar study, found that even in a perceptual identification task (i.e., an implicit memory task as in Leiphart et al., 1993), neither P300 nor behavioral indices could discriminate familiar vs. novel words. The literature on P300 in wakefulness and sleep also suggests that P300 disappears just when consciousness is lost (Harsh, Vos, Hall, Schreffer, & Badia, 1994).

It could be argued, however, that cerebral pathology (e.g., mild to moderate head injury) could disturb the normal association of P300 and consciousness outlined in the previous paragraph. This would render our methods of dubious value, again, in the target population (head-injured persons) for which they were developed. The answer to this potential line of criticism is that one would expect such major cerebral pathology to be associated with morphological signs (i.e., detectable with NMR, CT scans, MRI, etc), as in prosopagnosia. This condition, a seemingly excellent exemplar of a dissociation between conscious recognition and P300 signs of recognition (Renault, Signovet, DeBrulle, Breton, & Polgert, 1989) is a very rare disorder that *does* have distinct morphological correlates (Renault et al., 1989). In contrast, the type of subject typically suspected of malingering has no demonstrable lesion, and evidence of brain disfunction associated with head injury is circumstantial and/or based on personal report. Thus, there remains good reason to support the use of P300 methods in detection of malingering. Nevertheless, because of the logical possibility that P300 and conscious awareness could become dissociated following lesions too subtle to be visible, even using modern imaging diagnostics, we suggest caution in utilization of P300 evidence. Such evidence should always be utilized along with results of other diagnostic approaches.

In addition to the just discussed issues with ERP methods that pertain to both the autobiographical oddball, as well as to the matching-to-sample paradigm, there is a special set of issues that pertain more particularly to the latter: As noted earlier, if an individual scores at or significantly below chance on the behavioral indices of the ERP-enhanced HFCP (i.e., if the percent correct rate is less than or at 50%), and the P300 to the match is significantly greater than the P300 to the mismatch, the behavioral-physiological discrepancy should cast serious doubt on claims of cognitive deficit. A problem arises, however, if an individual scores (behaviorally) significantly greater than chance, but at a suspiciously low rate of accuracy, e.g., 75% correct. Guilmette et al. (1994) have suggested that less than 90% accuracy on the HFCP, particularly in potentially mild injuries, should raise suspicions about malingering. Could ERP data collected from the P300-enhanced HFCP help support the malingering implication of a 75% behavioral hit rate?

At first glance, the answer might appear to be negative. If a subject scoring a behavioral 75% hit rate showed a significant P300 amplitude difference, match vs. mismatch, there would be negligible discrepancy between behavior and brain wave: A plaintiff's attorney could argue, "Of course my client shows a significant difference between match and mismatch P300s. After all he *is* correctly discriminating these stimuli most (75%) of the time." Some reflection on the matter, however, in conjunction with a review of appropriate research, suggests a different conclusion.

There is ample evidence that as task difficulty and/or workload *increases*, P300 amplitude *decreases* (e.g., Johnson, 1988; Rosenfeld, Bhat, Miltenberger, & Johnson, 1992). Thus one might expect that persons having genuine difficulty with the HFCP task would show reduced P300s to all stimuli and little or no difference between P300 responses to matches and mismatches. There is also empirical support for the notion that if a secondary task is embedded within the P300-eliciting task, the effect will be *enhanced P300 amplitudes* to appropriate stimuli (Donchin, Kramer, & Wickens, 1986). A somewhat sophisticated litigating malingerer might well realize that performing at or below chance on an easy matching-to-sample task would raise suspicions, apart from availability of ERP data. Such a person might indeed elect to perform above chance but below 90% correct, i.e., at 75%–80% correct, (particularly if it became well-known that a 90% diagnostic cutoff score would likely be used). This person would have to attend especially carefully to match and mismatch stimuli, in order to achieve the 75% hit rate associated with simulated deficit. One might thus expect such an individual to show *enhanced* P300s to appropriate stimuli, in comparison to responses shown by honest test-takers who do not have to do the extra, embedded work of deception, but need only respond honestly on each trial.

All the notions just suggested have been empirically tested, and the results continue to support the use of P300-based methods in cases of suspected malingering: The first empirical question addressed concerned what effect a 75% hit rate would have on P300 amplitude in a group of simulators manipulated to perform at such a greater-than-chance, but lower-than-expected hit rate in an easy task. Two separate studies were undertaken by two different experimenters on two groups of subjects in two separate years (all detailed in Rosenfeld, Reinhart, Bhatt, Ellwanger, Sekera, & Sweet, 1997). In both studies, there were 2 groups of participants; a group in which each person was told to do his best (*Truth* group), and a group told to simulate malingering by producing an overall hit rate of about 75%; (*Malinger* group). In the first study, subjects were not specifically instructed as to how to achieve the 75% rate. Since in the course of debriefing for this experiment, it was discovered that many subjects had kept running track of error rates in order to achieve the target hit rate, we decided that for the second experiment we would try to discourage this possibly interfering extra task (Donchin et al., 1986). Thus, in the second study the subjects were told explicitly to not count their behavioral responses, but to "respond correctly like what feels close to a random 75% of trials".

The average hit rates for simulators in both studies were similar, at 71% and 70% for Experiments 1 and 2, respectively. These rates were significantly different ($p < .001$) than those of the *Truth* groups in both studies (98.4% correct, overall), indicating that the manipulations were successsful. Figure 4 shows the ERP results for Experiment 1 in Figure 4a and for Experiment 2 in Figure 4b; responses to match stimuli are shown. Each pair of traces shows the P300s of

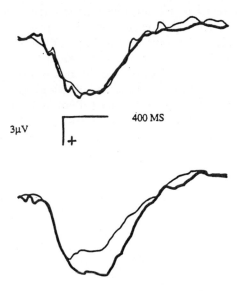

Fig. 4.   Group averaged match ERP waveforms (at Pz) for simulating malingerers (darker curve) and truthtellers in P300-HFCP study. Top (4a) is first study where some simulators kept track of numbers of matches and mismatches. Bottom (4b) is for second experiment.

Truth and *Malinger* groups, superimposed. It appears that there is no difference between P300s of *Truth* and *Malinger* subjects in Experiment 1; (this was confirmed in group t-tests where $p > 0.3$ for both amplitude and latency of P300). However, in within-individual tests, 69% of the Malinger group showed P300 match responses significantly greater (amplitude) than the P300s in response to mismatches (quite comparable to Rosenfeld et al. 1996, discussed above), whereas only 40% of the *Truth* group showed match-mismatch differences on this intra-individual basis. In Experiment 2, Figure 4b suggests that the P300 in the *Malinger* group was *greater* than that of the *Truth* group, and this was statistically confirmed ($p < .002$). Thus, as expected, a simulated hit rate that is greater than chance, but less than 90% on the P300-enhanced HFCP, will not reduce the match-mismatch amplitude difference for P300; if anything, such finely tuned simulation would appear to enhance the P300 match-mismatch difference. The plaintiff's argument that this difference is an indicator of reduced, but non-chance, differential responding to match and mismatch becomes less tenable. This follows from the literature-based expectation that real increases in task difficulty typically reduce P300 amplitude (Johnson, 1988; i.e., one would expect less of a match-mismatch difference, if any, in someone honestly scoring only 75% correct on the easy HFCP task). Presumably, norms for normal, match-mismatch differences in P300 could be developed, and someone scoring an honest 75% correct behaviorally would surely *not* be expected to have a match-mismatch difference in P300 *greater* than the norm. In contrast, one would expect a reduced match-mismatch difference. The results of the next study to be described convincingly support this expectation.

Ellwanger (1996), and Rosenfeld, Ellwanger, Bhatt, Hankin, Bermann, and Sweet (1995b) attempted to model the behavior of a person having real difficulty with a forced-choice procedure. We did this by continuing to use normal subjects, but in more difficult versions of the P300-enhanced HFCP. All subjects were run through a *standard* (easy) 3-digit, P300-enhanced HFCP. That is, all mismatches were different in all three digit positions from the three-digit samples. Subjects were also run through *difficult* conditions in which seven ($n = 10$) or nine ($n = 9$) digit strings were used both for sample and test stimuli. Moreover in these *difficult* conditions, mismatches differed from samples in only one digit position. (It was never the first position, but randomly varied over other positions throughout the run). We had learned in pilot studies that such tests would indeed produce accuracies of 70%–90% in subjects doing their best.) There were two behavioral measures of interest collected in these studies. One was accuracy and the other was subjectively perceived task difficulty on a scale of 1 to 100. Table 3 (from Ellwanger, 1996) shows the behavioral results. It appears that overall accuracy did not decline from the three-digit to the seven-digit task (99.9 to 98.2); however, if one looks at match stimuli only, the drop in accuracy is 99.2 to 93.3 ($p < .01$). Indeed, for accuracy, one should focus primarily on responses to rare match stimuli, since only these are expected to elicit P300. The increase in difficulty from three to seven-digit tasks was not significant at $p < .05$, however the increase from the seven to the nine-digit task was ($p < .01$). The drop in match accuracy from seven to nine-digit tasks was significant ($p < .001$), as was the loss in accuracy from three to seven digit tasks. Thus, our manipulation was able to elicit approximately 75% accuracy in the nine digit condition for honestly-responding, normal models of cognitive deficit.

ERP data are shown separately in Figure 5 for three, seven, and nine-digit conditions. The figure shows grand average match wave forms for the standard and two difficult conditions. There is an evident P300 to the match in the standard (easy) three-digit condition, but *no* P300s are evident for either the seven or nine digit (difficult) conditions. As comparison of P300 amplitudies for seven versus nine-digit conditions yielded no difference ($p > .9$), data for both difficult

Table 3. Mean Accuracy and Perceived Difficulty Ratings for the P300-enhanced HFCP.

Test (n)	Perceived difficulty (SD)	Match accuracy (SD)	Mismatch accuracy (SD)	Total accuracy (SD)
3 digit (19)	3.18 (3.13)	99.2 (1.9)	100 (0)	99.9 (0.3)
7 digit (10)	16.6 (21.4)	93.3 (7.5)	99.1 (1.4)	98.2 (2.3)
9 digit (9)	57.2 (23.2)	77.0 (8.1)	86.7 (15.3)	85.1 (12.2)

Note. Accuracy is the percentage of items correctly responded to ("yes" for a match, and "no" for a mismatch). Difficulty was rated on a scale from 1 (extremely easy) to 100 (extremely difficult). There were on average 18 match trials and 90 mismatch trials.

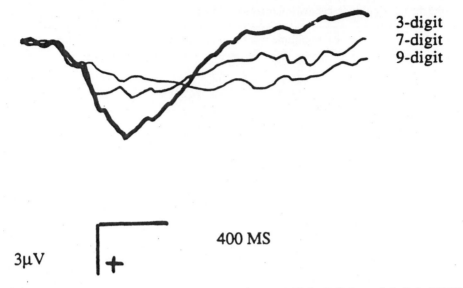

3-digit
7-digit
9-digit

400 MS

3μV +

Fig. 5.   ERP group averages (at Pz) to matches in 3-digit, 7-digit, and 9-digit HFCP paradigms.

conditions were combined. Figure 6 (from Ellwanger, 1996) illustrates clear match-mismatch differences in the standard (three-digit) task, but not in the difficult (seven and nine-digit combined) condition. Figure 7 (from Ellwanger, 1996) shows computer-calculated P300 match and mismatch amplitudes at the Pz site only as a function of task difficulty. It is evident that an interaction is present such that increased task difficulty removes the usual match-mismatch difference; (this interaction was significant, $p < .001$). The critical point is that even going from three to seven digit tasks in which actual overall accuracy drops only 1.7%, the P300 match-mismatch amplitude is obliterated totally (Table 3 and Figure 7). Thus, the problems discussed earlier with the P300-enhanced HFCP concerning the hypothetical low, but better-than-chance-scoring subject seem less of an issue; those who simulate 75% hit rates probably show *enhanced* P300 match-mismatch differences, whereas even slight, but genuine, difficulties with the task simply cancel out match-mismatch differences. Of course, our argument is loosely based on the assumption that a difficult task for normals is a reasonable model of a less difficult task for patients with true cognitive deficit related to brain dysfunction. This assumption requires verification. Yet it seems reasonable that whatever the source of discrimination difficulty, to the extent that the size of P300 match-mismatch amplitude difference represents this difficulty, our normal model of cognitive deficit may have considerable value. Indeed, Table 4 shows the correlation matrix for P300 amplitude, match identification accuracy and perceived task difficulty. Confirming results given earlier, these correlations were significant and further support the notion that as greater difficulty with the match-mismatch discrimination is experienced, the match-mismatch P300 amplitude difference decreases.

Table 4. P3 Amplitude-behavior Correlation Matrix.

	P3 amplitude	Accuracy	Perceived difficulty
P3	1.0		
Accuracy	.38*	1.0	
Perceived difficulty	−.45**	−.74***	1.0

Note. Number of observations = 34 (4 cases excluded that were missing a difficulty rating)
*p < .05; **p < .01; ***p < .001

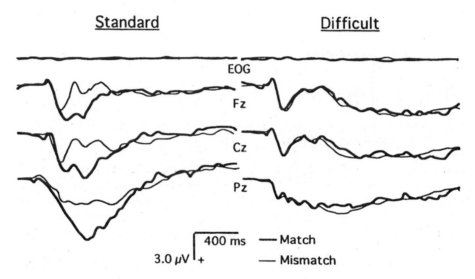

Fig. 6.   Superimposed match (darker) and mismatch grand average waveforms.
Standard is for 3-digit task; *Difficult* represents combined 7- and 9-digit tasks.
Fz, Cz, Pz are scalp site recordings. (Figure taken from Ellwanger, 1996.)

Combined behavioral and ERP methods

Although the main thrust of our work in detecting malingering has focused on
the P300 wave, we have tried to avoid the suggestion that P300-based methods
will ultimately come to replace standard behavioral instruments in neuropsychological testing. Indeed, we have previously argued in a different context that
if two or more testing instruments have some degree of orthogonality, then combining information from both instruments should afford greater diagnostic accuracy than that available from either index alone (Rosenfeld, 1995).

There are several approaches to the behavioral detection of malingered cognitive deficit, including identifying malingerers through the use of a battery of
neuropsychological tests, the use of a single neuropsychological test, or the use
of tests specifically designed to detect malingering (see Nies & Sweet, 1994 and
Chapter 9 for reviews). Behavioral tests of malingering are based on the premise

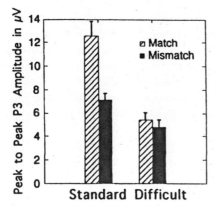 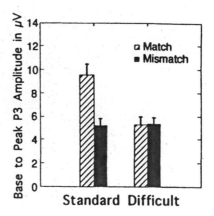

Fig. 7. Peak-to-peak (left) and base-to-peak (right) computer calculated mean P300 amplitude at Pz site. Error lines on top of bars are standard errors of the mean. (Figure taken from Ellwanger, 1996.)

that malingerers will perform differently than truly impaired subjects in terms of their overall level, or pattern of performance, on tests of cognitive function. Because the dependent measure in ERP tests of malingering is a direct measure of brain function, whereas, the dependent measure in behavioral tests does not directly measure brain function, the bases for deciding whether subjects are truly impaired or are malingering impairment are at least somewhat different for behavioral and ERP-based tests.

The goal of the study to be next described was to assess the relative effectiveness and the combined accuracy of two tests that have recently been developed for identifying patients who malinger neuropsychological deficits, the Booklet Category Test (BCT; DeFilippis & McCampbell, 1979), and the P300-based HFCP as described above. Tenhula & Sweet (1996) investigated methods to detect malingering based on a subject's pattern of errors across the seven subtests that make up the BCT. The BCT is a neuropsychological test of abstract reasoning and concept formation and its use in the assessment of malingering is a novel application developed by Tenhula and Sweet. In the presently described study, subjects were assigned either to a group simulating cognitive deficit, or to a control group. The experimental question was what the comparative sensitivities of the two tests would be when administered to the same subjects.

Research on assessment of malingering is complicated by the fact that it is very difficult to study actual malingerers, as they have no incentive to participate in research, and in most cases, suspicions that a given subject is malingering cannot be independently confirmed. Because of this, most research on malingering, including much of the research in our lab, has relied on subjects who are given instructions to simulate the behavior of malingerers. One reason we chose to compare ERP and behavioral tests of malingering within normal subjects, is the concern that because subjects are role playing, subject variables such as degree of effort, strategies for performance, knowledge concerning psychological testing, as well as situation variables, such as experimenter effects, will influence the accuracy of any test used to diagnose malingering. Therefore, although there may be questions regarding whether estimates of test characteristics based on subjects simulating cognitive deficits will be valid when the test is used in real world situations with actual malingerers, comparing different tests of malingering within (normal, simulating) subjects should allow for a reasonable estimate of test characteristics (such as test sensitivity) between tests, by controlling for many subject and situation variables.

The BCT and the P300-HFCP likely measure different mental processes: The BCT was designed as a challenging test to measure abstract reasoning and concept formation, whereas the match-to-sample test, that the P300-HFCP is based on (i.e., the MDMT), was designed specifically as a test of malingering. The MDMT was designed to be so easy that only malingerers (or subjects simulating malingering) perform poorly on the test. To the extent that the MDMT and P300-HFCP measure mental functioning, they can be thought of as tests of short term recognition and attention. Tenhula and Sweet (1996) proposed that the BCT can also be used as a standardized method for assessing the validity of a patient's neuropsychological performance. They developed discriminant functions and clinical decision rules that effectively discriminated between subjects simulating cognitive deficit from both normals and brain-injured non-simulators in a double cross-validation study. These discriminant functions and clinical decision rules were applied to the BCT performance of the subjects in the presently described study. Despite the different cognitive functions assessed by the BCT and P300-HFCP, both tests were used as decision tools designed to assess malingering, and therefore, it is possible to compare the relative ability of the two tests to identify this behavior. The characteristic of the BCT and P300-HFCP that was compared is sensitivity, or the percentage of simulating subjects correctly identified on the basis of the test.

The BCT consists of 208 cards on which there are various geometric shapes and figures. The cards are presented to the subject one at a time. For each card, the subject must decide what number (between 1 and 4) the shapes on the card suggest to him. Immediate feedback is given to the subject as to whether his response is correct or not. The subject's task is to integrate this feedback in order to generate the organizing principle for each subtest. There are seven separate subtests, the last of which is a recognition memory test for concepts taken from the cards of the first six subtests.

Classification of subjects as "malingerers" or "non-malingerers" as a result of their BCT performance was based on the discriminant functions and cut-off scores developed by Tenhula and Sweet (1996). In the double cross validation procedure, discriminant functions and cut-off scores were first developed in two studies separately, and were then cross validated by application to the subjects in the other study. In Study 1 of Tenhula and Sweet, there were 28 malingerers (simulators), 26 brain-injured subjects, and 34 normal controls; in Study 2 there were 17 malingerers (simulators), 24 brain-injured subjects, and 24 normal controls. Because the discriminant functions and cut-off scores developed from the two studies separately achieved similar hit rates, and there was minimal reduction in the hit rates upon cross-validation, optimal discriminant functions and cut-off scores were developed including all the subjects in both Study 1 and Study 2. It was these combined-groups discriminant functions and cut-off scores that were applied to the present study. Tenhula and Sweet found that the combined-groups discriminant functions correctly classified from 72.5% to 92.2% of subjects, and the combined-groups optimal cut-off scores correctly classified from 64.1% to 91.5% of subjects.

The results (Ellwanger, 1996) of these procedures involving combined BCT and P300 data were, first of all, that the simulation manipulations were successful. Simulators performed less accurately than truth-tellers in the P300-HFCP ($p < .001$). Likewise, on the BCT, simulators also performed less accurately than non-simulators (overall accuracy was 62% for simulators vs. 89% for non-simulators, $p < .001$). Figure 8 likewise indicates that match-mismatch P300 amplitude differences were the same in simulators as in non-simulators ($p > .85$). Using intraindividual analytic methods, it was found that neither the BCT nor the P300-HFCP by themselves yielded perfect discrimination between simulators and non-simulators, although sensitivities for both tests in detecting simulators were .8 (P300-HFCP) and .9 (BCT), which are similar for the small sample we used ($n = 10$ for simulators, $n = 11$ for non-simulators). If the outcomes of the two tests are used together such that the classification of subjects as "malingerers" is based on *either* a positive outcome in the P300-HFCP *or* a positive outcome in the BCT, then the rate of correct classification of malingerers = 100% (sensitivity = 1.0).

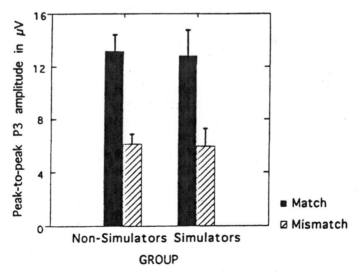

Fig. 8.   Mean, computer-calculated P300 (p-p) amplitude at Pz site. Error lines as in Fig. 7. (Figure taken from Ellwanger, 1996.)

Ellwanger (1996) did another study conceptually similar to the one just described, but using different behavioral and ERP paradigms. In this study, the behavioral test was the standard MDMT, the commercially available version of the HFCP. The ERP paradigm was the birthdate recognition paradigm, as described above. For the MDMT, Ellwanger adopted the criterion of Guillmette et al. (1994) that hit rates < 90% resulted in a diagnosis of simulation. For the ERP intraindividual analysis, he utilized bootstrapping methods as described above and in Ellwanger et al. (1996); Rosenfeld et al. (1996). In this presently described study, he utilized simulators only (i.e., no non-simulators), with $n =$ 25. Individually, the MDMT detected 17 of 25 (68%) simulators, whereas the P300 test detected 22 of 25 (88%) simulators. Again, *all* (100%) 25 simulators were correctly detected by at least one of the tests. Indeed, some simulators (32% of the sample) who were undetected by the MDMT were detected with the P300 test, but others (12%) who were undetected with the ERP test, were detected by the MDMT. It may thus be reasonable to conclude that these behavioral and ERP-based methods are tapping into somewhat different sources of information, and that the most profitable use of either type of information is, therefore, in conjunction with the other type.

Other ERP methods

Thus far, this review has mainly focused on one ERP dependent variable: P300 amplitude at a single scalp site (Pz). It should be acknowledged that there are other ERP components that have been studied over the past 30 years and that have been linked to psychological functions sometimes implicated in real, as well as malingered brain injury. Further, these other components' amplitudes and latencies might be best measured at sites other than Pz; (for reviews, see Andreassi, 1995; Rugg, 1995; or see Barcelo, Sanz, Molina, & Rubia, 1997 for a description of the use of ERPs in validation of frontal cortical assessment with the Wisconsin Card Sorting Test).

There is, however, another dependent variable ERP index that may be used with any ERP component and is at least to some extent independent of ERP amplitude at any single scalp site, or any linear or other combination of scalp sites. This index will be hereafter referred to as *profile*. It refers to the scaled *distribution* of ERP component amplitudes across various scalp sites. For example, the P300 component is usually largest at Pz, smallest at Fz, with amplitude at Cz somewhat less than at Pz though greater than at Fz. There is clearly, however, an infinitely large family of curves (P300 amplitude as f(site)) that would satisfy the basic attribute, Pz>Cz>Fz. Johnson (1993) has made a compelling argument that different psychological conditions will engage differing subsets of P300 neurogenerator neurons in the brain, the differential activations of which will be manifest in differing scalp amplitude distributions of the P300. Each distribution, however, has to be scaled within each condition so as to remove the potentially confounding effect of simple amplitude differences between conditions. These scaling methods have been discussed by McCarthy & Wood, 1985. A pertinent example will clarify this approach.

We have conceptualized truth-telling and deception (malingering) as two distinct psychological conditions, and thus hypothesize that truthful versus deceptive responses will be associated with differing scalp profiles (P300 amplitude distributions). We recently utilized our nine-probe P300-HFCP paradigm in a study of scaled scalp profiles associated with truthful versus malingered responses in simulated malingerers. This paradigm is like the one described above in some ways, but differs in that each sample number is followed by a series of nine probe numbers, only one of which is the correct match. (Then the next cycle of sample and probes begins; and so on.) ERPs are recorded during presentation of all probes, and are then sorted into four separate averages according to stimulus type/response type combination (i.e., match-honest response (RR), match-dishonest (RW), mismatch-honest (WW), and mismatch-dishonest (WR)). The four averages within each subject are then scaled within each of the four subtypes, using the vector length scaling method described by McCarthy & Wood, 1985, for reasons given therein. Figure 9 (Rosenfeld, Ellwanger,

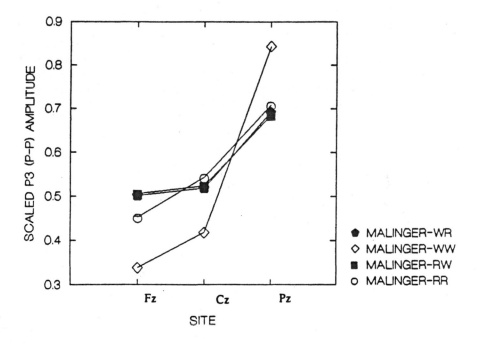

Fig. 9. Computer calculated, scaled P300 (p-p) amplitudes as a function of site within malingering group during 4 types of stimulus-response combination: match truthful (RR), match-deceptive (RW), mismatch-truthful (WW), and mismatch deceptive (WR). Note the superimposability of the deceptive responses, regardless of stimulus type.

Nolan, Bermann, & Sweet, in press) shows results within the malingering subjects, who were instructed to respond dishonestly about half the time, thereby yielding the four stimulus-response subtypes. Scaled P300 amplitude is plotted as a function of site for the four stimulus-response conditions. It is evident that the dishonest responses have the same profile, regardless of stimulus type, but that the honest responses show a stimulus condition by site interaction. This was confirmed by a 3-way ANOVA in which there was a significant interaction of site by condition; Truth = combined RR and WW versus Lie = combined RW and WR. While these results are for groups, and such interactions within individuals are not yet available, they strongly suggest that appropriate methods for use of profile as a novel ERP index in malingering situations may be forthcoming.

Challenges to further development of this approach for use in individuals are at least twofold:

(1) We have used the traditional midline set of scalp sites; Fz, Cz, Pz, but there may be a better (larger?) set to use for individual diagnosis. Research will be required to delineate the electrode set.

(2) Looking for condition by site interactions within an individual requires a known truthful (control) condition yielding the assuredly truthful-specific profile even in malingering suspects. This optimal procedure needs development.

Finally, it is noted that using the previously described, simple bootstrapping methods on unscaled P300 amplitude at one (Pz) site, we were able to identify 87% of our malingering simulators with this new, 9-probe P300-HFCP. The addition of other sites in the profile method, as well as integration of information from other ERP components and behavioral measures, should push detection sensitivity and specificity to very high levels.

Summary

The series of studies from our laboratory described in this chapter provide evidence that cognitive psychophysiological assessment can be used to detect malingered cognitive deficit. In particular, relevant to the need to evaluate potentially head injured individuals who are litigating, the use of P300 ERPs has been demonstrated to work well in discriminating this population from simulators feigning memory deficit. Behavioral measures of malingering, such as the HFCP, can be modified and used effectively in conjunction with P300 methodology. In so doing, sensitivity in detecting malingerers may be increased. Finally, administration of the BCT, a commonly used neuropsychological measure, and use of P300 techniques in the same individuals suggests that the most effective detection of malingering may result from employing neuropsychological and cognitive psychophysiological measures in conjuction with one another. Available evidence suggests that this area of inquiry can be fruitful and warrants further attention.

References

Allen, J.J., Iacono, W.G., & Danielson, K.D. (1992). The development and validation of an event-related-potential memory assessment procedure: A methodology for prediction in the face of individual differences. *Psychophysiology, 29*, 504–522.

Andreassi, J.L. (1995). *Psychophysiology: Human behavior and physiological response* (3rd ed.). Hillside, NJ: Lawrence Erlbaum.

Barcello, F., Sanz, M., Molina, V., & Rubia, F.J. (1997). The Wisconsin Card Sorting Test and the assessment of frontal function: A validation study with event-related potentials. *Neuropsychologia, 35*, 399–408.

Campbell, K., Houle, S., Lorrain, D., Deacon-Elliot, D., & Proulx, G. (1986). Event-related potentials as an index of cognitive functioning in head-injured outpatients. In W.C. McCallum, R. Zapolli, & F. Denoth (Eds.), *Cerebral psychophysiology: Studies in Event-related potentials* (pp. 486–488). Amsterdam: Elsevier.

Deacon, D., & Campbell, K.B. (1991). Effects of performance feedback on P300 and reaction time in closed-head-injured patients. *Electroencephalography and Clinical Neurophysiology, 78*, 133–141.

DeFillippis, N.A., & McCampbell, E. (1979) *The Booklet Category Test: Research and clinical form: Manual*. Odessa, FL: Psychological Assessment Resources.

Donchin, E., & Coles, M.G.H. (1988). Is the P300 component a manifestation of context updating? *Behavioral and Brain Sciences, 121*, 357–374.

Donchin, E., Kramer, A., & Wickens, C. (1986). Applications of brain event-related potentials to problems in engineering psychology. In M. Coles, S. Porges, & E. Donchin (Eds.), *Psychophysiology: Systems, processes and applications* (pp. 702–718). New York: Guilford.

Ellwanger, J.W., Rosenfeld, J.P., & Sweet, J.J. (1997). The P300 event-related brain potential as an index of recognition response to autobiographical and recently learned information in closed head injury patients. *The Clinical Neuropsychologist, 11*, 428–432.

Ellwanger, J.W. (1996). *The validity and utility of a p300 event-related potential based method for indentifying malingered memory deficit*. Doctoral Dissertation, Northwestern University, Evanston, IL, U.S.A.

Ellwanger, J., Rosenfeld, J.P., Sweet, J.J., & Bhatt, M. (1996). Detecting simulated amnesia for autobiographical and recently learned information using the P300 event related potential. *International Journal of Psychophysiology, 23*, 9–23.

Farwell, L.A., & Donchin, E. (1991). The truth will out: Interrogative polygraphy ("lie detection") with event-related potentials. *Psychophysiology, 28*, 531–547.

Guilmette, T.J., Hart, K.J., Guiliano, A.J., & Leineger, B.E. (1994). Detecting simulated memory impairment: Comparison of the Rey fifteen-item test and the Hiscock forced-choice procedure. *The Clinical Neuropsychologist, 8*, 238–294.

Harsh, J., Voss, U., Hull, J., Schrepfer, S., & Badia, P. (1994). ERP and behavioral changes during the wake/sleep transition. *Psychophysiology, 31*, 244–252.

Hiscock, M., & Hiscock, C.K. (1989). Refining the forced-choice method for the detection of malingering. *Journal of Clinical and Experimental Neuropsychology, 11*, 967–974.

Jennett, B. (1983). Scale and scope of the problem. In R. Rosenthal, E. Griffith, M. Bond, & J.D. Miller (Eds.), *Rehabilitation of the head injured adult*. Philadelphia: F.A. Davis.

Johnson, R. (1993). On the neural generators of the P300 component of the event-related potential. *Psychophysiology, 30*, 90–97.

Johnson, R., Jr. (1986). A triarchic model of P300 amplitude. *Psychophysiology, 23*, 367–384.

Johnson, R., Jr. (1988). The amplitude of the P300 component of the event-related potential. In P.K. Ackles, J.R. Jennings, & M.G.H. Coles (Eds.), *Advances in psychophysiology* (Vol. 2; pp. 69–138). Greenwich, CT: JAI Press.

Johnson, M.M., & Rosenfeld, J.P. (1992). A new ERP-based deception detector analog II: Utilization of non-selective activation of relevant knowledge. *International Journal of Psychophysiology, 12,* 289–306.

Leiphart, J., Rosenfeld, J.P., & Gabrieli, J. (1993). Event-related potential correlates of implicit priming and explicit memory tasks. *International Journal of Psychophysiology, 15,* 197–206.

McCarthy, G., & Wood, C.C. (1985). Scalp distributions of event-related potentials: and ambiguity associated with analysis of variance models. *Electroencephalography and Clinical Neuropsychology, 62,* 203–208.

Nies, K., & Sweet, J. (1994). Neuropsychological assessment and malingering: A critical review of past and present strategies. *Archives of Clinical Neuropsychology, 9,* 501–552.

Paller, K., & Kutas, M. (1992). Brain potentials during memory retrieval provide neurophysiological support for the distinction between conscious recollection and priming. *Journal of Cognitive Neuroscience, 4,* 375–390.

Pankratz, L. (1983). A new technique for the assessment and modification of feigned memory deficit. *Perceptual and Motor Skills, 57,* 367–372.

Renault, B., Signoret, J.L., De Bruille, B., Breton, F., & Polgert, F. (1989). Brain potentials reveal covert facial recognition in prosopagnosia. *Neuropsychologia, 27,* 905–912.

Rosenfeld, J.P. (1997). Method and system for detection of memory deficiency utilizing brain waves. U.S. Patent. No. 5,622,181.

Rosenfeld, J.P., Reinhart, A.M., Bhatt, M., Ellwanger, J.W., Sekera, M., & Sweet, J.J. (1997). P300 correlates of simulated malingered amnesia on a simple matching to sample task: Topographic analysis of deceptive vs. truthful responses. *International Journal of Psychophysiology, 28,* 233–247.

Rosenfeld, J.P., Ellwanger, J.W., Nolan, K., Bermann, R.G., & Sweet, J.J. (in press). P300 scalp amplitude distribution as an index of deception in a simulated cognitive deficit model. *International Journal of Psychophysiology.*

Rosenfeld, J.P., Sweet, J.J., Chuang, J., Ellwanger, J., & Song, L. (1996). Detection of simulated malingering using forced choice recognition enhanced with event-related potential recording. *The Clinical Neuropsychologist, 10,* 163–179.

Rosenfeld, J.P. (1995). Alternative views of Bashore and Rapp's (1993) alternatives to traditional polygraphy: A critique. *Psychological Bulletin, 117,* 159–166.

Rosenfeld, J.P., Ellwanger, J., & Sweet, J. (1995a). Detecting simulated amnesia with event-related brain potentials. *International Journal of Psychophysiology, 19,* 1–11.

Rosenfeld, J.P., Ellwanger, J., Bhatt, M., Hankin, B., Berman, R., & Sweet, J.J. (1995b). Simulated malingering on an easy matching-to-sample test increases P3 amplitude: Genuine difficulty reduces the amplitude. *Psychophysiology [Abstract], 32* (Suppl. 1), s65.

Rosenfeld, J.P., Bhat, K., Miltenberger, A., & Johnson (1992). Event-related potentials in the dual task paradigm. *International Journal of Psychophysiology, 12,* 221–232.

Rosenfeld, J.P. (1992). Method and apparatus for the analysis, display, and classification of event-related potentials by interpretation of P3 responses. U.S. Patent. No. 5, 113, 870.

Rosenfeld, J.P., Angell, A., Johnson, M., & Qian, J. (1991). An ERP-based control-question lie detector analog: Algorithms for discriminating effects within individual waveforms. *Psychophysiology, 28,* 320–336.

Rosenfeld, J.P., Cantwell, B., Nasman, V.T., Wojdac, V., Ivanov, S., & Mazzeri, L. (1988). A modified, event-related potential-based guilty knowledge test. *International Journal of Neuroscience, 42,* 157–161.

Rosenfeld, J.P., Nasman, V.T., Whalen, R., Cantwell, B., & Mazzeri, L. (1987). Late vertex positivity as a guilty knowledge indicator: A new method of lie detection. *International Journal of Neuroscience, 34,* 125–129.

Rugg, M.D. (1995). ERP studies of memory. In M.D. Rugg & M.G.H. Coles (Eds.), *Electrophysiology of mind* (pp. 132–170). New York: Oxford.

Rugg, M.D., Cowan, C.P., Nagy, M.E., Milner, A.D., Jacobson, I., & Brooks, D.N. (1988). Event related potentials from closed-head injury patients in an auditory "oddball" task: Evidence of dysfunction in stimulus categorization. *Journal of Neurology, Neurosurgery, and Psychiatry, 51,* 691–698.

Tenhula, W.N., & Sweet, J.J. (1996). Double cross validation of the booklet category test in detecting malingered traumatic brain injury. *The Clinical Neuropsychologist, 10,* 104–116.

Warrington, E. (1984). *Recognition Memory Test manual.* Los Angeles, CA: Western Psychological Services.

Wiggins, E.C., and Brandt, J. (1988). The detection of simulated amnesia. *Law Human Behavior, 12,* 57–78.

PART III

RELEVANT POPULATIONS

At present, diagnostic issues pertaining to several clinical populations are over-represented among cases likely to be sent to a neuropsychologist as forensic referrals. Of these, the vast majority will pertain to traumatic brain injury. Coverage of all relevant populations is not feasible. Therefore, the intent within the next section is to impart salient information pertinent to select populations that can be expected to be represented recurrently. Although traumatic brain injury is most often at issue in forensic neuropsychology cases at present, of these, a disproportionately large number will concern possible mild traumatic brain injury. Ronald Ruff discusses some of the important aspects of mild traumatic brain injury cases in Chapter 11. Another substantial population likely to be involved in litigation and frequently producing referrals to neuropsychologists consists of toxin-exposed individuals. In Chapter 12, David Hartman presents valuable information relevant to neuropsychological toxicology. To conclude that forensic neuropsychology is an exclusively adult-oriented undertaking would be mistaken. To the contrary, this section concludes with consideration of pediatric issues that may result in adversarial activity likely to involve neuropsychologists. Rudy Lorber and Helene Yurk, in Chapter 13, identify and elaborate on the unique knowledge base and characteristics of pediatric cases involved in litigation, especially pertaining to due process hearings within school systems.

Chapter 11

MILD TRAUMATIC BRAIN INJURY

Ronald M. Ruff
St. Mary's Medical Center, San Francisco, California

Ann M. Richardson
Private Practice, San Francisco, California

The majority of cases that neuropsychologists encounter in the forensic arena are individuals who may have sustained traumatic brain injuries. Within the subclassifications of traumatic brain injury, mild traumatic brain injuries (MTBIs) comprise well over half of all individuals who sustain TBI. Indeed, although estimates vary, evidence suggests that each year in the United States alone, over 1,300,000 individuals sustain an MTBI (Kay, Newman, Cavallo, Ezrachi, & Resnick, 1992). Many of these injuries occur as a result of motor vehicle accidents, and a majority of the MTBI patients are young men between 16 and 35 years of age. Although a majority will make a full recovery, a minority suffer from more chronic difficulties (McLean, Temkin, Dikmen, & Wyler, 1983; Rutherford, Merrett, & McDonald, 1978). However, the size of the minority who become chronically symptomatic varies from one study to another, from 7–8% (Binder, 1997), to 10–20% (Alexander, 1995), to an even higher estimate of approximately one third (Rimel et al. 1981). These numbers are estimates, and hopefully better technical research in the future will result in a more definitive range. Nonetheless, agreement exists in the literature that a minority of MTBI patients present with persistent symptoms, and over the past

15 years there has been a drastic increase in neuropsychologists being consulted by attorneys, to evaluate whether indeed these problems are related to brain damage or not.

The intent of this chapter is not to review the rapidly growing literature on MTBI. Comprehensive summaries are available on this subject, and examples include: Levin, Eisenberg, and Benton (1989); Gualtieri (1995); Rizzo and Tranel (1996); Kibbey and Long (1997), and for children and adolescents, Satz, Zauch, McClery, Light, Asarnow, and Becker (1997). Instead, this chapter raises the following four key questions that neuropsychologists face: (1) What are the challenges for diagnosing MTBI? (2) Can we rely on objective evidence? (3) What logic is used to sort through the symptoms? and (4) What role does the examiner's viewpoint play? In the subsequent four parts, each of these questions is dealt with first by addressing the extent of the challenges involved, then followed by possible resolutions.

Part I: What are the challenges for diagnosing MTBI?

Extent of problem
Any diagnosis is dependent on the criteria set forth in its definition. Does a uniform definition exist for MTBI?

Neurosurgeons have introduced the schema of the Glasgow Coma Scale (GCS; Teasdale & Jennett, 1974; Jennett & Teasdale, 1981), which allows for a classification between severe, moderate and mild brain injuries along a scale ranging from 3 to 15. The three target behaviors that are rated are eye-opening, motor response, and verbal response. The GCS is a useful scale to separate severe (scores 3–8), moderate (9–12) and mild TBI (13–15). Even within the range of mild TBI (GCS 13–15), there can be differences between patients. Patients are at a greater risk of having acute radiographic abnormalities with a score of 13 than 15 (Kraus & Nourjah, 1989). Also a greater incidence of poor outcome has been found with a GCS 13 versus 15, as well as between patients who have a GCS of 15 with versus without acute radiographic abnormalities (Hsiang, Yeung, Yu, & Poon, 1997). Moreover, many MTBI patients are never seen in an emergency trauma setting, and no GCS is administered. Indeed, for clinicians who evaluate patients weeks or months post-accident, the GCS score will always be 15. It can be difficult to establish a reliable GCS score retroactively, and thus GCS scores are not sufficiently sensitive to establish a classification system of MTBI for all disciplines.

Rehabilitation specialists have offered a definition of MTBI which was set forth by the American Congress of Rehabilitation Medicine (ACRM; Mild Traumatic Brain Injury Committee, 1993) to include *at least one* of the following criteria: (1) any period of loss of consciousness; (2) any loss of memory for events immediately before or after the accident; (3) any alteration in mental state at the time of accident (e.g., feeling dazed, disoriented, or confused); and/or (4) focal neurological deficit(s) that may or may not be transient. The following three exclusion criteria for MTBI are also delineated and include: (1) loss of consciousness exceeding approximately 30 minutes; (2) after 30 minutes, a GCS falling below 13; (3) post-traumatic amnesia persisting longer than 24 hours.

Psychiatrists have also for the first time included a definition for the postconcussional disorder when updating DSM-III-R to DSM-IV (American Psychiatric Association, 1994). This definition implies that a cerebral concussion is manifested through a loss of consciousness, post-traumatic amnesia, and less commonly, post-traumatic onset of seizure. DSM-IV further suggests the loss of consciousness should last for five or more minutes.

Having raised the issue that these three definitions are not uniform, let us also consider the following question: Who is the crucial source for diagnosing a cerebral concussion? Some clinicians insist that a police report or the medical records containing direct observations by paramedics and physicians should be the principal sources for diagnosing MTBI. However, in most cases the paramedics and ER physicians also have to rely on information obtained from the patient. In fact, a large portion of MTBI patients sustain a brief loss of consciousness that spans just seconds to minutes, and most often no observers, particularly emergency medical technicians, are present during that time. Moreover, some patients have only an altered sense of consciousness, and even if an observer is present, it is difficult to observe whether a patient is dazed, disoriented, or confused without asking the patient specific questions. Thus, the patient as a rule is the principal source, except for those very few cases in which a third party is present and is capable of determining a loss of consciousness or change of mental state. However, it seems illogical and inappropriate to diagnose MTBI *only* in those patients who are directly observed, since the initial availability of a competent observer at the scene is a random occurrence.

What time parameters are crucial for diagnosing an MTBI? If the ER records note "no loss of consciousness", does this necessarily rule out that the patient was unconscious at the scene of the accident? Does a GCS score of 15 obtained in the ER one hour after an accident indicate the absence of an MTBI?

One cannot really grasp the complexity of MTBI if one believes that most brain injured patients can give a definite response with respect to their own loss of consciousness or mental changes. Indeed, as a rule, individuals with brain injuries suffer not only from a loss of consciousness, but also concurrently from a loss of memory (post-traumatic amnesia; Levin, Benton, & Grossman, 1982). This leads to the following question: How is a patient who has no memory for a period of time after an impact able to remember whether *initially*, after the impact, there was a loss of consciousness? By attempting to answer this question, we enter murky waters, and it should come as no surprise that different clinicians and attorneys, who rely on medical information obtained from the patient at different time intervals, will *honestly* come up with different conclusions with respect to loss of consciousness.

It is too simplistic to conclude that if the patient's response was vague, that the paramedics' and emergency room documentations are more reliable. Indeed, these records are, as a rule, also based on questions posed to the patient who may have just had a traumatic insult to the brain. This creates two different dilemmas; the plaintiff is faced with the challenge of confirmation, whereas the defendant needs to ascertain whether the patient report can be fully accepted. These dilemmas have no perfect solutions for either side.

In summary, the diagnosis of MTBI is a challenge, because (1) no uniform diagnosis has been agreed upon, and (2) the patient is the typical source for a description of an altered level of consciousness, while at the same time the patient may be suffering from memory loss for the events surrounding the accident.

Resolution: Establish a definition

Given that multiple definitions exist in the literature, it is the clinician's responsibility to select and establish his or her own definition. In litigation, multiple clinicians will evaluate the same patient, and if different definitions for MTBI are used, this should be clearly delineated so that misunderstandings can be avoided.

The definitions developed by the ACRM and DSM-IV are applicable retroactively, but their criteria are also discrepant. Thus, we propose a subdivision of MTBI that incorporates the definition by the ACRM and the implicit definition of DSM-IV. In doing so, we propose a continuum, rather than multiple parallel definitions of MTBI (see Table 1).

Type I in Table 1 is consistent with the lowest entry criteria established by the ACRM, and Type III is basically aligned with the DSM-IV classification. Type II bridges the gap between the two discrepant definitions. These three proposed sub-types are presently being evaluated in ongoing research protocols to establish whether the symptom presentation varies between the three types of severity within TBI.

An important point is that the severity of MTBI must be defined by the acute injury characteristics, and *not* the severity of symptoms at some random point subsequent to the accident. Moreover, it must be emphasized that brain traumas

Table 1.   Classifications for MTBI.

	Type I	Type II	Type III
LOC	altered state or transient loss	definite loss with time unknown or < 5 minutes	loss 5–30 minutes
PTA	1–60 seconds	60 seconds–12 hours	> 12 hours
Neurological symptoms	one or more	one or more	one or more

are due to contact forces, or due to acceleration/deceleration trauma, and in essence the duration of unconsciousness is brief, usually seconds to minutes, and in some cases there is no loss of consciousness per se, but rather a period of altered or changed consciousness. If a GCS has been administered, it must be between 13 and 15, if indeed it is administered in the early phases. However, a GCS of 15 obtained at some random point after the accident does not exclude a diagnosis of MTBI.

Determining whether the patient is briefly unconscious or confused, or in a state of post-traumatic amnesia immediately following the accident, is still the key issue that needs to be addressed when evaluating the likelihood of a concussion (Alexander, 1995). To solicit this information, we primarily rely on questions posed to the patient and have found that the administration of the Galveston Orientation and Amnesia Test (GOAT), which retrospectively estimates both loss of consciousness and post-traumatic amnesia, can be most helpful (Levin, Benton, & Grossman, 1982). However, as noted earlier, such retrospective evaluations are challenging and their reliability can come into question (Gronwall & Wrighton, 1981). Particularly once the case enters litigation, we have found that some additional information needs to be provided for patients. When asking the patient to recall the first events before and after the accident, we typically preface the question by stating the following:

> Since the accident, you have discussed various aspects with your lawyers, friends, families, and various examiners, and you may also have had the opportunity to review various records, including police reports and other medical records. For the questions we are about to ask you, we would like you to rely on your *independent* recall without any reference to information you have learned after the accident.

Then, we typically proceed with the specific questions on the GOAT. Thereafter, we have the patient go over the accident in his or her own words, relaying all the information that they know about the accident, and at times we also have them identify the various sources upon which they have relied (e.g., talking with attorneys, medical records, police reports, etc.). Nonetheless, it is essential to review the medical records as well, and to conduct a thorough clinical interview to reach reasonable conclusions.

Even if a uniform subclassification for MTBI were established, further confusion around nomenclature exists that has not yet been addressed. As Kay et al. (1992) have pointed out, there needs to be a distinction made between "head injury" (which refers to an injury to the head, and may or may not include a brain trauma), "mild traumatic brain injury" (which infers damage to the brain), and "postconcussive disorder" (which refers to interactive sequelae subsequent to a concussion). These terms should *not* be used interchangeably; rather, MTBI should be used to describe the initial brain trauma, and a "postconcussive disorder" should refer to the residua that follow. As with MTBI, the term postconcussive disorder is also applied differently by clinicians. It is beyond the scope of this chapter to refine the criteria for a postconcussive disorder, however, DSM-IV has proposed research criteria that we hope will be refined in future diagnostic and statistical manuals (i.e., DSM-V).

In summary, the presently used classifications of MTBI are broad and can include, on the one hand, a patient who has a 30-minute loss of consciousness and 12 hours of post-traumatic amnesia and, on the other hand, a patient who has an altered sense of consciousness for seconds with a brief loss of memory. It would be beneficial if uniform definitions could be developed with sufficient differentiation to allow for clinically relevant subclassifications for MTBI and postconcussive disorder.

Part II: Can we rely on objective evidence?

Extent of problem
Given the subjective nature of the patient's presenting complaints, it is understandable that certain clinicians prefer to rely on objective data for their diagnosis. Thus, neuroimaging, such as CT or MRI, may be ordered to determine whether brain damage can be objectified. However, it has become recognized that the diffuse axonal injuries possibly associated with MTBI are typically not visible on static neuroimaging, such as computerized tomography (CT) and magnetic resonance imaging (MRI) scans (Crooks, 1991; Gennarelli et al., 1982; Povlishock, 1993). Preliminary evidence suggests greater sensitivity with the dynamic neuroimaging techniques of positron emission tomography (PET; Ruff et al., 1994) and single photon emission computerized tomography (SPECT; Nedd et al. 1993; Varney et al. 1995). However, when PET or SPECT scans are available, one must be cautious, because in litigation the point can be argued that the American Academy of Neurology (1996) has determined that due to insufficient research, we are not yet at a stage at which an MTBI can be conclusively diagnosed with dynamic neuroimaging. If the plaintiff's attorney goes to the trouble and expense of using a PET or SPECT scan as one of the diagnostic tools that might shed some light on this challenging issue, it is not uncommon that the defense attorney will hire an expert to testify as to the lack of experimental data to provide the certainty upon which to rely in formulating an opinion. EEGs have also been used to assist in the quantification of MTBI (Thatcher, Walker, Gerson, & Geisler, 1989); however, EEGs, including brain mapping and computerized EEGs, are also not recognized as definitive (American Congress of Neurology, 1989).

The answer to our question can therefore be clearly stated: At the present time, objective evidence in the form of neuroimaging or brain mapping has not been proven to be conclusive for diagnosing MTBI. Nonetheless, objective evidence should be gathered and can be of assistance; however, the absence of objective data does not allow one at this time to rule out that the patient did not sustain an MTBI (i.e., the absence of proof is not equal to the proof of absence – especially if the methods for proof are themselves flawed). On the other hand, this argument must be balanced, since *not every concussion causes permanent neurobehavioral damage*.

Resolution: Subjective vs. objective evidence

The principal neuropathology of MTBI is thought to be diffuse axonal injury (DAI) caused by the forces to the head which can result in shearing of the axons (Gennarelli et al. 1982). In addition, vascular injury can disrupt small veins, producing small hemorrhages or edema. Most of this research is based on animal literature, and the primary distribution of the injury seems to be in the parasagittal white matter spreading from the cortex to the brain stem. However, a few humans who, after accidents, died from systemic injuries sustained along with MTBI, were studied by Oppenheimer (1968) for the magnitude of their DAI changes. The neuropsychopathology noted in these humans correlated substantially with the animal data. In a similar study, Blumbergs et al. (1994) found multifocal DAI in five humans with MTBI resulting in particular damage to the fornices.

In a recent study, the CT scans of a relatively large sample of 1,405 consecutively imaged MTBI patients were evaluated, and abnormalities were detected in 8.2% of the sample (Borczuk, 1995). Thus, over 90% of the patients diagnosed by ER physicians with an MTBI demonstrated no brain damage on CT scans. Given that even CT and MRI scans typically have an insufficient sensitivity to demonstrate DAI in patients with MTBI, we cannot rely on these objective measures to quantify MTBI. Instead, the diagnosis remains dependent on clinical examination, history and measures of loss of consciousness and post-traumatic amnesia (Alexander, 1995). The lack of conclusive objective data can result in different interpretations, however, extremes must be avoided. On the one hand, defense experts should not imply the absence of MTBI on the basis of negative CT and MRI scans, when there is clinical agreement that static neuroimaging studies are, as a rule, negative in most MTBI patients. On the other hand, this does not give plaintiff experts the liberty to interpret all nuances in the neuropsychological profile as reflecting brain dysfunction. Even if there are PET or SPECT deficits on scans in MTBI cases, it is important to evaluate the clinical significance according to neuropsychological tests (Levin et al., 1992). As noted by Williams, Levin, and Eisenberg (1990), MTBI cases complicated by intracranial lesions or depressed skull fracture are at a greater risk for a more negative outcome from a neurobehavioral persective.

In summary, objective abnormalities do occur in a minority of MTBI patients. The absence of positive neuroimaging is not proof that the patient did not sustain a concussion. Given that objective medical diagnostic techniques are not yet proven as conclusive, a dilemma emerges in litigation. The defense attorney must challenge an "expert" who believes that every bump to the head causes DAI which results in permanent brain damage in every case. However, a plaintiff attorney also has to challenge an "expert" who only diagnoses brain damage in MTBI patients when CT or MRI images are positive. Since we cannot, at the present time, rely on objective evidence, we must depend on clinical interviews and quantifications of behavior on neuropsychological testing.

Part III: What logic is used to sort through symptoms?

Extent of problem

Given that patients are the critical source for diagnosing MTBI, and also given that objective neuroimaging techniques are not conclusive, lawyers and experts are left with both their experience and a system of logic to sort through symptoms that MTBI patients present. Let us first explore two systems of logic that are prevalent, particularly in the context of litigation.

Lawyers, without claiming to be unbiased, are typically working to present their case in a logically *consistent* fashion. Thus, both sides search for those aspects that are either consistent or inconsistent. In doing so, an implicit assumption is made that if the patient's statements within a deposition and across time to different examiners are consistent, then the information can be trusted. However, if lawyers and experts can uncover inconsistencies, then the patient's credibility can be challenged in proportion to the inconsistencies that have been uncovered. This approach is ultimately founded on the assumption that only patients with consistent symptom presentation have sustained MTBI.

Let us examine this logic. We all agree that the color of the sky is blue. Such a statement is typically accepted as "true". However, it is also correct that at night or given certain circumstances (moderator variable – time of day), the sky is dark gray. Now let us return to MTBI. If, for example, a patient states, "My attention is terrible", and we assess with a test in the morning that the patient's attention is indeed average, does this mean that the patient is a liar? If the patient's attention is "terrible" in the afternoon (moderator variable – excessive fatigue), then his statement is true. To the patient, his attention is not the same as it was before the accident. Wittgenstein (1958) has elegantly exemplified that language describes uses in its own context. Thus, in different contexts, the same statement has different meanings. Thus, symptoms in different contexts (e.g., testing situation vs. occupational setting) may not necessarily be proof of inconsistency.

Given that we have cautioned against an over-application of consistencies within symptom presentation, an expert must flag those inconsistencies that contraindicate the presence of MTBI. Let us give a few examples. Without a localized lesion or unusual complication, there is no relevant literature suggesting that a lateralized sensory or motor complaint should be associated with MTBI. Moreover, despite the fact that attentional dysfunctions are the most common cognitive impairments following an MTBI (Binder, Rolling, & Larrabie 1997; Levin, Eisenberg & Benton, 1989), gross discrepancies between the expected degree of attentional deficit (based on the literature) and a far greater impairment on the individual's test results cannot be fully explained by MTBI (Mittenberg et al. 1993, 1995). Notable discrepancies may also occur even across domains: for example, a patient may be unable to repeat more than 4 digits forward on an attention test – but 20 minutes later on an oral arithmetic test, is able to correctly subtract 20 from 115 (i.e., process 5 digits). These types of discrepancies must be flagged.

In the context of litigation, one needs to question the degree of the reported symptoms. For example, if a patient claims that there were absolutely no problems in his life prior to the MTBI, and thereafter, he reports a plethora of problems in all dimensions of life, then the validity of complaints can be called into question. For example, it is unlikely that an individual prior to an accident never had *any* headaches or problems with sleep, and now after the MTBI is suffering from chronic, severe migraines and never sleeps. Moreover, one must be cautious when, on a personality questionnaire, all scales are elevated. The same applies for the neuropsychological testing, since not all test scores across domains should be impaired. In summary, there are different degrees of inconsistency, and it is important to evaluate the context for these consistencies to determine the potential explanations and probability levels for the inconsistencies.

Exclusionary logic is a second system of logic frequently used by attorneys. Recently, while testifying in a court case, the first author was asked the following question: "Do you believe the patient ever lost consciousness?" At that point, the testimony was that the patient had a probable loss of consciousness at the scene of the accident. The defense attorney then had read from the emergency room record that the patient was not unconscious, and wanted the witness to agree or disagree with these medical records. An apparent conflict was created given the two different contexts that were presented. When attempting to explain, the witness was asked to answer with "yes" or "no". Eventually, of course, the witness was given a chance (by the plaintiff's attorney on re-direct) to explain that indeed, 30–40 minutes had passed between the time of the accident and the time when the patient was admitted to the emergency room. In these medical records it also appeared that the emergency room physicians were reporting the patient's status at the time of admission. Thus, the apparent inconsistency of the central question regarding loss of consciousness was subjected to a set of questions that did not explain the sequence of events that had transpired over time. In this case it was reasonable to conclude that the patient had a loss of consciousness at the scene of the accident for a few seconds, because a passenger in the same car who was not brain-injured was unable to awaken the patient, then observed during the ambulance ride that he became more alert, and then in the emergency room observed the patient to be fully conscious.

Though this example refers to exclusionary logic used by a defense attorney, plaintiff attorneys are equally likely to use such logic. For example, the first author has been asked by a plaintiff's attorney whether "objective evidence" existed that a patient's forehead was lacerated by 3 inches, since it required sutures in the emergency room. The response was "yes". This was followed by the question, "Then it is true that the patient sustained a head injury?" Again, my response was, "Yes, the patient did sustain an injury to his head". Questioning was stopped, and later on, an inference was drawn between an objective injury to the head being equivalent to a brain trauma. The two can go together, but this is *not* always the case, since one can have an injury to the head without a brain trauma.

In summary, it is important to avoid black-white logic, especially if conflicting situations are created. Moreover, inconsistencies should be examined in their

Table 2. Potential Indicators of Insufficient Effort.

Premorbid indicators
- Antisocial personality traits
- Borderline personality disorder
- Prior incapacitating injuries
- Prior claims for injury
- Poor work record or job satisfaction
- Intolerable life conflicts

Behavioral indicators during examination
- Uncooperative or inconsistent cooperation
- Suspicious, aloof
- Ill at ease, unfriendly
- Evasive

Test performance
- Frequently responds with "I don't know"
- Missed random items in a test comprising items with increasing difficulty
- Easily gives up on more difficult items
- Spurious results (e.g., unexplained outlier)
- Inconsistent profile within test battery or between tests across battery
- Malingering-like performance on direct measures of malingering

Patient's presenting postmorbid complaints
- Description of events surrounding accident in great detail
- Endorsement of an unusually large number of symptoms
- Absurd or preposterous reporting of symptoms
- Nonselectivity of physical, emotional, or cognitive symptoms
- Overidealized functioning or lack of reasonable difficulties before accident
- Inconsistencies in symptom reporting itself

Reported or observed activities of daily living
- Engaged in activities not consistent with reported deficits
- Discrepant capacity between work and recreation
- Refuses employment with partial disability

For personal injury litigants
- Significant financial stressors
- Resistance or lack of seeking reasonable remedies
- Lack of reasonable follow-through on available treatments
- Attributing all life's problems to accident
- Blaming others for all life's problems
- Seeking unnecessary examinations and treatments without consulting experts

---

Note. See Chapter 9 for differential diagnosis pertaining to these behaviors.

context, and *only* exposed as exaggerations or feigned symptoms *if* no reasonable moderator variables can provide a more parsimonious explanation. Partial questioning framed in exclusionary logic and potentially incorrect inferences also needs to be rebutted by the opposing attorney in questioning, whenever possible. For a list of indicators including inconsistencies that could raise the possibility of malingering, see Table 2 (Ruff, Wylie, & Tennant, 1996), and for a more detailed discussion, see Chapters 9 and 10.

Resolution: A reasonable framework

How can we adopt a reasonable framework for *consistency?* Let us start by admitting that it is a challenge to separate inconsistencies that are part of life from those that point to non-injury factors or even malingering. As mentioned, the hallmark cognitive sequelae subsequent to MTBI are attentional deficits. These attentional deficits actually result in behavioral fluctuations. Thus, fluctuations of performances and behavior can account for varying neuropsychological test performances. It would be incorrect to conclude that all deficits subsequent to MTBI must present themselves in a static fashion. However, extensive fluctuation, unusual deviations or generalized impairments must be flagged as clinically significant. In such cases it becomes imperative that the evaluation include a careful examination of: (1) pre-injury functioning levels, and (2) comorbid symptom interactions.

Pre-injury estimate

Estimating the patient's premorbid risk factors is essential in order to relate the present functioning levels to his or her estimated pre-existing functioning levels. Only this relative comparison allows one to answer the key question: is there a decline in functioning between the estimated premorbid functioning and the present functioning levels attributable to the incident in question? Among the *physical* risk factors, one must evaluate the medical history of the patient's family, rule out genetically transmitted weaknesses or deficits, and evaluate the developmental milestones. One also needs to rule out pre-existing neurological histories, including prior MTBIs, major alcohol or substance abuse, or other major medical illnesses that could have an impact on cerebral integrity. From an *emotional* perspective, it is important to determine longstanding psychiatric illnesses, and it has also been suggested that particularly vulnerable personality styles may make the symptom presentation more pronounced subsequent to an MTBI (Kay et al. 1992; Ruff, Camenzuli, & Mueller, 1996). For examples of vulnerable personality styles, see Table 3. From a *cognitive* perspective, one should evaluate pre-existing strengths and weaknesses of intellectual functioning. The degree of educational influence and opportunity should also be established. From a *psychosocial* perspective, it is essential to appraise the family structure, their ability to cope with a crisis, and to evaluate how they resolve psychosocial conflicts. *Vocational* and financial issues also need to be explored to evaluate the importance that the patient assigns to his or her work or vocational status. Given that in litigation financial gain may play a role, the patient's premorbid financial status may be considered. For a list of questions that can be posed when ascertaining the premorbid history, see Table 4. For a more comprehensive review of estimating premorbid functioning levels, see Ruff, Mueller, and Jurica, 1996, and Chapter 3 within this text.

Table 3. Vulnerable Personality Styles.

Style	Premorbid psychological traits	Postmorbid reactions
Overachiever	Sense of self derived from driven accomplishments, which is frequently accompanied by obsessive-compulsive traits	Catastrophic reaction if drop in performance is perceived
Dependency	Excessive need to be taken care of, frequently leading to submissive behaviors and a fear of separation	Paralyzed by symptoms, if a critical erosion of independence occurs
Borderline personality characteristics (not disorder)	Pattern of instability in interpersonal relationships and self-image with fear of rejection or abandonment	Exacerbation of personality disorganization, including despair, panic, impulsivity, instability, and self-destructive acts
Insecurity	Weak sense of self, which can include shame, guilt, and dependency needs	Magnification of symptoms
Grandiosity	Overestimation of abilities and inflating accomplishments, can include need for admiration and lack of empathy	Minimization or denial of symptoms. If failure results, crash of self-esteem can result in catastrophic reaction
Antisocial behavior	Tendency to be manipulative or deceitful, temperamental, impulsive and irresponsible, lacks sensitivity to others	Possible exaggeration or malingering, increased risk taking, irritability, takes little responsibility for recovery
Hyperactivity	Restless, unfocused and at times disorganized	Attentional difficulties and impulsivity may be compounded, possible oppositional behavior
Depression	Mood fluctuations dominated by negative affect	Increase of depressive symptoms, despondency
Histrionic style	Emotionality and attention-seeking behavior	Dramatic flavor to symptom presentation, blaming behavior
Somatic orientation	Preoccupation with physical well-being, reluctance to accept psychological conflicts	Endorsement of multiple premorbid physical symptoms intermixed with new or changing TBI residua
Post-Traumatic Stress Disorder	Prior stressors produced an emotional reaction of fear and helplessness	Decreased coping ability, cumulative effect of traumas with exaggerated reaction to current crisis

Table 4.   Clinical Questions for Estimating Premorbid Functioning after Traumatic
Brain Injury.

Physical-medical status

- What is the family medical history (e.g., cause of death of parents, grandparents, etc.)?
- Are there genetically-transmitted deficits or conditions?
- Were there complications in utero or in the birth process?
- Are there pre-existing neurological illnesses (e.g., seizure disorders, meningitis, exposure to toxins, drug abuse, or malnutrition)?
- What are the major illnesses or injuries that had been sustained (e.g., polio, diabetes, migraines, back pain)?
- What treatments have been provided (e.g., medication, surgery, ECT, etc.)?

Emotional-psychosocial status

- Did the patient suffer from psychiatric illnesses prior to the accident (e.g., affective disorders including depression, schizophrenia, personality disorders, post-traumatic stress disorder)?
- In the absence of psychiatric illnesses, does the individual have a particularly vulnerable personality style which existed prior to the accident (e.g., over-achievement, insecurity, grandiosity, quick to anger)?
- How would the clinician have expected this individual to react emotionally to a TBI (i.e., what were the key defense mechanisms by which such a crisis would have been handled)?

Cognitive-neuropsychological status

- What were the strengths and weaknesses within the individual's pre-existing cognitive functioning? Among the weaknesses, were there particular cognitive disabilities (e.g., learning disabilities, dyslexia, dyscalculia)?
- Was there a premorbid history of serious alcohol or substance abuse which may have impaired cognition?
- Were there any medical illnesses which may have compromised cognition prior to the TBI?
- Were there significant emotional or psychosocial problems which compromised premorbid cognitive functioning levels (e.g., prolonged periods of depression, social withdrawal, antisocial traits)?
- To what degree has education influenced the level of cognitive functioning (i.e., "over-achiever," "underachiever," lack of education or opportunity, availability of supportive education)?
- Did cognitive factors, such as finances or illness in the family, influence the educational levels obtained?

Psychosocial status

- In what sort of family structure was the TBI patient raised? Was the family dysfunctional, and were abuses tolerated?
- How did the family cope with crisis situations, both individually and as a unit, prior to the TBI?
- What roles were assigned to the various family members? How were conflicts resolved? Was the family, as a system, open to input from the outside?

Table 4 continues

Table 4 (continued).

- How did the TBI patient resolve psychosocial conflicts prior to the trauma? Did premorbid trends exist to focus on certain problems while ignoring other? Was the patient active vs. passive, responsible vs. irresponsible, etc?
- Were physical vs. emotional vs. cognitive problems dealt with differently? What labels or attitudes would be applied to TBI deficits?

Vocational and financial status

- What degree of importance did the patient assign to his/her work or vocational status? Were the work habits erratic or inconsistent, or was the patient a workaholic who lived to work?
- What was the financial situation of the patient at the time of the TBI? Was the patient living with unreasonable debt, or too close to the edge without financial buffers?

Co-morbid symptoms

In differentiating among the potential contributors that can compound the concussive symptoms, it is important to evaluate from a *physical* standpoint whether the accident has resulted in additional medical problems. That is, patients with a cerebral concussion can also sustain other head injuries such as eye damage, jaw fractures, etc. In addition, there may be orthopedic injuries to the neck and back, etc. which can compound the pain symptoms (Alexander, 1995). The *emotional* reaction to the accident must also be evaluated, since certain patients may have, for example, a sense of anger subsequent to being a victim in an accident. Having to deal with the inconvenience of seeing multiple doctors, and being unable to return to specific sports or activities that may once have provided stress-release, also need to be considered as potential contributors. The most common *cognitive* difficulties typically include attentional and short-term memory difficulties that can also affect learning and executive functioning (Dikmen, McLean, & Temkin, 1992; Levin, Benton, & Grossman, 1962). These in turn can also become emotional and vocational stressors. *Psychosocial* difficulties have also been frequently reported, since patients may become more irritable and have lower levels of energy, which can lead to conflicts and avoidance of certain social interactions. *Vocational* and *financial* problems can also be manifested subsequent to MTBI (Rimel, Giordani, Barth, & Jane, 1981).

Once the premorbid and comorbid aspects have been evaluated within the various disciplines, the analysis of interactions among the symptoms remains a key challenge. The patient does not live within a specific discipline, but rather lives within his or her dynamic life, and the interactions that interfere with specific activities need to be addressed. Kay et al. (1992) have pointed out that certain dysfunctional loops such as pain syndromes and cognitive difficulties can become self-sustaining, and can be of such a magnitude that they even mask neurological gains that take place over time. Note that even these dysfunctional loops are not to be viewed as static; rather, they fluctuate over time. Finally, evaluations within the first few weeks after the injury are frequently unable to pick up the emotional, psychosocial, vocational and financial consequences, since these typically develop over time.

In summary, a logical framework needs to be established that not only addresses discipline-specific inconsistencies, but explores the interactions of pre-injury, co-morbid and post-injury functioning in the context of the individual's daily life.

Part IV: What role does the examiner's viewpoint play?

Extent of problem

Given that each examiner is an expert within his or her discipline, the information that is elicited has to be relevant to that discipline, and information that falls outside of that discipline is typically viewed by that expert as less relevant. Thus, the focus of a discipline brings with it a specific viewpoint. For example, the orthopedic surgeon examining a knee that was bruised in the accident will not focus on the patient's personality characteristics that may emphasize the complaints of pain. A neuropsychologist who has delineated attention and memory difficulties also needs to be sensitive to the influences of certain pain medications on the patient's test performances. The point is that patients are seen in a discipline-specific context, and the patient's symptoms may be interpreted within the domain of a single specialty without relating symptom interactions between disciplines.

Postconcussive symptoms must also be analyzed in the context of everyday life. For example, neuropsychologists focus primarily on psychometrically quantified strengths and weaknesses of cognitive functions. Given that tests must be standardized, the examiner provides very clear-cut instructions, minimizing distractions by testing in a quiet room. Patients may perform fairly well on, for example, attention tests in such an artificial testing environment. However, in everyday life, the patient may be working in an office cubicle with lots of distracting noises occurring all day, and the demands on attention in such an environment may be substantially different. Thus, test scores do not always translate perfectly to the difficulties experienced in everyday life. For discussion on ecological validity, see Chapter 8.

The first two issues raised, namely discipline-specific viewpoints and the challenges of ecological validity, apply to the evaluation of all patients. Yet, as we have pointed out so far, specific challenges emerge with MTBI: (1) multiple definitions are being used, (2) no conclusive objective evidence has yet been developed for the diagnosis, and (3) the symptoms can be interactive and influenced by pre-morbid and co-morbid factors. Thus, clinicians must rely on their interpretations, and this can lead to reasonable disagreements. As Kay et al. (1992)

have pointed out, clinicians tend to become polarized between those who have a belief that MTBIs can, in fact, result in permanent residua, vs. those who are more skeptical and operate with the assumption that MTBI patients fully recover, and that if any residuals persist, these are a psychological reaction. Thus, the believers are convinced that certain individuals with MTBI can have permanent brain-based residua. These clinicians are open when examining patients to consider brain damage as a diagnostic possibility. Non-believers typically point to the absence of positive CT or MRI findings. Some non-believers point out that patients who present with lasting symptoms represent a similar portion of our population that present with a variety of complaints, even without a concussion (e.g., Youngjohn, Burrows, & Erdal, 1995). These clinicians are, as a rule, very skeptical that neurological damage can account for the patient's complaints subsequent to an MTBI. In summary, the biases of each discipline combined with the biases of the particular professionals can affect and influence how the facts are interpreted. This has, of course, not gone unnoticed by attorneys who may select experts based on their belief systems.

In addition to the above biases of the experts, it should be noted that the patients, as well, have a particular viewpoint. Finally, jurors can also have biases, which may lead them (for example) to classify women as being more "hysterical", and men as having a greater tendency to be "malingerers". In summary, the astute clinician needs to be aware of multiple layers of bias, and must also be willing to reflect on his or her own biases. These biases get wrapped around the facts, and from the outset can color the clinical interactions between the expert and the patient.

Resolution: How to deal with one's biases

Beyond a discipline-specific approach

As stated earlier, clinicians must become aware of their own biases. Clinicians should be able to explicitly state their belief systems. The present authors have the following biases: The majority of MTBI patients recover, yet a minority do not. It is a challenge to classify this "Miserable Minority." Pre-existing personality traits can color the patient's psychological reactions. Thus, not all post-concussive syndromes are driven primarily by neurological changes. However, some patients suffer from long-term neurologically-based residua secondary to MTBI.

We also believe that a more global approach is needed to understand the Miserable Minority. In addition to the psychodiagnostic and neuropsychological evaluations, the clinician should attempt to explore interactions. Interactions of problems within and between these domains can lead to crisis reactions that are greater than the sum of the deficits (Ruff, Camenzuli, & Mueller, 1996). For a model of such interactions, see Figure 11.1. In this model, the emotional, cognitive and physical domains overlap and are of equivalent importance for the

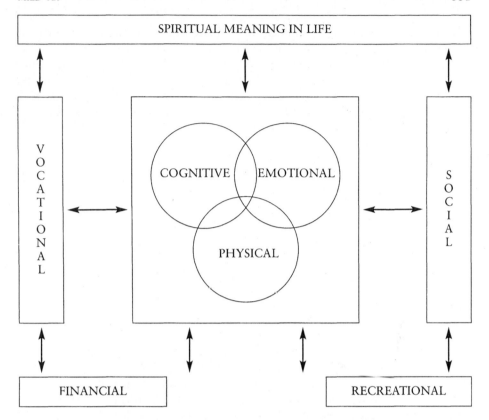

Fig. 11.1. A model depicting the interactions of domains important to the understanding of persistent problems in the "miserable minority".

interpersonal dynamic. This interpersonal dynamic, however, is interactive within the social and recreational network, which is also dependent upon vocational and financial components. One's meaning in life or spiritual direction can round out the picture. If, for example, an accident can result in loss of employment, this can in turn affect the finances, which can result in a limited recreational outlet or social life, which in turn can negatively influence the emotional well-being of the individual MTBI patient. In another example, when an individual's concussion results primarily in emotional lability and depression, this can lead to a lack of initiation for social reactions, which in turn can reduce the patient's confidence or motivation to return to work. A multitude of interactions needs to be described on a phenomenological level to understand the Miserable Minority.

The above model is not viewed as the "truth", but rather as a schema or construct that allows the clinician to explore potential interactions which may negatively influence an individual's outcome subsequent to an MTBI. Obviously, this approach does not negate or replace a discipline-specific evaluation.

However, it encourages the clinician to evaluate interaction and, conversely, to reject the notion that any information outside of one's own discipline is relegated to "noise". The discipline-based bias can lead to a myopic view, which most likely will not assist in the understanding of the Miserable Minority. We further believe that it is extremely critical to be selective regarding the type of information presented to the patient. As neuropsychologists, it is important to avoid *iatrogenic* effects in patients.

Test-battery specific bias

A further bias that we want to discuss is the selection of neuropsychological tools. Most neuropsychologists are particularly wedded to those tests upon which their clinical experience rests. Regrettably, this can become fertile ground for rigidity and biases. For example, some neuropsychologists who use the Halstead-Reitan batteries display a certain attitude of superiority due to the fact that they use the oldest and most commonly applied test battery. They can point to a 50 year history filled with countless examinations that have been based on this battery. Moreover, courts recognize traditions, and commonly-used techniques tend to be looked upon more favorably than newer techniques. However, being the "oldest" and most commonly used battery does not make it more scientifically valid. In the end, science is neither based on "longevity" nor "popularity", but rather on the validation of theories on the basis of empirical findings.

MTBI patients have been studied with multiple test batteries, and the Halstead-Reitan battery has *no* empirically proven advantages. Indeed, the tests for the Halstead-Reitan battery were developed prior to the recent advances made in the cognitive neurosciences (Halstead, 1947). With MTBI, attentional and memory difficulties are particularly pronounced, and the core Halstead-Reitan battery is not particularly sensitive to detection of learning and memory difficulties. While almost all clinicians using the Halstead-Reitan battery will expand it with additional tests, the impairment indices nevertheless rely on the core battery. We have been confronted with the argument that this impairment index presents a scientifically-based advantage. Let us examine this notion.

An *impairment index*, in essence, assumes that the sum total of deficits across a test battery is sensitive for capturing brain damage. If a battery is designed to capture the degree of diffuse damage, this assumption may have validity; however, no index has been designed for the specific deficits caused by MTBI. MTBI is not thought to result in a generalized deterioration of equal magnitude within and between the hemispheres. The assumption underlying an impairment index is reminiscent of Lashley's principle of mass action (1950), i.e., quantity of brain damage is the critical feature. However, both Lashley's principles of mass action and equipotentiality have been convincingly challenged (Pribram, 1971). Applied to the impairment index, brain damage should not be solely based on the amount or quantity of test scores that are deficient. It is the cluster of deficits that is essential; a cumulative approach is not wrong, but far less powerful. Thus, a pattern analysis, across a test profile, that emphasizes the evaluation of attention, memory, learning and executive functioning appears to us as equally or more sensitive, since these are the cognitive domains that are particularly affected after MTBI.

Like the Halstead-Reitan users, the first author also has a bias for using a certain neuropsychological test battery (Ruff & Crouch, 1991), as do all clinicians. These biases should be evaluated for what they are; a clinician should not insinuate that there is scientific proof of having a superior battery compared to the opposing neuropsychologist, *unless* there exists sound reasoning combined with empirical research to back such a claim. When assessing MTBI patients, it is essential that the comprehensive evaluation include a range of attention tests, and failing to include a sustained or continuous performance test would not be keeping up with the state of the art at this time. A subsequent focus should be on the assessment of memory, learning and executive functioning.

Bias of the times

Having worked with MTBI patients for over 20 years, we have encountered various shifts between generally accepted beliefs. In the 1970s and early 1980s, the defense and plaintiff arguments centered on the notion that if CT images were negative, then the diagnosis of MTBI was disputed and the symptoms were explained as psychogenic. Based on research with MRI scans, EEGs, SPECT and PET scans, and especially the experimental animal literature, the field slowly accepted that MTBI could occur even in the absence of positive neuroimaging scans. Thereafter, it appeared that the loss of consciousness became the hallmark feature. If patients did not sustain a definitive loss of consciousness, then again, the defense and plaintiff's sides would argue over whether an MTBI did indeed occur. However, this controversy also became modified, since in the early 1990s it became acceptable that a change in consciousness is sufficient to indicate an MTBI (Mild Traumatic Brain Injury Committee, 1993). The above changes were primarily based on changing technology and knowledge relevant to MTBI, and also involved a certain dynamic process between the plaintiff and defense strategies.

A range of interpretations

Presently, there are multiple trends occurring, and in the following we will identify three interpretations that are being used. In a first approach, group data of MTBI patients are analyzed to question the reliability of "true neuropsychological deficits" (i.e., group-effect argument; Binder, Rohling, & Larrabee, 1997). Second, some neuropsychologists point out that MTBI symptoms reported by the Miserable Minority are similar to the complaints expressed by a minority of the general population who have not sustained MTBI (i.e., base rate argument; Youngjohn, Burrows, & Erdal, 1995). A third interpretation by neuropsychologists is to adhere to the notion that in many patients who are part of the Miserable Minority, the concussion resulted in diffuse axonal injury; thus, in part, some of the postconcussive symptoms can be attributed to chronic brain-based changes.

Let us examine the strengths and weaknesses of these three interpretations, beginning with the *group effect argument*. Given that, in the literature, it has become accepted that only a small portion of the overall MTBI sample suffers from more permanent residua, one needs to first address the question of whether group analysis is a reasonable way to evaluate whether MTBI can result in "true neuropsychological deficits". Most clinicians would agree that the population of MTBI is heterogenous. At one extreme, MTBI can occur after a patient is merely dinged during a collision on the football field. At the other extreme, an individual can be diagnosed as having MTBI following a post-traumatic amnesia of 12 hours and a positive CT scan. This raises the issue of comparability among studies, since their samples may be skewed to either extreme. This may explain why the rate for the poor outcome varies between 7–8% to over 20% among the investigators.

Having patients with different pre-injury levels can also influence a group analysis. For example, in the three-center study (Ruff et al., 1989), significant variations occurred not only among the patients, but also among the control groups in the three geographical regions. Indeed, on some of the tests (e.g., Paced Serial Auditory Addition Test), the performances for normals recruited in the Bronx were equivalent to the mild TBI patients at either the Galveston or the San Diego centers.

While group studies can be susceptible to various sampling biases, more group studies are needed. Indeed, it is our impression that too many case reports and opinion papers have emerged in the last 3–4 years, with only a handful of researchers studying well-controlled groups of TBI patients.

Nonetheless, group means cannot be the sole basis for determining neuropsychological deficits for individual patients (Marshall & Ruff, 1989). For example, it is conceivable that if someone had been functioning above the 90th percentile, and after MTBI this declines to the 50th percentile, the test results would still fall in the average range, i.e., 50th percentile. Thus, for each individual a reduction is always relative, and an "average" result does not necessarily prove that a specific patient has not experienced a decline.

Second, let us briefly address the *base-rate argument*. If, in a normal or non-MTBI population we find that 30% suffer from heightened anxiety, 20% from headaches, and 10% from attentional difficulties, then if within the MTBI population we find similar percentages, can we conclude that these are basically the same individuals? The notion that they are the same individuals is of course nothing but an assumption. To empirically evaluate this, one would have to first establish base-rates that are compatible with the MTBI populations. Thus, taking base-rates from overall populations would not be acceptable, since MTBI patients are typically younger and male. Even if one were to establish matched populations on demographics who would have identical rates of endorsed symptoms, the question that the clinician must answer would not be fully answered.

Obviously, baseline arguments are of interest and should be further researched, however, these group comparisons can never substitute for case-by-case diagnoses by an experienced clinician, who has to determine whether a patient was indeed one of the premorbid complainers. Future researchers will also need to carefully deal with the reasons behind the under-reporting or overreporting by certain MTBI patients of their symptoms. Moreover, it is important to evaluate whether the base-rate complaints are due to trait vs. state variables. In other words, a very technically tight research needs to be advanced, and still, one can fall into a logical trap.

Let us review an example. If in the base-rate population 50% of married couples divorce, and secondly, if we were to find that in women who are diagnosed with breast cancer the divorce rate is also 50%, we can conclude that the breast cancer diagnosis had no effect on the divorce rate. However, this would not allow us to further conclude that cancer had no effect on marriages. Similarly, equal base rates between normals and MTBI patients for symptoms such as headaches, depression or anxiety do not allow us to conclude that the MTBI did not adversely affect a patient. Let us, in an example, assume that 20% of the 16–35 year old males report a base rate of 3–4 headaches a week, and let us further assume that this is identical to the rate of an MTBI population matched in age and gender. Are these the same type of headaches with equal intensity and duration? Should we not also find out whether these headaches were spontaneous, vs. the result of a hangover, etc.? Thus, more refined qualitative and quantitative data need to be gathered to further this base-rate argument.

Thirdly, let us consider the *diffuse axonal injury argument*. Animal and human studies can be cited in which acceleration/deceleration has resulted in DAI, even when neuroimaging techniques were insensitive to establishing objective neuropathology. Given that our patient has an altered sense of consciousness or post-traumatic amnesia, then the inference is made that chronic brain trauma is the explanation for the patient's persistent and chronic deficits. The weakness with this particular viewpoint is the lack of lower limits. Does every blow to the head result in cell loss? Even if there were some cell loss, how relevant is it, since it is assumed that we lose cells every day as part of the aging process? This type of interpretation must be seen in the context of its vulnerability, especially if no stringent definitions for alteration of consciousness and posttraumatic amnesia are used. Our suggested subtypes may assist in determining whether there are differences in the type of chronic problems that are presented by patients with Type I, Type II, and Type III MTBIs (see Table 1). Moreover, systematic group studies that evaluate dynamic neuroimaging techniques are called for to further evaluate the prevalence of neuropathology. To study these new technologies, it is important that future studies go beyond case analysis and evaluate consecutively-admitted emergency room patients with, for example, SPECT or PET images independent of their outcome and whether or not they are in litigation.

Bottom line

As clinicians, we have accepted that a portion of MTBI patients will seek expert opinion with respect to their persistent and chronic residua. Out of these patients, a sub-group will present with multiple positive symptoms and complaints. The clinician is faced with a case-by-case decision; the patient's problems have to be diagnosed as being (1) neurogenic, (2) psychogenic, (3) based on the potential for financial gain, or (4) any combination of these. In doing so, one should select a specific definition for MTBI and, based on clinical interviews and a review of the medical records, gather as much information as is helpful. The first decision that should be reached is whether there was a concussion or MTBI, *per se.* Thereafter, one should evaluate pre-existing risk factors and evaluate co-morbid and post-morbid symptoms. Interactions among and between the postconcussive symptoms need to be explored in order to differentiate the degree to which these symptoms are based on interactions vs. the concussion per se. Finally, one has to recognize that changes over time are likely and that some fluctuation should be expected. Despite these fluctuations, determination of the reliability of neuropsychological findings can be enhanced by repeated testing over time.

In the future, the study of MTBI definitely will benefit from refinement in objective technologies that will be able to more definitively diagnose the existence and extent of brain pathology. However, even if such techniques were established, the clinical relevance still needs to be determined for each individual patient. Even if we set the criteria that five million diffuse axonal injuries would have to take place before we call it a mild TBI, this loss would affect different individuals at different rates. Thus, even with the advancement of neuroimaging techniques, we anticipate that clinical evaluations, including neuropsychological testing, will be important in determining whether and how a patient's brain functioning has been altered.

As the history of science has repeatedly demonstrated, one interpretation does not always win over another. Instead, a combination of two opposing interpretations may actually persist. Although a clinician, researcher, or lawyer may favor a certain interpretation at this time, the future may well prove that for different MTBI patients, different interpretations apply.

References

Alexander, M.P. (1995). Mild traumatic brain injury: Pathophysiology, natural history, and clinical management. *Neurology, 45,* 1253–1260.

American Academy of Neurology, Therapeutics and Technology Assessment Subcommittee (1989). Assessment: EEG brain mapping. *Neurology, 39,* 1100–1102.

American Academy of Neurology, Therapeutics and Technology Assessment Subcommittee (1996). Assessment of brain SPECT. *Neurology, 46,* 278–285.

American Psychiatric Association (1994). *Diagnostic and statistical manual of mental disorders* (4th ed.) Washington, DC: Author.

Binder, L.M. (1997). A review of mild head trauma. Part II: Clinical Implications. *Journal of Clinical and Experimental Neuropsychology, 19,* 432–457.

Binder, L.M., Rohling, M.L., & Larrabee, G.J. (1997). A review of mild head trauma. Part I: Meta-analytic review of neuropsychological studies. *Journal of Clinical and Experimental Neuropsychology, 19,* 421–431.

Blumbergs, P.C., Scott, G., Manavis, J., Wainwright, H., Simpson, D.A., & McLean, A.J. (1994). Staining of amyloid precursor protein to study axonal damage in mild head injury. *Lancet, 344,* 1055–1056.

Borczuk, P. (1995). Predictors of intracranial injury in patients with mild head trauma. *Annals of Emergency Medicine, 25*, 731–736.

Crooks, D.A. (1991). The pathological concept of diffuse axonal injury in head trauma. *Journal of Pathology, 165*, 5–10.

Dikmen, S., McLean, A., & Temkin, N. (1986). Neuropsychological and psychosocial consequences of minor head injury. *Journal of Neurology, Neurosurgery, and Psychiatry, 49*, 1227–1232.

Gennarelli, T.A., Thibault, L.E., Adams, J.H., Graham, D.I, Thompson, C.J., & Marcincin, R.P. (1982). Diffuse axonal injury and traumatic coma in the primate. *Annals of Neurology, 12*, 564–574.

Gronwall, D., & Wrightson, P. (1981). Memory and information processing after closed head injury. *Journal of Neurology, Neurosurgery and Psychiatry, 44*, 889–895.

Gualtieri, C.T. (1995). The problem of mild brain injury. *Neuropsychiatry, Neuropsychology, and Behavioral Neurology, 8*, 127–136.

Halstead, W.C. (1947). *Brain and intelligence.* Chicago: University of Chicago Press.

Hsiang, J., Yeung, T., Yu, A., & Poon, W. (1997). High-risk mild head injury. *Journal of Neurosurgery, 87*, 234–238.

Jennett, B., & Teasdale, G. (1981). *Management of head injuries.* Philadelphia: F.A. Davis Company.

Kay, T., Newman, B., Cavallo, M., Ezrachi, O., & Resnick, M. (1992). Toward a neuro-psychological model of functional disability after mild traumatic brain injury. *Neuropsychology, 6*, 371–384.

Kibbey, M.Y., & Long, C.J. (1997). Effective treatment of minor head injury and understanding its neurological consequences. *Applied Neuropsychology, 4(1)*, 34–42.

Kraus, J.F., & Nourjah, P. (1989). The epidemiology of mild head injury. In H.S. Levin, H.M. Eisenberg & A.L. Benton (Eds.), *Mild head injury.* New York: Oxford University Press.

Lashley, K.S. (1950). In search of the engram. *Symposia of the Society for Experimental Biology, 4*, 454–482.

Levin, H.S., Amparo, E., Eisenberg, H.M., Williams, D.H., High, W.M. Jr., McArdle, C.B., & Winer, R.L. (1987). Magnetic resonance imaging and computerized tomography in relation to the neurobehavioral sequelae of mild and moderate head injuries. *Journal of Neurosurgery, 66*, 706–713.

Levin, H.S., Eisenberg, H.M., & Grossman, R.G. (1989). *Mild head injury.* New York: Oxford University Press.

Levin, H.S., Williams, D.H., Eisenberg, H.M., High, W.M., & Guinto, F.C. (1992). Serial MRI and neurobehavioral findings after mild to moderate closed head injury. *Journal of Neurology, Neurosurgery and Psychiatry, 55*, 255–262.

Lezak, M.D. (1995). *Neuropsychological assessment* (3rd ed.) New York: Oxford University Press.

Marshall, L.F., & Ruff, R.M. (1989). Neurosurgeon as a victim. In H.S. Levin, H.M. Eisenberg & A.L. Benton (Eds.), *Mild head injury* (18), 276–280. New York: Oxford University Press.

McLean, A., Temkin, N. R., Dikmen, S., & Wyler, A. R. (1983). The behavioral sequelae of head injury. *Journal of Clinical Neuropsychology, 5*, 361–376.

Mild Traumatic Brain Injury Committee (1993). Definition of mild traumatic brain injury. *Journal of Head Trauma Rehabilitation, 8*, 86–87.

Mittenberg, W., DiGuilio, D.V., Perrin, S., & Bass, A.E. (1992). Symptoms following mild head injury: Expectation as aetiology. *Journal of Neurology, Neurosurgery, and Psychiatry, 55*, 200–204.

Mittenberg, W., Tremont, G., Zielinski, R.E., Fichera S., & Rayls, K.R. (1996). Cognitive-behavioral prevention of postconcussion syndrome. *Archives of Clinical Neuropsychology, 11*, 139–145.

Nedd, K., Sfakianakis, G., Ganz, W., Uricchio, B., Vernberg, D., Villanueva, P., Jabir, A.M., Bartlett, J., & Keena, J. (1993). ^{99m}Tc-HMPAO SPECT of the brain in mild to moderate traumatic brain injury patients: compared with CT – a prospective study. *Brain Injury, 7, 469–479.*

Oppenheimer, D.R. (1968). Microscopic lesions in the brain following head injury. *Journal of Neurology, Neurosurgery and Psychiatry, 31, 299–306.*

Povlishock, J.T. (1993). Pathobiology of traumatically induced axonal injury in animals and in man. *Annals of Emergency Medicine, 22, 41–47.*

Rimel, R.W., Giordani, B., Barth, J.T., Boll, T.J., & Jane, J.A. (1981). Disability caused by minor head injury. *Neurosurgery, 9, 221–228.*

Rizzo, M., & Tranel, D. (1996). *Head injury and postconcussive syndrome.* New York: Churchill Livingstone.

Ruff, R.M., Camenzuli, L.F., & Mueller, J. (1996). Miserable minority: Emotional risk factors that influence the outcome of a mild traumatic brain injury. *Brain Injury, 10, 551–565.*

Ruff, R.M., & Crouch, J.A. (1991). Neuropsychological test instruments in clinical trials. In E. Mohr & P. Brouwers (Eds.), *Handbook of clinical trials: The Neurobehavioral approach* (Chapter 5; pp. 89–119). Amsterdam: Swets and Zeitlinger.

Ruff, R.M., Crouch, J.A., Tröster, A.I., Marshall, L.F., Buchsbaum, M.S., Lottenberg, S., & Somers, L.M. (1994). Selected cases of poor outcome following a minor brain trauma: Comparing neuropsychological and positron emission tomography assessment. *Brain Injury, 8, 297–308.*

Ruff, R.M., Levin, H.S., Mattis, S., Marshall, L.F., Eisenberg, H.M., Tabaddor, K., & High, W.M. Jr. (1989). Recovery of memory after mild head injury: A three center study. In H.S. Levin, H.M. Eisenberg, & A.L. Benton (Eds.), *Mild head injury (12),* 176–188. New York: Oxford University Press.

Ruff, R.M., Mueller, J., & Jurica, P.J. (1996) Estimating premorbid functioning levels after traumatic brain injury. *Neurorehabilitation 7, 39–53.*

Ruff, R.M., Wylie, T., & Tennant, W. (1993). Malingering and malingering-like aspects of mild closed head injury. *Journal of Head Trauma Rehabilitation, 8, 60–73.*

Rutherford, W.H., Merret, J.D., & McDonald, J.R. (1978). Symptoms at one year following concussion from minor head injuries. *Injury, 10, 225–230.*

Satz, P., Zauch, K., McClery, C., Light, R., Asarnow, R., & Becker, D. (1997). Mild head injury in children and adolescents: A review of studies (1970–1995). *Psychological Bulletin, 122, 107–131.*

Teasdale, G., & Jennett, B. (1974). Assessment of coma and impaired consciousness: a practical scale. *Lancet, 2, 81–84.*

Thatcher, R., Walker, R., Gerson, I., & Geisler, F. (1989). EEG discriminant analyses of mild head trauma. *Electroencephalography and Clinical Neurophysiology, 73,* 94–106.

Varney, N.R., Bushnell, D.L., Nathan, M., Roberts, R., Rezai, K., Walker, W., & Kirchner, P. (1995). NeuroSPECT correlates of disabling mild head injury: Preliminary findings. *Journal of Head Trauma Rehabilitation, 10(3),* 18–28.

Williams, D., Levin, H., & Eisenberg, H. (1990). Mild head injury classification. *Neurosurgery, 27, 422–428.*

Wittgenstein, L. (1958). *Philosophische Untersuchungen.* New York: McMillan.

Youngjohn, J.R., Burrows, L., & Erdal, K. (1995). Brain damage or compensation neurosis? The controversial post-concussion syndrome. *The Clinical Neuropsychologist, 9,* 112–123.

Chapter 12

NEUROPSYCHOLOGY AND THE (NEURO)-TOXIC TORT

David E. Hartman
Rush University, Chicago, Illinois

Introduction

Merging the *science of poisons* (toxicology) with the *science of brain-behavioral relationships* (neuropsychology) has produced the hybrid behavioral science of *neuropsychological toxicology,* the study of human neurological, behavioral, cognitive and emotional concomitants of toxic and neurotoxic exposure. Within the past 25 years, there has been an explosion of research demonstrating that acute and chronic exposures to neurotoxic substances are capable of causing observable impairments in brain function on standardized measures.

Such impairments are a function of direct or indirect damage to the nervous system by a toxic agent or *toxicant*. Exposure may occur by inhalation, ingestion, skin absorption or any combination of these routes. The exact number of neurotoxic substances among the 50,000–100,000 industrial substances and at least 2,000,000 mixtures of commercial chemicals is not known, but Tilson et al. (1995) suggest that from 3–28% percent may be neurotoxic. Approximately 1925 million pounds of substances already listed as neurotoxic are released annually into the air (*Neurotoxicity*, 1990).

Toxic torts

As recognition of the scope of neurotoxic effect has become clear, clinical and epidemiological exploration of toxic injuries has been paralleled by a demand for legal redress of those injuries. The resulting form of legal action has been called a "toxic tort", defined as a legal cause of action after a plaintiff develops an injury or illness in response to a chemical, industrial product, metal, drug, or other exogenous physical agent. The earliest, so called "toxic tort" was that of *Borel v. Fibreboard*, which, in 1973, established liability for an insulation worker who was employed for 33 years installing asbestos-containing insulation. The worker developed asbestosis and malignant mesothelioma as a result of exposure. A jury agreed that the asbestos manufacturers were negligent and strictly liable for failing to warn workers of risks in handling asbestos. On appeal, the Fifth Circuit ruled that the manufacturer of a product is assumed to possess expert knowledge about that product and is therefore responsible for a duty to warn of relevant dangers. A variety of other toxic agents may trigger illness and subsequent tort redress, including brain or testicular cancer from exposure to police radar guns, lead-related mental retardation or learning disability, or toxic encephalopathy from chronic solvent exposure.

Obviously, only a subset of toxic torts addresses areas that are within the purview of the neuropsychological expert; those that directly or indirectly address brain injury or psychological damages from toxic exposure. While not identified as such by the legal community, these legal actions can more accurately be called *neurotoxic* torts, reflecting the specific loci of injury and implying the necessity for neuropsychological expertise. There is no dearth of such materials, and hundreds of group and individual case studies exist that detail the effects of exposure to metals, solvents, pharmaceuticals, abused drugs, pesticides, and other well known neurotoxic substances (e.g., carbon monoxide, see Hartman, 1995). Familiarity with these substances and accompanying research is an essential prerequisite to working with clinical or forensic cases of neurotoxic exposure.

Metals

"Excessive exposure to metals . . . has been implicated in a remarkable range of adverse signs and symptoms involving the central nervous system (CNS) and behavior" (Weiss, 1983, p. 1175). Exposure to almost any metal can cause CNS dysfunction, and peripheral nervous system (PNS) toxicity has also been noted for a number of common metals (e.g., lead, arsenic). Several metals produce neurotoxic effects on the extrapyramidal system, damaging the caudate nucleus, putamen, globus pallidus and their connections to the thalamus, substantia nigra, and cerebellum (e.g., iron) (Feldman, 1982a). Many other metals can damage both the central and peripheral nervous system. According to an Environmental Protection Administration (EPA) list, the metals lead, arsenic, mercury, cadmium and chromium occupy five of the top twenty most hazardous substances, with lead, arsenic and mercury finishing first, second and third (Delany, Bateman, & Harman, 1996).

Lead exposure

Present-day extent of lead exposure in children has been a frequent source of toxic tort litigation, perhaps reflecting a large population of children with elevated blood lead levels. One survey suggests that 3.9% or almost 700,000 U.S. children have blood levels at or above 30 μg/dl (a, Annest, Roberts, & Murphy, 1982), which is 20 μg/dl above the current EPA cutoff for significant lead exposure. Racial and socioeconomic factors strongly influence the prevalence of lead elevation, with only 2% of white children showing elevated blood lead, as contrasted with up to 18% of poor, inner-city, black children. City living itself may be an additional risk. The U.S. Agency for Toxic Substances and Disease Registry found elevated blood lead levels in excess of 15 μg/dl in 17% of city dwelling U.S. children (Agency for Toxic Substances and Disease Registry, 1988). Landrigan and Curren (1993) estimate that 68% of inner city children in the United States have elevated blood lead levels, the larger number perhaps reflecting increased risk from living in old, poorly maintained buildings, and exposure to soil contaminated by factory waste or automobile exhaust.

Mercury exposure

Mercury poisoning has also been the subject of toxic tort. The metal is contained in dental amalgam, pesticides and fungicides, battery manufacturing, and many other industrial manufacturing processes, including plastic and paper production. Delany and Cusack (1996) cite the case of a 44 year old female who alleged chronic mercury poisoning while working as a dental assistant for 18 years. The case was settled for $50,000 with defense denying causation, and claiming that the plaintiff actually had multiple sclerosis. A two million dollar verdict was reached in New York for a 13 year old boy who sustained loss of cognitive function and personality changes among other symptoms, when a hospital and pediatrician failed to diagnosis his symptoms as due to mercury poisoning from a traditional Chinese remedy (Delany & Cusack, 1996).

Solvents

Organic solvents are chemical compounds or mixtures used by industry to "extract, dissolve or suspend" non-water-soluble materials including fats, oils, resins, lipids, cellulose derivatives, waxes, plastics, and polymers (*Organic Solvents*, 1985, p. 3). Solvents are *lipophilic* – having a special affinity for fat, and are disproportionately absorbed by brain and tissue having high fat content. The amount of solvent retained depends on a complex interaction between blood and tissue solubility, *a priori* toxicity (Astrand, 1975) and other individual variations including metabolic cycles, alcohol use, and body weight. For example, solvents are reported to have longer biological half lives in obese individuals than in thin persons (Cohr, 1985). Solvents are retained in multiple sites within the body and bioaccumulation can occur in cases of repeated exposure, depending upon frequency and extent (Ashley & Prah, 1997). All solvents depress CNS activity via anesthetic action. Solvent-exposed individuals are rendered progressively less sensitive to stimuli as a function of solvent concentration and/or exposure duration until unconsciousness, coma, or death occurs.

Early solvent cases entering the courts were not neurotoxic torts, but asserted the relationship of leukemia to solvent (benzene) exposure (Willman & Whitson, 1996). However, in 1995, a non-jury trial verdict was made in favor of a plaintiff alleging "toxic encephalopathy, mild organic brain syndrome. . . anxiety . . . and depression". The court's opinion did not address testimony about the scientific basis of benzene exposure and toxic encephalopathic symptoms (Willman & Whitson, 1996). The Louisiana Court of Appeals upheld the trial judge's decision to admit plaintiff's experts, including those who testified about neuropsychological damage, and thus may have been the first decision "upholding a ruling in favor of a plaintiff . . . that a solvent caused neurophysiological or neuropsychological damage" (Willman & Whitson, 1996, p. 22).

Pesticides

Over 4 billion pounds are applied worldwide, with the United States using 1.1 billion pounds of that total annually (Lang, 1993). Pesticides are applied to over 900,000 farms in the United States. Numbers of individuals exposed to pesticides are quite high, and include an estimated 340,000 pesticide production workers, 2.5–2.7 million migrant workers, and 1.3 million certified pesticide applicators. Pesticide workers are one of the largest occupational risk groups in the world (Davies, 1990; Moses, 1983).

Pesticides are a diverse group of toxicants that are grouped together by their pest-killing properties, rather than their chemical structure. Metals, hydrocarbons and other substances have been employed as pesticide. Most of the more common pesticides, including chlorinated hydrocarbons (e.g., DDT), organophosphates (e.g., Malathion, Diazinon, Ronnel), and carbamates (e.g., Baygon, Maneb, Sevin, Zineb) are lethal to insects via neurotoxic action (Morgan, 1982). Studies using higher mammals have also demonstrated pesticide neurotoxicity (e.g., Aldridge & Johnson, 1971; DuBois, 1971; Vandekar, Plestina, & Wilhelm, 1971). It is not surprising, therefore, that neurotoxic effects of pesticides are also found in human exposure victims, and that both cognitive and emotional functions are affected.

Delany, O'Brian, Barr and Caliendo (1996) cite a variety of recent pesticide case law, but most cases are related to property damage. However, several allowed testimony related to neurotoxic symptoms and pesticide causation.

Prescription drugs

Drugs are another class of potentially neurotoxic substances; their number is also not known, but a partial compilation occupies 347 pages of a recent text

(Brust, 1996). Drug-induced neurological disorders involve approximately 1–2% of neurological admissions or consultations (Jain, 1996). Lawsuits involving drug toxicity likely would be actionable as medical malpractice on the individual level. However, torts involving drugs or medical devices damaging to the nervous system might still be termed toxic torts.

Hazardous waste exposure

Many toxic exposures do not fall under a single category. For example, there are an estimated 32,000 hazardous waste sites, with an average of 6000 people living nearby each one, in proximity to solvents and heavy metals. It has been estimated that more than 2,000,000 children, women of childbearing age, and elderly persons live near waste sites designated as national priorities for cleanup (Amler & Lybarger, 1993).

Risk and type of injury

Given the numbers and quantities of substances involved, their variety and ubiquitous distribution in the environment, it should not be surprising that the number of individuals at risk for neurotoxic injury is quite high. In 1987, there were between 1023 and 1707 deaths that could be definitively linked to workplace neurotoxic exposure in the United States (Hessl & Frumkin, 1990). Neurotoxicity is considered by NIOSH (National Institute for Occupational Safety and Health) to be one of the ten leading causes of work-related disease and injury (Neurotoxicity, 1990). The number of disabilities or nonfatal impairments has not been tallied and would almost surely be underestimated, especially if diseases of currently unknown etiology are linked to neurotoxic exposure. Although controversial, some studies have linked amyotrophic lateral sclerosis, parkinsonism, some Alzheimer-type dementias and lupus to chronic, insidious effects of neurotoxic agents (Kardestuncer & Frumkin, 1997; Tilson, 1990; Wechsler, Checkoway, & Franklin, 1991).

Neurotoxic injury can occur on a continuum ranging from DNA damage to death. Damage is often measurable at multiple points in the organism, including "behavioral, neurophysiological, neurochemical and neuroanatomical levels" (Tilson et al., 1995, p. 364). In humans, all primary neurobiological systems are susceptible to neurotoxic effect, including disturbances of affect, autonomic impairments, damage to higher and lower cortical function, cerebellar disease, and peripheral nervous system damage. Assay of behavioral output from these damaged systems has been considered one of the most sensitive indicants of injury (Eskenazi & Miazlish, 1988).

Neuropsychological contributions to the toxic tort

Attorneys who seek to show, prove, or refute, damages in cases of neurotoxic exposure have increasingly called upon neuropsychological experts to quantify or question the nature of damages. Neuropsychological methods directly address the primary theory of damage in a toxic tort case; the claim of present injury or disease manifestation. A plaintiff cannot recover damages that are considered uncertain, speculative or conjectural (*Lavelle v. Owens-Corning Fiberglass*, 1981). In conventional cases of traumatic brain injury, the burden of taking the injury out of the realm of conjecture can rely on more direct, easily understandable information; the physical facts of the injury and subsequent surgical and radiological investigations may establish the foundation for the tort. In contrast, the bar to establish injury is higher in neurotoxic injuries; many exposures produce subtle or insidious damage rather than discrete, easily discernible injury. Moreover, many neurotoxic exposures operate at a neurochemical or microscopic cellular level that is not visible to the naked eye or common neuroradiological investigations. In these features, at least, neurotoxic injuries are similar to more conventional traumatic brain injury effects, with similar diagnostic and forensic approaches required (see Ruff, Chapter 11). It can be argued, however, that neuropsychological methods may be even more crucial for proving brain injury than in traumatic brain injury claims, where there potentially is a more easily understandable "fact" of auto accident, concussion, etc.

In their ability to uncover subtle functional impairments, neuropsychological methods can demonstrate a variety of direct and indirect manifestations of toxic physical injury, much as they are able with traumatic brain injury. Such demonstration is a necessity for showing damages and thereby taking the injury out of the realm of conjecture. Further, the neuropsychologist, functioning in the capacity of a clinical psychologist, can quantify *reactive* distress or *fear of disease,* considered in the context of a toxic tort as negligent infliction of emotional distress (Marrs, 1992).

Scientific versus clinical expertise

Neuropsychologists may function in the expert roles as scientists, clinicians, or both. The neuropsychologist with both scientific and clinical expertise holds an advantage over the pure scientist or clinician in the ability to integrate the science behind an observed clinical impairment that may or may not have been induced by toxic exposure. Marino and Marino (1995) point out there are differences in reasoning for purely scientific versus purely clinical experts and "it would be no more reasonable to presume that a physician was an expert in . . . scientific inference than it would be to expect a scientist to diagnose and treat disease" (p. 13). They caution that expert "clinical" reasoning based entirely on anecdotal knowledge accrued from seeing patients cannot, by itself "serve as a substitute for the scientific reasoning the plaintiff must provide to sustain his case in toxic tort" (Marino and Marino, 1995, p. 13). The neuropsychologist, an expert trained in the scientist-practitioner model of mainstream neuropsychology, is in an ideal position to circumvent problems noted by Marino and Marino, by integrating scientific data analysis with ipsative clinical observation.

Expertise vs. advocacy: The neuropsychologist as expert witness

Several basic criteria can be outlined for the neuropsychologist seeking to serve as an expert witness in toxic torts. First, since an "expert" is an individual with specialized knowledge that will assist the court, the neuropsychological expert in toxic torts presumably has pre-litigation knowledge about neurotoxic exposure-related issues. "Experts" with no previous experience in neurotoxic disorders who then acquire such experience as a result of the litigation in question should expect to have their credentials challenged.

Second, the neuropsychologist who reviews scientific data and evaluates patient performance must remain tied to the state of the literature to be considered a scientific expert, in addition to being a clinical expert. As a plaintiff's expert, the neuropsychologist may explain to the court that a collection of scientific studies supporting evidence of neurotoxic damage are competently performed, and related to plaintiff's claim of damage. Conversely, the defense neuropsychologist who believes it the case that there are objective errors in available research, may show that conclusions favoring the defense logically flow from a scholarly review of available studies.

Both the plaintiff and defense experts must avoid becoming *advocates* and maintain their dispassionate expertise. As Marino and Marino (1995, p. 7) warn, "if the meanest scientific data led a scientist to posit a causal link, or if the strongest possible data did not do so, the scientist would not be a proper expert witness because the scientist would no longer be acting as an scientist but as an *advocate*" (italics mine). Experts who have a history of testifying for only the plaintiff or defense, regardless of the data, and/or who are funded by groups which support or oppose specific neurotoxic disorders have their expertise overpowered by their position as advocates, with subsequent loss of professional credibility.

Both plaintiff and defense neuropsychological experts who seek to integrate the literature on a particular neurotoxic agent with data from a particular patient must be alert to poor experimental technique and literature biases that may obscure or exaggerate neurotoxic deficits. In a recent study reviewed by this author, for example, there was purportedly a finding that chronically-exposed lead workers were unaffected compared to controls. Among other problems with the study were the use of a very small subject population, no medical screening for substance abuse and especially the fact that "control" subjects may have been exposed to solvents in heavy industry and actually tended to perform worse than lead-exposed subjects. Reliance on such a study, regardless of whether it appears to support plaintiff or defense, damages the credibility of the neuropsychological expert.

Biases

Methodological biases

Needleman (1993) lists several concerns about biases in methodology at the expense of clinical judgment. While Needleman is clearly criticizing those who use statistics to *disprove* neurotoxic relationships, his critique remains relevant to forensic review of neurotoxic exposure research:

Arbitrary worship of the p value of 0.05

Significance levels of .05 or less have become the mark of "causality" when the actual meaning of this probability level is simply that there is one possibility in twenty that results were due to chance. The tradition of imputing special significance to a particular (level may have more to do with the fact that Fisher was permitted to reprint only two lines from the table of P values than to any intrinsic cut-off for causality at 0.05 (Needleman, 1993).

Overcontrol of "confounds"

Control of confounding influences should clarify, rather than obscure "real" relationships between neurotoxic exposure and subsequent neurobehavioral impairment. It is possible to overcontrol for confounding influences and subtract out variance properly attributable to the particular neurotoxic exposure. For example, removing variance for diagnoses of hyperactivity, developmental delay, or mental retardation in childhood lead studies may improperly remove their influence from subjects who were affected by lead severely enough to merit these additional diagnoses.

Ignoring the possibility of multi-generation toxic exposures

For example, poor parenting ability or low parent IQ may be due to parents living or working under the same neurotoxic conditions as their progenitors, a possibility that should be investigated before removing parenting or similar factors from the statistical model.

Accepting the null hypothesis from studies with inadequate power

Power analyses are rarely published, and findings which have accepted the null hypothesis may be using small data sets, or overanalyzing the data. Needleman gives an example of a study with 48 subjects that used a multiple regression analysis with 17 covariates – producing a power "between zero and .30" to reject the null hypothesis (p. 164).

Underestimating the epidemiological significance of a true, but "small" effect

In childhood lead studies, an IQ difference of 6 points "predicts a four-fold increase in the proportion of significantly impaired children" (p. 165).

Demands for proof of causality
A common, but meaningless methodological criticism utilizes the unrealistic expectation that all variables must be properly controlled and accounted for. Clinical research can be reanalyzed using infinite combinations of variables and yet bring the researcher no closer to a secure definition of "causality". The purpose of group studies is to accrue data from which causality is then inferred. Causality, therefore is an inference *made from* data; it is not *produced by* data analysis.

Evaluating isolated studies
When multiple studies are available, simple "balloting" or tallying of positive or negative studies does not constitute an adequate review of data. Meta-analysis or other quantitative review techniques, where applicable, are preferred.

Testimonial biases
After the evidence in a toxic tort is made available for expert review, the neuropsychological expert should strive to provide objective testimony that can be of assistance to the court. Experts may consciously, or otherwise, provide a less than adequate review or leave the realm of scientific objectivity to provide testimony. Neuropsychologists must be alert to many types of testimonial bias in toxic tort matters. For example, Marino and Marino (1995) list the following biases:

Failing to review or consider studies whose conclusions conflict with planned testimony
The expert who only considers research that supports his own, or discounts contrary research simply because it disagrees with planned testimony, violates the tenets of scientific objectivity.

Arguing that a negative study disproves the possibility of causality inferred from positive studies
Research rarely duplicates exactly a pre-existing study, especially in human clinical studies. Minor variations in methodology or subject selection may produce differing conclusions. A negative study says only that the particular study did not show a neurotoxic effect.

Suggesting that without a mechanistic understanding of the illness, no conclusions are possible
It is not necessary to know or demonstrate the complete chain of molecular events capable of causing neurotoxic brain injury to prove that a cause and effect relationship exists between an exposure and the end results of that exposure.

Other toxic tort testimonial biases

Stating that a pre-existing but unrelated condition makes it impossible to determine if neurotoxic brain injury exists
This author has seen statements in forensic reports stating that because an individual had been treated with antidepressants in the past, his moderately impaired neuropsychological test scores, but non-depressed mental status, made it impossible to obtain a "clean" relationship between exposure and subsequent symptoms. The trier of fact does not require a perfect relationship between exposure effects and symptoms. Human beings typically have imperfect medical or psychological histories. However, the expert must consider the nature and degree of their influence, rather than discount the possibility of developing an opinion because the relation is not unifactorial.

Arguing that unexposed populations report similar complaints
This appears to be a principal defense-related argument of Dunn et al., (1995). Many exposure-related complaints are non-specific in character, due to the subtle, diffuse nature of nervous system impairment. Showing that an unexposed population has similar complaints says nothing about the relationship between exposure and *the plaintiff in question*. The plaintiff must serve as his or her own control with respect to pre-and-post-morbid symptom comparison. It is scientifically illegitimate to compare an individual recently fallen ill with pesticide-related fatigue and loss of cognitive efficiency and no previous history of same, with a large group of unexposed patients, some of whom will have life-long histories of these symptoms for well known psychiatric reasons.

Arguing there is no clear "dose-response" relationship of a particular neurotoxic substance to human brain injury
Dose-response relationships, especially in cases of chronic exposure, may be uniform in studies of genetically similar laboratory animals, but are rarely so neatly found in human clinical studies of neurotoxic exposure. If a dose-response curve is not found, the alternative is not "no relationship", but rather that a "more complex relationship" must be identified to show cause and effect (Hill, 1965). Additional factors, such as age, alcohol use, use of prescription drugs which influence bioavailability of the toxic substance, or synergistically worsen its effect, are possible. Being exposed to cigarette smoke, for example, may cause and/or worsen atherosclerotic deposits, when combined with a high-fat diet (Zhu et al., 1993).

Using "Threshold Limit Values" (TLVs) as evidence for safety or neurotoxicity
Roach and Rappaport (1990) and Ziem and Castleman (1989) argue that TLVs are not true safety thresholds, but rather levels that were acceptable to industry (see Hartman, 1995). Roach and Rappaport concluded that some TLVs were blatantly "influenced by vested interests" and that the rest "are a compromise between health-based considerations and strictly practical industrial considerations with the balance seeming to strongly favor the latter" (p. 741).

Clinical biases: The "healthy worker," the "exploited" worker and the malingering worker

The evaluation of neurotoxicity is complicated from neurobiological as well as neuropsychological perspectives by the problem of individual difference variability in subject response to neurotoxic substances. Systematic biases in neurotoxic sensitivity may be introduced by what has been termed, the "healthy worker effect". This phenomenon involves the observation that employed workers will, on average, be healthier than the general population. Healthy workers do not, by definition, include the unemployed, retirees, the aged, the chronically ill, and other groups who will show higher mortality and poorer health than the working population. As part of the healthy worker effect, workers exposed to neurotoxic substances will selfselect for tolerance to those substances. Employees who show initial sensitivity to solvents or metals, for example, become ill and either voluntarily leave or are terminated from employment due to incapacity. Those who remain at work may form a select group with decreased sensitivity to a given neurotoxicant. Unfamiliarity with this effect may lead to a conclusion that a particular toxic agent has minimal effects in a working population; a possible *underestimate* of a true effect, although perhaps true in a given worker who remains on the job.

A bias that may skew the clinician toward *overestimates* of neurotoxic effect might be termed the *"exploited worker"* effect. Those employees with the lowest level of premorbid intellectual function also may be the most likely to be taken advantage of and assigned to dangerous exposure employment. These workers would be the least knowledgeable and least able by virtue of their low skills to find a safer job. Neuropsychological evaluations in a population of such workers exploited for their minimal skills could show impairments resembling neurotoxic damage that are actually related to *premorbid* capabilities.

Another potential area of bias is the inclusion of subjects who are currently litigating damage awards for neurotoxic poisoning. It is important for both clinical and research neuropsychologists to be sensitive to this source of variance in their neuropsychological assessments. Where there is potential for monetary or psychological secondary gain, evaluation of neuropsychological malingering (see Chapters 9 and 10) is strongly advised.

Determining cause and effect relationships in neurotoxic exposure

Even after the completion of copious record review and extensive neuropsychological testing, the neuropsychologist is responsible for integrating available data into a conclusion that addresses causality. *Epidemiology*, which studies pattern and spread of disease in groups, may or may not be helpful, but cannot address the plight of toxic exposure effects and damages in a single individual. Suitable exposure groups for neurotoxic injury symptoms can be difficult to obtain, since a study requires the cooperation of both potential exposure victims and their employers who may be responsible for their exposure. In addition, the statistical averaging of symptom expression of large groups is of less interest for clinical and legal purposes than the possible *range* of injury expression and factors that predispose an individual to be particularly vulnerable to such injury. A substance capable of widespread but small effects may not present for clinical evaluation in comparison to substances which may be capable of being a serious neurotoxic poison for some and having minor effects for others (e.g., lead).

The influence of neurotoxic exposure to symptom expression is complex, and it is naïve to insist that neurotoxicants show the same rigid cause and effect relationships proposed for infectious agents. Contrasting these so called "Koch Postulates" (Delany & Cusack, 1996) with neurotoxic exposure effects outlines the limitations inherent in applying infectious disease-based methods of epidemiology to toxic injury.

A more sophisticated analysis of environmental disease causation has been outlined by Hill (1965), that far better addresses issues related to the complexity of toxic injury. Hill proposed the following set of criteria to be considered when evaluating the association between an environmental, occupational hazard and human injury. Particularly notable is a relative de-emphasis on strict, so-called "dose-response" experimental causation and substitution of multifactorial clinical and experimental judgment.

1. *Strength*: Relevant information can be gathered by comparing diseases in occupations with toxic exposure to occupations not having such exposures, or with occupations that are just beginning to use particular toxic chemicals.

2. *Consistency*: An association between illness and exposure that has been observed repeatedly in different persons, places and times, validates and strengthens the scientific certainty of that association, especially when similar results are found in different avenues of investigation.

3. *Specificity*: Hill tends to downplay specific patterns of damage or disease in building a case for toxic injury, suggesting that lack of a pattern of disease does not necessarily invalidate the possibility of illness or injury. While specific patterns that link exposure to the injury may exist, they are not necessary to making a conclusion. If specificity exists, "we may be able to draw conclusions without hesitation; if it is not apparent, we are not thereby necessarily left sitting irresolutely on the fence" (p.297).

4. *Temporality*: One of the more important facets of causation is that there must be some reasonable relationship in time between exposure and injury.

5. *Biological gradient*: The fabled "dose-response curve" should be sought, but if it is not found, then the relationship between substance and effect is "weakened, though not necessarily destroyed". Rather, if a dose-response curve is not found, the alternative may be a "more complex relationship to satisfy the cause and effect hypothesis". Additional factors, such as age, co-existing drug use, etc., may influence the effect of common neurotoxic substances. Smoking, for example, may cause and/or worsen atherosclerotic deposits, when combined with a high-fat diet (e.g., Zhu, et al., 1993).

6.   *Coherence:* Exposure data should "not seriously conflict with the gener-
      ally known facts of the natural history and biology of the disease" (p.
      298). However, for toxic injuries, the "generally known facts" sometimes
      lag behind clinical observation. A common example was seen in the early
      understanding of lead poisoning, when "generally known facts" were that
      lead was absorbed through the skin, and body cleanliness was emphasized
      above environmental management of fumes and dust, e.g., "I remember
      the head surgeon of a great Colorado smelting company saying in a pub-
      lic meeting: 'It is not the lead a man absorbs during his work that poisons
      him but what he carries home on his skin'" (Hamilton, 1985, pp. 3–4).
      Detailed studies of lead's respiratory absorption were needed (Aub,
      Fairhall, Minot, & Reznikoff, 1926) before science caught up with the
      clinical reality of lead exposure effects.
7.   *Experiment:* Hill believes that experimental methodology is of "occasion-
      al" use in support of causation, however, experimental animal exposures

Table 1. Koch Postulates vs. Neurotoxic Exposure Effects.

Koch postulates	Neurotoxic exposure effects
The organism responsible for the disease must be present in every case of the disease.	Neurotoxic exposure symptoms are often non-specific and present in other illnesses. There is no "organism" and toxic agents may have been metabolized after causing damage, leaving no further chemical "signature."
The organism must be capable of being isolated and grown in pure culture.	While exposure parameters may be known, many neurotoxicants cannot be directly measured in vulnerable central nervous system tissue, and have short half-lives in fat stores.
The organism must be grown in pure culture and must be able to cause the disease.	Disease expression in neurotoxic exposure can vary widely according to individual ability to clear the material from the body, age, nutrition and other factors.
The organism must be recovered again from the experimentally infected host.	Many highly toxic agents, e.g., solvents, cannot be measured in human brain tissue without biopsy of brain tissue (an impractical and potentially dangerous procedure). In addition, many solvents and other neurotoxicants have short half-lives, thereby producing damage but leaving no long-term evidence of their presence.

rarely duplicate typical exposure conditions in humans and may have limited application if the injury includes subtle impairments in intelligence, memory and other neuropsychological functions that are difficult to test in animal models.

8. *Analogy:* Hall reports that in some cases "it would be fair to judge by analogy, with the effects of thalidomide and rubella before us, we would surely be ready to accept slighter, but similar, evidence with another drug or another viral disease..." (p. 299). Given the almost infinitely creative capacity of industry to create neurotoxic materials and the limited funding available to empirically investigate every creative variation, inference of neurotoxicity by analogy is often required. In this author's experience, it can also provoke a highly contentious defense counter-argument, particularly as a plaintiff's expert for an individual who has developed solvent neurotoxicity after exposure to a solvent variant that has not been subject to study, which is chemically related to other known neurotoxic solvents. Defense attorneys argue that since the exact solvent variant at contention has not been studied, that no conclusions can be reached about neurotoxic damages. I have termed this somewhat concrete defense counter-argument, the "Green Hammer-Blue Hammer-Yellow Hammer" approach (i.e., an attempt to prove that just because one knows that being struck in the head by a green hammer or blue hammer will be painful, there is no data to "prove" that being struck on the head by a yellow hammer will produce a similar effect).

Expert strategies

Several steps can be taken by neuropsychologists involved in toxic tort litigation to ensure that their opinions are as accurate as possible. One way to potentially increase diagnostic accuracy may be to incorporate the opinions of a multidisciplinary diagnostic team. This recommendations follows from the same type of observation noted in regard to discipline-specific assessment in cases of traumatic brain injury (see Chapter 11). Because neurotoxic substances may cause a variety of organ system problems, analysis of neurotoxic damage can often be outlined by a variety of specialists in their particular clinical domains. Neuropsychologists who see patients outside of the hospital/occupational health clinic milieu can consider recreating an equivalent multidisciplinary diagnostic team in the private sector in the service of either the defense or the plaintiff of a toxic tort. Close collaboration with occupational medicine physicians, neurologists, and toxicologists can maximize the likelihood of accurate diagnosis, effective treatment, and accurate expert opinion-making. Development of a system perspective is also important in ruling out alternative diagnoses or impairments and determining whether other system impairments have occurred in conjunction with toxic exposure. For example, many neurotoxins also cause hepatic or pulmonary damage that may interact or (as Tarter et al., 1988 have shown) largely account for neuropsychological dysfunction.

A second strategy to increase diagnostic accuracy is to obtain all available data. A skilled neuropsychologist, like all other health care providers, is only as effective a diagnostician as the data allow. It is not unusual for a clinician to be presented with incomplete data by an overenthusiastic (or Machiavellian) attorney in an effort to obtain his or her cooperation in a case. Nevertheless, neuropsychologists must make a reasonable attempt to obtain complete medical, occupational, psychological and educational records to determine whether a conclusion about neuropsychological injury is possible and justifiable. One must assume that all such data will eventually be disclosed by the adversarial process. However, a proactive search will reveal the information in a way that allows its early integration into clinical evaluation or record review. There are few situations more uncomfortable to the individual and unflattering to the profession of neuropsychology than being surprised by new information at the time of deposition or on the stand that forces retraction of otherwise carefully worked out conclusions. The axiom "knowledge is power" must guide the neuropsychologists participation in a toxic tort case, as in other adversarial proceedings.

The chain of inference

The neuropsychologist who evaluates a patient with neurotoxic exposure must attempt to develop a chain of logical causation that considers the influence of toxic exposure in the context of a broad life history. Accordingly, the first link in the chain of neuropsychological inference is provided by a review of the patient's pre-injury status, from a developmental, educational, medical, psychological and occupational viewpoint. The aim is the construction of a context within which to compare and contrast changes attributed to neurotoxic exposure.

The next step in the chain of inference is the verification of the fact of toxic exposure. What constitutes a "toxic" exposure will vary considerably from substance to substance and patient to patient; *the examiner cannot simply conclude exposure post-hoc from symptoms,* but must make a comprehensive effort to determine whether significant exposure took place. The services of a toxicologist or industrial hygienist to compute exposure levels may be useful in this regard.

Once the reasonable possibility of an exposure is confirmed, the next link in the chain of inference involves determination as to whether symptoms are credible from a toxicological perspective. Based upon the toxic properties of the substance in question, is it likely that exposure was damaging and that present symptoms are of the type and degree expected? For example, the toxicological and clinically observed effects of toluene exposure indicate that brief, low level, open air exposure is unlikely to produce neurotoxic damage. Alternatively, serious if not fatal neurotoxic brain damage, may occur from large, single exposures to such substances as carbon disulfide, carbon monoxide, cyanide, or solvent exposure sufficient to cause anesthesia and respiratory arrest. Some neurotoxic substances may have sustained effects on brain and behavior, even as

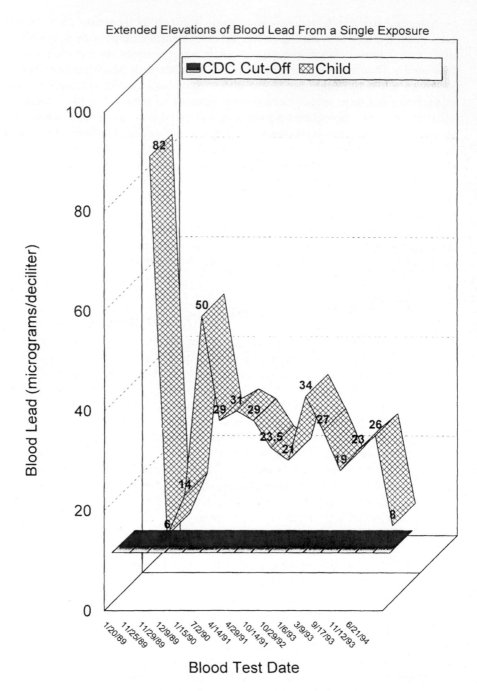

Fig. 12.   Extended elevations of lead from a single exposure compared with Center for
Disease Control cut-off's.

the result of a single exposure. The following graph, for example, documents extended blood levels of lead induced as a result of a single *pica* exposure. Despite multiple attempts to pharmacologically strip the blood of free lead (chelation), blood levels rise repeatedly, reflecting a characteristic heavy body burden of lead deposited into porous bone structures, and gradually diffusing into the bloodstream.

The clinical expert who is aware of the fact that over 90% of a child's lead burden may be contained in bone may help the trier of fact understand how a single exposure can persist in elevating blood levels over the course of several years.

The next link in the chain of causation is the determination through patient evaluation and record review of the relationship between exposure and symptom development or exacerbation. If symptoms developed *prior* to neurotoxic exposure and did not worsen afterward, they cannot be of neurotoxic etiology, no matter how similar and "textbook" they may appear. The patient who claims intellectual deterioration after solvent exposure, but who had long-standing difficulties with academic performance (see Matarazzo, 1990), cannot have academic deficits attributed to neurotoxic insult, unless they have worsened substantially after exposure. Ideally, there should be no other illnesses, drugs, stresses or other psychological diagnoses that might account equally well for symptoms.

Neurotoxic injuries do not have the same external validation criteria that diagnoses of cancer or heart disease have; unless there is moderate to severe damage, there may be no abnormal cells or blocked blood vessels to observe. Common imaging techniques (e.g., CT, MRI) may be unable to display correlates of toxic brain injury because the injury may be microscopic or neurochemical, rather than structural. Symptoms produced may not be unique to toxic exposure, but may be found in unrelated disease states, psychological illnesses, drug or alcohol abuse, or prescription drug use, among others. Failure to consider the influence of non-neurotoxic-based medical disorder can distort conclusions to the point of foolishness. This author has reviewed plaintiff neuropsychological reports that purported to find "toxic impairment-based" neuropsychological test results in patients with history of surgical removal of brain tumor, chronic alcoholism, lupus, and other potent confounds.

In an illustrative case, a record review and clinical evaluation was conducted on a female in her 40's who had demanded and received Workers' Compensation payments for "multiple chemical sensitivity". The patient claimed that she developed symptoms after working for several months in a framing shop, using solvents to clean glass and glues to fasten picture frames. At the time of record review and interview, the patient claimed she had no problems that could not be explained as a function of multiple chemical sensitivity and reported her intention to proceed in a civil suit against the owner of the frame shop.

However, review of the patient's medical history framed an entirely different picture. Her medical records documented a first psychotic break, accompanied by hallucinations, delusions and inpatient hospitalization several years before she began work at the frame shop. Both before and after her several month employment, medical records detailed repeated and unrelated psychotic episodes. Neuropsychological screening at the time of her interview was normal. The patient volunteered to demonstrate her multiple chemical sensitivity experimentally by allowing herself to breathe in odors from substances to which she claimed sensitivity, including perfumes and detergents. No observable cognitive, behavioral, emotional or neuropsychological changes could be elicited in this manner. Her actual diagnosis was within the domain of Schizophrenia, Paranoid subtype. There was no evidence for solvent neurotoxicity or chemical sensitivity. It was unclear to what extent her belief that she had been poisoned by solvents was delusional, as opposed to a more "rational" motivation for continued and expanding financial compensation for her injury.

In a similar case, this author was retained by a plaintiff's attorney in a case alleging neurotoxic exposure to hydrocarbon solvents. The individual in question had a history of beliefs in Satanism, other bizarre delusional beliefs and had been living in his car for an extended period to avoid toxic exposure. It was explained to the attorney that the closed environment of a car would itself constitute a box of outgassing plastics and other industrial materials, and that the plaintiff's choice of residences coupled with a premorbid belief system that was clearly psychotic did not work in the plaintiff's favor. The case was settled before trial.

The next link in the chain of inference is to determine whether there is internal consistency of neuropsychological test results. Neuropsychological test data must make sense with respect to both internal consistency within the test battery, as well as ecological validity (i.e., when comparing test behavior to real world capabilities). Within a test battery, the patient with complaints of severe attentional deficit should not complete the PASAT without error, but fail Speech Sounds Perception and Seashore Rhythm. The patient who struggles to produce 10 taps in a 10 second interval on the Finger Oscillation Test should not be able to wiggle his hand rapidly in the air nor play the piano in a bar on weekends. There should be a convincing and reasonable relationship between degree of impairment on test performance and degree of impairment in everyday life.

There is a substantial relationship between neuropsychological test performance and everyday capabilities in variety of brain-injured populations (Sbordone & Long, 1996). When this tenet is violated, there is good reason to suspect deliberate distortion of test results. This author once contested the ecological validity of a set of test results in which the claimant's grossly impaired neuropsychological test performance on tasks involving attention and concentration conflicted with his ability to perform supervised flying lessons in a small aircraft. In a similar case, neurotoxic attentional abnormalities were alleged, but the claimant's abnormal test results were not compatible with his continued squash playing at the local health club. A third litigant claimed severe attention, concentration and motor control deficits that precluded any form of work involving these capabilities. In the meantime, he passed the time at home, assembling plastic car models as a hobby (a hobby obviously requiring concentration and fine motor control).

Secondary gain in the form of large damage awards is an obvious influence on patient behavior in these types of cases. Only careful correlation of patient history and clinical observation with test results will allow the neuropsychologist to differentiate between valid psychometric indications for brain damage, and elaborate exaggerations in the service of secondary gain. Invalid patterns on the MMPI-2, wildly inconsistent performance within or across similar tests and/or batteries, and exaggerated performance on forced-choice tests of malingering (See Chapters 9 & 10) may all provide evidence of less than credible cooperation and performance. Personality test distortion and cognitive exaggeration may be completely independent of one another, depending upon whether the claimant wishes to emphasize cognitive or emotional "damage".

Clinical conclusions

There are at least five types of clinical conclusions possible in the evaluation of any patient with neurotoxic exposure, or for that matter, of any patient with symptom complaints in the context of an adversarial, litigated matter:

Symptoms may be *primary behavioral, cognitive, or affective concomitants of structural or neurochemical lesion*. Personality change as a result of lead poisoning, dementia from chronic heavy solvent exposure or slowed, confused thinking and depression from pesticide exposure are examples. Primary symptom relationship to toxic exposure is the strongest, most direct form of injury and preferred from a plaintiff's legal perspective, since it indicates direct physical injury and may have the least optimistic profile for patient recovery (leading to possibly larger damage awards).

Symptoms may be *reactive to consequences of real neurological disability*. An example would be an educated professional rendered unconscious from carbon monoxide exposure as the result of faulty heater installation who subsequently becomes depressed as a psychological reaction to loss of ability to continue in that profession. An individual with carbon disulfide-induced cerebral vasculopathy may react catastrophically to perceived deficiencies in cognition. In both of these cases, real neurological disability may coexist with, or be exacerbated by affective components and psychological reactions may be considered secondary to and aggravating of the direct toxic injury. There is also research that supports synergistic damage to the nervous system and other organ systems when organisms are exposed to both psychological stress and neurotoxic substances. A review of such studies concluded that elevated levels of neurotoxic damage were produced in animal subjects subjected to experimental stress, in addition to toxic exposure (Vogel, 1993). It is worth considering, therefore, whether highly stressed human subjects suffer from increased neurotoxic exposure effects; even if such stress is due to knowledge of exposure implications or is unrelated to exposure.

A third possibility is that symptoms are *reactive to psychosocial stressors of exposure*, rather than to structural or neurochemical abnormality. For example, an individual might develop neuropsychological symptoms after working in a factory that has been found to be in violation of OSHA toxic exposure standards. Fears of possible exposure, job loss, non-neurological medical consequences of exposure (e.g., cancer), legal proceedings, and financial hardships may produce "learned helplessness" and other reactive effects. These reactions, in turn, may produce abnormal neuropsychological test behavior. The patient, family, friends, and co- workers may view toxic exposure as more frightening and threatening than other illnesses, which reinforces the patient's use of medical services (Bolla, 1991). Knowledge of toxic exposure effects and fears of injury can occur in workers, regardless of literacy or formal education (Vaughan, 1993). Pavlovian-like conditioning of fear or other psychological responses may also occur (Bolla, 1991). Elevated levels of post-traumatic stress disorder-like symptoms have been seen, especially in patients with chronic solvent exposure (Morrow et al., 1989), with anxiety, depression, derealization and various cognitive problems, although it is not clear what proportion of symptoms is the result of direct organic involvement of limbic system structures and which symptoms are psychologically reactive.

Psychological reactions to the fear of toxic exposure are not as frequently seen as a sole cause of action in toxic tort litigation, but similar cases have reached the courts in so-called "cancerphobia" torts. Fear of the possibility of becoming ill is considered compensible and courts have "almost uniformly" allowed such cases to proceed (Gale & Goyer, 1986). Some neurotoxic substances are also carcinogens, and may justifiably trigger cancerphobia directly in a susceptible individual. Other forms of toxic exposure may produce a more general "nosophobic", or general, fear of disease response. Courts have been divided over what constitutes the necessary criteria for a true 'fear of disease' reaction as a form of damages. Some have required physical injury to have occurred, others have been satisfied with pure psychodiagnostic criteria. One court has required a perhaps psychologically unrealistic standard that there must be a 50% or greater probability of developing a future illness for there to be a reasonable grounds for fearful reaction. Donath (1995) suggests that a middle ground in such emotional damages suits may be a "reasonableness" statute in which a "reasonably constituted person" may reasonably fear developing cancer, even if the actual probabilities remain low. He suggest that while .5 percent increased risk over the general population is unreasonable to fear, an 8 percent increase over general levels "is substantial enough to make the plaintiff's fear reasonable". Mental health professionals, will, of course, dispute any such attempts to quantitate "reasonableness", noting that individual reaction can be highly subjective and variable, depending upon the background and psychological organization of the individual. Nevertheless, neuropsychological experts may be asked to testify in such cases, in particular, to help determine whether a plaintiff is a "reasonably constituted person" and whether fear of disease is reasonable under the circumstances of exposure. Differentiation of related psychological disorders tied to toxic exposure may be required (e.g., mass psychogenic illness, post-traumatic stress disorder, chronic technological disaster effects and others; see Hartman, 1995, Chapter 9).

Fourth, symptoms may be *unrelated*, linked to difficulties that existed prior to toxic exposure, or to stress that occur coincidentally with toxic exposure, (e.g., continuation of premorbid alcoholism or personality disorder, death of a parent or spouse.) This is a preferred toxic tort defense; that symptoms, if they exist, have no connection with the exposure, but can be explained as a function of premorbid facts.

Finally, symptoms may be *malingered*. Individuals in civil litigation, in particular, stand to gain significant compensation as a function of claimed damages (neuropsychological and otherwise). Ruling out the possibility of voluntary symptom production should be a part (albeit regrettable) of the clinical rule-out process for all forensic neuropsychological evaluations.

Strength of conclusions

The inference of scientific or clinical causation is exactly that, a rational judgment based on the best available data. It is the position of the expert to *infer* causation from statistical, historical and clinical data. The defense argument that statistics do not prove causality is true in this limited respect; it is the job of the expert to infer such causality in all cases.

Because such inference is necessarily indirect and probabilistic, the strength of conclusion for a neuropsychological expert may vary. Expert conclusions range from "possible" through "reasonably certain" or "to a reasonable standard of neuropsychological certainty". "Possible" relationships are the weakest, saying only that a relationship is not impossible, (i.e., that the likelihood of a relationship between a toxic agent and subsequent impairment is a probability greater than zero). This is not a very satisfactory conclusion and is essentially useless to the judge or jury in their fact finding task. "Possible" relationships would likely not meet the hurdles set by the *Daubert* decision (discussed below).

The next stronger level of inference, one of "more probable than not", is acceptable to the court; it is the assertion that the link between toxic exposure and subsequent injury is greater than 50 percent certain. The expert is saying that more of the evidence supports the link between toxic exposure and injury than refutes it. In a neurotoxic tort, where an injury may be multi-factorial, influenced by premorbid health state and interactive with age, intelligence or genetic susceptibility, an assertion of "more probable than not" is not an admission of uncertainty, but an acknowledgment of scientific complexity.

The final level of "reasonable degree of neuropsychological (medical, scientific, etc.) certainty" is the strongest level of certainty possible in a claim of inferential causality. Any greater assertion of causality usurps the role of the judge and jury in the legal process. The implication is that the expert is certain to the standard of the discipline, over and above a simple "preponderance of the evidence" that an agent has caused injury to the plaintiff. It is important to note that this level of certainty should not be confused with "statistical" certainty at a probability of .05 or less, which may be present, but not required, for an expert's judgment of reasonable certainty.

Standards of neuropsychological expert testimony in toxic tort claims: *Frye* & *Daubert*

Two standards addressing validity of scientific testimony are relevant to neuropsychological conclusions in toxic torts, the *Frye* rule and *Daubert v. Merrill Dow Pharmaceuticals, Inc,* commonly referred to as the *Daubert* standard (also discussed in Chapters 14 and 15). *Frye* articulated a "general acceptance standard" in which a scientific deduction "must be sufficiently established to have gained general acceptance in the particular field in which it belongs". The use of neuropsychological methodology to describe toxic brain injury certainly has gained such general acceptance in the field of clinical neuropsychology, so *Frye* does not pose an undue burden to the use of neuropsychological consultation *per se*. Alternatively, *Frye* seems to limit the opinions of a neuropsychologist who might seek to testify on a newly developed toxicant for which there is little research and therefore little "general acceptance" by the neuropsychological community of its neurotoxicity. In neurotoxic torts, *Frye* tends to favor the defendant, since testimony regarding an agent's toxicity, or disease-causing potential, particularly if the toxicant is newly developed or the subject of little research, will usually be met by contrary testimony from the defendant's expert (Marino & Marino, 1995). The controversy itself can be construed as a lack of "general acceptance" that therefore causes the plaintiff's argument to fail the test of the *Frye* rule.

Superceding *Frye* in the Federal courts was the 1993 decision by the United States Supreme Court, which ruled in *Daubert v. Merrell Dow Pharmaceuticals,* a toxic tort instituted by plaintiffs who sought to prove that birth defects were related to taking the morning sickness drug, *Bendectin*. Under the standard of *Daubert,* a court may consider four additional factors other than "general acceptance", including (1) whether the scientific theory or technique can be and has been tested, (2) whether it has been subject to published peer review, (3) the known or potential error rate and, (4) the existence and maintenance of standards controlling the techniques operation. The end result of *Daubert*, therefore is to reduce the court's automatic reverence for scientific experts and make the judge a "gatekeeper" who controls the quality of scientific evidence that is permitted. While ostensibly tightening the standard of scientific testimony (which would be a welcome outcome), *Daubert* places the burden of such scientific gatekeeping on judges who may have limited scientific backgrounds. Thus, an unintended effect of *Daubert* may be to make case law involving toxic torts and neuropsychological testimony more unpredictable, as judges attempt to rule on scientific methodologies (Delaney & Cusack, 1996). Use of *Daubert* standards also may favor a defense outcome; with defendants winning approximately 75 percent of recent Federal cases (Delaney & Cusack, 1996). While it is possible that this lopsided ratio reflects, not so much the additional hurdles of scientific proof required of the plaintiff, but the quality of the plaintiff's case (e.g., cases of multiple chemical sensitivity), it is clear that plaintiffs and experts must more completely address issues of basic science in *Daubert* than under *Frye*.

On one hand, application of clinical neuropsychological methods to a human plaintiff may be more acceptable under *Daubert* than an inference of neurotoxicity based solely on animal studies, since the issue at hand is related to human neurotoxic effects. Alternatively, *Daubert* criteria can and have been abused in the service of the adversarial process. As an example, a recent brain injury case decided for the defense on testimony of Dr. Ralph Reitan, who apparently persuaded the judge that his battery, which was admitted as "medical" scientific evidence, was the sole standard within neuropsychology capable of satisfying *Daubert* criteria (Reed, 1996). While limited at this time to the federal courts, the ruling could be used as a precedent to force a neuropsychological expert to adhere to a battery that is neither optimal for testing of neurotoxic disorders, nor the choice of the majority of practicing clinicians according to surveys of the field (Sweet, Moberg, & Westergaard, 1996; Guilmette, Faust, Hart, & Arkes, 1990).

Daubert also has been interpreted as requiring every step of the expert's analysis to pass the full set of standards. Thus, "any step that renders the analysis unreliable under the *Daubert* facts renders the expert's testimony inadmissible . . . whether the step completely changes a reliable methodology or merely misapplies that methodology" (Delaney & Cusack, 1996, p. ix–16). It is therefore up to the neuropsychologist who uses test batteries that include tests of unproven sensitivity in neurotoxically-exposed individuals to prove that they have correctly applied such methodology in order to prevent their entire battery and subsequent testimony from being rendered inadmissable.

Preparing for *Daubert*

Neuropsychological experts are likely to have their scientific and clinical expertise vigorously challenged under *Daubert*. Preparation for litigation challenge begins when a case is accepted for evaluation and should include personal research in the toxicant area at issue. To avoid becoming an "advocate", a "neutral" and open-minded approach to diagnosis and testimony is optimal. In the interest of maintaining that approach, the neuropsychological decision-making process should be both inductive and deductive; evidence must point to exposure as the primary etiology, while alternative explanations are systematically ruled out if a plaintiff's claim is to be supported. The neuropsychological report or review should seek to integrate testing data with medical, occupational and psychological history.

If there is a finding of exposure-related effect, it must be based on more than just temporal connection with exposure. Toxicants cannot be linked to neuropsychological test abnormalities by a simple *post hoc* decision rule. In other words, there must be a higher standard of proof than, "patient was in the vicinity of a toxicant . . . patient shows abnormal neuropsychological test results . . . therefore . . . patient shows neurotoxic brain damage". The clinician who bases testimony primarily on temporality will fail the hurdle of *Daubert*.

In addition, the neuropsychologist should:

1. Be able to justify why he or she is considered an expert in the field of neuropsychology generally, and neurotoxic injuries, specifically.
2. Be able to explain in plain language how conclusions were derived.
3. Be ready to summarize the results of neuropsychological testing, and how they compare or contrast with results from other experts.
4. Have peer reviewed studies available in support of an opinion. Avoid making conclusions that are (a) without substance from a scientific perspective, i.e., that lead exposure does *not* cause impairments in young children, (b) when there is no known scientific disease process, (e.g., fibromyalgia), or (c) if there are no studies to reasonably suggest human neurotoxicological effects, (e.g., dental amalgam).
5. Be able to summarize relevant scientific literature and integrate conclusions into the known body of human neurotoxic research on a particular subject.
6. Be specific about sources of information relied upon. Try not to rely heavily upon possibly "tainted" sources of information, (e.g., other parties to the lawsuit, the husband or wife of the plaintiff), but rather on sources who would not benefit from the outcome of the case. Even patient self-report, by itself, is not the most valid clue to toxic exposure, and self-report neuropsychological questionnaires, by themselves, may be unable to distinguish between neurotoxic exposure profiles and patterns indicative of high levels of personal distress for other reasons (Dunn, Lees-Haley, Brown, Williams, & English 1995).
7. Be able to discuss which alternative diagnoses were entertained, ruled out and why. The expert who is seen as having considered other options carefully is less likely to be, or be perceived as an advocate or a "hired gun" with an opinion that is purchased, rather than scientifically derived.
8. Avoid using neurotoxic exposure as a throw-away or "scattershot" diagnosis in the service of an unrelated complaint to increase the "seriousness" of damages, (e.g., plaintiff was harmed not only by being rendered comatose from auto accident, but by breathing fumes from the fire extinguisher used to put out the car fire.) Unless the neuropsychologist is prepared to back up such a claim scientifically and clinically, such an approach detracts from credibility and may diminish perceived expertise in the more important aspects of the case.

Test battery construction

To realistically satisfy *Daubert* criteria, a conservative and complete approach to neuropsychological assessment is recommended. Not all neuropsychological instruments are useful clinical tools. Tests designed for large group epidemiologic neurotoxicological research, for example, may be without clinical norms and thus problematic for use with individual patients. Tests with relatively low sensitivity to neurotoxic disorders (e.g., the Luria-Nebraska Neuropsychological Battery) would be a poor choice for accurate measurement of neurotoxic symptoms.

Because neurotoxicants often produce diffuse degradation of brain function, tests and batteries that emphasize cognitive efficiency and sustained, whole brain integration will be more sensitive to neurotoxic disorder than more specific and brief Luria-like tasks. In this regard, many components of the Halstead-Reitan Neuropsychological battery would be appropriate measures to include in an evaluation for neurotoxic symptom expression, including the Category Test, Tactual Performance Test, and Trailmaking. Unfortunately, the HRB is notably weak in other areas commonly found to be impaired as a function of neurotoxic exposure, including memory function and fine motor coordination of the type assessed by the Grooved Pegboard. The HRB also fails to include other tests of brain function that have been found useful in detecting other concomitants of neurotoxic exposure (e.g., continuous visual attentional vigilance to rapidly displayed information, and color discrimination).

Regarding the former, research by Morrow, Robin, Hodgson, and Kamis (1992) indicates that the use of a continuous performance test (i.e., scanning a computer screen and responding to a degraded image of the number "9") may be particularly sensitive to neurotoxic solvent-related effects. Similar CPT's already in common use (e.g., Gordon Diagnostic System) are particularly useful in assessing neurotoxins that leave the brain with limited neurochemical reserves. While initial behavior is adequate, as reserves are depleted or not restored, behavior degrades – an effect opposite that of healthy individuals, who may actually exhibit facilitation with continuous performance.

With respect to color discrimination, Mergler and colleagues in several studies suggest the use of color vision discrimination tests to assess early effects of neurotoxic exposure (Mergler, Belanger, DeGrosbois, & Vachon, 1988; Mergler & Blain, 1987; Mergler, Bowler, & Cone, 1990; Mergler, Huel, Bowler, Frenette, & Cone, 1991). For example, they found impaired desaturated color vision on the Lanthony D-15 color panel in 17 out of 21 (81%) patients presenting with exposure to solvents, Freon, or pesticides at an occupational health clinic (Mergler et al., 1990). Type II color vision loss (concurrent red-green and blue-yellow) which "most often results from damage to the optic nerve" was found in 52% of patients with diagnosed cognitive impairment (p. 671). While desaturated color perception testing is relatively uncommon among neuropsychologists at present, its research efficacy may encourage its inclusion in clinical and forensic evaluations.

Neuropsychologists who seek to implement a battery capable of satisfying both *Daubert* federal court precedent and research precedent should consider combining the strengths of tests from a well researched battery like the Halstead-Reitan with additional validated tests of memory (e.g., MAS, WMS-III), motor coordination (e.g., Grooved Pegboard) and continuous performance would appear to provide a "best of all possible worlds" approach by maximizing both *Daubert* acceptability and clinical sensitivity. Common, well researched objective tests of personality (e.g., MMPI-2; PAI), are appropriately included, as are measures of cognitive symptom exaggeration and standardized life history assessment via medical history questionnaire.

Conclusion

The growing need for neuropsychological testimony in cases of personal injury reflects recognition of the unique role of the neuropsychologist as *the* forensic expert who can straddle the worlds of scientific data analysis and clinical service, and in doing so, disentangle complex symptom formations. Nowhere is the need for such testimony greater than in the area of toxic tort litigation. Evaluation of patients with possible toxic exposure requires the best of what neuropsychology has to offer – psychological evaluation of subjective symptomatology, developmental clinical review of patient status to properly view symptoms in the context of life history, objective assessment of neurocognitive status, research review and scientific justification of test results, and finally, framing of final conclusions in a clear, comprehensive manner. Neuropsychological participation within the toxic tort arena also serves the judicial process by allowing appropriate cases to be compensated for injury as well as saving society the cost of rewarding claimants for ill-founded or falsified complaints. By maintaining high levels of accuracy in scientific and clinical diagnosis in the forensic arena, neuropsychologists also provide evidence for appropriate regulation of neurotoxic substances, thus indirectly improving environmental standards, increasing worker safety, and improving awareness of toxic safety issues as well as expanding the science of psyche-soma relationships.

References

Aldridge, W. N., & Johnson, M. K. (1971). Side effects of organophosphorus compounds: Delayed neurotoxicity. *Bulletin of the World Health Organization, 44*, 59–63.

Amler, R. W., & Lybarger, J. A. (1993). Research program for neurotoxic disorders and other adverse health outcomes at hazardous chemical sites in the United States of America. *Environmental Research, 61*, 279–284.

Ashley, D. L., & Prah, J. D. (1997). Time dependence of blood concentrations during and after exposure to a mixture of volatile organic compounds. *Archives of Environmental Health, 52*, 26–33.

Astrand, I. (1975). Uptake of solvents in the blood and tissues of man. A review. *Scandinavian Journal of Work Environment and Health, 1*, 199–218.

Aub, J. C., Fairhall, L. T., Minot, A. S., & Reznikoff, P. (1926). *Lead poisoning (Medicine Monographs 7)*. Baltimore: Williams & Wilkins.

Becking, G. C., Boyes, W. K., Damstra, T., & MacPhail, R. C. (1993). Assessing the neurotoxic potential of chemicals A multidisciplinary approach. *Environmental Research, 61*, 164–175.

Bolla, K. I. (1991). Neuropsychological assessment for detecting adverse effects of volatile organic compounds on the central nervous system. *Environmental Health Perspectives*, 95, 93–98.

Brust, J. C. M. (1996) *Neurotoxic side effects of prescription drugs*. Boston: Butterworth Heinemann.

Cohr, K.-H. (1985). Definition and practical limitation of the concept organic solvents. In Joint WHO/Nordic Council of Ministers Working Group (Eds.), *Chronic effects of organic solvents on the central nervous system and diagnostic criteria (Document 5)*. Copenhagen: World Health Organization, Regional Office for Europe.

Davies, J. E. (1990). Neurotoxic concerns of human pesticide exposures. *American Journal of Industrial Medicine*, 18, 327–331.

Delany III., J. J., Bateman, D. P., & Caliendo, C. (1996). Multiple Chemical Sensitivity. In J. J. Delany III (Ed.), *Toxic tort law & science manual* (pp. XIV-1–XIV-8), Wayne PA: Mealey Publications.

Delany III, J. J., Bateman, D. P., and Harman, V. S. (1996). Metals, particulates and fibers. In J. J. Delany III ((Ed.), *Toxic tort law & science manual* (pp. XIV-1–XIV-10), Wayne PA: Mealey Publications.

Delany III, J. J. & Cusack, A. (1996). Overview of the basic tools needed for litigating toxic torts. In J. J. Delany III (Ed.), *Toxic tort law & science manual* (pp. XIV-1–XIV-32), Wayne PA: Mealey Publications.

Delany III, J. J., & Cusack, A. (1996). Daubert and toxic torts. In In J. J. Delany III (Ed.), *Toxic tort law & science manual* (pp. XIV-1–XIV-26), Wayne PA: Mealey Publications.

Delany III, J. J., O'Brian, W. J., Barr, B. D., and Caliendo, C. Pesticides. In (J. J. Delany III (Ed.), *Toxic tort law & science manual* (pp. XIV-1–XIV-30), Wayne PA: Mealey Publications.

Donath, G. (1995). Curing cancerphobia phobia: Reasonableness redefined. *University of Chicago Law Review*, 62, 1113–1137.

DuBois, K. P. (1971). The toxicity of organophosphorus compounds to mammals. *Bulletin of the World Health Organization*, 44, 233–240.

Dunn, J. T., Lees-Haley, P. R., Brown, R. S., Williams, C. W., & English, L. T. (1995). Neurotoxic complaint base rates of personal injury claimants: Implications for neuro- psychological assessment. *Journal of Clinical Psychology*, 51, 577–584.

Eskenazi, B., & Maizlish, N. (1988). Effects of occupational exposure to chemicals on neurobehavioral functioning. In R. Tarter, D. Van Thiel, & K. Edwards (Eds.), *Medical neuropsychology: The impact of disease on behavior*. New York: Plenum Press.

Feldman, R. G. (1982). Central and peripheral nervous system effects of metals: A survey. *Acta Neurologica Scandinavica*, 66 (Suppl. 92), 143–166.

Gale, F. J., & Goyer, J. L. (1986). Recovery for cancerphobia and increased risk of cancer. *Defense Law Journal*, 35, 443–464.

Guilmette, T., Faust, D., Hart, K., & Arkes, H. (1990). A national survey of psychologists who office neuropsychological services. *Archives of Clinical Neuropsychology*, 5, 373–392.

Hamilton, A. (1985). Forty years in the poisonous trades. *American Journal of Industrial Medicine*, 7, 3–18.

Hanninen, H. (1982b). Psychological test batteries: New trends and development. In R. Gilioli, M. G. Cassitto, & V. Foa (Eds.), *Neurobehavioral methods in occupational health* (pp. 123–130). New York: Pergamon Press.

Hartman, D. E. (1995). *Neuropsychological toxicology*. New York: Plenum Press.

Hessl, S., & Frumkin, H. (1990). *Beyond neglect: The problem of occupational disease in the United States*. Chicago: Workplace Institute.

Hill, A.B (1965). The environment and disease: Association or causation? *Proceedings of the Royal Society of Medicine*, 295–300.

Jain, K. K. (1996). *Drug-induced neurological disorders.* Seattle: Hogrefe & Huber.

Kardestuncer, T., & Frumkin, H. (1997). Systemic lupus erythematosus in relation to environmental pollution: An investigation in an African-American community in north Georgia. *Archives of Environmental Health, 52,* 85–90.

Landrigan, P. J., & Curran, A. (1992). Lead A ubiquitous hazard. *Environmental Research, 59,* 279–289.

Lang, L. (1993). Are pesticides a problem? *Environmental Health Perspectives, 101,* 578–583.

Lavelle v. Owens-Corning Fiberglass Corp. (1981), 507 N.E.2d 479 (Ohio Com. Pl. 1986) citing 30 Ohio Jurisprudence 3d 24, Damages Section 14.

Mahaffey, K. R., Annest, J. L., Roberts, J., & Murphy, R. S. (1982). National estimates of blood lead levels: United States 1976–1980: Association with selected demographic and socioeconomic factors. *New England Journal of Medicine, 307,* 573–579.

Marino, A. A., & Marino, L. E. (1995). The scientific basis of causality in toxic tort cases. *University of Dayton Law Review, 2,* 2–62.

Marrs, S.D. (1992). Mind over body: Trends regarding the physical injury requirement in negligent infliction of emotional distress and fear of disease cases, *28, Tort and Insurance Law Journal,* 1.

Matarazzo, J. D. (1990). Psychological assessment versus psychological testing: Validation from Binet to the school, clinic and courtroom. *American Psychologist, 45,* 999–1017.

Mergler, D., & Blain, L. (1987). Assessing color vision loss among solvent-exposed workers. *American Journal of Industrial Medicine, 12,* 195–203.

Mergler, D., Belanger, S., De Grosbois, S., & Vachon, N. (1988a). Chromal focus of acquired chromatic discrimination loss and solvent exposure among printshop workers. *Toxicology, 49,* 341–348.

Mergler, D., Blain, L., Lemaire, J., & Lalande, P. (1988b). Colour vision impairment and alcohol consumption. *Neurotoxicology and Teratology, 10,* 255–260.

Mergler, D., Bowler, R., & Cone, J. (1990). Colour vision loss among disabled workers with neuropsychological impairment. *Neurotoxicology and Teratology, 12,* 669–672.

Mergler, D., Huel, G., Bowler, R., Frenette, B., & Cone, J. (1991). Visual dysfunction among former microelectronics assembly workers. *Archives of Environmental Health, 46,* 326–334.

Mergler, D., Huel, G., Bowler, R., Iregren, A., Belanger, S., Baldwin, M., Tardif, R., Smargiassi, A., & Martin L. (1994). Nervous system dysfunction among workers with long-term exposure to manganese. *Environmental Research, 64,* 151–180.

Morgan, D. P. (1982). Pesticide toxicology. In A. T. Yu (Ed.), *Survey of contemporary toxicology* (p. 136). New York: Wiley.

Morrow, L. A., Robin, M. N., Hodgson, M. J., & Kamis, H. (1992). Assessment of attention and memory efficiency in persons with solvent neurotoxicity. *Neuropsychologia, 30,* 911–922.

Morrow, L. A., Ryan, C. M., Goldstein, G., & Hodgson, M. J. (1989). A distinct pattern of personality disturbance following exposure to mixtures of organic solvents. *Journal of Occupational Medicine, 31,* 743–746.

Moses, M. (1983). Pesticides. In W. Rom (Ed.), *Environmental and occupational medicine* (pp. 547–571). Boston: Little, Brown.

Needleman, H. L. (1993). The current status of childhood low-level lead toxicity. *NeuroToxicology, 14,* 161–166.

Neurotoxicity: Identifying and Controlling Poisons of the Nervous System (1990). Office of Technology Assessment. OTA-BA-436, GPO No. 052–003–01184–1.

Reed, J. E. (1996). Fixed vs. flexible neuropsychological test batteries under the Daubert standard for the admissibility of scientific evidence. *Behavioral Science and the Law, 14,* 315–322.

Roach, S. A., & Rappaport, S. M. (1990). But they are not thresholds: A critical analysis of the documentations to threshold limit values. *American Journal of Industrial Medicine, 17*, 727–753.

Sbordone, R. J., & Long, C. J. (1996). *Ecological validity of neuropsychological testing.* Delray Beach FL: St. Lucie Press.

Spencer, P. S. (1990). *Neurotoxicity. Identifying and controlling poisons of the nervous system.* Washington, DC: Office of Technology Assessment, OTA-BA-436.

Sweet, J., J., Moberg, P. J., & Westergaard, C. K. (1996). Five-year follow-up survey of practices and beliefs of clinical neuropsychologists. *The Clinical Neuropsychologist, 2*, 202–221.

Tarter, R. E., Van Thiel, D. H., & Edwards, K. L. (Eds.). (1988) *Medical neuropsychology.* New York: Plenum Press.

Tilson, H. A., MacPhail, R. C., & Crofton, K. M. (1995). Defining neurotoxicity in a decision-making context. *NeuroToxicology, 16*, 363–376.

Vandekar, M., Plestina, R., & Wilhelm, K. (1971). Toxicity of carbamates for mammals. *Bulletin of the World Health Organization, 44*, 241–249.

Vaughan, E. (1993). Chronic exposure to an environmental hazard: Risk perceptions and self- protective behavior. *Health Psychology, 12*, 74–85.

Vogel, W. H. (1993). The effect of stress on toxicological investigations. *Human and Experimental Toxicology, 12*, 265–271.

Wechsler, L. S., Checkoway, H., & Franklin, G. H. (1991). A pilot study of occupational and environmental risk factors for Parkinson's disease. *NeuroToxicology, 12*, 387–392.

Weiss, B. (1983). Behavioral toxicology and environmental health science. *American Psychologist, 38*, 1174–1187.

Willman, R. K., & Whitson, K. E. (1996). Exposure to solvents. In J. J. Delany III (Ed.), *Toxic tort law & science manual* (pp. XIV-1–XIV-40), Wayne PA: Mealey Publications.

World Health Organization (1985). *Organic solvents and the central nervous system.* Copenhagen: WHO.

Zhu, B. Q., Sun, Y. P., Sievers, R. E., Isenberg, W. M., and Glantz, S. A. (1993). Passive smoking increases experimental atherosclerosis in cholesterol-fed rabbits. *Journal of the American College of Cardiology, 21*, 225–232.

Ziem, G. E., & Castleman, B. I. (1989). Threshold limit values: Historical perspectives and current practice. *Journal of Occupational Medicine, 31*, 910–918.

Chapter 13

SPECIAL PEDIATRIC ISSUES:
NEUROPSYCHOLOGICAL APPLICATIONS AND CONSULTATIONS IN SCHOOLS

Rudy Lorber
Lake Shore Neuropsychological Services, Highland Park, Illinois

Helene Yurk
University of Chicago, Chicago, Illinois

Introduction

Traditional models of forensic neuropsychology have focused upon the presence or absence of cognitive and related deficits in the individual being examined. Even when the neuropsychologist is focusing upon an area of relative or absolute strength, this typically occurs in forensic situations for the purposes of describing the absence of an acquired form of neuropathology. The present chapter will focus upon the application of neuropsychological and related test measures, along with the unique expertise of pediatric neuropsychologists to issues arising from Federal and State laws that affect the education of children and adolescents, as well as related entitlements that these laws mandate. Inherent in the model described within this chapter is a comprehensive integrated assessment approach that runs counter to the purpose of solely documenting the presence or absence of neurologically-based deficits. Rather, a clinical model will be presented in which there is a need to document *all* aspects of the individual's cognitive and related strengths. Further, beyond the widely accepted general definition of neuropsychology as the study of brain/behavior relationships, pediatric neuropsychologists are also in a unique position to assist in the determination of the interrelationship between cognitive abilities/disabilities and an individual's academic, social, emotional, and behavioral functioning. By virtue of training and experience, the pediatric neuropsychologist is often in the best position to integrate the findings of the varied professional disciplines that may be involved with children in school settings (e.g., occupational therapist, physical therapist, and speech and language pathologist). In addition, school-age children who come to the attention of pediatric neuropsychologists are likely to have had a variety of medical evaluations, as well as other related assessments (e.g., audiological) that may not be readily interpretable by school personnel.

In making the transition from a deficit-based model to a remedially-based model of neuropsychological functioning, the pediatric neuropsychologist must either evaluate the "whole" child, or do so in a coordinated approach with other qualified professionals. Although many of the areas of functioning listed below will be familiar, it is the ability to relate these underlying processing skills to an individual's functional performance and school-related needs, that places the pediatric neuropsychologist in a unique position. Deficit-based implications derived from neuropsychological testing typically will be of no practical use to school personnel. Rather, a comprehensive and generally understandable description of cognitive processing skills, and when appropriate the relationships of these skills to academic, social, emotional, and behavioral functioning, is what is required to communicate complex issues within the school environment.

Part I. The developing nervous system and major neuropsychological functions

Given the multiple etiologies for learning disabilities and related cognitive processing deficits, both congenital and acquired (e.g., Cohen, 1991; Gaddes & Edgell, 1994), it is essential to have a basic understanding of the development of the nervous system, as well as the physiological correlates of disorders of learning. Therefore this chapter will depart briefly from the format of this text to review relevant aspects of nervous system development.

Development of the nervous system
During the third week of gestation, a thickened area near the dorsal surface of the embryo begins to develop (Martin, 1996; Spreen, Risser, & Edgell, 1995). This thickened area of the dorsal ectoderm is referred to as the neural plate. In the center of the neural plate, cells on the inner and outer edges become narrower, and form a longitudinal neural groove, which deepens and eventually folds over onto itself. The neural groove begins to close, forming a tube, starting at the midpoint and extending in both rostral and caudal directions. The process of the tube closing and becoming a fluid-filled central canal is generally completed by the fourth week of gestation (Rourke, Bakker, Fisk, & Strang, 1983). The posterior portion of the neural tube forms the spinal cord, whereas the anterior portion becomes the brain.

If the above described process of early nervous system development is defective, developmental anomalies may occur (Berg, 1996; Volpe, 1977). For example, failure of the anterior portion of the tube to develop properly will result in defects of the overall structure of the brain such as anencephaly (absence of cerebral hemispheres (Martin, 1996)). This defect usually occurs between 24–26 gestational days (Berg, 1996; Hynd & Willis, 1988). Furthermore, failure of the posterior neural tube closure (myelomeningocele) occurs at approximately 26 days gestation. Myelomeningocele, (incidence from one to six per 1,000 live births) is often defined by such brain abnormalities as the Arnold-Chiari malformation, agenesis of the corpus callosum, heterotopias, polymicrogyria, and disordered lamination (Gilbert, Jones, Rorke, Chernoff, & James, 1986). Difficulties with motor and bladder/bowel control may also occur (Wills, 1993). Individuals with myelomeningocele may have obstructive hydrocephalus. The neuropsychological profile of children with myelomeningocele and/or hydrocephalus may be remarkable for nonverbal learning disabilities along with impairment in mathematical ability, visual-motor, and motor skills (Friedrich, Lovejoy, Shaffer, Shurtleff, & Beilke, 1991; Hurley, Dorman, Laatsch, Bell, 1990; Snow, Prince, Souheaver, Ashcraft, 1994; Wills, 1993; Wills, Holmbeck, Dillon, & McLone, 1990).

During the fourth week of gestation, three primary vesicles appear from the anterior portion of the neural tube, which include the prosencephalic, mesencephalic, and rhombencephalic vesicles (Martin, 1996; Rourke, Bakker, Fisk, & Strang, 1983). The prosencephalic vesicle eventually becomes the forebrain, while the mesencephalic and rhombencephalic vesicles become the midbrain and hindbrain, respectively. During the fifth week of gestation, the prosencephalic vesicle divides into the telencephalon and diencephalon. The telencephalon forms the cerebral hemispheres, which are separated by the longitudinal fissure and cover the remainder of the prosencephalon (i.e., the diencephalon). The diencephalon eventually develops into the epithalamus, thalamus, hypothalamus, subthalamus, and metathalamus. The rhombencephalic vesicle divides into the metencephalon and the myelencephalon. During the sixth week, the pontine flexure forms and divides the rhombencephalic vesicle into the myelencephalon (which becomes the medulla) and the metencephalon (which becomes the pons and cerebellum).

The ventricular system develops out of the large cavities within the cerebral vesicles (Martin, 1996). The forebrain cavity develops into the two lateral ventricles along with the third ventricle. The lateral ventricles are connected with the third ventricle by the interventricular foramen (of Monro). The fourth ventricle develops from the cavity within the hindbrain, and it is connected to the third ventricle by the cerebral aqueduct. The ventricular system contains cerebrospinal fluid (CSF), which is produced mainly by the choroid plexus. Various pathological processes, such as constricting effects of brain tumors, inflammation, or swelling from trauma can block the interventricular formina or the cerebral aqueduct. Thus, the flow of CSF out of the formamina of Lushka and Magendie may be blocked. The interventricular formamina and the cerebral aqueduct are especially vulnerable to such pathological processes due to their narrow size. Furthermore, hydrocephalus may develop if ventricular size increases before the bones of the skull are fused together (Berg, 1996).

Children shunted for hydrocephalus have been found to have lower overall intellectual and academic abilities, impaired nonverbal abilities, and variable skills in the areas of executive functions, attention/concentration, deficient social maturity along with motor/tactual abilities (Billard, Santini, Gillet, Nargeot, & Adrien, 1986; Dennis et al., 1981; Fletcher, Bohan et al., 1992; Fletcher, Francis et al, 1992;).

The development and placement of neurons is an essential process in early brain formation (Baron, Fennell, & Voeller, 1995; Spreen, Risser, & Edgell, 1995). Neuroependymal cells divide in the ependymal layer or ventricular zone and migrate to the outer layers including the mantle and marginal layers. Cell generation occurs by mitosis within the ventricular zone, according to a fixed sequence between the second and fourth months of gestation. As a result of this process, neuroblasts, or nerve cell precursors, and glioblasts, or glial cell precursors, are formed. There are various pathological processes that can occur during the proliferation stage, which may result in developmental or learning problems in the young child. Specifically, microcephaly, which is due to a disturbance in cell proliferation, is remarkable for lowered intellectual functioning and small brain size (Rourke, Bakker, Fisk, & Strang, 1983; Volpe, 1977). There are familial and sporadic cases (e.g., alcoholism, diabetes) of microcephaly vera. Macrocephaly is another disorder of proliferation consistent with a well-formed brain, but of larger than normal size. Development and intellectual abilities are compromised in approximately half of these cases (Volpe, 1977).

Upon completion of the proliferative stage, the neuroblasts move to permanent locations within cellular zones of the developing cortex. This process of migration determines the ultimate destination of neurons. The cortex is composed of four embryonic regions including the ventricular, marginal, intermediate, and subventricular zones (Martin, 1996; Spreen, Risser, & Edgell, 1995). The neural tube is lined by the ventricular zone which contains mitiotic cells, and the marginal zones which contain the cellular processes. As proliferation continues, an intermediate zone of neurons begins to form. By the eighth to tenth week of gestation, the intermediate zone expands to include the cortical plate, from which the cortex will develop, and the subventricular zone. The subventricular zone or the subependymal zone is a secondary zone in which cell proliferation may continue after birth. Golgi type II neurons along with glial cells may be generated from this zone. The process of migration involving the cortical plate occurs in an inside-out pattern. As such, neurons migrate to the deepest layer of the cortex (i.e., the sixth layer) and subsequent neurons must migrate past these cells to higher levels. Thus, the top layer of the cortex is the last layer to be formed. Conversely, migration in the cerebellum occurs in an outside-in pattern (Spreen, Risser, & Edgell, 1995). Importantly, the processes of cell proliferation and migration vary across stages and areas of cortex during the development of the nervous system.

There are various disorders of migration, which occur secondary to either a genetic disturbance or extrinsic factors, such as a viral infection. Schizencephaly is the most severe of the migration disorders, and involves agenesis of a portion of the cerebral wall (Aniskiewicz, Frumkin, Brady, Moore, & Pera, 1990; Volpe, 1977). As such, the cortical plate is remarkable for large neuronal heterotopias (i.e., misplaced or displaced areas of gray matter within the ventricular walls or white matter). Neuronal heterotopias are the least severe and most common type of migration disorder. Neuronal heterotopias usually are present in the other migration disorders and typically result in intellectual deficiencies. Children with schizencephaly generally have seizures, spasticity, and severely delayed neurological development. There are several other disorders of migration, which involve abnormalities in the development of gyri and sulci, including lissencephaly (few or no gyri), pachygria (abnormal, broad gyri), and polymicrogyria (mulitude of small gyri) (Baron, Fennell, & Voeller, 1995). All of these disorders are remarkable for impaired neurological development.

After migration, neurons undergo neuronal specification and differentiation. First, neurons aggregrate or form major cellular masses with neighboring cells. After aggregation, the neurons begin to differentiate, which is a four-step process (Spreen, Risser, & Edgell, 1995). This process includes the development of the cell body, selective cell death, axonal and dendritic development, and synaptogenesis. Various pathological processes may occur at any stage of differentiation. For example, programmed cell death is critical for the normal development of the nervous system. Many more neurons are produced by the nervous system than are required, and only a limited number survive, which results in a "fine-tuning of neural wiring" (Spreen, Risser, & Edgell, 1995). During axonal and dendritic development, connections between neurons are made in order to establish communication within the system. Abnormalities in cell death and/or the growth of axons and dendritic spines generally result in abnormal development.

Myelination, which involves the formation of myelin sheaths around axons, begins in the spinal cord by the third month of gestation and continues until adulthood (Rourke, Bakker, Fisk, & Strang, 1983). Within the cortex, the sensory areas of the brain myelinate before the motor areas, which parallels the order of the functional development of sensory and motor systems (Rosenzweig & Leiman, 1989). During the fetal period, the rate of myelination correlates directly with an increase in brain weight (Lemire, Loeser, Leech, & Alvord, 1975). Although myelination continues into adolescence, the bulk of the myelination process occurs shortly after birth (Rosenzweig & Leiman, 1989). Myelination serves to increase the conduction velocity, lowers the action potential threshold, and increases a neuron's ability to carry impulses, thereby increasing the efficiency of the overall neuronal system (Spreen, Risser, & Edgell, 1995). The process of myelination is correlated with major developmental milestones. At birth, only a few tracts are myelinated, including those that mediate the sucking reflex. As the infant matures, he or she will be able to lift chin (1–2 months), sit (7–8 months), crawl (9–10 months), stand and walk (12 months), depending upon the myelination of various motor tracts (Dodge, 1975). During early childhood, the child will eventually be able to dress independently and have voluntary sphincter control.

During birth, many complications or traumatic events can occur (e.g., prematurity, low birth weight, infections, toxic damage, nutritional disorders, anoxic episodes, traumatic brain injury) which may disrupt normal development (Spreen, Risser, & Edgell, 1995). The weight of the brain increases fourfold in weight and size between birth and adulthood (Rosenzweig & Leiman, 1989). Specifically, the brain becomes 1 kg heavier during the first 15 years of life (Rourke, Bakker, Fisk, & Straing, 1983). Brain growth is facilitated by an increase in myelination, the formation of synapses and dendrites (increase in length of dendrites), production of neurons after birth (only smaller nerve cells and those around the subventricular zones), and the formation of glial cells (Rosenzweig & Leiman, 1989). Additionally, the cortex becomes more developed and specialized during this time with increases in cortical thickening and cortical fissuration (Rourke, Bakker, Fisk, & Strang, 1983). Primary sensory and motor areas are the first areas to develop, followed by the progressive development of sensory, parietal, and temporal association areas, respectively (Spreen, Risser, & Edgell, 1995). The development of the brain is influenced during this time by both instrinsic (genetic control) and extrinsic factors (environmental factors). As such, during this early stage, the developing brain of the infant is susceptible to environmental toxins, malnutrition, and other negative influences.

The above developmental course may be significantly altered if a child sustains an injury or trauma during his or her first few years. The age of the child, or more importantly, the developmental level at the time of injury, affects the behavioral outcome secondary to the trauma (Kolb, 1989). Brain plasticity, or the ability of the brain to modify neural mechanisms in response to a brain injury, can alter the expression of the trauma (Bach-y-Rita & Bach-y-Rita, 1990). According to Chelune and Edwards (1981), various factors impact a child's outcome secondary to a traumatic injury, including age or maturational level of the brain; type of lesion and how it affects the brain's growth processes; biological and behavioral mechanisms of recovery available to the individual at the time of injury; premorbid level of ontogenetic interactions with the environment; residual strengths and weaknesses that affect the acquisition and attainment of future developmental experiences; and environmental conditions likely to facilitate or impede the developmental process. With regard to age effects, Kolb and Whishaw (1996) reported that there are three critical age periods with children less than 12 months of age demonstrating greater functional impairment than lesions incurred at a later age. Specifically, children between the ages of 1–5 years have a greater chance for reorganization of brain function, while children older than 5 years have less of a chance for successful reorganization. For example, if a child sustains a traumatic injury in the left temporal area at age 3, some aspects of language development may be severely affected. During the ages of 2–5, children generally experience a significant increase in average sentence length and use morphological, syntactic, and linguistic rules for constructing and understanding sentences (Paul & Cohen, 1982). As such, trauma at this stage may interfere with the normal development of expressive and receptive language. However, if the injury occurs before language is fully developed, undamaged tissue in the right hemisphere may assume the functions of the left damaged tissue, even if at some expense to the replacement areas (Chelune & Edwards, 1981). For example, visual-spatial or visual-perceptual areas, which are normally mediated by the right hemisphere may be compromised if the right hemisphere assumes responsibility for the development of language functions (Kolb & Whishaw, 1996).

Implications of neurodevelopment for assessment of major neuropsychological functions

Following is a description of crucial areas of neuropsychological, cognitive, and academic processing, that need to be assessed when undertaking functional analyses of children for the purpose of working within the school setting. The final analysis and integration of these processing areas can be presented in a variety of hierarchies, depending upon the specific results obtained and the role and background of the consumer.

Motor functioning

Assessment of lower level abilities, such as motor speed and strength, tactile discrimination, and rapid fine-motor manipulation skills, are necessary in the comprehensive neuropsychological evaluation of children. Not only are these skills required at a very basic level for functional daily living skills (e.g., dressing or meal preparation), but also for the wide range of graphomotor tasks required by school-aged children. In the early school years, sensory/motor deficits can affect preschool and kindergarten functions ranging from cutting and pasting shapes to learning the formation of printed cursive letters. Continued deficits in this area can impact the speed of test taking or the rapid production of lecture notes. Congenital or acquired lesions to the sensory and motor regions of the brain can result in deficits that range from the global motor impairment indicative of cerebral palsy to more subtle sensory/motor deficits

Visual-perceptual and spatial functioning

Spatial ability is defined as the "ability to represent and organize the surrounding environment into a coherent spatial framework, to integrate visual percepts, to attend to specific locations in space, and to manipulate objects, either visually or tactually" (page 5, Morrow & Ratcliff, 1988). There is a wide range of manifestations of spatial disorders, including visuoperceptual, visuospatial, and visuoconstructive disorders (Benton & Tranel, 1993). Impairments in any one of these areas may adversely affect a child's ability to learn or function within the environment. For example, a child may experience a visual field cut or neglect, or impaired visual scanning or figure-ground differentiation that may impair his or her ability to read or copy information from the blackboard. A child may not be able to compute written arithmetic problems presented in a horizontal or vertical orientation due to a visual-perceptual deficit affecting the organization of the printed materials. Furthermore, a child may not be able to perceive or differentiate between various letters or arithmetic symbols. Individuals with these types of visual-perceptual deficits often perform poorly on tasks requiring quick and accurate visual scanning, visual form discrimination and form matching. These skills especially impact upon early acquisition of pre-academic concepts. Furthermore, a child may have difficulty recognizing or knowing the meaning of objects despite normal visual perception and language, which is referred to as a visual agnosia (Damasio, 1985).

Children may also demonstrate visuospatial or visuoconstructive disorders, which may compromise their ability to learn or function independently within the environment. For example, a child may have difficulty describing the spatial arrangement of a particular room, describing the travel routes to certain places, or indicating the locations of cities on a map, secondary to either a posterior right hemisphere or bilateral lesion (De Renzi, 1997). They may also demonstrate impaired directional orientation or localization of points in space, a devastating disability in a child who has to navigate the corridors of a complex school building. Children with visuoconstructive disorders may be impaired in their ability to copy various figures or line drawings, as well as adversely affected in their ability to write or draw. Therefore, in addition to formal measures of one's spatial-constructional and graphomotor abilities, naturalistic examples of functional writing and drawing skills should also be examined.

Children with the above disorders are, with proper documentation, entitled to appropriate supports, services, accommodations, and adaptations within their learning environment. Such supports can be as simple as moving a page within the child's visual field. An example of a more complex case involved a 15 year old who, while attending the ninth grade, was failing biology. Neuropsychological testing revealed superior verbal intelligence in the presence of severely impaired spatial constructional skills, finger agnosia, and fine tactile discrimination deficits. Analysis of handwriting also revealed the presence of dysgraphia. Examination of all of the aspects that were contributing to overall grades in biology class yielded a potential functional relationship with her neuropsychological profile. Specifically, her grade in this class was comprised of a fifty percent contribution of quizzes, midterm, and final test scores, and a fifty percent contribution of laboratory projects. Although this youngster was receiving scores in the upper ninety percent to the one hundred percent range on quizzes and tests (i.e., comprised of multiple choice, fill in the blank, and short written answer questions), she was also consistently receiving failing grades for laboratory projects. These projects, as one might guess, were typically of the form that included examining, under a microscope, a slide of some type of cell or organism, and then drawing what was observed and labeling its component parts. It was this primary requirement of laboratory assignments (i.e., reproducing the observed slide) that resulted in failing this aspect of class, and thus reduced her overall grade. Appropriate educational adaptations and modifications resulted in a primary weighing of her exceptional test performances, along with modified laboratory assignments (e.g., labeling the components of a cell or organism from a pre-printed magnification of the specimen in question).

In terms of direct supports and services to a student such as described above, although not able to directly determine the educational need for a specific amount, frequency or form of physical, occupational, or vision therapy, the pediatric neuropsychologist can initiate a referral for such evaluations, and support or refute the recommendations derived from such assessments. Thus, pediatric neuropsychologists can be in the position to make the logical/empirical link between motor, visual/perceptual, and graphomotor skills, and the array of related functional, adaptive, and academic skills that derive therefrom.

Attention

Disturbances in attention may differentially affect a child's academic progress depending upon the type of attentional problem he or she is experiencing. These disturbances may take the form of difficulty with sustained attention, divided attention, or selective attention (Gaddes & Edgell, 1994). Children with poor sustained attention have difficulty maintaining their attention for a specified period of time while completing a particular task. Visual or auditory continuous performance tasks are often used to assess sustained attention. For these tasks, visual or auditory stimuli are sequentially presented on a screen or tape recorder, respectively, over a period of time, with instructions for the recipient to indicate in some way (e.g., press button, raise hand) when a given number or letter is perceived (Barkley, 1990; Lezak, 1995). On divided attention tasks, a child may have difficulty maintaining his or her attention between two tasks or entertaining multiple cognitive concepts simultaneously, such as alternating between number and letter sequences. Finally, on tasks of selective attention, a child is directed to focus attention on a specific stimulus within his or her environment in the presence of distracting stimuli. Selective attention is usually measured by cancellation tasks or visual search tasks in which a child must cross out a target letter or number within an array of letters and numbers (Lezak, 1995). Furthermore, many children with attentional disorders frequently have difficulties in the areas of executive functioning including difficulties in planning and organization, sequencing, and inhibition of inappropriate response tendencies (Denckla, 1989).

As outlined above, there are different types of attention, mediated by various systems within the brain. The reticular activating system, which is located within the brainstem, is responsible for maintaining a general level of arousal or consciousness (Kolb & Whishaw, 1996). From the brainstem, fibers extend upward to the thalamus or to other areas of the cortex, regulating various aspects of attention (Mesulam, 1985). Thus, specific brain regions are interconnected and organized into systems, and a lesion or injury to any part of the system can result in different expressions of attentional problems. The attentional system appears to be hierarchically arranged in a vertical direction with the subcortical system being responsible for general arousal states, whereas the cortex, which is at the top of the hierarchy, is responsible for the fundamental control of attention (Fisher, 1998; Mesulam, 1985). Anterior areas of the brain (prefrontal cortex) have been implicated with motor functions and are involved in response preparation and initiation of action, whereas the posterior brain systems are involved with sensory selection (Fisher, 1998; Kolb & Whishaw, 1996; Mesulam, 1990).

Traumatic brain injury or other medical problems can disrupt the "normal" operation of the attentional systems within the brain, resulting in varied expressions of attention disorders. Thus, children may have an attention disorder secondary to some type of traumatic injury or medical illness or they may have a developmental form of an Attention-Deficit/Hyperactivity Disorder (ADHD). As described within the Diagnostic and Statistical Manual of Mental Disorders-Fourth Edition (DSM-IV), a diagnosis of ADHD is defined by persistent difficulties in the areas of attention, impulsivity, or hyperactivity, that exceed that typically observed in individuals at a comparable level of development. Presently, ADHD is divided into three categories: predominantly inattentive type; predominantly hyperactive-impulsive type; and combined type. For a diagnosis of ADHD, symptoms must have been present since at least age 7, occur across at least two different settings, and significantly interfere with social, academic, or occupational functioning.

The prevalence of ADHD ranges from 3–5% of school-aged children, and it is more common in males as well as the first-degree biological relatives of children with ADHD (DSM-IV). Barkley (1995) has proposed that ADHD is a disorder of response inhibition and executive function involving deficient self-regulation, impaired cross-temporal organization of behavior, impaired directing of behavior toward the future and away from the moment, and diminished social effectiveness and adaptation. Many different etiologies for ADHD ranging from neurological (brain dysfunction, neurochemical abnormalities, neurological immaturity), genetics (familial-hereditary transmission, genetic anomalies), biochemical factors, psychosocial (chronic stress, poor child management, family dysfunction), diet and allergies, and toxins (prenatal exposure to alcohol, nicotine and lead) have been investigated (Barkley, 1995; Frank, 1996). The disorder is believed to have a genetic influence as documented by a postive family history of ADHD in family studies (Biederman et al., 1986; Cantwell, 1972). Twin studies and adoption studies also provide evidence for the heritability of ADHD (Goodman & Stevenson, 1989; Morrison & Stewart, 1973). Some researchers also view ADHD as being a biochemical disorder involving an imbalance of neurotransmitters (Zametkin & Rappaport, 1987). Finally, the cortex, especially the pre-frontal area, has been implicated in ADHD. At the present time, it appears that given the ongoing research in the area of ADHD, we are likely to continue to learn more about the various subtypes and associated comorbid conditions, and thus better understand the complex interaction between the genetics, neurochemical actions, and associated brain regions underlying this disorder (Pennington, 1997).

Children with attentional problems often have difficulty within the classroom, as they are unable to maintain their attention for extended periods of time or they become easily distracted by extraneous stimuli within their environment. As such, they may be unable to remain focused while completing or learning new lessons. Consequently, these individuals may fall behind their peers in their learning and have gaps in their knowledge base. These children may disrupt the classroom by talking out of turn, interrupting classmates, introducing irrelevant topics of conversation to the class discussion, or engaging in off-task behaviors (Ylvisaker, Szekeres, & Hartwick, 1994). Furthermore, these children may have difficulty with those organizational skills necessary to complete short-term and long-term classroom projects or homework assignments. They may also have difficulty organizing their ideas within a story or report. Another potential functional problem is the inconsistency often exhibited in children diagnosed with this disorder.

Finally, as described by Lorber (1994), one or more of the overt symptoms associated with the presence of attentional difficulties displayed by children may be the result of related neuropsychological, cognitive, or emotional factors. Overt inattention may, for example, result from either a central auditory processing disorder (see below) or a significant degree of underlying anxiety. Neurologically-based inattention may also be misdiagnosed in children with severe memory disorders or language-based disabilities. Therefore, the diagnosis and recommended treatment approaches within school settings must also be accompanied by comprehensive assessment of those related factors whose direct expression, or behavioral reaction to their existence, can lead to overt manifestations of behaviors interpreted as inattention or distractibility.

Auditory processing

To better understand the complexity, impact, and remedial interventions for central auditory processing disorders (CAPD), an area of study not necessarily familiar to pediatric neuropsychologists, the recent work of Ferre (1997) entitled *Processing power: A guide to CAPD assessment and treatment* can be extremely helpful. Ferre describes a three stage model of central auditory processing skills, which includes auditory attention and arousal, auditory sensory reception, and output planning related to auditory processing skills. Specifically, Ferre characterizes auditory attention and arousal to include "selective attention or the ability to attend to a target in background or competing noise; arousal or attention to a new auditory signal; and the ability to localize a sound source in space". Auditory sensory reception is described to include "signal detection, short-term storage, discrimination, recognition, identification, sensory (acoustic) analysis, perception, information association, integration, and coordination". Finally, Ferre describes the process of output planning related to auditory processing skills as including "integration, long-term memory, recall, verbal retrieval, and information organization and sequencing". Although neuropsychologists are typically not able to formally diagnose the presence of either a peripheral hearing loss or CAPD, screening of these and related functions are essential in the school-aged individual for three major reasons. First, such screening is an excellent means for determining the need for a referral to a consulting audiologist for a comprehensive audiological evaluation. Assessment by an audiologist skilled in the diagnosis and treatment of CAPDs, is essential in the appropriate treatment of individuals for whom this disability comprises even a small part of their neuropsychological profile (Bellis & Ferre, 1996).

Neuroanatomically, the auditory system can be divided into two components, the peripheral and the central mechanisms. The peripheral system accounts for the detection of auditory signals and the transformation of the signals into electrical impulses. The central auditory system begins at the level of the brainstem, with ascending fibers that travel both ipsilateraly and contralaterally to the primary auditory cortex (i.e., Heschl's gyrus). It is this combination of ipsilateraly and contralaterally fibers that, in part, account for timing and intensity differences that allow for the localization of sounds. The secondary auditory cortex includes Wernicke's area, and with its associated linguistic functions (see description to follow), is in part responsible for associating meaning to the auditory signal. In addition, the accurate processing of auditory information involves Broca's area and the issula, both of which aid, via connections to the primary and secondary auditory cortices, with the integration of auditory information. Finally, fibers within the corpus callosum allow for the integration of interhemispheric and multimodal information.

As discussed later within this chapter, a detailed patient history can be crucial, for example, given that the presence of chronic ear infections in the young child has been related to the later presence of CAPDs, even when individuals are able to "pass" pure-tone audiometric evaluations (Fredricks, 1969). Therefore, the fact that a child has received and passes a "hearing screening" at school does not negate the possibility of the presence of a significant CAPD.

Ferre (1997) describes a number of overt and observable characteristics that have been identified in children with CAPD, including: difficulty understanding speech in noisy environments; difficulty hearing in groups; difficulty listening; short attention span; appearing anxious or stressed when required to listen; distractibility; difficulty following directions; seeming to hear but not understanding what people say; difficulty remembering information presented aurily; impaired speech and language skills; impaired phonetic decoding and encoding (spelling); poor reading skills; impulsive behaviors; disorganization; poor peer relationships; and poor self esteem.

Memory

There is limited research on developmental memory disorders within a pediatric population. Pennington (1991) suggested that researchers have not focused upon this aspect of empirical study because children with a pervasive amnestic syndrome would likely present as being mentally impaired due to their inability to learn or retain novel information. Pennington also suggested that brain areas implicated in memory functioning are the last regions of the brain to develop, and as such, these brain areas can escape early trauma or insult. Due to the limited information regarding developmental memory disorders, such research in children has focused upon acquired memory deficits secondary to such traumatic injuries as a closed head injury, seizure disorder, infection, tumors, anemia associated with sickle cell disease, or stroke (Squire & Shimamura, 1996; Pennington, 1991).

When children are referred for a neuropsychological assessment, memory problems are not a common referral question, although they may be the significant underlying problem. Therefore, comprehensive testing should include measures of verbal and nonverbal memory functions, and especially in children, the ability to store and retrieve novel and old-learned information. It is this latter function that is rarely assessed in typical school-based evaluations. According to Pennington (1991), common symptoms associated with the presence of an underlying memory disorder can include attentional problems, impulsivity, emotional lability, language problems, declining intellectual functioning, school failure, confabulation, poor self-esteem, and learning disabilities.

Similar to adults, children attend to and process information in their environment through a sensory register (Gaddes & Edgell, 1994; Mesulam, 1990, 1985). Information that reaches the sensory register is basically unanalyzed information. Unlimited information can be processed within the sensory register, although the child will divert his or her attention to the most salient information. A breakdown in memory can occur at this stage if a child has an attentional disorder, as the proper information will not be encoded into the register for later processing (Squire & Shimamura, 1996). From the sensory register, information is transferred to short-term memory or immediate memory if it is recognized as familiar or meaningful. Information is transformed into different formats, either auditory or visual, when it reaches the short-term memory stores. A limited amount of information may remain in a child's short-term memory store, although this range of information increases with age, especially between ages eight and nineteen. Most individuals can remember seven digits, plus or minus two, and four or five words (Gaddes & Edgell, 1994). Information is lost from the short-term memory store within a few seconds, if it is not rehearsed. Polymodal sensory areas of the posterior parietal cortex, posterior temporal cortex, and frontal lobe appear to be implicated in short-term memory functioning.

From short-term memory, information is transferred to long-term memory. However, information is processed separately and simultaneously within the short-term and long-term memory stores (Kolb & Whishaw, 1996). The transfer of information from short-term to long-term memory is mediated through elaborative rehearsal, and the information is later retrieved from long-term memory through retrieval cues. The more efficiently the information is initially organized or processed, the easier it will be to recall the information at a later point in time. Long-term memory is divided into two categories including procedural memory (i.e., memory for skills such as riding a bicycle) and declarative memory (i.e., memory of facts) (Bauer, Tobias, & Valenstein, 1993). Declarative memory is divided into episodic (autobiographical information) and semantic memory (knowledge of meaning and facts). It appears that the cerebellum and basal ganglia are implicated in procedural memory, while the hippocampus, amygdala, rhinal cortex, and prefrontal cortex are purported to mediate declarative memories (Kolb & Whishaw, 1996).

As outlined above, there are different types of memory systems that are mediated by specific networks or processes within the brain. For example, if a child experiences a traumatic injury to the medial temporal cortex, memory problems affecting the acquisition of novel information or anterograde amnesia can result. Conversely, damage to the insula of the temporal lobe and the medial frontal cortex can result in disordered memory for previously learned information, or retrograde amnesia (Kolb & Whishaw, 1996; Squire, 1986). Furthermore, a child may experience memory deficits for either verbal or nonverbal information, depending upon the lesion site. Generally, children who are right-handed or left hemisphere dominant for language experience verbal memory deficits secondary to a left-sided lesion. A child may experience a breakdown in memory at various junctures, ranging from the sensory register, short-term memory store, long-term memory store, or recognition memory. However, research suggests that most children experience short-term and/or long-term memory problems secondary to faulty encoding or processing of information (Gaddes & Edgell, 1994).

It is essential for educators and neuropsychologists to have a good understanding of the various types of memory disorders, and their potential impact upon a child's ability to learn, as they directly affect the establishment of appropriate educational accommodations, adaptations, and modifications within the learning environment. For example, a child with a memory disorder may have difficulty completing assignments, fail to bring necessary materials to class, or forget to turn in homework assignments. As such, compensatory strategies to address such organizational difficulties can include use of a memory notebook or assignment book. However, given that a child can just as readily forget to employ this compensatory strategy, it is often equally important to establish an appropriate behavioral intervention to help create the habit of employing the technique. School personnel, parents, and other caretakers need to work together to optimize a consistent approach to addressing a child's needs in this aspect of their functioning. For example, the child's teacher may have to check in with the child after each class period and at the end of the day to ensure that assignments were recorded properly and/or the child takes the assignment book home at the end of the school day. A child may also have difficulty learning school rules, routes within the school, or remembering the names or faces of their peers or teachers (Ylvisaker, Szekeres, & Hartwick, 1994). Furthermore, a child may fail to retain basic concepts or specific aspects of the curriculum, which could adversely affect academic progress. This difficulty can result in isolated gaps in acquired knowledge, or lack of acquisition of whole aspects of this curriculum, if sequential building blocks of component concepts or skills are required.

Memory accommodations can include "multi-sensory" learning approaches. This technique can include the simultaneous presentation of concepts or skills via visual, auditory/verbal, and even tactile modalities. Such an approach can facilitate a child's learning by providing different neural sites to draw upon when attempting to recall the desired information. For example, if a child is experiencing visual-spatial memory problems, learning of novel visual information could be facilitated by either relating the information to personal life experiences or using semantic knowledge of basic scripts (Ylvisaker, Szekeres, & Hartwick, 1994). Similarly, a young child who is experiencing difficulty remembering how to form letters, can associate letter forms with animal shapes to facilitate their production (Connell, 1985). Further, classroom and testing accommodations for a child with impaired memory skills can include alternative methods of testing a child's knowledge. For example, the use of multiple-choice questions provides recognition cues to assist in the retrieval process, rather than requiring unaided recall.

Language

In order to develop a basic understanding of childhood language disorders, it is essential to review normal language development. Generally, children acquire language skills in a predictable pattern, although a variety of factors (e.g., biological, psychological, social, and environmental) may alter the rate and expression of language acquisition in a particular child (Rice, 1989). According to Paul and Cohen (1982), infants between the ages of zero to two months respond to sounds within the speech frequency zone, and have an early preference for speech over other rhythmic sounds. Between nine and twelve months, attempts at nonverbal communication are made through the use of gestures and sounds. The young child uses his or her first recognizable words between 12 to 18 months. Between 18 to 24 months, two words are combined to form simple sentences, and thus the child begins to use language to request information. Average sentence length increases from two to four words or more between the age of two and five years. Furthermore, the child begins to master morphological and syntactic rules within sentences and use linguistic rules for understanding sentences. At this level, children begin to employ language for conversational purposes. Between ages five and twelve, language becomes increasingly more complex, including the development of metalinguistic awareness (e.g., an internalized understanding of the more complex aspects of language).

Abnormalities within this hierarchical pattern of linguistic development are classified as being either a developmental or an acquired language disorder. Developmental language disorders occur when a child either fails to develop normal language skills or is delayed in language acquisition, whereas acquired aphasia of childhood refers to an abnormality in language secondary to a traumatic event or disease process after a relatively normal period of language development (Benson & Adila, 1996). Developmental language disorders may be secondary to an anomaly or trauma within the left hemisphere of the developing brain during the prenatal, perinatal, or postnatal period before the development of formal linguistic skills. Research has documented either reversed asymmetry or lack of asymmetry in posterior perisylvian areas of the brain in children with delayed language, whereas in most individuals, the left temporal plane is larger on the left as compared to the same area on the right (Geschwind

& Levitsky, 1968; Plante, Swisher, & Vance, 1989; Plante, Swisher, Vance, & Rapcsak, 1991). Galaburda and Kemper (1979) found evidence of heterotopias in the left perisylvian areas of adult brains in post-mortem studies of individuals with significant language impairments and dyslexia. It has also been documented that children with left hemisphere dysfunction often have specific language deficits including poor speech production and difficulty with semantic and syntactic comprehension (Rankin, Aram,& Horowitz, 1981). Furthermore, children with a history of a left hemispherectomy have been found to have problems with syntactic comprehension and production (Dennis, 1980). It has also been found that children with developmental language disorders have a positive family history of such disorders (Stromswold, 1997). Specifically, Stromswold (1997) indicated in her review of 12 studies of children with language disorders that 30–70% of immediate family members and 9–42% of relatives had a higher percentage of language disorders as compared to controls. Finally, developmental language disorders are often seen in conjunction with other disorders (e.g., hearing loss, mental retardation, chromosomal abnormalities, emotional disturbance, autism, epilepsy) as well as secondary to environmental factors (e.g., social deprivation, inadequate instruction) (Robinson, 1991; Spreen, Risser, & Edgell, 1995).

Children with developmental language disorders may display such language abnormalities as impaired expression, reception, phonology, central processing, or pragmatics. Disorders of expressive language are remarkable for limited speech output and range of vocabulary, difficulty recalling words, difficulty producing sentences with developmentally appropriate length or complexity, and simplified grammatical structures and sentence types (DSM-IV). Children with expressive language disorders may also experience poor speech motor skills which may adversely affect the child's ability "to sequence, temporally organize, and control movements and postures of the peripheral speech mechanism" (Hodge, 1991 p. 69). Children with poor receptive language experience difficulty understanding words, sentences, or specific types or classes of words. Receptive language disorders could also be related to a hearing loss, poor discrimination between auditory signals, and poor decoding of the meaning of the phonological aspects of speech (Spreen, Risser, & Edgell, 1995). Expressive language difficulties usually accompany receptive language disorders, as the development of expressive language skills is dependent upon the acquisition of receptive language skills (DSM-IV).

In cases of acquired childhood aphasia, in which the child had some degree of normal language function prior to an injury, the prognosis and course of the disorder is dependent upon injury severity and location, as well as level of premorbid language development. If a child sustains an injury in the speech areas of the left hemisphere at an early age, during the first year or two, there is a strong possibility that language will proceed to develop in other areas of the brain (Locke, 1992). For example, linguistic skills can develop in the right hemisphere or in both hemispheres due to the plasticity of the developing brain of a young child. As such, the child may demonstrate only a mild language disorder or may eventually display normal language function in spite of injury site. Benson and Ardila (1996) emphasized the importance of the age of the child at the time of injury, reporting that the earlier the age of the trauma, the greater the likelihood for the nondominant hemisphere to assume language function. As such, there is an increased probability for a better and more complete recovery.

Much of our knowledge of acquired aphasias of childhood is drawn from extrapolation of studies of adult patients who have identified localized lesions. However, most classification systems have been questioned, as specific language deficits do not always correlate with lesion location. Nevertheless, it is generally accepted that a spoken word is relayed to Wernicke's area from the auditory cortex, and then transmitted to Broca's area for speech production (Geschwind, 1972). In reading, information is relayed from the primary visual areas to the angular gyrus, where the visual pattern is transformed into the auditory form of the word. When spelling a spoken word, the auditory pattern is transmitted from the auditory cortex to the angular gyrus, where the auditory pattern is transformed into the visual form of the word. This model of language function has been instrumental in understanding developmental and acquired language disorders of children (Hynd, & Willis, 1988). However, due to the plasticity of the developing brain, the model is not necessarily directly applicable for individuals with brain conditions acquired in childhood. Furthermore, the severity and location of the trauma, along with level of premorbid language development, greatly affect the manifestation of childhood language disorders.

A comprehensive assessment of language function involves the measurement of speech, as well as receptive and expressive language abilities. Generally, following a comprehensive screening of linguistic skills, if deficits are found formally or informally, referral to a speech and language pathologist may be warranted. The quality of a child's speech is analyzed by the speech pathologist for articulatory imprecision, phonatory weakness, hypernasality, prosody, and slow or rapid rate of speech (Ylvisaker, Szekeres, Haarbauer-Krupa, Urbanczyk, & Feeney, 1994). Language should be examined for conversational language, repetition of spoken language, comprehension of spoken language, and word-finding ability (Benson & Ardila, 1996). Verbal output is further assessed according to expressive vocabulary level, phonology, morphology, syntax, semantics, and pragmatics. Impairment in any of the above areas can adversely affect learning, including reading, writing, and arithmetic functioning as well as social/emotional functioning and self-esteem. Montgomery and Levine (1995) reported that children with developmental language disorders experience processing deficits in other areas of cognitive functioning (i.e., auditory perceptual processing, problem-solving, and verbal short-term memory) which negatively impacts upon their learning as they progress in school.

Abstract reasoning and concept formation
According to Piaget's theories (1960), children develop higher-order forms of cognitive thinking as they mature. They progress through six stages of cognitive development, beginning with the sensorimotor stage through formal operations.

At the earliest level, children's thinking is independent of overt behavior (Spreen, Risser, & Edgell, 1995). However, they eventually begin to develop the ability to mentally represent actions and anticipate future experiences (Wood, 1981). At approximately age 12, children achieve the formal operations stage, that is the ability to think abstractly, problem-solve through deductive reasoning, and infer causal attributions (Piaget, 1960). The development of higher level thinking or abstract reasoning in early adolescence is correlated with the slower maturation of the frontal lobe as compared to other areas of the brain (Kolb & Whishaw, 1996).

Children under age 12 would not be expected to perform at a level commensurate with older individuals on test measures assessing higher-level problem-solving, abstract reasoning, response inhibition, cognitive flexibility, sequencing, and associative learning. Hence, the necessity of age-appropriate norms. However, as they mature, they slowly develop the necessary higher-level thinking to solve such problems. Importantly, if a child sustains an injury to areas of the brain that mediate such cognitive processes (e.g., the prefrontal cortex, including the dorsolateral prefrontal, inferior prefrontal, and medial frontal cortices), the proper development of higher level cognitive thinking may by disrupted. Furthermore, due to the fact that neighboring areas of the cortex may already be committed to the development of other functions, the possibilities for functional reorganization may be quite limited.

Individuals with frontal lobe dysfunction generally do well on general intellectual and basic linguistic measures, given that the breakdown of functioning occurs in application of higher order behavioral strategies or in approach to problem solving, rather than due to a lack of knowledge. Frontal lobe dysfunction has been associated with difficulties in the areas of regulatory functions (ability to initiate, modulate, or inhibit ongoing mental activity), executive functions (planning, goal-setting, and controlling behavior), and social discourse (Dennis, 1991). Furthermore, personality changes have been reported with frontal lobe damage including altered mood (exaltation or depression), lack of insight, apathy, indifference, distractibility, decrease in spontaneity and initiative, lack of judgment, irritability, slowness in thinking, and impulsivity (Stuss & Benson, 1984; Stuss, Gow, & Hetherington, 1992; Trimble, 1990).

Deficits in the above areas of cognitive processing can therefore interfere with all higher level aspects of a child's academic functioning. For example, the child may be able to learn to read, but not be able to comprehend the passage, as it may be beyond his or her level of abstract thinking or inferential skills (Ylvisaker, Hartwick, Ross, & Nussbaum, 1994). Furthermore, deficits in inductive and deductive reasoning can interfere with a child's ability to make appropriate generalizations about a set of principles, or infer causes from effects, respectively (Ylvisaker, Hartwick, Ross, & Nussbaum, 1994). These types of processing deficits can adversely affect a child's ability to acquire higher level mathematical concepts, as well as verbal concept formation. Furthermore, impairments in self-monitoring, along with higher level reasoning deficits, can negatively affect a child's ability to self-correct, acquire or demonstrate a skill or concept, engage in social problem-solving and follow school or societal rules. In addition, if a child is rigid and inflexible in approach to problem-solving, he or she may have difficulty considering alternative hypotheses or generating alternative courses of action when confronted with even common problems. Furthermore, children with abstract reasoning and logical problem-solving deficits frequently have difficulty engaging in goal-directed activities and generating appropriate planning strategies to achieve desired goals. They can also lack an appropriate level of initiative to execute the goals that they have generated (Ylvisaker, Hartwick, Ross, & Nussbaum, 1994).

Numerous forms of educational intervention have been proposed to address this form of cognitive processing deficit. As an example, for the child who experiences poor self-monitoring skills, educators could implement an "alarm" or "self-check" system, signaling the child to stop working and check over his or her output (Ylvisaker, Szekeres, Hartwick, & Tworek, 1994). Similarly, for those children with deficient problem-solving skills, a problem-solving guide could be employed to assist the child's learning. For example, Ylvisaker, Szekeres, Hartwick, and Tworek (1994) suggested that a child be instructed in "self-questioning" for alternative responses. Furthermore, these researchers proposed that children should be taught to look at possible solutions from at least two different perspectives, scan environmental cues for appropriate behavioral responses, and set specific times or places for situation specific-behaviors. As outlined above, it is readily apparent that difficulties in verbal abstraction and higher-level concept formation underlie all aspects of learning, with increasing impact as the academic curriculum moves from rote memorization to the synthesis and integration of novel higher level concepts and skills. Clinically, it is not unusual to observe children with such deficits succeeding fairly well at the lower grade levels, only to begin experiencing increasing difficulty as the academic curriculum becomes more complex.

Reading

Reading disabilities are the most common and frequently diagnosed learning problem among school-age children (Beitchman & Young, 1997; Lyon, 1996). The exact prevalence of reading disabilities is unclear due to the comorbidity of other learning and behavioral disorders, but it is estimated to occur in approximately four percent of the population of school-age children (DSM-IV). The definition of learning disabilities, including reading disabilities, and the eligibility criteria for receiving special education services, are typically relatively well documented at the level of State Boards of Education (e.g., Illinois State Board of Education, 1990). However, issues related to the etiology of reading disorders along with underlying concomitant neuropsychological processes are frequently overlooked when establishing the appropriate form of remedial intervention to address a given child's disability. This is especially important given that different empirically derived reading intervention strategies build upon and/or remediate different underlying cognitive processing issues (e.g., visual, auditory, rote memory). Therefore, the unique expertise and comprehensive perspective provided by a pediatric neuropsychologist can readily augment both the understanding of, and services provided to, a child in his or her learning environment, especially when evaluating children with acute or chronic medical disease. While most schools are quite proficient in diagnosing deficient reading skills, an understanding of the neurological and neuropsychological substrates of the disorder can be crucial in obtaining the appropriate type and intensity of special education services necessary to assist the child.

Reading disabilities can occur in those who have either sustained some type of traumatic injury after the age of 12 months, which may have affected various language or speech centers within the brain, or those children who experience developmental dyslexia (Gaddes & Edgell, 1994). Although developmental dyslexia has had many definitions, this term generally reflects partial or complete inability to acquire reading skills despite adequate instruction, underlying cognitive processing ability, and social, cultural, and environmental support and opportunity to do so. In addition, children with various medical disorders such as neurofibromatosis or Tourette's Syndrome, or those undergoing therapeutic regimens for the treatment of various medical disorders (e.g., irradiation therapy) can experience specific learning disabilities that may interfere with their academic progress (Cohen, 1991).

According to the DSM-IV, a reading disability is defined as "Reading achievement, as measured by individually administered standardized tests of reading accuracy or comprehension, is substantially below that expected given the person's chronological age, measured intelligence, and age-appropriate education". The disturbance must "...significantly interfere with academic achievement or activities of daily living that require reading skills..." and "If a sensory deficit is present the reading difficulties are in excess of those usually associated with it" (p. 50). Educators, school psychologists, pediatric neuropsychologists, other related mental health workers and school personnel, can all potentially diagnose reading disabilities according to specific guidelines established by state or federal agencies. The specific nature of these guidelines vary from state to state, but all conform to the general guidelines contained within federal regulations. Most state agencies have chosen to employ a discrepancy score model typically comparing a child's cognitive potential (e.g., intellectual functioning) to his or her reading achievement. This discrepancy score may be calculated according to a regression formula. Some agencies further require documentation of a specific processing deficit, which may underlie the reading disorder. Examples of such processing deficits include impairments in the areas of attention, memory, perception, visual-motor, conceptualization, organization, processing speed, or social perception (Illinois State Board of Education, 1990). The complexity that arises in the calculation of this apparently simple comparison of potential to level of achievement, and this is true for any area of academic achievement, is what specific measures of "potential" and achievement are employed. Specifically, previous research, as well as the process of psychometric test construction, have identified that global measures of intelligence and academic achievement tests were both designed, in part, to create stable results of a child's functioning over time. As such, both of these types of test measures can be insensitive to specific neuropsychological deficits (e.g., Taylor, Fletcher, & Satz, 1984). This is, in part, due to high intercorrelations within subtests and between individual test scores, indicating that the different tasks are to some degree measuring the same construct. For example, many individual intelligence test subtests measure multiple cognitive skills that can be processed and mediated both verbally and nonverbally. The fact that the same areas of cognitive impairment that determine academic deficits may also impact upon a child's performance on measures of intellectual functioning, greatly limits the use of an IQ score/achievement test discrepancy "formula" as a determinant of a learning disability (Doehring, 1978). Although impaired scores on intellectual and achievement tests may speak to one or more areas of generalized or specific cognitive deficit, they may not be able independently to identify specific areas of underlying neuropsychological difficulty or processing strength that account for a child's test performances. Therefore, test measures employed to determine cognitive potential may vary depending upon the child's underlying neuropsychological profile. As in the case of a child with a coexisting language disorder and fine-motor deficit affecting speed of manipulating objects, administration of a nonverbal intelligence test measure that does not require a timed motor response can be a more appropriate indicator of cognitive potential, than the typical intellectual test measures administered in schools.

Before one can fully understand the neuropsychological processes underlying reading disorders, it is essential to briefly review the neurobiology of reading disorders. Reading disorders are divided into two categories. Children with a developmental reading disorder are assumed to have a congenital defect affecting the process of reading, while children with an acquired reading disorder have sustained some type of traumatic injury after age 12 months which has interfered with their ability to learn to read (Gaddes & Edgell, 1994). Early studies of traumatically induced dyslexia within the adult population have led to a better understanding of reading disorders in the pediatric population. Research with adults has focused upon correlational studies between lesion location and functional impairment in reading. These early studies have implicated lesions within the splenium of the corpus callosum, primary visual cortex of the left hemisphere, and the left angular gyrus in acquired reading disabilities (Adams, Victor, Ropper, 1997; Albert, & Helm-Estabrooks, 1988; Damasio & Damasio, 1983; Geschwind, 1972). Specifically, visual information or words are presented to either the right or left visual field, and the information is subsequently processed in the left and right occipital areas, respectively. The visual information travels from the primary visual cortices to the left temporal lobe via the left angular gyrus. The angular gyrus is regarded as a "way station" in which abstract word forms or orthographic units are phonetically analyzed, which arouses the corresponding auditory form of the word in Wernicke's area. Orthographic units are also connected to semantic units, which allows for the comprehension of written words. According to Friedman, Ween, and Albert (1993), a model for normal reading is mediated by two primary systems: "One links orthographic information directly to the phonological system. The other accesses phonology from orthography via semantics" (p. 41).

In disorders of acquired dyslexia secondary to a severe traumatic injury within the adult patient, the reading process is disrupted from an already established premorbid level of functional reading ability. For example, if an adult patient sustains a lesion within the splenium of the corpus callosum, that patient may experience hemialexia, which is a rare condition, defined as the ability to read written text only when it is presented within the right visual field (Gaddes & Edgell, 1994). Furthermore, if a lesion occurs within the left occipital lobe and the splenium, word blindness or alexia without agraphia will result. Importantly, the ability to write, speak, or identify words by tactile and auditory modes is still intact, as Wernicke's area still maintains connections with the sensory and motor areas of the brain. Aphasic alexia commonly occurs in adults with acquired reading disabilities. These disorders range from deep dyslexia (semantic paralexias), surface dyslexia (errors of regularization or the pronunciation of irregular words with regular pronunciation), word-form dyslexia (words are recognized only after naming each letter of the word aloud), and phonological dyslexia (pseudoword reading is impaired) (Benson & Ardilla, 1996; Coslett, 1997; Friedman, Ween, & Albert, 1993).

In contrast to the acquired reading disabilities within an adult population, developmental reading disabilities and traumatically-induced reading disabilities in children are much more varied and nonspecific in their expression. The processes underlying developmental dyslexia are presumed to be much different than those of traumatic dyslexia. Multiple theories of underlying processes have been entertained as etiological factors in reading disabilities, including visual disturbances, cerebral dominance, perceptual-motor or auditory perceptual dysfunction, language disorders, and multi-sensory deficits. Currently, the most widely accepted view of reading disabilities is linked to a language deficiency, with a primary deficit in phonological coding (Mann & Brady, 1988; Stanovich, 1982; Vellutino, 1987; Wagner & Torgesen, 1987). As part of Velluttino's theory, a specific reading disability is associated with deficient phonemic segmentation and poor vocabulary development. Furthermore, Mann & Brady (1988) documented that poor readers have a faulty phonetic coding system, which adversely affects verbal memory, speech perception, naming ability, and sentence comprehension.

There are many neurobiological substrates, which underlie developmental dyslexia. Specifically, various genetic studies have reported that dyslexia has a familial risk of between 35% to 40% of first degree relatives, heritability of approximately 50%, and is heterogeneous in terms of the mode of transmission (e.g., polygenetic, recessive, and autosomal dominant; Pennington, 1995, 1991). Furthermore, dyslexia has been linked to chromosome 15 (Smith, Kimberling, Pennington, & Lubs, 1983). Developmental dyslexia has also been linked to aberrant neurological development during fetal development such as deviations related to asymmetry of the planum temporale and morphological anomalies (Foundas, Leonard, Gilmore, Fennell, & Heilman, 1994; Hynd & Hiemenz,

1997). Research based upon post-mortem studies has documented cases of symmetrical planum temporale within the superior posterior surface of the temporal lobe, along with focal cortical dysplasias in the perisylvian areas of the left hemisphere (Galaburda & Kemper, 1979; Galaburda, Sherman, Rosen, Aboitiz, & Geschwind, 1985; Humphreys, Kaufmann, & Galaburda, 1990; Hynd, & Semrud-Clikeman, 1989). Furthermore, MRI and CT studies have documented a greater symmetry of the length and area of planum temporale in dyslexics as compared to controls (Hynd, Semrud-Clikeman, Lorys, Novey, & Eliopulos, 1990; Larsen, 1990). These neurodevelopmental anomalies have been related to faulty processes during neuronal genesis and migration, occurring during fetal gestation. However, despite the wealth of knowledge regarding the correlational studies between asymmetry of the planum temporale and morphological anomalies, the underlying processes mediating this relationship are still not clearly understood.

Various neuropsychological processes underlie the nature and presentation of reading disorders. In the past, visual-perceptual disorders, such as defective scanning or visual spatial deficits, were regarded as primary explanations for children displaying reading disabilities. This was based upon the observation that children with reading disorders presented with letter reversals and mirror writing (Orton, 1925). In addition, poor readers were often thought to have eye-movement defects, including difficulty with binocular coordination, eye tracking, and directional scanning. It was hypothesized that these visual problems interfered with a child's ability to visually process written text. However, as previously mentioned, more recent research argues against the belief that visual-perceptual difficulties account for the majority of reading disabilities (Vellutino, 1987). Vellutino postulated that reading disabilities are due to a subtle language deficiency with a primary deficit in phonological processing. However, more recently, Eden, Stein, Wood and Wood (1995) provided support for the fact that a reading disability may, to some extent, result from dysfunction of the visual and oculomotor system.

There are various other processes associated with reading difficulties, ranging from defective auditory processing, short attention span, memory problems, poor sequencing, and abstract reasoning difficulty (Beitchman & Young, 1997). At the sight-reading level, a child with poor auditory processing skills can have significant difficulty discriminating, perceiving, and manipulating sounds, as well as in sustaining his or her attention in a distracting auditory environment. Verbal memory deficits have also been related to reading problems due to difficulties in employing phonologically-based codes to store verbal information (Torgesen, 1988, 1985). As such, the child may understand a passage, but have difficulty retaining the information from the passage, while considering questions about the material (Ylvisaker, Hartwick, Ross, & Nussbaum, 1994). If a child evidences difficulty with higher level thinking skills or verbal abstraction abilities, he or she may understand the basic nature of information provided in a reading passage, yet have difficulty answering questions requiring inferences (Ylvisaker, Szekeres, & Hartwick, 1994). As outlined above, a careful analysis of a child's reading, working memory, and reasoning skills is necessary, as the nature of the deficit will dictate the treatment course. It is essential for the pediatric neuropsychologist to educate school personnel regarding the nature and/or etiology of a child's reading disability, when such information has a direct bearing upon type and/or intensity of the proposed intervention services. Potential interventions may range from retraining impaired cognitive functions, teaching compensatory strategies, creating environmental compensations, or choosing appropriate instructional procedures (Ylvisaker, Szekeres, & Hartwick, 1994).

Writing

Disorders of written expression are generally diagnosed in combination with other learning disabilities. According to the DSM-IV, a writing disorder is defined as "writing skills, as measured by individually administered standardized tests (or functional assessments of writing skills), are substantially below those expected given the person's chronological age, measured intelligence, and age-appropriate education" (p. 53). This difficulty "significantly interferes with academic achievement or activities of daily living that require the composition of written texts (e.g., writing grammatically correct sentences and organized paragraphs)" (p. 53). A child's writing disorder may be secondary to an acquired or traumatic injury or may be due to a developmentally-based disorder. The term agraphia is used when there is a loss of the ability to produce written language, after a period of relatively normal functioning, secondary to brain damage (Benson & Ardila, 1996). The term dysgraphia is typically employed when the writing impairment is of a milder form. In the childhood form of the disorder in which the child fails to develop age-appropriate writing skills, the term developmental agraphia has been used (Gaddes & Edgell, 1994).

Precise criteria for the diagnosis of a writing disorder are unavailable due to the wide expression of potentially deficit skills. For example, individuals diagnosed with a writing disorder can display varied problems, including poor handwriting, spelling, grammar usage, punctuation and capitalization, and sentence structure. Additionally, development and organization of ideas for written narratives can be poor. Often, children with writing disorders experience concomitant difficulties in the areas of language, motor functioning, visual-motor/spatial skills, memory functions, and/or attentional problems. These problems may underlie the child's writing disorder or coexist with it. For example, children with visual-motor and spatial problems may have difficulty with the act of forming letters, or their writing may be remarkable for poorly defined margins, sloping lines, and inadequate or excessive spacing (Harris, 1995). Furthermore, children with language difficulties may produce extremely legible writing, the content of which is extremely impoverished or disconnected due to word retrieval problems or expressive/receptive language difficulties. For example, if a child is unable to correctly and spontaneously name correct words, written text will be remarkable for paraphasic errors or near-miss responses.

Benson & Ardila (1996) divided the agraphias into two main categories, including those involving a linguistic deficit or aphasic agraphia, and those involving spatial/motor impairments or mechanical agraphia. The characteristics of the aphasic agraphias mimic the subtypes generally found in aphasic disorders (e.g., Broca's, Wernicke's, Conduction, and Global; Benson & Ardila, 1996). There are also many subtypes of the mechanical agraphias, involving either a motor or spatial defect.

In contrast to agraphia seen in an adult population secondary to a traumatic brain injury, children may manifest a developmental type of agraphia or failure to develop age appropriate writing skills. Gaddes and Edgell (1994) reported that a child may experience awkward or slow handwriting due to faulty early development of the left motor or premotor areas of the brain, whereas more posterior dysfunction can result in poor letter and word recognition. Furthermore, anomalies within the left temporal area may produce aphasic-like symptoms and/or poor phonetic discrimination. Sandler et al. (1992) reported four different categories of writing disorders for children, including those with (1) fine motor and linguistic deficits, (2) visual-spatial deficits, (3) attention/memory deficits, and (4) sequencing deficits. Subtype 1 reportedly demonstrated reading problems (decoding and comprehension) and difficulties in the areas of spelling, punctuation, capitalization, and slow written output. They also demonstrated fine-motor deficits, including finger agnosia, mirror movements, fine motor dyspraxia, and poor imitation of opposition sequences. Expressive language deficits were also found among the children in this first subtype. Specifically, they demonstrated difficulty on tasks of oral sentence formulation, rapid naming, and picture naming. In subtype 2, children displayed visual-spatial deficits. This category is similar to the adult version of spatial agraphia as described above. In subtype 3, affected individuals often demonstrated attention and memory deficits. They also demonstrated poor phonetic decoding and numerous spelling errors (errors of encoding), while legibility and the mechanics of written language were relatively intact. The last subgroup was described as displaying significant visual and fine-motor sequencing deficits. Affected individuals also demonstrated difficulty on tasks of finger agnosia and fine-motor sequencing. Academic testing indicated that reading skills were intact, while mathematical computation was deficient. Finally, difficulties in the areas of spelling, legibility, mechanics, and letter formation were noted.

When evaluating a child with a writing disorder, it is important to screen for processing deficits that may underlie the child's writing difficulties and/or coexist with the disability. For example, a child mistakenly thought to have an attentional deficit may be viewed by his or her school as having writing difficulties solely due to "sloppiness", and thus deemed not appropriate to receive Special Education Services to address this writing deficiency.

Mathematics and quantitative functioning
There appears to be far more limited information regarding learning disabilities in the area of mathematics, as compared to reading disorders. There are few

studies addressing mathematical disorders alone; previous research has focused upon the investigation of mathematical disorders in combination with other learning or behavioral disorders. According to the DSM-IV, the prevalence of mathematical disorders is estimated at approximately one percent of school-age children. A learning disability in mathematics is defined as "mathematical ability, as measured by individually administered standardized tests, is substantially below that expected given the person's chronological age, measured intelligence, and age-appropriate education" (DSM-IV, p. 51). Furthermore, the disturbance must significantly interfere with academic achievement or activities of daily living and be in excess of a sensory deficit, if one is present.

Despite having a formal definition of a mathematical disorder, as outlined in the DSM-IV, underlying processes defining the nature of the disorder are unclear. Gaddes and Edgell (1994) described some basic processes underlying mathematical skills, including number concept, relative value, spatial skills, accurate reading and writing of numbers, retrieval of numbers, facts, and procedures, and calculation skills. Mastery of a number concept involves the ability to formulate a meaningful mental image of a number quantity, while relative value involves the ability to estimate relative size, distance, or spatial relationships. Basic spatial skills appear to be strongly related to proficiency in arithmetic. As Gaddes and Edgell (1994) stated, "Multidigit numbers must be written horizontally from left to right and spaced evenly. Addition sums are usually written in vertical columns, and the answer is measured from the decimal to the left, the units, tens, hundreds, and thousands having their value indicated by their spatial position. Multiplication and long division demand not only horizontal and vertical spacing, but oblique spacing to the left in multiplication and to the right in long division" (p. 418). Accurate reading and writing of numbers is also essential for success in mathematics. Individuals with aphasia may have a verbal paraphasia, which would impede their ability to verbalize a numerical response. Finally, success in arithmetic depends upon intact retrieval of mathematical facts and computational procedures.

Disorders of mathematical calculation can be divided into two categories, those involving a loss of mathematical ability (acalculia) secondary to a traumatic injury, versus failure in the development of mathematical abilities (developmental dyscalculia). Much of our knowledge about mathematical skills again comes from studies of adult patients who have sustained some type of traumatic injury or illness, resulting in the loss of mathematical ability. The various types of acquired mathematical disabilities commonly seen within an adult population are described below.

In contrast to acalculia, children who fail to develop age-appropriate arithmetic skills are classified as having a developmental dyscalculia. Developmental dyscalculia is generally not diagnosed until the end of the second or third grade, or sometimes even later in the child's educational career, given that children in the earlier grades are not at the appropriate developmental level for the mastery of certain higher order mathematical concepts (DSM-IV). Geary (1993) proposed three subtypes of mathematical disabilities, with the first two involving verbal processes, and the third involving spatial deficits. The first subtype involves impairment in the ability to represent and retrieve arithmetic facts in semantic memory, involving impairment in the left posterior area of the brain. Children with this problem often have difficulty learning and automatizing basic arithmetic facts and procedures. Children also often have accompanying language and reading deficits. The second subtype involves deficient use of arithmetic procedures resulting in difficulty acquiring basic number concepts and is related to left cerebral hemisphere dysfunction. The final subtype involves visual-spatial deficits with lesions within the right posterior hemisphere. Furthermore, Kosc (1974) proposed various subtypes of developmental dyscalculias ranging from impaired reading and writing of mathematical symbols, to difficulty in understanding mathematical concepts and performing calculations, and an inability to name numbers.

Children with developmental dyslexia may also experience comorbid reading and language problems, which may alter the presentation of their mathematical disorder. Jordan, Levine, and Huttenlocher (1995) reported that children with language problems have more difficulty on verbally-mediated tasks of arithmetic ability, while they performed within normal limits on nonverbal tasks of arithmetic ability. Thus, mathematical word problems are likely to be extremely difficult for children with this form of the disorder. Conversely, children with low intellectual ability demonstrated generalized impairment across arithmetic problems secondary to compromised higher order cognitive processes. Ackerman and Dykman (1995) reported that when children have comorbid arithmetic and reading problems, they can experience underlying deficits in the areas of processing speed, paired-associate learning, and analogical reasoning.

Disorders in mathematics have also been implicated as part of a larger syndrome – a right hemisphere nonverbal learning disorder. Myklebust (1975) originally defined a nonverbal learning disability as involving an inability to comprehend the significance of various aspects of the environment, difficulty anticipating events, failing to learn and appreciate nonverbal gestures, and difficulty acquiring the ability to determine the significance of basic nonverbal aspects of daily living. Other associated features have included disturbed social relationships; poor self-help skills; right-left confusion; difficulty learning to tell time, reading maps, or follow directions; disturbances in mathematics; and deficits in learning the meaning and actions of others. Rourke (1987, 1988, 1989) postulated a model for the syndrome of nonverbal learning disability based upon the theory of Goldberg and Costa (1981). Rourke (1988) described this syndrome as involving deficits in the areas of visual-spatial organization, psychomotor and tactile-perceptual skills (greater on left side of body), and conceptual problem-solving abilities. Deficient arithmetical skills and poor social reasoning skills were also hallmark signs of the disorder. Rourke (1987) documented the presence of the nonverbal learning disability in a group of children with severe mathematical disorders and in more specialized populations of medically ill children, such as those with head injuries, hydrocephalus, cancer survivors with a history of irradiation therapy, and congenital absence of the corpus callosum.

Rourke (1987) postulated that the syndrome of nonverbal learning disabilities was due to a disruption of white matter within the right hemisphere, resulting in impaired intermodal integration. Based upon Goldberg and Costa's model (1981), there is more white matter than gray matter in the right hemisphere, as compared to the left hemisphere. Furthermore, the right hemisphere is remarkable for interregional integration, resulting in a greater neuronal capacity for dealing with informational complexity. Conversely, the intraregional pattern of connectivity of the left hemisphere is better equipped to process simple, unimodal information, and execute discrete motor acts. Thus, a disorder in mathematics can be part of a larger syndrome secondary to a disruption of white matter within the right hemisphere. This disruption interferes with intermodal integration within the right hemisphere, resulting in impairment of novel information processing and informational complexity.

Understanding the exact nature of the underlying processing deficits that accounts for a learning disability in the area of mathematics is essential. Knowledge of the type of mathematical errors that a child produces, along with an understanding of the available intact processing skills upon which to build, will help determine the type and intensity of the educational strategies to be employed. For example, an educator could adjust the mode of presentation of remedial stimuli based upon the child's strengths and limitations. Specifically, for a child with visual/spatial and motor strengths, information could be presented via actual objects or "manipulatives". Other adaptations include orally stating the problem, presenting the problem in written or symbolic form, or presenting the problem in a fixed visual display (Jones, Wilson, & Bhojwani, 1997). If a child experiences visual-perceptual, spatial or visual-motor deficits, he or she may not be able to complete written mathematical problems on a worksheet if the worksheet is filled with other problems. The educational technique of having a class complete as many mathematical problems on a series of worksheets within one minute, can be especially devastating for a child with this form of underlying processing deficit.

Furthermore, if the child experiences organizational problems, he or she may have difficulty organizing the numbers on the page, despite intact calculation skills (Ylvisaker, Szekeres, & Hartwick, 1994). Deficits in the areas of reasoning and verbal abstraction may interfere with the application in solving word problems or understanding more abstract algebraic relationships despite proficiency with basic mathematical skills. The child may also have difficulty identifying critical information in mathematical word-problems (Ylvisaker, Szekeres, & Hartwick, 1994). Also, memory problems may impede a child's ability to learn and retain basic mathematical concepts. As such, children may not master procedural rules, such as multiplication or borrowing. Attentional deficits while executing arithmetical procedures in working memory may also contribute to poor computational skills (Geary, 1993).

Part II. Applications of pediatric neuropsychology to educational due process hearings and overview of laws and entitlements that apply to school age children

Following will be a discussion of the pertinent aspects of the Federal laws that govern the provision of entitlements for school-aged children with a variety of disabilities under special education and related laws. The information presented is not meant to be a legal primer, but rather a general overview to help facilitate a working knowledge of the so-called "due process" educational hearings. Such knowledge can greatly assist pediatric neuropsychologists who may become involved as expert witnesses.

Unique involvement in "due process" hearings

The reader will note a distinction between the involvement of a pediatric neuropsychologist in this type of procedural process as compared to other adversarial proceedings described within this book. Because of the unique nature of due process hearings within school settings, pediatric neuropsychologists not employed or regularly consulting with a public school system can become involved at the request of either parents or school district personnel. A resulting evaluation may determine the need for educational programming not already in place or a change in a child's educational placement or services. If parents or a school district cannot agree upon the appropriate services or educational placement, either party, as mandated by Federal law, may request a due process hearing. An impartial due process hearing officer, who has received training pertaining to relevant laws and hearing procedures, administers this hearing. The hearing officer's role is to determine which party to the hearing has presented the more appropriate recommendation, placement, supports, services, or accommodations necessary to meet a disabled child's educational needs. In this sense, when supported by objective evaluation results, the neuropsychologist can be viewed as advocating for particular intervention(s), services, or placements. Throughout the remainder of this chapter, the term *advocacy* will describe such scenarios, in which the pediatric neuropsychologist's role is often one that facilitates meaningful integration of data collected by all relevant sources, including educational experts employed by the school system, all of which should be working in the best interests of the student.

Relevant laws

The Education of All Handicapped Children Act of 1975, or P.L. 94–142, when passed into law by Congress, began to more formally address the growing concerns regarding the number of children who were either excluded from receiving public education due to the nature of their disabilities, or who were receiving inadequate or inappropriate forms of education to address aspects of their disabilities. Central to this law was the declaration that all disabled children should receive

> . . . a free appropriate education which emphasizes special education and related services designed to meet their unique needs, to assure that the rights

of handicapped children and their parents or guardians are protected, to assist States and localities to provide for the education of all handicapped children, and to assess and assure the effectiveness of efforts to educate handicapped children. (P.L. 94–142, Sec. 601[c])

Not only has this law been through numerous revisions, but also many State and Federal court cases have provided important interpretations of the law as written. Several important historical facts, as well as components of the present law will be discussed. In 1990, following several revisions, the name and some of the content of P.L. 14–142 was changed to the Individuals with Disabilities Education Act, or as it is more commonly known IDEA (P.L. 101–476). Of note in the 1990 revision, was that all terms related to the word "handicapped" were changed to the term "disabled". At the time of writing this chapter, President Clinton had recently signed into law P.L. 105–17, which after much input from professionals, parents, and individuals with disabilities, further amends IDEA. However, as this text goes to press, the specific regulations of P.L. 105–17 are still in draft form awaiting ratification, and thus some of the wording of the law as described below may change.

As partially defined under the pertinent laws, for a child or adolescent to qualify as having a "disability" and thus be eligible to receive Special Education and Related Services, he or she must have been evaluated per the guidelines contained within the law. Such evaluations are described generally as being conducted by a multidisciplinary team, which includes school personnel, parents, and independent professionals with expertise and "knowledge in the area of suspected disability". Pediatric neuropsychologists can potentially add a significant degree of understanding regarding an individual's cognitive functioning, learning strengths and weaknesses, and the need for individualized remediation, adaptations, supports, and services. As described within the law (34 CFR 300.532), the overall evaluation process should employ measures that are:

(1)   . . . administered in the child's native language or other mode of communication, unless it is clearly not feasible to do so;
(2)   Have been validated for the specific purpose for which they are being used; and
(3)   Are administered by trained personnel in conformance with the instructions provided by their procedures.

Numbers two and three above, can certainly become issues of contention if school personnel attempt to employ neuropsychological test measures without proper training, or if a variety of general test measures are employed and then proposed to represent a comprehensive neuropsychological test battery.

Also contained within 34 CFR 300.532, are the following general descriptions of the evaluation measures:

(1)   Tests and other evaluation materials include those tailored to assess specific areas of educational need and not merely those which are designed to provide a single general intelligence quotient.

(2)   No single procedure is used as the sole criterion for determining an appropriate educational program for a child.

(3)   The child is assessed in all areas related to the suspected disability, including, where appropriate, health, vision, hearing, social and emotional status, general intelligence, academic performance, communicative status, and motor abilities.

(4)   The tests are selected and administered so as best to ensure that when a test is administered to a child with impaired sensory, manual, or speaking skills, the tests accurately reflect the child's aptitude or achievement level or whatever other factors the test purports to measure, rather than reflecting a child's impaired sensory, manual, or speaking skills (except where those skills are the factors which the test purports to measure).

This latter stipulation can be a crucial factor in the identification or misidentification of an individual's functioning, such as in the case of a child who on one isolated measure appears to demonstrate global mental retardation, when in fact the child is challenged by severely impaired receptive and expressive language skills, as well as fine-motor abilities. Alternative neuropsychological and related test measures may be required to reflect more accurately a child's cognitive potential when specific processing deficits are demonstrated to negatively bias performances on certain test measures (e.g., measures of global intellectual functioning). For example, measures of nonverbal intellectual functioning or adaptive functioning may better reflect an individual's cognitive potential when they are challenged by a significant language disorder.

Once an appropriate set of evaluation procedures is undertaken, either by a pediatric neuropsychologist alone, or as part of a multidisciplinary team, the child must meet the criteria for one of thirteen "exceptional characteristics", and as a result of that disabling condition require "special education and related services". The Federal Register, in 300.22, defines related services as:

> . . . transportation and such developmental, corrective, and other supportive services as required to assist a child with a disability to benefit from special education, and includes speech-language pathology and audiology services, psychological services, physical and occupational therapy, recreation, including therapeutic recreation, early identification and assessment of disabilities in children, counseling services, including rehabilitation counseling, orientation and mobility services, and medical services for diagnostic and evaluation purposes. The term also includes school health services, social work services in schools, and parent counseling and training.

Following are the thirteen areas of disability as defined in the Federal Register. They are listed in their entirety because, although they may generally conform to accepted diagnostic criteria, they may vary somewnat from DSM-IV. Therefore, the pediatric neuropsychologist who advocates for patients within the school setting must be familiar with the following definitions, given that it is these, and *not* necessarily the DSM-IV criteria, that must be met for a child to be qualified to receive services.

(1)   *Autism* means a development disability significantly affecting verbal and nonverbal communication and social interaction, generally evident before age 3 that adversely affects a child's educational performance. Other characteristics often associated with autism are engagement in repetitive activities and stereotyped movements, resistance to environmental change or change in daily routines, and unusual responses to sensory experiences. The term does not apply if a child's educational performance is adversely affected primarily because a child has an emotional disturbance, as defined in paragraph (b) (4) of this section.

Note: If a child manifests characteristics of the disability category "autism" after age 3, that child still could be diagnosed as having "autism" if the criteria . . . in this section are satisfied.

(2)   *Deaf-Blind* means concomitant hearing and visual impairment, the combination of which causes such severe communication and other developmental and educational problems that they cannot be accommodated in special education programs solely for children with deafness or children with blindness.

(3)   *Deafness* means a hearing impairment that is so severe that the child is impaired in processing linguistic information through hearing, with or without amplification, that adversely affects a child's educational performance.

(4)   *Emotional Disturbance* is defined as follows:
      (i)   The term means a condition exhibiting one or more of the following characteristics over a long period of time and to a marked degree that adversely affects a child's educational performance.
             (a)   An inability to learn that cannot be explained by intellectual, sensory, or health factors.
             (b)   An inability to build or maintain satisfactory interpersonal relationships with peers and teachers.
             (c)   Inappropriate types of behavior or feelings under normal circumstances.
             (d)   A general pervasive mood of unhappiness or depression.
             (e)   A tendency to develop physical symptoms or fears associated with personal or school problems.
      (ii)   The term includes schizophrenia. The term does not apply to children who are socially maladjusted, unless it is determined that they have an emotional disturbance.

(5)   *Hearing Impairment* means an impairment in hearing, whether permanent or fluctuating, that adversely affects a child's educational performance but is not included under the definition of deafness in this section.

(6)   Mental Retardation means significantly subaverage general intellectual functioning, existing concurrently with deficits in adaptive behavior and manifested during the developmental period, that adversely affects a child's educational performance.

(7) *Multiple disability* means concomitant impairments (such as mental retardation-blindness, mental retardation-orthopedic impairment, etc.), the combination of which causes such severe educational problems that the problems cannot be accommodated in special education programs solely for one of the impairments. The term does not include deaf-blind.

(8) Orthopedic impairment means a severe orthopedic impairment that aversely affects a child's educational performance. The term includes impairments caused by congenital anomaly (e.g., clubfoot, absence of some member, etc.), impairments caused by disease (e.g., poliomyelitis, bone tuberculosis, etc.), and impairments from other causes (e.g., cerebral palsy, amputations and fractures or burns that cause contractures).

(9) *Other Health Impairment* means having limited strength, vitality or alertness, due to chronic or acute health problems such as a heart condition, tuberculosis, rheumatic fever, nephritis, asthma, sickle cell anemia, hemophilia, epilepsy, lead poisoning, leukemia, or diabetes, that adversely affects a child's educational performance.

Note: A child with attention deficit disorder (ADD) or attention deficit hyperactivity disorder (ADHD) may be eligible. . . if the child's condition meets the disability categoriese. . . and because of that disability the child needs special education and related services. Some children with ADD or ADHD who are eligiblee. . . meet the criteria for "other health impairments". . . These children would be classified as eligible for services under the "other health impairments" category if (1) the ADD or ADHD is determined to be a chronic health problem that results in limited alertness, that adversely affects educational performance, and (2) special education and related services are needed because of the ADD or ADHD. The term "limited alertness" includes a child's heightened alertness to environmental stimuli that result in limited alertness with respect to the educational environment.

Other children with ADD or ADHD may be eligible. . . [for special education] . . . because they satisfy the criteria applicable to other disability categoriese. . . For example, children with ADD or ADHD would be eligible for services under the "specific learning disability category" if they meet the criteria . . . or under the "emotional disturbance" category if they meet the criteria . . . Even if a child is not eligible for services under [special education law] the requirements under Section 504 of the Rehabilitation Act of 1973 and its implementing regulations at 34 CFR Part 104 may still be applicable.

(10) *Specific learning disability* is defined as follows:

    (i) General. The term means a disorder in one or more of the basic psychological processes involved in understanding or in using language, spoken or written, that may manifest itself in an imperfect ability to listen, think, speak, read, write, spell, or to do mathematical calculations, including such conditions as perceptual disabilities, brain injury, minimal brain dysfunction, dyslexia, and developmental aphasia.

> (ii)   Disorders not included. The term does not include learning problems that are primarily the result of visual, hearing, or motor disabilities, mental retardation, or emotional disturbance, or environmental, cultural, or economic disadvantage.

(11) *Speech or language impairment* means a communication disorder, such as stuttering, impaired articulation, a language impairment, or a voice impairment, that adversely affects a child's educational performance.

(12) Traumatic brain injury means an acquired injury to the brain caused by external physical force, resulting in total or partial functional disability or psychosocial impairment, or both, that adversely affects a child's educational performance. The term applies to open or closed head injuries resulting in impairments in one or more areas, such as cognition; language; memory; attention; reasoning; abstract thinking; judgment; problem-solving; sensory, perceptual, and motor abilities; psychosocial behavior; information processing; and speech. The term does not apply to brain injuries that are congenital or degenerative, or brain injuries induced by birth trauma.

(13) Visual impairment including blindness means an impairment in vision that, even with correction, adversely affects a child's educational performance. The term includes both partial sight and blindness.

As mentioned above, another of the Federal Laws that apply to school-aged children is Section 504 of the Rehabilitation Act Of 1973. The published definition of this Act states that:

> No qualified individual with disabilities, shall, solely by reason of his or her disability, be excluded from the participation in, be denied benefits of, or be subjected to discrimination under any program or activity receiving Federal financial assistance.

Under Subpart D of Section 504, this act "applies to preschool, elementary, secondary, and adult education programs and activities that receive or benefit from Federal financial assistance and to recipients that operate, or receive benefit from Federal financial assistance for the operation of such programs". This Act also mandates a "free appropriate public education", with an appropriate education defined as ". . . the provision of regular or special education and related aids and services that . . . are designed to meet individual educational needs of handicapped persons as adequately as the needs of nonhandicapped persons". For example, under Section 504, children diagnosed with an Attention Deficit Hyperactivity Disorder, who do not otherwise qualify for Special Education, may receive such modifications as spreading the administration of a test over a greater length of time, and inserting specified break periods within the test administration. Such modifications can allow the individual who experiences motoric restlessness to get up and move around during these specified intervals. However, as a result of this modification, their actual time of exposure to the individual test items is the same as all of the other students.

For many of the above disabling conditions, the pediatric neuropsychologist may not be the primary diagnostic clinician. However, neuropsychologists either alone or collaboratively with other professionals are often crucial in the diagnosis of subtle comorbid conditions that can easily be masked by one or more of the above conditions. For example, a child who is challenged by a peripheral hearing impairment may be viewed by the school system as not displaying any other disabling condition (i.e., all of his or her educational difficulties are the direct result of the diagnosed hearing impairment). However, the child may have a coexisting learning disability or speech and language disorder. Due to the typically complex nature of many children with disabilities, often states employ a system of determining one or more primary and secondary disabling conditions, when the category of multiple disability does not apply.

Following the undertaking of comprehensive neuropsychological and related evaluations, the pediatric neuropsychologist is faced with a number of clinical decisions pertinent to patient care in general, as well as to aspects of special education and related laws. Probably, the most immediate care issue is whether the child is ready to return to school if they were recently discharged from a hospital or rehabilitation facility following an acute injury. Most clinicians are aware of patients who are discharged from medical facilities with stable medical conditions, yet whose cognitive status is in a state of ongoing recovery. For such children, their acute status may not allow them to readily benefit from a mainstream or even a typical special education setting. If a situation exists in which a patient is staying at home for some extended period of time, they may be entitled to receive special education and related services in their home setting.

Following the undertaking of a comprehensive evaluation, the parents /guardian of the child or school personnel must then decide if they wish to pursue either special education services or services under Section 504 of the Rehabilitation Act. If parents or the public school chooses to initiate this process, the first step typically involves the school undertaking a case study evaluation. Although different states may comprise the case study evaluation team in different ways, a typical constellation of school personnel is likely to include the child's mainstream teacher if they are currently in school, a school psychologist, a special education teacher, an administrative representative of the school district (e.g., a special education director), a school social worker, and a school nurse. In some cases a speech and language pathologist, and/or occupational and physical therapists may also be members of the evaluation team. If psychometric evaluations have been conducted by the school system, they are likely to include standardized measures of intellectual functioning, measures of academic achievement, and possibly measures of linguistic functioning. Vision and hearing screenings are typically conducted by a school nurse, and a social-developmental study is typically conducted by the school social worker (via an interview with parents and the child). These latter two assessments are conducted to rule-out exclusionary factors described above (e.g., vision problems, economic or environmental factors, or native language issues) that may be accounting for a child's poor educational performance. A review of a child's educational history, including prior grades and performance on group administered tests (if they have been in school), is often presented, along with the results of an assessment of the child's learning environment (e.g., observation in the class-room or rehabilitation setting, and interviews with teachers and therapists).

When additional specialized evaluations have been conducted by outside individuals or agencies (i.e., independent evaluations), public schools are man-dated to consider those assessments in the determination of the presence or absence of any of the above thirteen disabling conditions that qualify a child for the provision of special education and related services. Although public schools are not mandated to accept every independent evaluation that they receive, the school must show good faith in considering all outside evaluations. A child's eli-gibility for special education and related services must be re-assessed every three years, ". . . or more frequently if conditions warrant or if the child's parent or teacher requests an evaluation" (34 CFR 300.534).

Following the determination that a child does qualify for special education, a document entitled an "Individualized Education Program" or IEP must be cre-ated. This document, although fluid in nature, governs the disabled child's edu-cation over the course of the following school year. This document must contain certain components, and be created by the multidisciplinary team, with parental input. Although the creation of an IEP is technically undertaken within a sepa-rate meeting from that which established a child's eligibility for special educa-tion, these two meetings typically occur concurrently. The type or amount of services to be provided to a child, and the setting in which those services are to be delivered to a child (i.e., the "Placement") cannot be determined until after a child's IEP is created. It is at this point important to state the expectations of special education law for the creation of a disabled child's IEP, specifically, it should reflect a "free appropriate education" in the "least restrictive environ-ment". Both parents' and professionals' expectations in developing an IEP can-not be for the creation of the "best" or "ideal" educational program, but rather for an "appropriate" one. However, the determination of what is in fact appro-priate, can range from immediate consensus to deadlocked disagreement.

Required for the creation of an IEP, is a written document containing a num-ber of necessary components. First, the IEP must contain documentation regard-ing a child's "present levels of educational performance". This statement should reflect the nature in which the determined disability or disabilities impact upon a child's education, both academically, and if appropriate, socially, emotionally, behaviorally, and also from a daily living skills standpoint. Although this state-ment should reflect the objective data collected as part of the case study evalu-ation, it should be written in a way that all participants, including parents, can readily understand. Of extreme significance at this initial point in the creation of an IEP, is that each and every aspect of a child's disability be described, because it is this initial component that will drive the creation of the rest of this document. Specifically, each area of educational impact must then be addressed via the creation of an "annual goal", which is then supported by one or more "short term instructional objectives". These goals and objectives, for the pur-pose of evaluating the efficacy of the IEP, should be written in as objectively measurable a format as is possible. The annual goals for a disabled child should reflect each area of educational impact. Therefore, goals may be written to address academic skills (e.g., To improve Jon's reading comprehension skills), social skills (e.g., To improve Jon's social problem-solving skills), as well as daily living skills (e.g., Jon will independently use the toilet without accidents). These annual goals should reflect, via the creation of appropriate objectives, what gains might reasonably be expected to occur within an academic school year. The short-term objectives should establish the intermediate steps towards the attainment of the annual goal. For example, under goal "To improve Jon's reading comprehension skills", the following objectives might be written:

1. Given a sixth grade reading passage, Jon will be able to state the main idea, and three specific details.
2. Given a sixth grade reading passage, Jon will be able to answer literal comprehension questions.
3. Given a sixth grade reading passage, Jon will be able to answer inferential comprehension questions.
4. Given a sixth grade reading passage, Jon will be able to write a study guide containing the important factual details in sequential order.

Also contained within the IEP are the specific evaluation procedures and methods for determining if the individual objectives, and thus the annual goal, are being met (e.g., "Jon will receive a grade of 80 percent or better on both literal and inferential reading comprehension questions").

Following the creation of the above components of a child's IEP, the type and amount of special education services needed to address the established goals and objectives are determined. Special education and related services are never provided based upon availability or policy, but rather must be provided based upon each individual child's needs (i.e., how long will it take to implement all of the stated goals and objectives). As such, statements by school personnel such as "we only provide so much occupational therapy" or "our speech and language pathologist is only here for two days per week", if they are designed to limit the amount of service provision, may be a violation of the disabled child's rights. Therefore, the IEP may contain statements such as "300 minutes per week of learning disabilities services", "60 minutes per week of occupational therapy", "60 minutes per week of psychotherapy", or "60 minutes per week of adaptive physical education". Also contained within the IEP document is the amount of time that a child will be included in regular education programming and activities. Transportation to and from special education programs is also considered a related service for a disabled child, and is therefore included in the IEP.

IEPs should also contain a listing of necessary accommodations, adaptations, and modifications, along with any necessary assistive technology needed to allow a child to benefit from instruction. For example, *accommodations* for a child with dyslexia might include reading test questions aloud, and transcribing answers dictated by the child orally. *Adaptations* might include having a child who demonstrates memory retrieval difficulties taking tests either with extended time or non-timed, or to only be tested using multiple-choice questions or open book exams. *Modifications* might include having a child with a central auditory processing disorder take tests in a quiet room that is separate from the rest of the class. Finally, assistive technology devices are defined under 34 CFR 300.6 as " . . . any item, piece of equipment, or product system, whether acquired commercially off the shelf, modified, or customized, that is used to increase, maintain, or improve the functional capabilities of children with disabilities". Assistive technology services are described as ". . . any service that directly assists a child with a disability in the selection, acquisition, or use of an assistive technology device". Thus, if a child with cerebral palsy is unable to communicate orally, yet has the ability to communicate via augmentative communication technology, provision of such technology is the responsibility of the school system, *if* it is deemed appropriate for the child's use.

It is at this point in the IEP process that a child's placement is determined. In essence, where services will be provided, and who will provide them, is the last step in the development of an IEP. As stated in the law, each disabled child's educational placement:

1. Is determined at least annually,
2. Is based on his or her individualized education program, and
3. Is as close as possible to a child's home.

Although number 3 above, speaks to the requirement to consider the least restrictive environment, all appropriate options must be considered, and documented in the IEP. The range of options can include everything from a minimal amount of special education services provided within a regular education classroom setting, through the provision of placement in a 12-month residential therapeutic setting at public expense. Also included within this range of services can be home-based or hospital-based instruction for a medically ill child, or institutional instruction for an incarcerated child. Although placement in a child's neighborhood school is always preferable, it is not always appropriate.

Several court rulings have determined that the degree of severity of a child's disability may require more time than afforded in the normal school year to achieve an appropriate education. In such circumstances, "extended school year" (ESY) instruction (i.e., instruction during the summer) has been deemed appropriate to meet a child's educational needs. Initially, two factors were determined to be required for the provision of ESY services. First, it was expected that a child would have to demonstrate empirically substantial regression of acquired skills during the summer months of non-instruction. Secondly, given that all children may not retain all of the information they acquired during the previous school year, it would have to be demonstrated empirically that the child could not readily recoup the previously acquired skills within a reasonable time period. However, when school systems began establishing formulas for determining issues of cognitive regression and periods of recoupement, irrespective of the specific criteria that they employed, it became quite clear that a child would have to have already demonstrated this loss of previously acquired knowledge or skills before they were entitled to receive assistance to keep this regression from happening. As a result, several court rulings determined that it was inappropriate to require a child to have to endure such a situation during the previous school year, before they could be determined eligible to receive ESY services in the year to follow. In fact, courts have stated that compelling predictive data, supported by expert professional documentation of the above two factors, could suffice to qualify a child for ESY instruction. Therefore, neuropsychological test data can be extremely important in helping to qualify an individual for ESY services without the need for a loss of functioning to first have occurred. Specifically, measures of verbal memory functioning, including issues of consistent retrieval and delayed recall of novel information can be most helpful in this regard. However, in some cases, it has been necessary to establish short-term learning paradigms to demonstrate this phenomenon. Such situations can include instruction in a specific skill on a Friday, and then evaluating its maintenance on the following Monday. Similarly, pre-post testing over the winter vacation period may demonstrate not only loss of a specific skill, but also the amount of time necessary to recoup the lost function.

One related issue in the law addresses extended school day instruction. Several court cases have determined that after-school remedial instruction can be reimbursed to parents if the standard school day program is not sufficient to meet a child's educational needs. In addition, extended school day instruction has also been deemed appropriate if providing such extended services has precluded the need to place a child in a more restrictive educational setting (e.g., a residential program). It is possible, for example, to infer that placement in a therapeutic day school, whose therapeutic program lasts beyond the early afternoon, can be appropriate to meet a child's emotionally-based educational needs, if it precludes the need for a 24-hour a day residential school setting.

Another important area in which the pediatric neuropsychologist's expertise regarding brain/behavior relationships can be crucial, concerns the disciplinary codes of schools and school districts. Virtually all schools maintain disciplinary codes for their pupils, and many may have specific predetermined consequences for violation of those codes. Such punitive action can range from minor consequences or reprimands, to suspensions of varying lengths, through expulsion. For nondisabled children, under most circumstances, the school policy will take precedence. However, for a child with an identified or suspected disability, the law may allow a potentially different course of action. For a child currently receiving special education services, especially if it has already been determined their disability impacts upon their overt behavior, children cannot be denied their special education and related services. When disabled children violate school rules, the child's multidisciplinary special education team must meet to undertake a "manifestation determination", specifically to determine whether or not the rule violation is a direct result of the presence of a disability. Issues to be considered include whether the child was capable of understanding the rules, understanding the impact and/or consequences of their behavior, as well as their ability to control their behavior. To make this determination, the educational team must consider all relevant evaluation results and expert opinions. If it is determined that a child's behavior is not a manifestation of their disability, they may be disciplined according to the school's general disciplinary code. However, if it is determined that the rule violation is a result of a disability, limitations can be placed upon the degree and nature of the imposed consequences, although special education services cannot be denied.

Given the complexity of these behavioral issues, it is crucial for pediatric neuropsychologists, when appropriate, to make the clinical connection between the presence of cognitive, neurological, and/or severe socially-based disabilities (e.g., Asperger's Disorder), and their potential link to a child's impaired ability to comprehend, appreciate, or control their behavior and thus conform to school rules. Anticipating in advance the relationship of patients' severe neuropsychological deficits (e.g., in the area of memory or abstract reasoning, global level of cognitive functioning, or disorders of impulse control), can greatly assist the process of manifestation determination, or more importantly potentially prevent the occurrence of such behaviors. The issue of potential prevention of extreme behavioral difficulties and rule violations can be addressed through the creation of a "Behavior Intervention Plan" as part of the child's IEP. Therefore, not only can this issue be addressed proactively by a pediatric neuropsychologist, but also his or her presence at a school meeting can greatly set the stage for an understanding of the complex nature of the interaction of neuropsychological deficits and overt behavior.

The final, and often most important aspect of a comprehensive evaluation designed to assist an individual in receiving an appropriate education is how clinicians present their results and recommendations for their patients. In this time of managed care and restrictions placed upon the amount of clinical time available to each patient, assessments undertaken with advocacy in mind, necessarily fly in the face of the trend of briefer and more efficient test reports. For the purpose of advocacy, as suggested above, especially when clinicians cannot be present to "stand behind" their findings and recommendations, the evaluation report must stand alone in presenting the strategic ramifications of the results of testing, and the recommendations derived therefrom.

The evaluation report typically contains four main components: The reason for referral and relevant past history; the summary and explanation of the test results; clinical impressions within the context of school law; and specific recommendations. Within this model of report writing, the initial section presents all parental and/or referral source descriptions of the referral issues, parents' perceived strengths and concerns regarding their child, relevant medical history, and a summary and integration of all relevant previous evaluation data, both formal and informal. This section of the evaluation report is the venue for presenting the often complex history of the patient leading up to the current issues at hand. It is not unusual for there to exist numerous fragmented pieces of data in need of integration, thus allowing for a more comprehensive understanding of the child. It is not unusual for school personnel not to be aware of all of the obvious and subtle issues that comprise the child that they are currently working with. It is this component of the written report that typically presents all relevant information regarding your patient to the reader in a logical and integrated fashion.

As with any evaluation report, the audience can include parents, physicians, school personnel, and other related mental health, allied health, and therapeutic professionals, all of which may benefit from an understanding of the complex history of a child. Further, this integrated summary also lays the groundwork for the strategic presentation of the historical record. Read by school personnel, Due Process Officers, and attorneys, this section presents the earliest determinants of the patient's disability, as well as confirming and disconfirming data and perceptions of the clinician's diagnostic impressions. If the patient is already engaged in an adversarial process, this section should also address, via the available data, the clinical hypotheses proposed by others.

Presentation and integration of neuropsychological test data poses an especially difficult challenge given the multiple audiences who may ultimately read the evaluation report. In the case of child advocacy, the writer must attempt to demystify and present test results such that individuals not familiar with such data can better understand the individual test results and their implications. However, test descriptions should also illuminate, where appropriate, the complexity and interrelated nature of the patient's disabilities, as well as clearly define their strengths. The challenge to all pediatric neuropsychologists working with a school-age population, is to go beyond descriptions of neuropathology to descriptions of functional expressions of test results. Although many alternative wordings can be found for any specific subtest, the clinician should attempt to employ functional descriptions that relate the obtained results to observable behaviors and measurable cognitive and academic skills.

The next section of the evaluation report in the current example contains clinical impressions. This section is not only the clinician's opportunity to summarize the patient's cognitive strengths and impairments, but also potentially accomplishes two additional purposes. First, this section should, as defined by federal and state school law, present the determination of the specific disabling conditions that may qualify a child for special education and related services. Although states and school districts vary regarding the determination of one or more primary and secondary exceptional or disabling characteristics, the clinician should present impressions of the category or categories in which a child qualifies for assistance.

Finally, the evaluation report should contain as many, and as specific recommendations as the clinician feels is appropriate to meet a child's needs. Although the independent clinician cannot, and should not attempt to dictate to schools the specific methodology that they should employ with a child, the written report should document recommended interventions, supports, services, accommodations, and adaptations for consideration by the public school, as well as by parents and private therapists. Further, this section of the evaluation report should also contain recommendations for additional evaluations that may be helpful in further clarifying the overall clinical picture, as well as determining any additional needs. Evaluations such as speech and language, and occupational and physical therapy will typically be provided by the school system at no charge if an appropriate rationale is provided. Although courts have ruled that schools do have a choice in the methodology that they employ, this does not mean that they can employ inappropriate or ineffective techniques to remediate or treat a child. For example, for a child with a central auditory processing disorder and impaired memory retrieval skills, who is having significant difficulty acquiring phonetic decoding skills, an example of an individualized empirically-based multisensory phonics instruction program is appropriate to include as a recommendation. Such a recommendation is not dictating methodology, if the child is currently receiving a rote sight-word based instructional program that has not been demonstrated to be effective after a reasonable period of time. In addition, recommendations should help to enlighten parents, mental health, and/or educational professionals to better understand the impact of the child's disability upon their interactions with that child or the child's functional status. For example, a special education teacher may not be aware of the impact that a central auditory processing disorder or a severe nonverbal learning disability has upon attempts to provide educational remediation to a child.

When disagreements occur between school personnel and parents regarding issues of proper diagnosis of their child under the special education categories, appropriateness of the IEP, provision of related services, or educational placement, procedural safeguards exist to help determine if a child is receiving a free appropriate public education in the least restrictive environment. Schools are required to give parents a clearly written and easily understandable "procedural safeguard notice", informing them of all of their rights and corrective procedures available under the law. This should include parents' rights to an independent evaluation, prior written notice of meetings, the need for parental consent, access to documentation collected about their child, and the procedures involved in requesting a due process hearing. This administrative procedure is one of the major mechanisms for resolving disputes between parents and schools. Therefore, parental notification of rights will also include directions on how to file for a due process hearing, issues related to the child's placement pending the outcome of the hearing, including, but not limited to, possible interim placements, the appeals process, attorney's fees, and, the availability of an impartial mediation process. Although when due process hearings are undertaken it is most common for parents to have filed a request, schools can and do file due process hearings against parents. This latter situation occurs when, for example, schools feel that a child should be evaluated and parents do not, or when schools feel that a child requires more intensive services or a more restrictive educational placement, and parents disagree.

The newest revision of the Special Education Law requires that school districts attempt to resolve disagreement through mediation, whenever a due process hearing is requested by either party. In this situation parents are provided with a list of qualified mediators, and the full cost of this process is born by the State. This process is, however, voluntary. The mediation process cannot be used to deny or delay a due process hearing, and the results of the mediation are not binding. In fact, discussions that occur during mediation are confidential, and are thus not admissible in either due process hearings or civil litigation.

All states maintain a pool of trained due process officers to preside over the process. These individuals, who may come from numerous backgrounds, are extensively trained in special education law, and are neither affiliated with the school district in which the hearing is taking place, nor have any other personal interest in the child or issues at hand. Due process officers will typically conduct a pre-hearing conference to more formally establish each parties' issues in the matter, and lay out the ground rules for the hearing. For example, the due process officer will establish which of the two parties will present their case first, and ask parents to determine whether the hearing is "open" or "closed". Specifically, if parents decide that only individuals directly involved in hearing itself (e.g., witnesses or relevant school personnel) can be present, it is considered a "closed" hearing. If, as has happened, parents invite friends or family for support, members of advocacy groups or graduate students for the purpose of observation only, or the news media, then the hearing is considered "open". If the hearing is deemed open, then the school district can also invite appropriate individuals to observe the process (e.g., a school psychology intern).

Although a due process hearing is not the same as a civil court proceeding, many of the same rules and procedures exist. Due process officers can, at either parties' request, issue subpoenas for witnesses, request documents, and directly question any of the participants in the hearing. Although the law requires that, in general, a child's educational placement not be changed pending the outcome of the due process ruling, there are certain circumstances in which an immediate change of placement is necessary until an appropriate placement is determined (e.g., when children are dangers to themselves or others, or if it is determined that the current environment is detrimental to the child). In these circumstances if the two parties cannot agree upon an interim placement, the due process officer will make that determination. All documents and evidence to be used in the due process hearing, including all evaluation reports and recommendations for the child, must be disclosed at least five business days prior to the onset of the hearing. Not doing so can result in the exclusion of that information unless both parties agree to allow it to be admitted. Similar to civil litigation cases, expert witnesses are examined and cross examined, typically by attorneys from both sides, regarding their qualifications as an expert, the findings of their evaluation, and the nature of the resulting recommendations.

Unlike civil litigation, in which expert witnesses typically are excluded from observing the ongoing proceedings, in the case of due process hearings experts can be allowed to sit through the entire process, and thus serve several roles for either party. In addition to presenting the results of their comprehensive evaluation and the recommendations contained therein, pediatric neuropsychologists can hear the testimony of other professionals for the purpose of rebuttal. In our experience, it is unfortunately not uncommon for individuals who do not have training as neuropsychologists, to nonetheless attempt to discount, refute, or

reinterpret neuropsychological test results. Within this venue, the pediatric neuropsychologist can be allowed to listen to the testimony of others, and then
develop direct issues of rebuttal to be presented at a later point in the hearing.
Although many of the attorneys, by virtue of their specialization in school law,
have acquired fairly extensive knowledge regarding issues of assessment and
intervention, they typically do not possess the unique expertise of a pediatric
neuropsychologist. Therefore, neuropsychologists can, within the setting of the
due process hearing, provide attorneys with questions to be employed with other
witnesses. Thus, the pediatric neuropsychologist can become an integral member of either parties' team in assisting in the presentation of their case in regard
to the issue of a child's entitlement to a free appropriate public education in the
least restrictive environment.

Following the completion of the due process hearing, the due process officer
will produce a written ruling based upon the evidence provided within the hearing. This document will typically provide a background of the issues that resulted in the request for the due process hearing, finding of facts based upon the
available evidence, findings of law (i.e., how the available evidence and legal
arguments are in accord with the law and previous court rulings), and then,
most importantly, a specific order regarding what the school should do in terms
of diagnosis, intervention, and placement, to insure that a child is receiving an
appropriate education. This recommendation can be based upon either parties'
arguments, or a synthesis of both parties' recommendations. In some cases, due
process officers have crafted a third alternative of what they believe would be
appropriate to meet a child's educational needs.

Conclusions

Evaluating children for the purpose of educational advocacy can be an extremely time consuming process. Some approaches to neuropsychological testing and
presentation of test results, although necessary and extremely valuable for
patient care in medical settings, are not necessarily conducive for the purpose of
assisting parents, school systems, and independent service providers, in providing an appropriate education to disabled children. Further, when more direct
advocacy within the school setting is required, such as in the case of a due
process hearing, pediatric neuropsychologists must be able to adapt to the
requirements of the complex legal issues that impact the roles of all professionals working with disabled children. Although a tremendous amount of clinical
time is required to adequately address the educationally-based needs of complex
disabled children, the outcome of such an endeavor can be highly rewarding. It
is not unusual in the professional role as a child advocate to follow individual
patients for many years, assessing their progress and refining their educational
and therapeutic needs. As we have tried to outline, the laws that govern the entitlements of children in educational settings through the age of 21 are extensive,
and ongoing court rulings continuously change and modify the interpretations

of the applicable laws. Pediatric neuropsychological practice in this area requires a commitment to a broader base of knowledge than is typically taught in academic neuropsychological programs (e.g., educational issues). Nonetheless, pediatric neuropsychologists can serve potently in a role that no other member of a child's educational team can, especially when it comes to the issues of neurologically challenged children. Rather than approach a child in a fragmented fashion as only a collection of isolated evaluation data and historical facts, pediatric neuropsychologists can serve the unique role of synthesizing all of the available data, and thus assisting in the development of a truly integrated and systematic appropriate IEP for their patients. Given the broader perspective of the positive impact upon society of each disabled individual reaching their potential, the rewards for all should be evident. It is typically not that schools do not want to appropriately educate a child; rather, more often, when a disabled child is not receiving appropriate education, it is due to a lack of understanding of the true nature and complexities of his or her disability.

References

Ackerman, P. J., & Dykman, R. A. (1995). Reading disabled students with and without comorbid arithmetic disability. *Developmental Neuropsychology, 11*, 351–371.

Adams, R. D., Victor, M., & Ropper, A. H. (1997). *Principles of neurology* (6th ed.). New York: McGraw-Hill.

Albert, M. L., & Helm-Estabrooks, N. (1988). Diagnosis and treatment of aphasia, Part I. *Journal of the American Medical Association, 259*, 1043–1047.

Aniskiewicz, A. S., Frumkin, N. L., Brady, D. E., Moore, J. B., & Pera, A. (1990). Magnetic resonance imaging and neurobehavioral correlates in schizencephaly. *Archives of Neurology, 47*, 911–916.

American Association on Mental Retardation (1992). *Mental retardation: Definition, classification, and systems of support* (9th ed.). Washington, DC, Author.

American Psychiatric Association (1994). *Diagnostic and statistical manual of mental disorders* (4th ed.). Washington, DC: Author.

Bach-y-Rita, P., & Bach-y-Rita, E. W. (1990). Biological and psychosocial factors in recovery from brain damage in humans. *Canadian Journal of Psychology, 44*, 148–165.

Barkley, R. A. (1990). *Attention deficit hyperactivity disorder: A handbook for diagnosis and treatment.* New York: Guilford Press.

Barkley, R. A. (1995). Attention deficit hyperactivity disorder in children and adults. *ADHD Workshop Manual.*

Baron, I. S., Fennell, E. B., & Voeller, K. K. S. (1995). *Pediatric neuropsychology in the medical setting.* New York: Oxford University Press.

Bauer, R. M., Tobias, B., & Valenstein, E. (1993). Amnesic disorders. In K. M. Heilman & E. Valenstein (Eds.), *Clinical neuropsychology* (3rd ed.) (pp. 523–602). New York: Oxford University Press.

Beitchman, J. H., & Young, A. R. (1997). Learning disorders with a special emphasis on reading disorders: A review of the past 10 years. *Journal of the American Academy of Child and Adolescent Psychiatry, 36*, 1020–1032.

Bellis, T.J. & Ferre, L.M. (1996). Assessment and Management of CAPD in Children. *Educational Audiology Association Monograph*, December, 23–28.

Benson, D. F., & Ardila, A. (1996). *Aphasia: A clinical perspective.* New York: Oxford University Press.

Benton, A., & Tranel, D. (1993). Visuoperceptual, visuospatial, and visuoconstructive disorders. In K. M. Heilman & E. Valenstein (Eds.), *Clinical neuropsychology* (3rd ed., pp. 165–213). New York: Oxford University Press.

Berg, B. O. (1996). Developmental disorders of the nervous system. In B. O. Berg (Ed.), *Principals of child neurology* (pp. 665–690). New York: McGraw-Hill.

Biederman, J., Munir, K., Knee, D., Habelow, W., Armentano, M., Autor, S., Hoge, S. K., & Waternaux, C. (1986). A family study of patients with attention deficit disorder and normal controls. *Journal of Psychiatric Research, 20,* 263–274.

Billard, C., Santini, J. J., Nargeot, M. C., & Adrien, J. L. (1986). Long-term intellectual prognosis of hydrocephalus with reference to 77 children. *Pediatric Neuroscience, 12,* 219–225.

Cantwell, D. P. (1972). Psychiatric illness in the families of hyperactive children. *Archives of General Psychiatry, 27,* 414–417.

Chelune, G. J., & Edward, P. (1981). Early brain lesions: Ontogenetic-environmental considerations. *Journal of Consulting and Clinical Psychology, 49,* 777–790.

Cohen, B. H. (1991). Neurologic causes of learning disabilities. *Seminars in Neurology, Vol. II,* 7–13.

Connell, D.R. (1985). *Writing is child's play.* Circle Pines, MN: American Guidance Service.

Coslett, H.B. (1997). Acquired dyslexia. In T. E. Feinberg & M. J. Farah (Eds.), *Behavioral neurology and neuropsychology* (pp. 197–208). New York: McGraw-Hill.

Damasio, A. R. (1985). Disorders of complex visual processing: Agnosias, achromatopsia, Balint's Syndrome, and related difficulties of orientation and construction. In M.M. Mesulam, *Principles of behavioral neurology* (pp. 259–288). Philadelphia, PA: F.A. Davis Company.

Damasio, A. R., & Damasio, H. (1983). The anatomic basis of pure alexia. *Neurology, 33,* 1573–1583.

Denckla, M. B. (1989). Executive function: The overlap between attention deficit hyperactivity disorder and learning disabilities. *International Pediatrics, 4* 155–160.

Dennis, M. (1980). Capacity and strategy for syntactic comprehension after left or right hemidecontication. *Brain and Language, 10,* 287–317.

Dennis, M. (1991). Frontal lobe function in childhood and adolescence: A heuristic for assessing attention regulation, executive control, and the intentional states important for social discourse. *Developmental Neuropsychology, 7,* 327–358.

Dennis, M., Fitz, C. R., Netley, C. T., Sugar, J., Harwood-Nash, D. C. F., Hendrick, E. B., Hoffman, H. J., & Humphreys, R. P. (1981). The intelligence of hydrocephalic children. *Archives of Neurology, 38,* 607–615.

De Renzi, E. (1997). Visuospatial and constructional disorders. In T. E. Feinberg & M. J. Farah (Eds.), *Behavioral neurology and neuropsychology* (pp. 297–307). New York: McGraw-Hill.

Doehring, D. G. (1978). The tangled web of behavioral research on developmental dyslexia. In A. L. Benton & D. Pearl (Eds.), *Dyslexia: An appraisal of current knowledge.* New York: Academic Press.

Dodge, P., Prensky, A., & Feigin, R. (1975). *Nutrition and the developing nervous system.* St. Louis: C.V. Mosby.

Eden, G. F., Stein, J. F., Wood, M. H., & Wood, F. B. (1995). Verbal and visual problems in reading disability. *Journal of Learning Disabilities, 28,* 272–290.

Ferre, J.M. (1997). *Processing power: A guide to CAPD assessment and treatment.* San Antonio, TX: The Psychological Corporation/Communication Skill Builders.

Fisher, B. (1998). *Attention deficit disorder misdiagnosis: Approaching ADD from a brain-behavior/neuropsychological perspective for assessment and treatment.* New York: CRC Press.

Fletcher, J. M., Bohan, T. P., Brandt, M. E., Brookshire, B. L., Beaver, S. R., Francis, D. J., Davidson, K. C., Thompson, N. M., & Miner, M. E. (1992). Cerebral white matter and cognition in hydrocephalic children. *Archives of Neurology, 49*, 818–824.

Fletcher, J. M., Francis, D. J., Thompson, N. M., Brookshire, B. L., Bohan, T. P., Landry, S. H., Davidson, K. C., & Miner, M. E. (1992). Verbal and nonverbal skill discrepancies in hydrocephalic children. *Journal of Clinical and Experimental Neuropsychology, 14*, 593–609.

Foundas, A. L., Lenoard, C. M., Gilmore, R., Fennell, E., & Heilman, K. M. (1994). Planum temporale asymmetry and language dominance. *Neuropsychologia, 32*, 1225–1231.

Frank, Y. (1996). Attention deficit hyperactivity disorder. In Y. Frank (Ed.), *Pediatric behavioral neurology* (pp. 179–202). New York: CRC Press.

Fredricks, J. A. M. (1969). The agnosias. In P. J. Vinken, & G. W. Bruyn (Eds.), *Handbook of clinical neurology* (Vol. 4; pp. 207–240). Amsterdam: North Holland Publishing.

Friedman, R., Ween, J. E., & Albert, M. L. (1993). Alexia. In K. M. Heilman & E. Valenstein (Eds.), *Clinical neuropsychology* (3rd ed., pp. 37–62). New York: Oxford University Press.

Friedrich, W. N., Lovejoy, M. C., Shaffer, J., & Shurtleff, D. B., & Beilke, R. L. (1991). Cognitive abilities and achievement status of children with myelomeningocele: A contemporary sample. *Journal of Pediatric Psychology, 16*, 423–428.

Gaddes, W. H., & Edgell, D. (1994). *Learning disabilities and brain function* (3rd ed.). New York: Springer-Verlag.

Galaburda, A. M., & Kemper, T. L. (1979). Cytoarchitectonic abnormalities in developmental dyslexia: A case study. *Annals of Neurology, 6*, 94–100.

Galaburda, A. M., Sherman, G. F., Rosen, G. D., Aboitiz, F., & Geschwind, N. C. (1985). Developmental dyslexia: Four consecutive patients with cortical anomalies. *Annals Neurology, 18*, 222–233.

Geary, D. C. (1993). Mathematical disabilities: Cognitive, neuropsychological, and genetic components. *Psychological Bulletin, 114*, 345–362.

Geschwind, N. C. (1972). Language and the brain. *Scientific American*, 2–10.

Geschwind, N. C., & Levitsky, W. (1968). Human brain: Left/right asymmetries in temporal speech region. *Science, 161*, 186–189.

Gilbert, J. N., Jones, K. L., Rorke, L. B., Chernoff, G. F., & James, H. E. (1986). Central nervous system anomalies associated with the meningomyelocele, hydrocephalus, and the Arnold-Chiari malformation: Reappraisal of theories regarding the pathogenesis of posterior neural tube closure defects. *Neurosurgery, 18*, 559–564.

Goldberg, E., & Costa, L. D. (1981). Hemispheric difference in the acquisition and use of descriptive systems. *Brain and Language, 14*, 144–173.

Goodman, R., & Stevenson, J. A. (1989). A twin study of hyperactivity – II. The etiological role of genes, family relationships and perinatal adversity. *Journal of Child Psychology & Psychiatry, 5*, 691–709.

Harris, J. C. (1995). *Developmental neuropsychiatry, Volume II: Assessment, diagnosis and treatment of developmental disorders*. New York: Oxford University Press.

Hodge, M. M. (1991). Assessing early speech motor function. *Clinics in Communication Disorders, 1* (2), 69–86.

Humphreys P., Kaufmann, W. E., & Galaburda, A. M. (1990). Developmental dyslexia in women: Neuropathological findings in three patients. *Annals of Neurology, 28*, 727–738.

Hurley, A. D., Dorman, C., Loatsch, L., Bell, S. (1990). Cognitive functioning in patients with spina bifida, hydrocephalus, and the "cocktail party" syndrome. *Developmental Neuropsychology, 6*, 151–172.

Hynd, G. W., & Hiemenz, J. R. (1997). Dyslexia and gyral morphology variation. In C. Hulme & M. Snowling (Eds.), *Dyslexia: Biology, cognition, and intervention* (pp. 38–58). London: Whurr Publishers.

Hynd, G. W., & Semrud-Clikeman, M. C. (1989). Dyslexia and brain morphology. *Psychological Bulletin, 106* (3), 447–482.

Hynd, G. W., Semrud-Clikeman, M., Lorys, A. R., Novey, E. S., & Eliopulos, D. (1990). Brain morphology in developmental dyslexia and attention deficit disorder/hyperactivity. *Archives of Neurology, 47,* 919–926.

Hynd, G. W. & Wills, W. G. (1988). *Pediatric neuropsychology.* Boston, MA: Allyn and Bacon.

Illinois State Board of Education, Department of Special Education (1990). *Criteria for determining the existence of a specific learning disability.* Author.

Jensen, P. S., Martin, D., & Cantwell, D. P. (1997). Comorbidity in ADHD: Implications for research, practice, and DSM-V. *Journal of the American Academy of Child and Adolescent Psychiatry, 36,* 1065–1079.

Jones, E. D., Wilson, R., & Bhojwani, S. C. (1997). Mathematics instruction for secondary students with learning disabilities. *Journal of Learning Disabilities, 30,* 151–163.

Jordan, N. C., Levine, S. C., & Huttenlocher, J. (1995). Calculation abilities in young children with different patterns of cognitive functioning. *Journal of Learning Disabilities, 28,* 53–64.

Kolb, B. (1989). Brain development, plasticity, and behavior. *American Psychologist, 44,* 1203–1212.

Kolb, B., & Whishaw, I.Q. (1996). *Fundamentals of human neuropsychology* (4th ed.). New York: W.H. Freeman and Company.

Kosc, L. (1974). Developmental dyscalculia. *Journal of Learning Disabilities, 7,* 46–59.

Larsen, J. R., Hoien, T., Lundberg, I., & Odegaard, H. (1990). MRI evaluation of the size and symmetry of the planum temporale in adolescents with developmental dyslexia. *Brain and Language, 39,* 289–301.

Lemire, R. J., Loeser, J. D., Leech, R. W., & Alvord, E. C. (1975). *Normal and abnormal development of the human nervous system.* New York: Harper and Row.

Lezak, M. D. (1995). *Neuropsychological assessment* (3rd ed.). New York: Oxford University Press.

Locke, J. L. (1992). Thirty years of research on developmental neurolinguistics. *Pediatric Neurology, 8,* 245–250.

Lorber, R. (1994, October 9). Can neuropsychological tests help diagnose children referred for ADD/ADHD? [commentary] *The ADHD Report,* 2(5).

Lyon, G. R. (1996). Learning disabilities. *Future child, 6* (1), 54–76.

Mann, V. A., & Brady, S. (1988). Reading disability: The role of language deficiencies. *Journal of Consulting and Clinical Psychology, 56,* 811–816.

Martin, J. H. (1996). *Neuroanatomy text and atlas* (2nd ed.). Stamford, CT: Appleton & Lange.

Mesulam, M-M. (1985). Attention, confusional states, and neglect. In M.M. Mesulam, *Principles of behavioral neurology* (pp.125–168). Philadelphia, PA: F.A. Davis.

Mesulam, M-M. (1990). Large-scale neurocognitive networks and distributed processing for attention, language, and memory. *Annals of Neurology, 28,* 597–613.

Montgomery, J. W. & Levine, M. D. (1995). Developmental language impairments: Their transactions with other neurodevelopmental factors during the adolescent years. *Seminars in Speech and Language, 16* (1), 1–13.

Morrison, J. R., & Stewart, M. A. (1973). The psychiatric status of the legal families of adopted hyperactive children. *Archives of General Psychiatry, 28,* 888–891.

Morrow, L., & Ratcliff, G. (1988). Neuropsychology of spatial cognition: Evidence from cerebral lesions. In J. Stiles-Davis, M. Kritchevsky, & U. Bellugi, (Eds.), *Spatial cognition: Brain bases and development* (pp. 5–32). Hillsdale, NJ: Lawrence Erlbaum.

Myklebust, H. R. (1975). Nonverbal learning disabilities: Assessment and intervention. In H. R. Myklebust (Ed.), *Progress in learning disabilities* (Vol. III; pp. 85–121). New York: Grune and Stratton.

Orton, S. T. (1925). Word blindness in school children. *Archives of Neurology and Psychiatry, 14*, 581–615.

Paul, R., & Cohen, D. J. (1982). Communication development and its disorders: A psycholinguistic perspective. *Schizophrenia Bulletin, 8*, 279–293.

Pennington, B. F. (1991). *Diagnosing learning disorders: A neuropsychological framework*. New York: Guilford Press.

Pennington, B. F. (1995). Genetics of learning disabilities. *Journal of Child Neurology, 10* (Suppl. 1), 569–577.

Pennington, B. F. (1997). Attention deficit hyperactivity disorder. In T. E. Feinberg & M. J. Farah (Eds.), *Behavioral neurology and neuropsychology* (pp.803–807). New York: McGraw-Hill.

Piaget, J. (1960). *Psychology of intelligence*. Patterson, NJ: Littlefield Adams.

Plante, E., Swisher, L., & Vance, R. (1989). Anatomical correlates of normal and impaired language in a set of dizygotic twins. *Brain and Language, 37*, 643–655.

Plante, E., Swisher, L., Vance, R., & Rapcsak, S. (1991). MRI findings in boys with specific language impairment. *Brain and Language, 41*, 52–66.

Rankin, J. M., Aram, D. M., & Horwitz, S. J. (1981). Language ability in right and left hemiplegic children. *Brain and Language, 14*, 292–306.

Rice, M. (1989). Children's language acquisition. *American Psychologist, 44*, 149–156.

Robinson, R. J. (1991). Causes and associations of severe and persistent specific speech and language disorders in children. *Developmental Medicine and Child Neurology, 33*, 943–965.

Rourke, B. P. (1987). Syndrome of nonverbal disabilities: The final common pathway of white-matter disease/dysfunction. *The Clinical Neuropsychologist, 1*, 209–234.

Rourke, B. P. (1988). The syndrome of nonverbal disabilities: Developmental manifestations in neurological disease, disorder, and dysfunction. *The Clinical Neuropsychologist, 2*, 293–330.

Rourke, B. P. (1989). *Nonverbal learning disabilities: The syndrome and model*. New York: Guilford Press.

Rourke, B. P., Bakker, D. J., Fisk, J. L., & Strang, J. D. (1983). *Child neuropsychology: A introduction to theory, research, and clinical practice*. New York: Guilford Press.

Rosenzweig, M. R., & Leiman, A. L. (1997). *Physiological psychology*. New York: Ramdom House.

Sandler, A. D., Watson, T. E., Footo, M., Levine, M. D., Coleman, W. L., & Hooper, S. R. (1992). Neurodevelopmental study of writing disorders in middle childhood. *Developmental and Behavioral Pediatrics, 13*, 17–23.

Smith, S. D., Kimberling, W. J., Pennington, B. F., & Lubs, H. A. (1983). Specific reading disability: Identification of an inherited form through linkage and analysis. *Science, 219*, 1345–1347.

Snow, J. H., Prince, M., Souheaver, G., Ashcraft, E. (1994). Neuropsychological pattern of adolescents and young adults with spina bifida. *Archives of Clinical Neuropsychology, 9*, 277–287.

Spreen, O., Risser, A. T., & Edgell, D. (1995). *Developmental neuropsychology*. New York: Oxford University Press.

Squire, L. R. (1986). Mechanisms of memory. *Science, 232* (4758), 1612–1619.

Squire, L. R., & Shimamura, A. (1996). The neuropsychology of memory dysfunction and its assessment. In I. Grant & K. M. Adams (Eds.), *Neuropsychological assessment of neuropsychiatric disorders* (pp. 232–262). New York: Oxford University Press.

Stanovich, K. E. (1982). Individual differences in the cognitive processes of reading: 1. Word decoding. *Journal of Learning Disabilities, 15*, 485–493.

Stromswold, K. (1997). Specific language impairments. In T. E. Feinberg & M. J. Farah (Eds.), *Behavioral neurology and neuropsychology* (pp.755–772). New York: McGraw-Hill.

Stuss, D. T., & Benson, D. F. (1984). Neuropsychological studies of the frontal lobes. *Psychological Bulletin, 95*, 3–28.

Stuss, D. T., Gow, C. A., & Hetherington, C. R. (1992). "No longer Gage": Frontal lobe dysfunction and emotional changes. *Journal of Consulting and Clinical Psychology, 60* (3), 349–359.

Taylor, H. G., Fletcher, J. M., & Satz, P. (1984). Neuropsychological assessment of Children. In G. Goldstein & M. Hersen (Eds.), *Handbook of psychological assessment* (pp. 211–234). New York: Pergamon.

Torgesen, J. K. (1985). Memory processes in reading disabled children. *Journal of Learning Disabilities, 18*, 350–356.

Torgesen, J. K. (1988). Studies of children with learning disabilities who perform poorly on memory span tasks. *Journal of Learning Disabilities, 21*, 605–612.

Trimble, M. R. (1990). Psychopathology of frontal lobe syndromes. *Seminars in Neurology, 10*, 287–294.

Volpe, J. J. (1977). Normal and abnormal human brain development. *Clinics in Perinatology, 4* (1), 3–31.

Wagner, R. K., & Torgensen, J. K. (1987). The nature of phonological processing and its casual role in the acquisition of reading skills. *Psychological Bulletin, 101*, 192–212.

Wills, K. E. (1993). Neuropsychological functioning in children with spina bifida and/or hydrocephalus. *Journal of Clinical Child Psychology, 22*, 247–265.

Wills, K. E., Holmbeck, G. N., Dillon, K., McLone, D. G. (1990). Intelligence and achievement in children with myelomeningocele. *Journal of Pediatric Psychology, 15*, 161–176.

Wood, D. J. (1981). Cognitive development. In M. Rutter (Ed.), *Scientific foundations of developmental psychiatry* (pp. 230–244). Baltimore MD: University Park Press.

Ylvisaker, M., Hartwick, P., Ross, B., & Nussbaum, N. (1994). Cognitive assessment. In R. C. Savage & G. F. Wolcott (Eds.), *Educational dimensions of acquired brain injury* (pp.69–119). Austin, TX: Pro-Ed.

Ylvisaker, M., Szekeres, S. F., Haarbauer-Krupa, J., Urbanczyk, B., & Feeney, T. J. (1994). Speech and language intervention. In R. C. Savage & G. F. Wolcott (Eds.), *Educational dimensions of acquired brain injury* (pp.185–235). Austin,TX: Pro-Ed.

Ylvisaker, M., Szekeres, S. F., & Hartwick, P. (1994). A framework for cognitive intervention. In R. C. Savage & G. F. Wolcott (Eds.), *Educational dimensions of acquired brain injury* (pp.35–67). Austin, TX: Pro-Ed.

Ylvisaker, M., Szekeres, S. F., Hartwick, P., & Tworek, P. (1994). Cognitive intervention. In R. C. Savage & G. F. Wolcott (Eds.), *Educational dimensions of acquired brain injury* (pp.121–184). Austin, TX: Pro-Ed.

Zametkin, A. J., & Rappaport, J. L. (1987). Neurobiology of attention deficit disorder with hyperactivity: Where have we come to in 50 years? *Journal of the American Academy of Child and Adolescent Psychiatry, 26*, 676.

PART IV

PARAMETERS OF THE FORENSIC ARENA

With even minimal exposure to forensic cases, neuropsychologists immediately learn that there are important differences between the arenas of traditional clinical practice and forensic practice. Of the various adversarial proceedings that may involve neuropsychologists and other health care professionals, none are more intimidating and anxiety-producing than those held within courtrooms. These proceedings may involve civil or criminal law and may involve jury trials or bench trials. Whatever the nature of the case and independent of which side requests the deposition or trial testimony of the neuropsychologist, it is essential to possess a basic understanding of courtroom standards and procedures, as well as the court's and attorneys' expectations of expert witnesses. In Chapter 14, attorney J. Sherrod Taylor describes the general environment of court proceedings and delineates important legal decisions pertaining to neuropsychologists as expert witnesses. In the closing chapter of the book, Paul Lees-Haley and attorney Lawrence Cohen elaborate on the forensic process (including legal expectations, relevant controversies, and professional cautions) pertaining to clinical neuropsychologists, with specific attention to production of credible expert witness testimony.

Chapter 14

THE LEGAL ENVIRONMENT PERTAINING TO CLINICAL NEUROPSYCHOLOGY

J. Sherrod Taylor
Taylor, Harp & Callier, Columbus, Georgia

Introduction

The Greek philosopher Herakleitos tells us that change alone is unchanging. Although he possessed no understanding of the topics discussed in this chapter, the words of this ancient thinker ring clear today. Only one fact is certain, the legal environment pertaining to clinical neuropsychology is in a state of flux.

Before we proceed to examine this changing environment in detail, a few preliminary observations are in order. First, readers will probably notice immediately that this chapter is the only one in this volume solely authored by a practicing attorney. Specifically, the author is a plaintiff's trial lawyer who often represents individuals with traumatic brain injury (TBI) and their families during personal injury litigation. Moreover, the author has written a book reviewing the medicolegal aspects of TBI from the plaintiff's perspective (Taylor, 1997). Nonetheless, a concerted effort will be made to present a balanced view of the subjects discussed here. If, in rare instances, the plaintiff's perspective seeps in, readers are asked to forgive those lapses. Second, largely because of the author's background and professional interests, this chapter focuses upon personal injury cases dealing with brain damage, even though neuropsychologists may also serve as expert witnesses in a variety of other civil and criminal cases. This focus is intentional because it is within the context of personal injury litigation that clinical neuropsychologists play a dominant, if not a predominant, role. Finally, although this chapter considers both the historical and theoretical underpinnings of forensic neuropsychology in order to set the stage for a deeper understanding of the legal environment, most comments made here concern practical suggestions about how expert neuropsychologists may become effective participants in the civil justice system.

Overview of the legal environment

Although the legal environment pertaining to clinical neuropsychology is in a state of flux, it is undeniable that it has become more hospitable to practitioners within the past two decades. Court appearances by neuropsychological experts have increased dramatically. Indeed, neurospsychologists today commonly serve as experts during brain injury litigation. One commentator has suggested that practicing neuropsychologists will probably be summoned to testify about the status of a patient at least once during the course of treatment (Bigler, 1986).

To confirm the rapid ascendancy of neuropsychologists to the forensic stage, in April, 1997, the author searched for the word 'neuropsychologist' on a major online computerized database of reported court decisions. This research identified 401 state and federal decisions containing this term. With the exception of a few federal trial court opinions, the majority of these cases were from appellate courts. Although the first opinion mentioning the term 'neuropsychologist' appeared in 1954 (*Nichols v. Colonial Beach Oil Co.*), fifty-six percent of the cases were published within the past five years; 83% within the past decade; and 98% since 1980. Approximately seventy percent of these cases were civil actions. Thus, there can be little doubt that neuropsychologists have become important expert witnesses in recent years. Whether they are retained by plaintiffs or defendants, neuropsychologists provide important information to legal fact-finders about the consequences of the traumatic event which is the subject of the litigation (Kreutzer, Harris-Marwitz, & Myers, 1990).

Contributing factors

Clinical neuropsychologists are professionals who apply principles of assessment and intervention based upon the scientific study of human behavior, as it relates to abnormal and normal functioning of the central nervous system (Division of Clinical Neuropsychology, 1989). Neuropsychology – or neurological psychology – has deep historical roots. Essays about this subject have been published since the early 1850s. However, as we have seen, only recently has this discipline assumed paramount significance during litigation.

It has been asserted that within the past twenty years, more has been learned about the brain generally and about traumatic brain injury (TBI) specifically than in all the previous years of human history. When the United States Congress declared the 1990s to be 'The Decade of the Brain,' it spawned widespread interest in brain disorders among professionals from many disparate disciplines. Unprecedented interest in brain studies precipitated an era of great change in the understanding of brain-behavior relationships and in clinical neuropsychology. As a result, the legal environment pertaining to clinical neuropsychology also began to change. The recent emergence of neuropsychologists as litigation experts may be attributed to at least five significant factors.

Increasing TBI population

First, in the past two decades, the population of TBI survivors has increased, due to innovations in life-saving medical interventions and technology. The advent of emergency shock trauma units and the development of sophisticated neuroimaging procedures have led directly to the survival of numerous individuals injured by traumatic accidents. People who previously died from brain trauma now often live. Consequently, many legal actions, which only a few years ago would have sounded in 'wrongful death,' presently are predicated upon claims of 'personal injury.'

Although clinical neuropsychologists rarely testify in death cases, they commonly proffer evidence in injury cases. It is well known that a variety of traumatic events, such as motor vehicle accidents, falls, on-the-job incidents and violence cause brain injuries that may later lead to litigation. Since courts regularly require evidence regarding the cause, nature and extent of any claimed brain damage flowing from these accidents, they customarily turn to clinical neuropsychologists, seeking expert opinions on these subjects.

Advocacy organizations

Second, in 1980, TBI survivors and their families joined concerned health care providers and attorneys to form the National Head Injury Foundation (NHIF). Today this survivor advocacy organization is known as the Brain Injury Association (BIA). Since its inception, this group has strived to inform both the general public and practicing lawyers about the efficacy of presenting neuropsychological evidence during TBI litigation. In 1987, this organization held its first annual seminar for trial lawyers and emphasized the roles played by clinical neuropsychologists in court. Subsequent BIA conferences have underscored the importance of neuropsychological testimony.

Following the BIA's lead, other organizations have sponsored similar gatherings. These include the American Psychological Association, International Association for the Study of Traumatic Brain Injury, Association of Trial Lawyers of America, American Bar Association, International Neuropsychological Society, European Brain Injury Association and the International Brain Injury Association. Importantly, members of these various groups regularly author articles and make professional presentations which highlight the forensic aspects of clinical neuropsychology.

Advent of neurolaw

Third, in 1991, this author and his colleagues introduced a new field of medical jurisprudence, called *neurolaw*, to the neuropsychological (Taylor, Harp, &

Elliott, 1991) and legal communities (Taylor, 1991). Neurolaw will be the subject of a later section of this chapter. At this point readers only need to recognize that supporters of the new practice area strongly endorse the roles played by neuropsychologists in the legal setting.

Supply of neuropsychologists

Fourth, in recent years, the number of persons trained in clinical neuropsychology has increased substantially. This development has created greater visibility of this subspeciality and an increased availability of these expert witnesses. Relative abundance of qualified practitioners has led directly to an increased perceived need by attorneys and courts for neuropsychological expert services during litigation.

Response of the legal system

Fifth, since 1980, state legislatures and courts have begun to appreciate the evidence that competent clinical neuropsychologists are able to provide during litigation. By promulgating statutory and common law rules that facilitate the presentation of neuropsychological testimony in courts of law, these legal entities continue to encourage neuropsychologists to participate in the legal arena. In this regard, we should remember - the law encourages that which it permits.

Problem areas

Although courts customarily accept testimony offered by clinical neuropsychologists, these experts still must occasionally overcome some formidable hurdles. Not all persons in our society or within the juridical community recognize the significance of neuropsychological evidence. Not all courts admit testimony by these experts. Naysayers abound; and, the issues they raise demand our attention. For this reason, we shall now review some of the present challenges to neuropsychology within the legal setting and examine an illustrative case study which exposes many of the concerns existing in the current environment.

Challenges to neuropsychology

Futurists suggest that there exist today two primary groups of change agents (Theobald, 1996). One group of ideologues seeks to change the world so that it comports to their personal sense of what is right; the other group strives to help people cooperate with one another. Both groups are active within the realm of forensic neuropsychology. We shall examine the second group during our later discussion of 'neurolaw.' The remainder of this section discusses the first.

Certain members of the first group urge courts to refuse to admit neuropsychological evidence, arguing that clinical psychology is merely 'junk science' (Hagen, 1997). Others go a step further, contending that we live amidst a litigation explosion – fueled by America's therapeutic culture which encourages psychologists to join attorneys in bringing ever more victims of injury into the courtroom (Garry, 1997). However, court statistics do not support the existence of a litigation explosion. The National Center for State Courts (1994) reported that the volume of tort litigation has declined steadily since 1990.

Nonetheless, those persons who espouse antagonistic views regarding neuropsychological evidence have found at least some support among clinicians. For example, in their numerous publications, Ziskin and Faust (1988; Faust & Ziskin, 1988, 1989) frequently criticize the scientific underpinnings of neuropsychological testimony. And, although Brodsky (1989) called attention to the one-sided views of those authors, Matarazzo (1990), no doubt impelled by a strong sense of intellectual honesty, felt the need to disagree with some, but agree with other of their harsh opinions during his presidential address to the American Psychological Association.

Moreover, a minority of courts have also shown little regard for neuropsychological evidence. As we shall see in the case study that follows, those courts have been especially reluctant to approve testimony by neuropsychologists when they perceive that it constitutes a threat to the practice of medicine. Indeed, some litigants themselves attempt to pit neuropsychologists against physicians in ways that many court-watchers find extremely distasteful. By suggesting that persons holding the Ph.D. degree in clinical neuropsychology or psychology are not as competent to render opinions about brain-behavior relationships as people holding the M.D. degree, these ideologues strain both logic and the controlling rules of law.

Nonetheless, what we may call the 'R.D. (real doctor) versus Ph.D. confrontation' may impact adversely upon the outcome of litigation. Neuropsychologists who customarily work along side medical doctors in the clinical setting may well be surprised by the way some attorneys attempt to manipulate both groups in the legal environment. Those lawyers who use such unfair tactics during litigation have one singular goal: to demonstrate to courts and legal fact-finders (i.e., jurors) that the opinions of physicians are superior to and hence 'outweigh' those held by neuropsychologists. During the course of trial, this intellectually dishonest stance may create many serious problems for both courts and litigants alike.

Case study
To fully comprehend the current legal environment, clinicians need to be aware of the somewhat tortured history of neuropsychological evidence. Although most American courts now permit neuropsychologists to testify about the cause, nature and extent of brain injury, securing the legal right to present expert testimony has been a difficult undertaking. Issues surrounding proof about the *cause* of brain damage have been especially troublesome in some quarters. This section explores the causation question using appellate court cases arising in the states of Georgia and North Carolina to show how clinical neuropsychologists, attorneys and others have worked together in an effort to ensure the admissibility of this vital evidence. One particular case – *Drake v. LaRue Construction Company* (1994) – with which the author is intimately familiar, is illuminating. In the *Drake* case, the Georgia Court of Appeals held that a neuropsychologist was competent to render an opinion as to the cause of a workers' compensation claimant's brain injury. Initially, readers may find that holding to be less than

remarkable since today most jurisdictions allow such evidence. However, reviewing the events that preceded the decision in *Drake* is important because that case resolved a dispute existing in Georgia for several years and because the process of arriving at the *Drake* ruling delineates many issues which hound neuropsychologists even today (Taylor, 1995a).

From the outset, neither clinicians nor trial lawyers anticipated that presenting evidence on the cause of TBI would pose significant problems in *Drake*. The Georgia Court of Appeals previously had ruled in *Jacobs v. Pilgrim* (1988) that testimony of this sort was admissible. The plaintiff in the *Jacobs* case called two psychologists as expert witnesses to support her allegations of brain injury. One testified that the plaintiff suffered a coup-contrecoup type of head injury, resulting in memory loss, during the motor vehicle accident which was the subject of the litigation. Another psychologist opined that plaintiff suffered organic brain syndrome resulting from the wreck. However, the defendant offered testimony by a neuropsychologist who found that plaintiff's condition was not consistent with brain damage. Recovering a jury verdict for only $15,000, plaintiff appealed and contended that the trial court erred in allowing the defense witness to opine that her condition resulted from a psychological reaction to stress predating the accident.

However, the *Jacobs* court affirmed the jury's award, finding that the defense expert's testimony was relevant to cast doubt on the other psychologists' explanations of the causal relationship between the collision and plaintiff's health problems. Under the holding in the *Jacobs* case, most clinicians and attorneys concluded that thereafter neuropsychologists would be allowed to testify regarding the origin of brain disorders in Georgia courts. Interestingly, it should be noted that it was the defendant – rather than the plaintiff – who proffered the neuropsychologist's expert testimony in the *Jacobs* case.

The same appellate court, citing its prior *Jacobs* decision, again upheld the admissibility of neuropsychological evidence on the issue of causation in *Morris v. Chandler Exterminators, Inc.* (1991). However, the Supreme Court of Georgia – without referencing the *Jacobs* case – overruled the *Morris* decision, finding that a psychologist was not qualified to render an opinion 'concerning a diagnosis of a mental disorder when such mental disorder requires a professional opinion as to a physical disorder – here organic brain damage' (*Chandler Exterminators, Inc. v. Morris*, 1992). Essentially, this new ruling meant that only a medical doctor – and not a psychologist or neuropsychologist – could testify about the cause of brain injury in Georgia. All seven justices of the court concurred in the decision.

Since a unanimous court rendered this opinion, members of the Georgia Psychological Association, brain injury advocacy groups and trial attorneys quickly recognized that drastic measures would need to be taken, if neuropsychological testimony on causation were ever again to be admitted into evidence in the State of Georgia. Acknowledging that it would be highly unlikely to expect that state supreme court to overrule its *Chandler* decision, concerned individuals and organizations determined to seek a legislative, instead of judicial, remedy for this problem.

During the mid-winter meeting of the Georgia Psychological Association in 1993, clinicians joined members of several TBI advocacy groups and trial lawyers (including this author) in drafting proposed amendments to Georgia's Psychology Practice Act that were designed to rectify the evidentiary problems created by the *Chandler* decision. Subsequently, the Georgia General Assembly passed and the state's Governor signed a bill allowing psychologists and neuropsychologists to render a variety of professional opinions about brain damage – including opinions regarding its cause. The precise language used in these amendments appears later in this chapter in the section discussing important legal statutes.

These changes in the statutory law of Georgia became effective on July 1, 1993, and with them in place, observers of the legal scene again expected that both plaintiffs and defendants henceforward would be allowed to offer evidence regarding the cause of brain damage in Georgia courts (Taylor, 1993). Amazingly, this understanding turned out to be incorrect. After the amendments took effect, the Georgia Court of Appeals decided another case, *Handy v. Speth* (1993), which was pending at the time the modified statutes were enacted. In *Handy*, the court erroneously relied upon the Supreme Court's prior *Chandler* decision, holding that the trial court had properly precluded testimony by a neuropsychologist about the cause of plaintiff's brain injury.

Finally, in 1994, the Georgia Court of Appeals corrected this mistake through its decision in the aforementioned *Drake* case. There the court held:

> We discern that the intent of the amendment was to legislatively overrule *Chandler*, at least to the extent that a neuropsychologist is not qualified to render an opinion concerning the diagnosis of the pathology of organic brain disorders and brain damage. Under the amendment, [the neuropsychologist] is not incompetent to testify that in his opinion Drake's cognitive deficits were caused by his head injury.

Later, the Georgia Supreme Court denied all efforts to further review the *Drake* decision. Now it is clear that neuropsychological evidence pertaining to the cause of brain injury is admissible in the State of Georgia.

Regrettably, the *Drake* decision came too late to be of assistance to the plaintiff in the *Handy* case. Addressing this anomalous situation in the *Drake* opinion, the court observed that in *Handy* no issue was raised as to whether the judgment there should have been reversed and a new trial ordered on grounds that the neuropsychologist's testimony had *become* admissible under the law in effect when the *Handy* appeal was decided. By omitting note of the amended statutes, both plaintiff's counsel and the appellate court contributed to an unjust result in the *Handy* case.

This case study is important for several reasons. First, it illustrates how an appellate court may go awry when it fails to take notice of prior decisional law. After all, the Georgia Supreme Court in *Chandler* originally could have avoided the problems described here, if it had simply followed the correct statements of law announced in *Jacobs* that allowed testimony by neuropsychologists about the cause of TBI. Similarly, that court could have simply affirmed the *Morris* decision and this, too, would have produced the same result. Second, this *Drake* case study reveals how neuropsychologists, trial lawyers, TBI survivors and advocacy groups may join together to correct erroneous court decisions through legislation. Third, this case study demonstrates the need for clinical neuropsychologists, as well as courts and lawyers, to stay abreast of changes in the law. If plaintiff's neuropsychologist, involved attorneys or the appellate court had recognized earlier the impact of the amendments to Georgia's Psychology Practice Act, the plaintiff's case in *Handy* could have been salvaged. Finally, the *Drake* decision confirms that Georgia courts now follow the majority rule that allows psychologists and neuropsychologists to testify about the cause, nature and extent of brain injury. Although these are critically important lessons, some other courts have not yet learned them.

Readers, perhaps, would naturally expect that this saga would end at this point. But, more remains to be said. Approximately two and one-half years following the *Drake* decision, the Court of Appeals of North Carolina rendered its opinion in *Martin v. Benson* (1997). In that case, the court held that, when a neuropsychologist expressed 'an opinion that the plaintiff did not suffer a closed head injury' as a result of an accident, he 'testified as to medical causation.' Then, this appellate court found that 'this testimony invades the field reserved for the practice of medicine' in North Carolina. Thus the court refused to allow the neuropsychologist's opinion about causation. In reaching this surprising conclusion, this appellate court cited Georgia's *Chandler* case as support, even though *Chandler* had been found to have been legislatively overruled in the *Drake* decision.

When the *Martin* case came before the North Carolina Supreme Court, that court failed to squarely address the issue of causation of brain injury in that state. Instead, the supreme court reversed the court of appeals on the ground that plaintiffs had failed to object to the defense neuropsychologist's testimony in the trial court. In taking that position, the supreme court found that plaintiffs had waived their right to appellate review of that testimony.

Subsequently, in *Curry v. Baker* (1998), the North Carolina Court of Appeals upheld a $900,000 jury verdict in favor of a man injured in a vehicular accident, even though a neuropsychologist had testified that the plaintiff sustained TBI caused by the collision. Again skirting the question of admissibility of that evidence, the court affirmed the trial court's judgment because defendant failed to properly object to the clinician's testimony at trial. Thus, the current rule applicable in North Carolina appears to be that neuropsychologists (a) are competent to testify about psychological and emotional conditions of patients without expressing, opinions as to the organic causes of those conditions and (b) are also competent to testify that the psychological and emotional conditions of patients are not consistent with other patients who have been medically diagnosed with brain injuries. Of course, this North Carolina rule remains outside the mainstream position taken by other courts that generally allow neuropsychologists, to render opinions about the physical and organic causes of TBI. Therefore, as long as a few courts continue to rule that neuropsychologists tread on medical turf when they testify about brain injury, the legal environment will remain in a state of flux.

Legal bases of neuropsychological evidence

Although some neuropsychologists may believe that matters of law should be left wholly to consideration by the judges and attorneys involved in particular cases, this is an untenable point of view. Clinicians themselves need to be cognizant of the legal principles underpinning their testimony if they are to avoid embarrassment during legal proceedings. Statutes and appellate court decisions constitute the sources of law forming the predicate for neuropsychological evidence. Statutes – commonly called 'code sections' by practicing attorneys and knowledgeable experts – articulate the public policy of every jurisdiction and establish the precise rules and procedures governing legal actions. Legislatures enact statutes in order to define the parameters of lawful conduct and to enunciate legal principles. On the other hand, appellate courts interpret statutes, review trial court judgments and otherwise author opinions setting forth common law rules used in trial proceedings. Since there are numerous statutory enactments and appellate court decisions that may impact upon clinical neuropsychologists, several of those which are especially significant will now be examined.

Statutes

To be sure, there are potentially many statutes dealing with neuropsychologists and the subjects about which they testify in court. These code sections principally involve the following topics: (a) the neuropsychologist's scope of practice, (b) legal definitions relevant to neuropsychological subjects, and (c) qualifications of expert witnesses. We shall now briefly review these key areas of interest.

Scope of practice

Before testifying in any legal action, it is essential for clinical neuropsychologists to become familiar with precisely how controlling statutes define the practice of neuropsychology in their own specific jurisdictions. Although some state laws carve out 'neuropsychology' as a professional specialty, others consider the practice of neuropsychology to be part of the practice of 'psychology' generally. From the legal perspective, it matters little whether particular statutes consider neuropsychologists as a distinguishable group or as members of the profession of psychology as a whole.

What is important, however, is the manner in which statutes define the scope of practice for neuropsychology or psychology. One state's law, which addresses both disciplines, is illustrative. The Official Code of Georgia Annotated, OCGA §43–39–1 (2)(1993), provides: "'Neuropsychology' means the subspecialty of psychology concerned with the relationship between the brain and behavior, including the diagnosis of brain pathology through the use of psychological tests and assessment techniques".

This code section is important because it clearly (a) establishes neuropsychology to be a subspecialty of psychology, (b) describes the subjects with which this practice is concerned, and (c) sets forth the methods through which this practice is pursued. Moreover, it also announces that neuropsychologists may diagnose 'brain pathology'. It was, of course, this latter provision that led directly to the decision of the Georgia Court of Appeals in the previously discussed *Drake* case.

However, even though the statute quoted above distinguishes 'neuropsychology' as a special area of practice, a later portion (i.e., OCGA §43–39–1(3)(1993)) lumps neuropsychologists together with psychologists generally, and pertinently provides:

> 'To practice psychology' means to render . . . to individuals . . . for a fee . . . any service involving the application of recognized principles, methods, and procedures of the science and profession of psychology, such as, but not limited to, diagnosing and treating mental and nervous disorders and illnesses, rendering opinions concerning diagnoses of mental disorders, including organic brain disorders and brain damage, engaging in neuropsychology. . .

Of course, that part of the statute is significant because it confirms that neuropsychologists and psychologists alike may testify in Georgia about brain damage and this language also played a vital role in the *Drake* decision described earlier in our case study.

By apprehending the legal scope of their practices, neuropsychologists preview how courts may later look at their professional expertise. When presenting their expert opinions to any tribunal, witnesses must always stay within the bounds of their specific areas of practice. Deviations will likely result in trial judges excluding their expert testimony.

Relevant legal definitions

Legal definitions of particular terms often become important during litigation. As a result, neuropsychologists must become familiar with those relevant to their expert testimony. Without doubt, the most important term which may be defined in a state's law code is 'traumatic brain injury'. The Official Code of Georgia Annotated, OCGA §37–3–1 (16.1) (1993), is illustrative: " 'Traumatic brain injury' means a traumatic insult to the brain and its related parts resulting in organic damage thereto which may cause physical, intellectual, emotional, social or vocational changes in a person . . .".

Currently, approximately a dozen American states define TBI in a similar manner in their codes. Importantly, these statutes enable neuropsychologists to testify about the organic aspects of brain damage and they allow neuropsychologists to avoid being perceived as intruding upon areas relegated, in some states, to the practice of medicine. By using the language found in definitional statutes, neuropsychologists become able to 'speak the language of the court' in their reports, records and testimony.

When relevant and desirable definitions are not available, proactive neuropsychologists may work with their state's psychological association to draft and otherwise encourage their own legislatures to enact them. As the previously described case study reveals, this affirmative action can be very important in some instances.

Qualifications of expert witnesses

Both federal and state laws describe generally the factors that enable certain persons, including neuropsychologists, to render opinion evidence during litigation. For example, Rule 702 of the Federal Rules of Evidence, used in federal courts, provides: "If scientific, technical or other specialized knowledge will assist the trier of fact to understand the evidence or to determine a fact in issue, a witness qualified by knowledge, skill, experience, training or education, may testify in the form of an opinion or otherwise".

State law codes also define experts as those who possess knowledge, skill, experience, training or education outside the ken of ordinary lay persons. By interfacing these general considerations with specific information found in statutes defining a professional's scope of practice, courts are able to determine whether an individual will be qualified to serve as an expert witness in a particular case.

Appellate court cases

The common law found in state and federal appellate court decisions provides neuropsychologists with specific guidance regarding (a) the admissibility of their expert testimony, (b) the subjects about which they may testify, and (c) the degree of accuracy they must employ when expressing their professional opinions in the court. However, since these decisions may vary widely, it is beyond the scope of this chapter to address particular holdings. In other words, to determine the contents of decisions governing their testimony, the neuropsychologist should discuss during a pre-testimony conference the relevant cases with the particular lawyer who plans to call him or her as an expert witness. Nonetheless, two important decisions will now be reviewed briefly.

Neuropsychologists who desire to obtain a comprehensive overview of the manner in which appellate courts have dealt with neuropsychological evidence may read the decision of the Iowa Supreme Court in *Hutchison v. American Family Mutual Insurance Company* (1994). That case offers a useful review of the myriad topics about which neuropsychological testimony has been both allowed and disallowed. However, some inaccuracies do appear in that decision; therefore, readers must exercise caution during their review. Several decisions cited therein have now been overruled. Moreover, since that decision was rendered, other cases have been decided. Notwithstanding these relatively minor problems, the *Hutchison* case remains a valuable introduction to neuropsychological evidence. Similarly this author and his colleagues have published a detailed discussion of many of these cases (Taylor, Elliott, & Harp, 1991). Other authors have followed this path (Stern, 1999).

Neuropsychologists, especially those who testify in federal court, should also become familiar with the United States Supreme Court's decision in *Daubert v. Merrell Dow Pharmaceuticals, Inc.* (1993). Although that decision does not specifically involve testimony by a neuropsychologist, it does contain important legal principles governing the admission of expert testimony. *Daubert* examines the methodology used by experts in reaching their opinions and requires the trial judge to ensure that an expert's testimony rests both on a reliable foundation and is relevant to the case at hand. However, this decision has proven to be highly controversial and subject to wide interpretation by lower federal courts. Some federal circuit courts have attempted to expand, while others have sought to limit, the holding in this case. For this reason, before testifying in federal court, neuropsychologists should consult with the attorneys who plan to call them as witnesses to ascertain how *Daubert* has been interpreted within the jurisdiction in which they plan to testify.

Neurolaw

Although clinical neuropsychologists are among the most important expert witnesses during brain injury litigation, they may find interfacing with the legal system to be confusing or frustrating. This confusion and frustration may flow in part from a lack of familiarity with the legal process (Penrod, 1996). To address those concerns, a new field of medical jurisprudence arose in 1991. Called 'neurolaw,' this area of trial practice was introduced to the neuropsychological and legal communities simultaneously (Taylor, 1991; Taylor, Harp, & Elliott, 1991). Neurolaw constitutes a synthesis of medicine, neuropsychology, rehabilitation and law which deals with the medicolegal ramifications of neurological injuries - notably, acquired brain damage.

The central thesis of neurolaw proposes that the financial resources secured through legal remedies offered by the civil justice system contribute to improving quality of life for people with neurological injuries and their families (Taylor, 1995b). When survivors of brain injury obtain compensation during personal injury litigation, they receive money damages that may be used to fund costly rehabilitation programs, replace earnings lost as a result of injury and compensate for attendant physical pain and mental suffering. These monetary awards, therefore, promote optimal recovery following injury and instill hope into the lives of those who are touched by acquired neurological maladies.

Cardinal principle

Neurolegal philosophy and practice are governed by four cardinal principles (Taylor, 1996). Each of these principles will now be described with greater specificity. In reviewing these individual principles, readers will note that each is designed especially to enhance the clinical outcomes of patients with brain injury and to increase the capability of expert witnesses to provide clinical testimony during litigation, while simultaneously reducing confusion or frustration that may arise in legal cases. Of course, not all cases that are considered in the beginning of litigation to be 'brain injury' cases will ultimately be determined to be such.

First principle

The first principle of neurolaw provides that better legal outcomes promote better clinical outcomes for persons with neurological injuries. Clinical neuropsychologists, as well as other health care experts, generally acknowledge that there is improvement in the functional abilities of patients following the application of neurorehabilitative care. Research studies reveal that early therapeutic intervention (Malec et al., 1993) and lengthy and intensive treatment programs (Spivak et al., 1992) are important factors that contribute significantly to the achievement of favorable clinical outcomes. Substantial compensatory damages awards derived through litigation may be utilized to support neurorehabilitative activities and, consequently, will contribute to improving quality of life for patients and their families.

Second principle
Neurolaw's second principle urges that success in neurolitigation is dependent largely on the quality and quantity of expert evidence. Therefore, neurolawyers – those attorneys who specialize in neurological injury cases – are highly motivated in their quest to proffer an abundance of expert evidence in brain injury litigation. This evidence generally comes from members of the patient's multidisciplinary treatment team, and, of course, clinical neuropsychologists are integral participants on that team. Although plaintiff's attorneys usually rely on treating or consulting neuropsychological experts to furnish this vital evidence, defense lawyers obtain this evidence from experts who have been retained to review the claimant's medical records and other documentation who appear in litigation under the auspices of the 'independent medical examination.' Regardless of whether neuropsychologists testify as plaintiff's or defendant's experts, their task is straightforward: to confirm the presence or absence of the claimed injury and to render opinions regarding the cause, nature and extent of the claimant's health condition. Since neurological injuries – especially those involving the human brain – are complex, neurolaw is considered to be the most 'expert intensive' form of personal injury practice.

Indeed, neurolitigation often becomes a 'battle of experts.' In TBI cases, it is not uncommon for two similar-looking neuropsychological evaluation patterns to have profoundly different implications, depending on the individual examined and the context of the situation involved. For this reason, it is in the interpretative function that forensically sophisticated neuropsychologists may become vital to the outcome of the litigation (Miller, 1992).

Third principle
The third principle of neurolaw holds that mutual cooperation among concerned professionals enhances the probability of successful neurolitigation. This principle is of vital significance because it has been suggested that the medicolegal system rarely works in harmony for the best interests of litigating patients (Packard, 1993). Consequently, dealing with brain injury in the legal environment requires teamwork. Neurolaw's supporters, therefore, encourage legal and health care professionals to strive to overcome barriers which historically have tended to divide them. Whether neuropsychologists serve plaintiff's or defendant's interests when they testify, they still have a legal duty to provide courts and juries with

accurate and authoritative evidence which is probative of the legal issues being tried. Moreover, attorneys have a professional duty to ensure that they assist expert witnesses in bringing this evidence before the tribunal in an efficient manner. Ideally, when experts and attorneys work together in harmony, much of the frustration and confusion that can arise during litigation disappears.

Fourth principle
Neurolaw's fourth cardinal principle suggests that to be successful, clinical and legal professionals require 'litigation literacy.' That term means possessing a basic awareness of the intricacies of the legal system. Specifically, this notion urges experts and the lawyers who proffer their testimony to become aware of the legal rules and procedures that govern the rendering of opinion testimony.

Practical considerations
General recognition by clinical neuropsychologists of the multiple roles they play in neurolitigation is seldom sufficient for them to be considered 'litigation literate.' Rather, these clinical experts need to know specifically what is expected of them. In other words, they need to fully comprehend the litigation project that lies before them. There have been several excellent papers authored by clinicians that have previously addressed this subject (Guilmette & Giuliano, 1991; Kreutzer et al., 1990; Miller, 1996; Miller, 1992; Rothke, 1992). This chapter adds the view of a practicing trial lawyer to that body of literature.

The litigation project
Neuropsychologists have two principal tasks within the legal environment. First, it is their job to confirm the presence or absence of brain injury in the plaintiff. Second, neuropsychologists strive to provide information relating to the cause, nature and extent of the litigant's health condition. When brain injury appears, these experts seek to determine how the brain injury has impaired the claimant's ability to function in the real world, what the plaintiff would be like in the absence of injury and whether neurorehabilitative treatments exist that may allow the plaintiff to compensate for any deficits (Miller, 1992).

Prior to presenting legal evidence, neuropsychologists must ascertain which specific rules govern the degree of certainty upon which their opinions should be based. Although a few jurisdictions require expert opinions in civil lawsuits to be predicated upon 'a reasonable degree of *certainty*' (i.e., greater than 95% chance), the vast majority of courts only require opinions to be expressed in terms of 'a reasonable degree of *probability*' (i.e., greater than 50% chance). Thus, in most instances, neuropsychologists possess a greater degree of flexibility with respect to their professional opinions than they may initially anticipate.

This legal construct of 'probability' is of crucial significance and is consistent with the burden of proof that civil litigants bear, called '*preponderance of the evidence*.' To establish a fact or an opinion by a preponderance only requires showing that the fact or opinion is more likely correct than not. This is the reason that trial judges commonly instruct lay jurors that this rule does not necessitate proof to an absolute certainty, since such proof is rarely obtainable. Therefore, when neuropsychologists fully accept the legal concept of 'probability', if that is appropriate under the governing rules of evidence, they take a giant step toward cooperating with courts and lawyers during neurolitigation. Moreover, application of the proper standard for expert testimony demonstrates that the neuropsychologist is literate in the ways of the law.

Cause of injury

Making a determination regarding the cause of the claimant's alleged condition is of paramount importance during litigation. If the traumatic event did not cause or contribute to causing plaintiff's condition, there can be no recovery of damages and the litigation will fail. For this reason, when clinical neuropsychologists testify, they must clearly articulate any opinions they have regarding causation. Neuropsychologists usually find it easy to testify that a particular incident produced a moderate or severe brain injury. They are rarely challenged when they present such testimony. However, if the plaintiff's alleged brain injury is classified as a mild traumatic brain injury (MTBI), then presenting definitive testimony as to the cause of subtle alterations in brain behavior relationships may prove more challenging. So-called MTBIs may not be readily apparent, may not be manifest on medical testing or neuroimaging and may even have been overlooked initially by other medical professionals. Sometimes there may exist a plethora of alternative explanations for the cause of plaintiff's condition.

In instances in which some confusion or doubt exists as to the cause of plaintiff's condition, neuropsychologists may find solace by applying the scientific principle of parsimony which suggests that no more causes should be assumed than are necessary to account for the facts presented by a particular case. Known as Ockham's razor, this maxim has guided scientific explanation for more than six hundred years and has been followed by those persons who refuse to multiply causes unnecessarily (Robinson, 1995). Following this maxim, one noted authority has observed that, in the absence of reliable data concerning an individual's premorbid state, abnormal behavior observed after head injury may be assumed to be the result of brain damage (Jennett, 1982).

At all times, neuropsychologists should remember the concept of legal 'probability' and should refrain from speculating about causation issues – especially if they arise in the context of alternative 'possible' causes. What is 'possible' is never probative in a legal case. Evidence must always be couched at least in terms of probability. Experts who fall into the trap of rendering professional opinions predicated upon mere possibilities fail to abide by the applicable rules of evidence. One legal writer has reminded us that testimony presented in terms of possibilities or alternatives may lead to jury speculation and, therefore, does not aid or assist the jury in its determination (Kolpan, 1989).

Nature of injury

Neuropsychological literature contains abundant information pertaining to the nature of brain injury (i.e., physical, intellectual, emotional, social, or vocational dysfunction). Making reference to this literature during expert testimony is acceptable in all courts. However, as we have seen, many state legislatures have codified specific definitions of 'traumatic brain injury' and similar conditions within state statutes. Neuropsychologists, therefore, should become familiar with the definitions that may apply in their own geographic areas and should employ them appropriately when they testify.

Extent of injury

A major purpose of neuropsychological evidence is to establish the extent, if any, of the claimant's alleged injury. More serious injuries create entitlement to more substantial monetary awards in meritorious cases. During neurolitigation, plaintiffs have two primary aims. First, they wish to prove that they have sustained significant personal injuries. Second, they desire to present evidence which demonstrates a specific plan (i.e., life care plan) under which neuropsychologists and others may address those injuries. Plaintiff's announced goals mesh well with both medical and neurorehabilitative paradigms. As neuropsychologists know, the medical model focuses on what is wrong with the plaintiff, while the neurorehabilitation paradigm focuses on increasing the patient's capacities through the use of intervention and treatment (Condeluci, 1992). By presenting evidence about the extent of plaintiff's injuries utilizing both paradigms, neuropsychologists contribute to the acquisition of compensatory damages awards needed to meet all of the claimant's future needs.

Of course, neuropsychologists who render testimony on behalf of defendants may be used by defense counsel legitimately to show either that plaintiff's problems were not caused by the traumatic event that is the subject of the litigation or seek to minimize the extent of the injuries claimed in order to negate or limit the amount of a verdict or settlement. However, regardless of the position taken by neuropsychological witnesses, great care must be taken to focus all testimony about extent of injury on issues relevant to plaintiff's impairment, disability and handicap, if any. These are the principle factors that impact on the plaintiff's future life.

Presenting neuropsychological evidence

After neuropsychologists complete the evaluations and assessments that form the bases of their professional opinions, they must then grapple with how they may effectively present their evidence during litigation. As we have previously mentioned, before presenting evidence at deposition or trial, neuropsychologists should meet with the attorneys who plan to offer their testimony (Kreutzer et al., 1990). During these preparatory conferences, neuropsychological experts outline their views about the case at hand, describe the testing protocols and results that have led to the formulation of their opinions and suggest potential questions which the attorneys may pose in order to elicit relevant testimony. Moreover, during these meetings, clinicians should ask attorneys to explain the procedural and evidentiary rules that will be used by the court to determine the admissibility of the evidence they intend to offer. The chief goal of these conferences is to facilitate an exchange of important information so that witnesses and lawyers alike will be comfortable in the legal environment.

At all times, neuropsychologists should impress upon attorneys that they view their role to be that of *educators, rather than advocates*. However, simultaneously, neuropsychologists should make it clear that they hold certain professional opinions relevant to the case and that they intend to stand by those opinions when they testify. In this manner, *experts become advocates for their own opinions, rather than advocates for particular litigants*. This is of vital importance, since attorneys rely upon the information provided by their experts when they construct their plans for trying lawsuits. Taking this approach, neuropsychologists confirm that they will express their opinions objectively, without speculation. And, they show their keen desire to be honest and truthful in their evidence, regardless of the party for which they testify (Gudjonsson, 1993). Upon completion of the preparatory conference, neuropsychologists should be prepared to present direct testimony and cross-examination testimony during depositions and in court.

Direct examinations

During direct examinations, attorneys ask neuropsychologists non-leading questions designed to obtain testimony favorable to the parties whom they represent. Customarily, direct examinations occur either during depositions prepared for use at trial or during 'live' presentations in court; direct examinations seldom occur during the discovery deposition taken by opposing counsel before trial. For this reason, direct examinations are non-threatening and afford experts unimpeded opportunities to articulate their professional opinions concerning the cause, nature and extent of the plaintiff's alleged condition.

During direct examinations, neuropsychologists strive to educate legal factfinders about the pertinent features of the case and to express their opinions in meaningful ways, without resorting to professional jargon. In these question and answer sessions, witnesses carefully review their own credentials, claimant's pre- and post-injury status, salient documentary evidence and the results of neuropsychological evaluations. To be successful in this endeavor, clinicians must make orderly presentations that demonstrate both the accuracy of their evidence and their own personal credibility.

Cross-examinations

During cross-examinations opposing attorneys ask neuropsychologists questions designed to precipitate responses favorable to the other litigating party. Cross-examinations may be conducted during discovery depositions, depositions taken for use at trial or in court. Unlike direct examinations, cross-examinations are frequently peppered with leading questions suggestive of the answers attorneys seek to secure. For this reason, neuropsychologists find these question and answer sessions to be somewhat threatening and capable of producing at least some anxiety. Much of this anxiety arises because most clinical experts appreciate that attorneys are rapidly becoming adept in pursuing vigorous cross-examinations. Consequently, experts perceive becoming increasingly vulnerable when they testify (Guilmette & Giulianio, 1991).

In order to prepare to meet these challenges, neuropsychologists must recall that the purpose of cross-examination boils down to the opposing lawyer's desire (a) to obtain helpful information, (b) impeach the expert's testimony in some fashion, or (c) to limit the scope and effect of the expert's testimony (Stern, 1995). Attorneys acquire helpful information when neuropsychologists make concessions regarding their opinions. Instead of questioning these witnesses about the particular case at hand, skillful cross-examiners commonly take a tangential approach by asking about general, rather than specific, issues. Additionally, attorneys may impeach neuropsychologists using a variety of tactics. Showing that experts have previously expressed contrary or inconsistent opinions, pointing out the existence of professional literature containing contrary information and attacking an expert's credentials are typical methods of impeachment. Moreover, by proving that neuropsychologists did not conduct a relevant or comprehensive evaluation of the plaintiff, attorneys seek to limit the testimony of experts.

However, neuropsychologists who are both competent and 'litigation literate' may thwart the goals of cross-examination. If they have conducted complete, appropriate and thorough clinical assessments, clinical witnesses may avoid embarrassment. If they apprehend the ways of the law, instead of fearing them, neuropsychologists may use cross-examinations as opportunities to bolster, rather than denigrate, their direct testimony. The examples of sample testimony that follow may help to clarify these points.

Cross-examination example number one
Question: Doctor, you have testified that the accident that is the subject of this case caused the plaintiff's brain injury, is that correct?
Answer: Yes, that's right.
Question: Doctor, isn't it true that matters involving the cause of brain injury lie outside the scope of the practice of neuropsychology?
Answer: No, that is not correct. The field of neuropsychology includes causes of brain injury and, in this state, neuropsychological practice is governed by a specific statute that permits me to render my opinion on this subject. And, I understand that this right has been upheld by our appellate courts.

Cross-examination example number two
Question: Doctor, is it your testimony that this plaintiff's brain injury involves physical, emotional, cognitive and employment deficits?
Answer: Yes, that is what I found when I conducted a comprehensive evaluation. And that's what I would expect to find, based on similar cases. Also, relevant professional publications indicate that these are the problems commonly encountered in brain injury cases.

Cross-examination example number three
Question: Doctor, you have expressed your opinions in terms of probability, not certainty, is that correct?
Answer: That is correct. It is my understanding that this is what I am supposed to do in this case. I might add, that my opinion is based upon the same degree of certainty that my own profession employs in cases like this one.

Potential pitfalls

A litigation pitfall is a trap that catches the unsuspecting witness by surprise. Most neuropsychologists recognize that several of these traps may await them in virtually every brain injury case, especially those involving subtle deficits. Three are especially pernicious. First, even though loss of consciousness is not required in order for an individual to sustain brain injury (Mild Traumatic Brain Injury Committee, American Congress of Rehabilitation Medicine, 1993), some attorneys persist in attempts to have neuropsychologists agree that it is prerequisite. Savvy experts assiduously refrain from agreeing with that proposition. Second, some lawyers strive to have clinical experts opine that brain injuries must appear on neuroimaging in order to be real. This, too, is a 'red herring' that neuropsychologists should refute during their testimony. Varney and Shepherd (1991) and numerous others (see Rizzo & Tranel, 1996) have confirmed that even under the most restricted definition of minor head injury, a patient's computed tomography (CT) scan, magnetic resonance image (MRI) and electroencephalogram (EEG) typically are normal. Clinical experts would do well to remember that the absence of a positive finding on any diagnostic test is not evidence of a lack of injury. Further, even more severely injured patients, for whom there is no question of injury may have negative neuroimaging scans (e.g., Rimel, Giordani, Barth, & Jane, 1982). Although an infrequent occurrence, it is possible for a comatose patient to have negative neuroimaging. Finally, some attorneys seek an agreement from neuropsychologists that brain injury claimants commonly prolong their own dysfunction because of litigation. This area of clinical research remains controversial, with some investigators, such as Rimel et al. (1981), reporting that pending litigation had a negligible effect, if any, on the recovery of patients with head injury, while others, such as Binder and Rohling (1996), have reported that financial incentives do affect presentations and outcomes. It would seem that there is no reason to pre-judge a particular claimant to be negatively affected by involvement in litigation, since researchers have shown that both no effects and negative effects are possible.

The future

As we have seen, for many years courts and legislatures perceived no acute need to address clinical neuropsychology in the forensic setting. Consequently, they refrained from doing so. This absence of necessity accounts for the dearth of statutes and court decisions dealing with neuropsychologists prior to 1980. In this respect, the law is much like a living organism that evolves only when confronted with a need for change. Even when society generally considers change to be desirable, the law often responds slowly. Using the legal doctrine known as *stare decisis*, courts tend to stand by previous precedents and do not often disturb settled positions. Thus, we may observe correctly that legal policy dictates stasis, absent necessity for change.

However, eventually legislatures and courts typically catch up with the ideas and concepts prevailing within society. To borrow a term from the field of paleobiology, we may say that the law is subject to periods of 'punctuated equilibrium' in which legal stasis is interrupted by marked shifts in public policy. As we have observed, during the 1980s and thereafter, many factors converged to encourage the participation of clinical neuropsychologists in the legal process. We have only to look at the dramatic increase in the number of court decisions mentioning neuropsychologists to confirm these experts' impact in recent years.

Testifying on behalf of plaintiffs, neuropsychologists provide evidence that may lead courts and juries to award compensation needed by survivors and their families to fund care programs, replace lost income and compensate for pain and suffering. Testifying on behalf of defendants, neuropsychologists provide evidence that may demonstrate the absence of brain injury or minimal effects upon quality of life. Under either circumstance, the task of the neuropsychologist is to make a careful and impartial assessment of the claimant and to identify genuine deficits, while also exposing any exaggeration of symptoms that may be present (McKinlay, 1992).

Although no commentator would likely argue that the legal roles played by neuropsychologists can continue to expand *ad infinitum*, as of this date, there appear to be few limits on the horizon. The same factors that led originally to the ascendancy of neuropsychologists in the forensic setting continue to operate in high gear today. If anything, they seem to be proliferating, especially when we consider that emerging neuroimaging techniques promise to confirm neuropsychological assessments in even greater detail in the future. Notwithstanding the multiple attacks now mounted against neuropsychologists, these are likely to constitute only minor obstacles to their expert testimony. All things considered, the future legal environment pertaining to clinical neuropsychology will probably remain very hospitable.

References

Bigler, E.D. (1986). Forensic issues in neuropsychology. In D. Wedding, A.M. Horton & J. Webster (Eds.), *The neuropsychology handbook* (pp. 526–547). New York: Springer.

Brodsky, S.L. (1989). Advocacy in the guise of scientific advocacy: An examination of Faust and Ziskin. *Computers in Human Behavior, 5*, 261–264.

Chandler Exterminators, Inc. v. Morris, 416 S.E. 2d 277 (1992).

Condeluci, A. (1992). Brain injury rehabilitation: The need to bridge paradigms. *Brain Injury, 6*, 543–551.

Curry v. Baker, 502 S.E. 2d 667 (N.C. Ct. App. 1998).

Daubert v. Merrell Dow Pharmaceuticals, Inc., 113 S. Ct. 2786 (1993).

Drake v. La Rue Construction Co., 451 S.E. 2d 792 (1994).

Divison of Clinical Neuropsychology (1989). Definition of a clinical neuropsychologist. *The Clinical Neuropsychologist, 3*, 22.

Faust, D., & Ziskin, J. (1988). The expert witness in psychology and psychiatry. *Science, 241*, 31–35.

Faust, D., & Ziskin, J. (1989). Computer-assisted psychological evidence as legal evidence. *Computers in Human Behavior, 5*, 23–36.

Garry, P.M. (1997). *A nation of adversaries: How the litigation explosion is reshaping America.* New York: Plenum Press.

Gudjonsson, G.H. (1993). The implications of poor psychological evidence in court. *Expert Evidence, 2* (3), 120–134.

Guilmette, T.J., & Giuliano, A.J. (1990). Taking the stand: Issues and strategies in forensic neuropsychology. *The Clinical Neuropsychologist, 5*, 197–219.

Hagen, M.A. (1997). *Whores of the court: The fraud of psychiatric testimony and the rape of American justice.* New York: Regan Books/HarperCollins.

Handy v. Speth, 435 S.E. 2d 623 (1993).

Hutchison v. American Family Mutual Insurance Company, 514 N.W. 2d 882 (1994).

Jacobs v. Pilgrim, 367 S.E. 2d 49 (1988).

Jennett, B. (1982). The measurement of outcome. In N. Brooks (Ed.), *Closed head injury: Psychological, social and family consequences* (pp. 37–43). New York: Oxford University Press.

Kolpan, K.I. (1989). Expert courtroom testimony. *Journal of Head Trauma Rehabilitation, 4*, 95–96.

Kreutzer, J.S., Harris-Marwitz, J., & Myers, S.L. (1990). Neuropsychological issues in litigation following traumatic brain injury. *Neuropsychology, 4*, 249–259.

Malec, J.F., Smigielski, J.S., DePompolo, R. et al. (1993). Outcome evaluation and prediction in a comprehensive integrated post-acute and patient brain injury rehabilitaion programme. *Brain Injury, 7*, 15–29.

Martin v. Benson, 481 S.E. 2d 292 (N.C. App. 1997), *reversed and remanded on other grounds,* 500 S.E. 2d 664 (N.C. 1998).

Matarazzo, J.D. (1990). Psychological assessment versus psychological testing: Validation from Binet to the school, clinic and courtroom. *American Psychologist, 45*, 999–1017.

McKinlay, W.W. (1992). Assessment of the head-injured for compensation. In J.R. Crawford, D.M. Parker, & W.W. McKinlay (Eds.), *A handbook of neuropsychological assessment* (pp. 381–392). Hillsdale, NJ: Lawrence Erlbaum.

Mild Traumatic Brain Injury Committee, American College of Rehabilitation Medicine (1993). Definition of mild traumatic brain injury. *Journal of Head Trauma Rehabilitation, 8*, 86–87.

Miller, L. (1992). Neuropsychology, personality, and substance abuse in the head injury case. *International Journal of Law and Psychiatry, 15*, 303–316.

Miller, L. (1996). Neuropsychology and pathophysiology of mild head injury and the postconcussion sydrome: Clinical and forensic considreations. *Journal of Cognitive Rehabilitation* (Jan./Feb.), 8–23.

Morris v. Chandler Exterminators, Inc., 409 S.E. 2d 677 (1991).

National Center for State Courts. (1994). *Examining the work of state courts, 1994* (Publication No. R-178). Williamsburg, VA: Author.

Nichols v. Colonial Beach Oil Col, 132 N.Y.S. 2d 72 (1954).

Official Code of Georgia Annotated § 37-3-1 (16.1) (1993).

Official Code of Georgia Annotated § 43-39-1 (2) (1993).

Official Code of Georgia Annotated § 43-39-1(3) (1993).

Packard, R.C. (1993). Mild head injury. *Headache Quarterly, 4*, 42–52.

Penrod, L.E. (1996). Medicolegal aspects of traumatic brain injury. In L.T. Horn & N.D. Zasler (Eds.), *Medical rehabilitation of traumatic brain injury* (pp. 227–248). Philadelphia: Hanley & Belfus.

Rimel, R.W., Giordani, B., Barth, J.T., Ball, T.J., & Jane, J. (1981). Disability caused by minor head injury. *Neurosurgery, 9,* 221–228.

Rimal, R., W., Giordani, B., Barth, J.T., & Jane, J. (1982). Moderate head injury: Completing the clinical spectrum of brain trauma. *Neurosurgery, 11,* 344–351.

Robinson, D.N. (1995). *An intellectual history of psychology* (3rd ed.). Madison, WI: University of Wisconsin Press.

Rothke, S. (1992). Expert testimony by neuropsychologists: Addendum to Schwartz and Satz. *The Clinical Neuropsychologist, 6,* 85–91.

Spivak, G., Spettel, C.M., Ellis, D.W., et al. (1992). Effects of intensity of treatment and length of stay in rehabilitation outcomes. *Brain Injury, 6,* 419–434.

Stern, B.H. (1995). Cross-examination of the defendant's neuropsychologist in a mild traumatic brain injury case. *Trial Diplomacy Journal, 18,* 137–144.

Stern, B.H. (1999). Admissability of neuropsychological testimony. *Trial Diplomacy Journal, 22,* 27–34.

Taylor, J.S. (1991). Neurolawyers: Advocates for TBI and SCI survivors. *The Neurolaw Letter, 1,* 1.

Taylor, J.S. (1993). Neuropsychological evidence in Georgia: New law. *Verdict, Fall,* 46–48.

Taylor, J.S. (1995a). Landmark case. *Georgia Psychologist,* Winter, 35–36.

Taylor, J.S. (1995b). Neurolaw: Towards a new medical jurisprudence. *Brain Injury, 9,* 745–751.

Taylor, J.S. (1996). Neurorehabilitation and neurolaw. *NeuroRehabilitation, 7,* 3–14.

Taylor, J.S. (1997). *Neurolaw: Brain and spinal cord injuries.* New York: ATLA Press/Clark Boardman Callaghan.

Taylor, J.S., Elliott, T., & Harp, J.A. (1991a). Neuropsychological evidence on appeal. In A.C. Roberts (Ed.), *Litigating head trauma cases* (pp. 18–1 through 18–42). New York: John Wiley & Sons.

Taylor, J.S., Harp, J.A., & Elliott, T. (1991b). Neuropsychologists and neurolawyers. *Neuropsychology, 5,* 293–305.

Theobald, R. (1996). Agents of change. In G.T. Kurian & G.T.T. Molitor (Eds.), *Encyclopedia of the future* (Vol. 1, pp. 10–11). New York: Simon & Schuster Macmillan.

Varney, N.R., & Shepherd, J.S. (1991). Minor head injury and the post-concussive syndrome. In J. Dywan, R.D. Kaplan, & F.J. Pirozzolo (Eds.), *Neuropsychology and the law* (pp. 24–38). New York: Springer-Verlag.

Ziskin, J., & Faust, D. (1988). *Coping with psychiatric and psychological testimony* (Vols. 1–3; 4th ed.). Marina Del Ray, CA: Law & Psychology Press.

Chapter 15

THE NEURO-PSYCHOLOGIST AS EXPERT WITNESS:
TOWARD CREDIBLE SCIENCE IN THE COURTROOM

Paul R. Lees-Haley
Woodland Hills, California

Larry J. Cohen
University of Michigan Law School, Ann Arbor, Michigan

The justification for expert testimony is that it can help the trier of fact understand the evidence or determine some fact in dispute: "If scientific, technical, or other specialized knowledge will assist the trier of fact to understand the evidence or to determine a fact in issue, a witness qualified as an expert by knowledge, skill, experience, training, or education may testify thereto in the form of an opinion or otherwise. (Rule 702, Federal Rules of Evidence (FRE))".

Experts testify in a variety of settings requiring such factual determinations: jury trials, bench (judge) trials, hearings on evidentiary motions, administrative hearings, arbitrations, and legislative hearings. Thus, the expert may be testifying to arbitrators, hearing officers, judges, juries, and others. For ease of discussion we will refer throughout to jury trials, and refer to the finder of fact as the jury, but with some relatively minor adjustment the same principles apply to all of these fact-finding settings.

A neuropsychologist seeking to accomplish the central purpose of assisting the jury should follow three guidelines:

1. Practice scientific and ethical neuropsychology.
2. Direct your testimony toward the goal of using your specialized knowledge to help the jury understand the evidence or determine facts in issue.
3. Speak in reasonable language, avoiding technical terminology not likely to be understood by the lay audience with which you are communicating.

As obvious as these keys to success may sound, many neuropsychologists are only marginally effective when they testify, and many fail altogether. We believe this happens largely because neuropsychologists do not understand these guidelines, because they are not motivated to follow them, or because they believe they have to accommodate their patients or lawyers who retain them as consulting or testifying experts.

One goal of this chapter is to help neuropsychologists better understand the forensic process and their place in it so they have a reasonable chance of being effective witnesses in depositions, hearings and trials. Another goal is to identify important problems neuropsychologists confront as witnesses and discuss how these problems interfere with testifying effectively. A third goal is to make recommendations for coping with the challenges of testifying effectively to the jury.

A prefatory word of caution is in order. For purposes of intellectual and professional growth and development, one's critics are one's best friends, not those who stop at complimenting one's good work. Toward the goal of making this chapter useful to practicing neuropsychologists, we focus on errors and criticisms, and on positive steps neuropsychologists can take to improve their performance. This pedagogical strategy should not be misconstrued as condemnation of the field in general, of particular schools of thought, or of any individuals (with the exception of the anti-science fringe). Quite the contrary, we sincerely compliment the high rate of peer-reviewed publications in neuropsychology, the conscientious concern most neuropsychologists demonstrate for improving the state of the art, and the willingness most neuropsychologists express to further their understanding of their profession. We also compliment the substantial efforts neuropsychologists are making to better understand the interaction between their profession and the legal process.

Forensic neuropsychology is a merit system, in which the participants conduct their activities in the open. Over the long run, what one knows, rather than who one knows or associates with, and how well one presents one's beliefs, rather than how much one accommodates to what others want to hear, determines who is rewarded with positive experiences and with further opportunities to participate in the forensic process.

Basic tensions

Some fundamental tensions threaten scientific objectivity in forensic neuropsychological work. The first tension arises when a treating neuropsychologist is asked to testify on behalf of the patient. The therapeutic nature of the psychologist-patient relationship is such that the neuropsychologist may feel an inclination to interpret the data and apply the literature in ways that help fulfill the patient's needs and goals. Despite efforts to achieve objectivity, a good psychologist is highly motivated to help the patient. The expert who agrees to serve as both treating health care provider and forensic expert is on thin ice ethically because of the potential conflict in these dual roles (e.g., see American Psychological Association, 1992 and Committee on Ethical Guidelines for Forensic Psychologists, 1991).

A second tension arises over conflicts of values, laws and ethics. At times the law and the applicable ethics are contradictory, or at least they appear so. For example, it is not uncommon to be told that the judge has ruled that the expert cannot mention certain important information in front of the jury, despite the expert's sincere opinion that the data are relevant clinically or scientifically. The expert is then sworn to tell the truth, the whole truth, and nothing but the truth. Does this oath mean the expert is to tell the whole truth, except as edited by the judge? What is the expert to do when the judge has suppressed some piece of information that the expert feels is essential to understanding the truth of the case?

Similarly, it is not uncommon to discover that the patient whose confidentiality one has protected has put his or her mental state at issue in a public forum, where complete disclosure, rather than confidentiality, is expected. What is the most responsible way simultaneously to respect the rights of the: (1) patient to responsible evaluation and treatment and privacy, (2) parties in the litigation to a truthful and accurate assessment of the damages in issue, and (3) legal system, including the judge and the jury, to see all the data so they can decide the case fairly and on its merits?

A related tension involves the disclosure of test data and test materials. While we believe that full disclosure is proper and desirable, we recognize that there are those in the field who think otherwise. Indeed, some psychologists have taken very aggressive stands against disclosure, citing professional ethics, copyright laws or concerns about misuse of the data and test materials as reasons for withholding them. Even if there were a consensus within the field of psychology to withhold some or all of the data or test materials, there would remain the tension arising from the belief widely held among lawyers that the advocacy system in this country cannot function properly when a party is deprived of the opportunity to cross-examine a witness, and especially an expert witness, about the bases for an opinion.

A fourth tension arises from the most fundamental of epistemological questions: Do we really know enough about brain function and mental health to offer conscientious, reliable opinions about the difficult forensic questions presented in the particular matter at hand? Some observers have argued that forensic neuropsychology does not yet exist as a scientific discipline and that neuropsychological testimony should not be allowed in the courtroom (e.g., see Faust, 1991; Faust, Ziskin, & Hiers, 1991; Ziskin, 1995). Other experts believe that there has been sufficient advancement to offer opinions based reasonably in the current state of scientific knowledge (Matarazzo, 1987, 1990).

Another tension arises from the competition between economics and objectivity. In many forensic cases the neuropsychologist is retained by an attorney

to review records, conduct an assessment in person, or otherwise engage in professional services for which the neuropsychologist is compensated. There is a
perception – unfortunately sometimes accurate – that if one wishes to receive
future business from that attorney or others with whom that attorney commu-
nicates, one needs to interpret the data and apply the literature in ways that
accomplish the attorney's goals. We believe that altering one's opinion in this
manner is not only a moral and legal problem, it is also a pragmatic mistake.
As Benjamin Franklin once observed, if scallywags realized what good business
it is to be honest they would reform.

Every neuropsychologist must make his or her own decision about how to
deal with this persistent problem. As far as the legal system is concerned, however, we encourage candid skepticism. Just as good scientists always suspect
error and bias in every investigator, and do not accept opinions as facts until
they are replicated by multiple investigators, we believe judges and jurors should
be skeptical of all expert witnesses.

We do not intend to imply anything unspoken here about the competence or
morality of individual expert witnesses. Rather it is our impression, based on
experience and observation, that incompetence and lack of integrity are both
sufficiently common to be substantive concerns. It is a mistake to presume that
credentials or prominence are narrow bandwidth predictors of ethical conduct.
Furthermore, all experts, however competent and well intentioned, are human,
meaning that they are all subject to limitations, bias and error.

The scientific skepticism we advocate should not be confused with cynicism
or hostility. On the contrary, scientific skepticism assumes nothing sinister; it is
healthy realism. For example, we have no critical feelings toward those who
imagine their children are the most wonderful people in the world; it is trans-
parently obvious that one's own children are the most wonderful people in the
world. Writers use computer spell checkers and accountants invented double
entry bookkeeping because it is normal behavior for humans to make errors.
Brilliant students take notes in class because they cannot remember what they
are told. Steps are painted bright yellow because we do not necessarily look
where we are going. As Wilson noted, in a discussion of research that applies
here, one can easily be deceived by personal interest when interpreting results,
and "Only the naïve or dishonest claim that their own objectivity is a sufficient
safeguard" (Wilson, 1990, p. 44).

The point is that no matter how competent and ethical the practicing neuropsychologist may try to be, the neuropsychologist who becomes involved in a
forensic context encounters an array of forces pulling in multiple directions, some
of which are divergent from principles of good scientific methodology and applicable ethics. We believe that regardless of the temptation in specific circumstances
to deviate from good scientific practice, the neuropsychologist who yields on this
count gambles with harming a number of people, including oneself. Distorting
the data exposes the neuropsychologist not only to embarrassment, but also to
the possibilities of damage to the expert's reputation and to ethics charges, and
perhaps even to the loss of one's license. Distorting the data may cause iatrogenic

or nocebo effects which may harm the patient, waste the resources of the courts, defraud the defendant or plaintiff, or sidetrack valuable health care resources from other patients. Yielding to these pressures often proves counterproductive in maintaining and developing business opportunities, and may bring about the very opposite of what the neuropsychologist is trying to accomplish.

It is important for forensic neuropsychologists to be aware of and to consciously address the problems arising from these tensions. Neuropsychologists who take seriously the recommendations we offer here may well find the business end of their practices improving over the long run, and may well discover they are better off losing clients and lawyers who demand, as the price for loyalty, that neuropsychologists surrender their honesty, credibility and integrity.

Unique features of neuropsychology

For purposes of this essay, we define neuropsychology primarily as the study of behavior of individuals and groups with neurological problems. We freely acknowledge that neuropsychology is a multi-disciplinary field, not the monopoly of psychologists, and we agree that descriptive and normative research and research on higher cognitive functions of persons with no known neurological problems are all part of the field.

In a forensic setting, neuropsychologists are typically asked to apply their training, research and experience to disputes over (1) the extent to which an individual presents certain abilities or deficits, (2) the real life consequences of those abilities or deficits for functioning, (3) the extent to which those abilities and deficits can be influenced through intervention, and (4) the likely maintenance of those abilities or persistence of those deficits over time. Neuropsychologists applying scientific method and the results of scientific research can bring unique assets to these disputes. For instance, rather than rely on the impressionistic, idiosyncratic experiences of one individual, neuropsychologists can rely on standardized norms. These norms permit them to give well-founded opinions about what is normal or abnormal, and what to expect from a person with particular characteristics.

At the same time, however, the field of neuropsychology is besieged with uncertainties and disputes about fundamental aspects of what neuropsychologists study and how they go about learning what they know in any given case. The persistence of these uncertainties and disputes bears directly on the confidence with which they can offer opinions on the various matters typically arising in litigation.

Controversies concerning underlying biology

Notwithstanding the remarkable progress of the past twenty years regarding the nature of brain functioning, there remain far more questions than answers about what happens to the brain as a result of trauma, toxic exposure or other neurological insult to the brain. This is not to discount what has been learned in certain areas, for example, about language function, human memory, intelligence, and visual processing. Rather, the point is to acknowledge how much remains uncertain or unknown about many aspects of cognitive functioning, and the resulting limitations this imposes on the conclusions that can be reached about such central forensic issues as causation, impairment and the potential for recovery.

For example, while a great deal is known about how certain toxins affect neurological functioning, there are many substances for which toxic consequences are anticipated, but not really known. Most chemicals have never been the object of scientific investigation by neuropsychological researchers. Neuropsychologists must be cautious when testifying about the neurobehavioral damage that may result from exposure to one of these substances, much less about the long term consequences of that exposure (see Chapter 12 by Hartman).

The scope and depth of understanding in neurological science is expanding at an extraordinary rate. Many of today's uncertainties will be tomorrow's established areas of knowledge. Neuropsychologists should be respectful of the evolving character of scientific thinking, and not press the borders of knowledge and understanding so far as to suggest they know more now than they really do about the biology of brain function.

Controversies concerning methodology

Methodology consists of the steps we take to reach conclusions. A useful exercise every neuropsychologist might consider before sending in a report or testifying to opinions is to review each point and ask the simple question, "How do I know that?" We urge this exercise not because we doubt the ability of neuropsychologists to make useful contributions to the causation and damages questions typically addressed to them by the attorneys, but to ensure that the answers the neuropsychologists give make sense in light of the pertinent peer-reviewed literature from a number of disciplines.

In this section we will explore common examples of fallacious answers to the "how do I know" question. The more general point, though, is that all too often in forensic cases the opinions and conclusions neuropsychologists offer are the result of excessive speculation, over-interpretation of limited data, *ad hominem* arguments, confusion of inverse probabilities, and over-reliance on self-report data. Testifying neuropsychologists may then try to hide the resulting weaknesses in their analyses by using language so vague as to be unintelligible or so technical as to be inaccessible to the lay audience. Attorneys are increasingly prepared to recognize such camouflage and demonstrate to the jury that, as the saying goes, the emperor has no clothes. Forensic neuropsychologists who deliberately consider the methodological foundation for opinions they offer not only may avoid the resulting embarrassment, but should make a genuine contribution to resolving the disputed issues at hand.

Excessive speculation often occurs when the neuropsychologist is pressing for conclusions that will support the interests of the patient or of the client who retained the expert's services. The problem here may not be that the neuropsychologist makes a particular mistake in interpreting the available data. Rather, the expert tries to make more of the data than is appropriate considering how the data were collected, the scope of the information available, or the appropriateness of the data for the matter in issue. This error can take many forms, (e.g., presumed validity, presumed reliability, the absence of normative standards, or a lack of supporting substantive information about the issue).

The practice of presuming validity or reliability has been rejected unequivocally by the American Psychological Association (APA) through its published ethical standards and by reference to other published professional standards. The APA guidelines for psychological testing are generally accepted by psychologists (American Educational Research Association (AERA), APA, & National Council of Measurement in Education (NCME), 1985). It is certainly true that many competent and honest psychologists disagree with some of the APA's positions, (e.g., what some consider a tendency to sacrifice science for practice values). However, APA members sign an agreement on their membership applications to abide by APA ethics. Accordingly, neuropsychologists who are APA members open themselves up to painful cross-examination when they do not abide by the APA standards.

Regardless of whether the neuropsychologist is an APA member, one factor considered by courts in evaluating the admissibility of an expert's testimony is whether the underlying methodology is generally accepted in the profession. Accordingly, the expert who relies on a novel methodology or who offers an idiosyncratic position may not be permitted to testify on the issue.

The *Standards for Educational and Psychological Testing* (APA, 1985) clearly place the responsibility for validity (the test measures what it purports to measure) and reliability (the test measures whatever it measures consistently) on the examining expert. However, many tests used in neuropsychological assessment do not meet the standards set forth in the *Standards for Educational and Psychological Testing*. Interpretation of these inadequately validated tests is speculative and misleading.

As a general proposition, we recommend using tests that have been evaluated in peer-reviewed literature for both their validity and their reliability. More specifically, we suggest considering the following points when selecting tests to employ in forensic neuropsychological evaluations: (1) whether the test has been the subject of repeated validation studies in peer-reviewed journals; (2) whether there are peer-reviewed studies demonstrating reliability of the test with the population and context similar to the current application; (3) whether there is a published test manual in compliance with the *Standards for Educational and Psychological Testing* (APA, 1985); (4) whether that manual provides information concerning error rates, reliability coefficients, validity studies, discussion of intended populations and applications of the test, statements of limitations, well-defined norms, and explicit instructions for administration and scoring; and (5) whether there are appropriate data concerning norms, taking into consideration that tests normed on small groups are unreliable, and that tests normed *for patients with one disease process or injury may be unreliable when applied* to patients presenting different diseases or injuries.

A pervasive source of excessive speculation is the absence of normative standards in some of the most fundamental areas of forensic neuropsychology. There is a lack of consensus, for example, about which combination of tests or which test battery is superior as a resource for forensic neuropsychological practice. Tests drawn from the Halstead-Reitan Battery are among the most widely used, but there is no battery used by a majority of forensic neuropsychologists (e.g., see Lees-Haley, Smith, Williams, & Dunn, 1995). As a result, neuropsychologists must be cautious indeed in making claims as to the general acceptance of their methods.

Likewise, there is widespread disagreement about many of the most basic definitions in the field. For example, there is no generally accepted definition of mild traumatic brain injury (MTBI) or post concussive syndrome (PCS) or the complaints associated with these terms. As noted in recent reviews, even the highest quality empirical research studies use different definitions of MTBI (e.g., see Binder, 1997; Binder, Rohling & Larrabee, 1997; Satz, Zauch, McCleary, Light, Asarnow, & Becker, 1997). Different professional interest groups propose different definitions of mild head injury. Experts give different answers in depositions when asked to define MTBI. Some experts testify that their definition of post concussive syndrome is generally accepted, ignoring a century of controversy so deep that the concept has not even been accepted as a diagnosis in the DSM-IV, the diagnostic manual most widely referenced by testifying experts in psychology, psychiatry, and neuropsychology.

Finally, there is a dearth of fundamental normative descriptive research about naturally occurring neuropsychological phenomena. For example, there is not one peer-reviewed study in the neuropsychology literature reporting naturalistic observations of spontaneous memory lapses for substantial samples of persons resembling typical plaintiffs. Although descriptive research is emerging concerning base rates of neuropsychological complaints, Reitan remains correct in the oft-repeated observation that neuropsychologists do not know what is normal (Reitan, 1992). Fundamental ecological research is needed as much today as it was when Roger Barker recommended it decades ago (e.g., see Barker, 1968).

Yet another reason for the excessive speculation that pervades forensic practice is that testifying neuropsychologists often do not have the background or correlative information they need to support their opinions. Sometimes the fault rests with the neuropsychologist for not collecting available information. Sometimes the fault rests with the client or with the client's attorney for not providing available information or, worse, for holding back relevant information.

The answer in all of these instances is that the forensic neuropsychologist should not extrapolate from insufficient evidence to a speculative conclusion.

Rather, the expert should be straightforward in describing the limitations of what is known and then either refuse to speculate altogether or at least be very clear about the limits of what can be said on the subject. This may not be satisfying in the sense that the forensic neuropsychologist may have to concede the limits of what is known or what is knowable. It is, however, fair to all concerned, in that it permits the trier of fact to base the ultimate decisions in the case on a valid appraisal of the situation in light of current knowledge and available data and evidence.

Turning next to *over-interpretation* of data, one common problem is the claim that neuropsychological testing can detect more subtle deficits than a clinical neurological examination. This statement is true as a general proposition, and is of interest for such research purposes as studying group differences. However, this statement can be highly misleading in the context of a particular case, such as when the neuropsychologist testifies that a few subtle deficits detected by neuropsychological testing mean the patient has pervasive neurological impairments, notwithstanding a normal neurological examination. A thoughtful inspection of this sort of interpretation yields such odd combinations of opinions as that the injury: (1) is so broadly incapacitating that the patient can never work again in any job, yet (2) so difficult to discern that a competent neurologist cannot detect it on a neurological examination, and (3) so subtle that often even the patient does not know that the impairment exists. It does little for our credibility to testify, in effect, that the patient did not notice being totally disabled for several months after the injury, and mistakenly went on working.

Another common over-interpretation of data is the conclusion that a small number of low test scores demonstrates the presence of an injury. Low tests scores should stimulate further inquiry and analysis of other relevant data, such as history and mental status information. However, statistical variation by itself may not be a sign of abnormality, and when it is, the presence of abnormal statistical deviation does not necessarily indicate an injury relevant to the case at hand. It only acquires meaning in the context of its interpretation together with other pertinent information bearing on the matter at hand.

Occasionally, a neuropsychologist will offer nothing more than test scores to support opinions on the question of whether there is a causal relationship between the alleged causal event and the reported symptoms and impairments. Neuropsychological tests are non-specific, however, in the sense that they cannot tell us the causes of the neuropsychological deficits. Accordingly, causal opinions can be derived, if at all, only by drawing together information from a variety of sources such as the clinical history, appropriate medical examinations, and appropriate tests evaluated in the context of a thorough history of the patient's pre- and post-injury background.

Ad hominem arguments take many forms. One example is the implicit claim that one's qualifications make one's opinions true. This elementary fallacy is familiar to anyone who recalls introductory science classes and appears in surprising places. Years of experience, offices held, board-certification, number of publications, and other fine and respect-worthy credentials are testified to as if they somehow proved an opinion about a particular patient in a particular case. Our view – which is consistent with scientific tradition and court precedent – is that credentials prove nothing. As the Court in *Viterbo v. Dow Chemical*, 826 F.2d 420, 421 (5th Cir. 1987) said so appropriately, "The mere fact that an expert says something does not make it so." The appellate court properly declined to be impressed by experts who expected to be heard, much less taken seriously, merely because of their credentials. Instead of replying in kind to *ad hominem* arguments, we recommend relying on evidence, reason, good science, and explicit disclosure of the basis for opinions. One should always be suspicious of the expert who brings up credentials when asked the basis for an opinion.

Confusion of inverse probabilities is a surprisingly common fallacy in expert testimony by neuropsychologists. This problem is closely linked with lack of awareness of base rates of psychopathology in general and brain injury symptoms in particular (e.g., see Dunn, Lees-Haley, Brown, Williams, & English, 1995; Fox, Lees-Haley, Earnest, & Dolezal-Wood, 1995a, 1995b; Gouvier, Cubic, Jones, Brantley, & Cutlip, 1992; Gouvier, Uddo-Crane, & Brown, 1988; Kessler et al., 1994; Lees-Haley & Brown, 1993; Regier et al., 1988; Robins et al., 1984). For example, although it is more likely than not that a patient who recently suffered a significant brain-injury will complain of problems such as memory, attention, or headaches, it is *not* more likely than not that a patient complaining of problems with memory, attention, or headaches has recently suffered a significant brain injury. Complaints about these sorts of problems occur with notable frequency in the general population, as well as among those suffering chronic psychological problems. Nonetheless, some experts conclude from the mere presence of subjective complaints that the patient must have suffered an organic injury.

This problem in logic is also related to *over-reliance on self-report data*. There are conflicting opinions on how much credence one should attach to various possible foundations for opinions (e.g., self reported complaints, documented history, collateral interviews, neuropsychological tests, and neuroimaging, especially functional imaging). Although properly developed tests are more reliable than interviews, and actuarial decision procedures superior to clinical judgment, it is not at all uncommon for testifying experts to discard test data and normal expectations based on controlled research and base their opinions, instead, on the self-reported complaints of the patient or on the opinions of lay family members or friends. It makes no more sense to rely on a mildly injured person's claims of severe injury than to rely on a severely brain damaged person's claims of total recovery.

Self report data are so prominently unreliable that concerns about their validity were the subject of a front page story in the *APA Monitor* in January, 1997. There was a related prominent story in the January issue of the *APS Observer* (Rowe, 1997), following up on the November, 1996 symposium on these problems sponsored by the National Institute of Health. Self report problems are compounded when the patient is administered a lengthy checklist of symptoms possibly experienced since the relevant incident. A competent forensic neuropsychologist makes efforts to obtain background records on the individual and correlative information to fully understand the injury in context.

The concerns we raise here about methodology are not intended to suggest that individual clinical evaluations or forensic examinations should be subjected to the same standards as a controlled scientific study. Rather, we are arguing that clinical or forensic evaluations that produce opinions inconsistent with other neurological findings or the prevailing peer-reviewed scientific literature have a special burden to produce evidence to account for the inconsistency. Thus, for example, if 95% of mild brain injury patients do not exhibit the alleged problem, more evidence is required to support the expert's opinion about the reported problem than borderline scores on a few neuropsychological tests or subjective lay reports from interested parties concerning perceived changes in performance. Extraordinary claims require extraordinary evidence, and experts claiming to have found exceptions to views based on controlled research owe an explanation to the scientific community and to the courts.

Controversies concerning probable outcomes

A prototypical question to be asked of a neuropsychologist in litigation is whether there is a causal relationship between the alleged causal event and the patient's reported symptoms. A second common question concerns what the patient will experience in relation to those symptoms in the future. Most neuropsychologists would concede that current testing and the overwhelming majority of the empirical research focus on diagnostic issues, rather than rehabilitation and recovery (Johnstone & Farmer, 1997). However, neuropsychologists who must respond to questions about causation must be cautious in offering opinions about the likely course of recovery and the likely persistence of impairments.

Unique features of the legal process

Those who enter the American legal system believing they are participating in a process designed to search for truth may be sadly disappointed with what they find. One could imagine a system of justice in which there is a concerted effort by all involved to ferret out the best available description of some historical event or to identify the best possible solution to some problem in light of all of the information that can be accumulated on the subject at hand. While the American legal system may indeed arrive at something that may reflect the truth or reach some result we might equate with justice, the underlying methodology is nothing like a scientific investigation. Rather, it is an adversarial system, in which truth and justice are expected to emerge from a process in which two or more opposing sides present partisan versions of some event and their respective preferences about what relief should apply, and then attack the foundations, integrity and credibility of their opponent's version and preferences.

In an effort to maximize procedural fairness, courts and legislatures in the various federal and state jurisdictions across the country have imposed on this process certain rules as to how the opposing sides may go about collecting information and presenting their respective positions. While some of these rules have the effect of expanding the breadth and depth of information available to each side, there are some important ways in which these rules tend to exclude information from the ultimate decision-making process. As a result, the most one may hope for in the end is not the best answer or fairest result in some abstract sense, but the best and the fairest in the more limited context of the information made available to the jury or other decision maker.

The system also proceeds on the assumption that each side is more or less equally capable of marshaling information and presenting its position. This is a patently false assumption, particularly when it comes to the lawyers who represent the respective sides. As a practical matter, lawyers vary in their skill and competence as much as do the members of any profession. Even if the opposing lawyers were equally capable in some abstract sense, the demands on their time and resources when the particular case is being developed or presented may be quite different, and may vary from one case to the next, resulting in a significant imbalance in what reaches the decision maker.

The point here is not to be cynical of or disparaging toward the American legal system. Rather, it is to put aside naive "civics class" assumptions about a human enterprise that is subject to the same general types of information acquisition, resource management and decision-making problems, as any other human enterprise. The expert witness needs to know the workings of this enterprise well enough that those factors do not interfere with the expert's presenting information and opinions in a reasonable and accurate manner under the circumstances. Thereafter, the expert who has been diligent and acted with integrity perhaps should recognize that the outcome is a product of the practical workings of this particular method of resolving disputes, and not always a completely true or fair statement about the immediate situation or about the way the world operates.

Gatekeeping rules

The formal legal process is set in motion by the filing of a Complaint in the civil arena and by the presentation of an Indictment in the criminal arena. These kinds of documents define the boundaries for inquiry and presentation as the case proceeds. There are rules that permit the expansion of these boundaries under certain circumstances, as in the civil arena with Rule 15, Federal Rules of Civil Procedure (FRCP), where a Complaint may be amended to add additional claims and Rule 19, FRCP, where a litigant may be allowed to join other parties. Such rules are not unlimited, however, and a party may be forced to proceed without being able to assert all potential claims or without being able to name all potentially responsible parties.

Sometimes, for example, a party or claim may not be included because of the expiration of a statute of limitation. Every jurisdiction limits the time frame within which an action may be initiated against a person or entity. These time limits vary considerably across the country, simply on the basis of a policy decision by the legislature or the courts about how much time should be allowed to seek redress of grievances. Some jurisdictions delay the start of the applicable time period until the one capable of asserting the claim discovers or reasonably should have discovered who caused the damage. Most jurisdictions also delay the start of the applicable time period while the person is incompetent.

Another reason a potentially responsible party may be excluded is that the party enjoys some immunity or other kind of protection from prosecution or liability. For example, in many jurisdictions there are restrictions on the right to pursue civil relief from the government. In virtually every jurisdiction, an employee who collects or is entitled to receive worker's compensation benefits may not pursue the employer for civil damages, except under very unusual circumstances.

The significance of such restrictions, and of the scope of the Complaint or Indictment for the forensic neuropsychologist is that the expert may feel that at least some of the focus of attention should be on the conduct of a person or an entity who is not involved in the case or on some conduct that is not formally in question. The expert may feel that the most important elements of the matter clinically or scientifically have been excluded legally. Paradoxically, under such circumstances, both sides may push the expert in uncomfortable directions. One of the lawyers may try to divert the expert's attention from the non-party or non-issue, even to the extent of trying to get the expert to disregard, or at least minimize the significance of, the non-party or non-issue. The other lawyer, meanwhile, may imply that the expert should assign all of the responsibility to, or at least maximize the significance of, the non-party or non-issue.

In any event, it may be that neither side shows much interest in the expert's desire to give what the expert may feel is a complete and accurate account of, for example, the source and nature of the injury causing event. It is important not to let the lawyers' strategies and tactics distract the neuropsychologist from conducting a careful and comprehensive analysis of the patient's functioning and future course, and then reporting and discussing the findings to the full extent permitted by the circumstances (e.g., in written reports or open-ended questions during testimony). The neuropsychologist retained solely to evaluate a patient for purposes of the litigation may be more constrained in the inquiry and analysis by the questions presented by the person who retained the neuropsychologist (e.g., is this person suffering a memory loss, is this person exaggerating symptoms). Even then, however, the neuropsychologist should be prepared in drafting any report and in responding to appropriate questions when testifying to give a complete and accurate account of the findings and conclusions in the case.

This is as good a place as any to make the important point, applicable to the entire forensic process, that the expert must resist both temptation and external pressure to allow the lawyer to influence the expert's reasoning or conclusions. The most obvious example is when the lawyer suggests that the expert modify the emphasis of some opinion. In extreme instances, the lawyer may attempt to persuade the expert to reverse or substantively alter an oral or written opinion. We urge treating and retained neuropsychologists to proceed with great caution when the proposed changes have to do with the substantive expert opinions. We recognize that there are times when an attorney can appropriately request changes (e.g., alterations having to do with format, legal requirements of report writing, or *de minimus* errors). For example, a report may be excluded because an expert fails to comply with regulations governing reports in certain workers' compensation jurisdictions, in Federal Court, or in other jurisdictions. In such circumstances, it would be entirely appropriate for the attorney to request that the expert alter the report to comply with applicable laws, legal procedures, or important administrative regulations. As a general rule, however, changing a report to satisfy a lawyer may be the first step on the road to loss of credibility in the case and in the community.

Discovery rules

There was a time when lawyers were not permitted much, if any, opportunity to learn about the other side's case before trial, but, rather, tried their cases based on what they could learn from their clients and from their own independent investigations. This remains much the practice in the criminal arena, especially on the prosecution side. In the civil arena, courts and legislatures have developed rules permitting lawyers to discover the other side's case, in the hope that this would encourage the resolution of cases on their merits, rather than on the basis of surprise or the differing trial skills of the lawyers involved. Unfortunately, discovery has developed into something of a contest in which the character and amount of the information exchanged depend on the lawyer's skill in framing and interpreting inquiries. Moreover, while the intent of the rules is to expand everyone's knowledge about the case, there are ways, ironically, in which these rules also function to restrict the availability of information that can be used at trial.

The challenge to the neuropsychologist is to resist being distracted and drawn off center by this process. The neuropsychologist should stay focused on providing the most professionally responsible analysis and presentation possible within the constraints of the information available and what the neuropsychologist is permitted to address in oral examinations.

The forensic neuropsychologist can insist that the retaining attorney provide all of the information and reports available concerning the patient and the litigation generally. The neuropsychologist who waits passively for the lawyer to provide all the relevant information voluntarily is making a mistake. Many lawyers do not appreciate the importance of this information for a proper neu-ropsychological evaluation. Others who do appreciate its importance may neglect to assemble it or may simply forget to pass it along, or may try to hide it from the expert. Accordingly, neuropsychologists should develop the habit of asking for what they need, and sometimes asking repeatedly to make sure they receive it. The essential problem is that an opinion based on inadequate infor-mation is deficient. However, it is also embarrassing to learn critically import-ant data for the first time during cross examination by an attorney intent on destroying one's credibility.

The neuropsychologist should ask to see whatever is available with respect to what other health care professionals, experts, lay witnesses and parties will have to say about anything relating to the neuropsychologist's analysis and con-clusions. This includes, for example, deposition testimony, reports, statements, investigation summaries and like materials. In addition to providing a founda-tion for expert opinions, these data may give the neuropsychologist a better understanding of the case and one's role in it, and potential sources of criticism of the neuropsychologist's work.

There is some material attorneys normally withhold to protect legal rights associated with privileged communications. In most jurisdictions virtually every-thing the lawyer gives the expert or discusses with the expert can be discovered by the opposing side simply by inquiring about such interactions and commu-nications. Accordingly, attorneys are cautious about what they say to an expert about their own thoughts and ideas about the case. Such thoughts and ideas are protected in the civil arena by what is called the "work product" privilege, but that privilege is waived when these thoughts and ideas are communicated to a third party, like an expert, who does not fall within the scope of that privilege. Likewise, communications between the lawyer and the lawyer's client are zeal-ously protected by the attorney-client communications privilege, and this too may be waived if the content of such communications is divulged to a third party.

Neuropsychologists may normally expect that all of their communications with the people they are evaluating will be protected similarly, based on a psy-chologist-patient privilege. Such privileges are created by the legislatures in most jurisdictions and by the courts in others. It is important to understand that these kinds of privileges, where they exist, tend to be interpreted by the courts nar-rowly, according to the strict letter of whatever statute, rule or case decision cre-ated the privilege. As with the attorney-client privilege, a psychologist-patient privilege can be waived by disclosure to a third party who does not share in the privilege. A patient who places his or her mental condition at issue in seeking some benefit by virtue of that condition, waives the privilege, at least to the extent of the benefit sought. Thus, previously confidential material may become public record.

It is important to understand that in most, if not all jurisdictions, the doc-tor-patient privilege arises only in the context of a therapeutic relationship. This is because the whole reason for a privilege in the first instance is to encourage

that level of open communications between a patient and a psychologist necessary to encourage a positive therapeutic relationship. Where there is no therapeutic relationship, as in an examination undertaken solely for purposes of a
forensic determination, there may not be any protection for the communications
between the neuropsychologist and the patient.

We urge neuropsychologists to acquire at least a general familiarity with the
discovery rules of the jurisdiction in which they are practicing. Further, neuropsychologists need to acquire an informed familiarity with the law relating to
the psychologist-patient privilege, including especially when the privilege arises,
who owns (and therefore can waive) the privilege, and the circumstances under
which the privilege is or may be waived.

Neuropsychologists should be aware that there are procedures available to
require delivery of documents the neuropsychologist would prefer withholding
and procedures for requiring a response to a question the neuropsychologist
would rather not address. The method may go by different names in different
jurisdictions, but almost invariably will involve the lawyer who wants the information filing a motion to compel production of the documents or response to a
question. Since the neuropsychologist is not a party to the case, most jurisdictions do not provide an obvious mechanism by which the neuropsychologist can
make his or her views on the subject known. Typically, the neuropsychologist
simply relies on the lawyer who is offering the neuropsychologist as a witness
to present the neuropsychologist's views. The neuropsychologist can be more
assertive, however, by doing as little as asking the court by letter to permit the
neuropsychologist to be heard directly on the subject, or retaining a lawyer to
assist with being heard, or going so far as to intervene in the case for the limited purpose of addressing the issue. The obvious determining factors are how
important the issue is to the neuropsychologist and how many resources the
neuropsychologist wants to invest in the issue.

Additionally, the neuropsychologist should be aware that the court usually
has the power to impose monetary, and perhaps more severe, sanctions on the
neuropsychologist who takes an unreasonable position. Accordingly, it may be
prudent for the neuropsychologist to solicit advice from an attorney before withholding records or refusing to answer a question before taking a final stand on
the issue. The possibility of sanctions being imposed should not be so great a
threat to the neuropsychologist, though, as to allow the lawyer to bully the neuropsychologist into doing something unethical, professionally irresponsible or
illegal. We recommend in such situations soliciting the guidance of a lawyer who
is not involved in the case; someone who is interested solely in what is best for
the psychologist, representing his/her best interests.

There is one substantive illustration of the question of withholding materials that requires specific comment. Some psychological experts strive to avoid
cross examination on their data, and go so far as to claim that only members of
their profession are competent to make meaningful inferences from their tests
and test data. They contend that judges, jurors, attorneys, and consultants from
other professions should not be allowed to see the data, making a variety of

protests that in effect rest on the assumption that these people are incompetent or will behave dishonestly or irresponsibly. Some experts claim that the normal standard practice of psychologists is not to permit attorneys or the public to have access to psychological test questions and answers.

We recognize that this is a hotly debated issue in psychology generally and in neuropsychology in particular. We agree with the attorney for the American Psychological Association who has suggested that those who argue for withholding these data are being paternalistic (Bersoff, 1995). In our view, consultants from many disciplines are competent to offer potentially valid criticisms of neuropsychological assessment instruments and procedures for deriving conclusions from neuropsychological test data, including survey researchers, questionnaire design experts, epidemiologists, biostatisticians, statistics experts, industrial psychologists, educational testing experts, neurologists, vocational rehabilitation researchers, ophthalmologists, audiologists, and attorneys, among others. Those who feel otherwise, and would withhold their test data and instruments, must be prepared to demonstrate why the benefits to the opposing side to cross-examine the expert about the expert's methodology and analysis are outweighed by whatever risk the expert believes arise from disclosing such information.

Neuropsychologists who refuse to disclose their tests and test data should be aware that they are bucking a recent trend toward greater openness in civil discovery. Judges and lawyers in jurisdictions across the country have become frustrated with the extent of the gamesmanship taking place in discovery proceedings. Public policy makers and clients are very concerned about the very high costs involved in the different discovery mechanisms (especially depositions, but also interrogatories, requests for production, and requests for admissions). Finally, there seems to be general consensus that greater sharing of information about each side's position is likely to encourage earlier and more productive discussions about settlement, and thus avoid the time and expense involved in litigation generally.

One result of these kinds of concerns is experimentation with disclosure requirements. Practices vary considerably across the jurisdictions experimenting with this reform, but the best way to understand the idea is that on those subjects about which one must make a disclosure, the disclosing party must reveal everything the other side would be entitled to request. The existence of this affirmative obligation to disclose means that neuropsychologists may find the attorneys with whom they are working prepared to give up the neuropsychologist's materials, reports and opinions rather than waiting to be asked for it. Likewise, the neuropsychologist may receive voluntarily from the other side much of the information and materials the expert would like to have for the evaluation. Neuropsychologists should take advantage of such circumstances to inform the lawyer with whom they are working about the kinds of information the other side should be disclosing, so that lawyer can make sure there is a full and proper disclosure.

Admissibility rules

While the rules of discovery are very broad, permitting inquiry into everything relevant or likely to lead to the discovery of relevant information, the rules of admission of evidence at trial are considerably more narrow. As a general proposition the rules require that the evidence be relevant to the matters in issue: " 'Relevant evidence' means evidence having any tendency to make the existence of a fact that is of consequence to the determination of the action more probable or less probable than it would be without the evidence (Rule 401, FRCP)".

There are many other restrictions on admission in the rules of evidence (e.g., character evidence or hearsay), and the neuropsychologist would be well served to have at least some global introduction to them, if only to get a sense of the overall environment.

The forensic neuropsychologist should have a working understanding of the rules of evidence specifically applicable to expert testimony. These rules are subject to some variation across the federal and state jurisdictions, but most jurisdictions use the federal rules, or something very similar. We cited Rule 702, FRE, earlier, the rule that defines the circumstances under which experts are permitted to testify in a civil case. There are three other rules of evidence specifically governing expert testimony:

> Rule 703, FRE: The facts or data in the particular case upon which an expert bases an opinion or inference may be those perceived by or made known to the expert at or before the hearing. If of a type reasonably relied upon by experts in the particular field in forming opinions or inferences upon the subject, the facts or data need not be admissible in evidence.

> Rule 704, FRE: (a) Except as provided in subdivision (b), testimony in the form of an opinion or inference otherwise admissible is not objectionable because it embraces an ultimate issue to be decided by the trier of fact. (b) No expert witness testifying with respect to the mental state or condition of a defendant in a criminal case may state an opinion or inference as to whether the defendant did or did not have the mental state or condition constituting an element of the crime charged or of a defense thereto. Such ultimate issues are for the trier of fact alone.

> Rule 705, FRE: The expert may testify in terms of opinion or inference and give reasons therefor without first testifying to the underlying facts or data, unless the court requires otherwise. The expert may in any event be required to disclose the underlying facts or data on cross-examination.

Courts in each jurisdiction are likely to have interpreted these rules under various circumstances and neuropsychologists who go to court frequently may be well served by reading the applicable literature in the jurisdiction concerning these rules, attending a seminar dealing with these rules, talking about the current state of the law with an attorney, or even reading the cases themselves.

Lawyers typically will challenge the admissibility of part or all of an expert's testimony at one of two times. The occasion most familiar to experts is during the expert's testimony at trial, when the opposing lawyer stands to assert an objection to the expert giving an opinion. Less familiar to many neuropsychologists, but even better for the lawyer is the motion *in limine*. This is a motion the attorney files before trial asking the court to consider excluding an expert's testimony on any of several grounds, such as the relevance of the opinion, the competence of the expert and even the need for expert testimony.

Another ground, one especially applicable to a field like neuropsychology in which there is considerable debate about methodology and explanatory theories, is that the proposed opinions do not have a sufficiently strong scientific basis to permit the expert to present them to the jury. Lawyers often refer to this as the junk science motion. The neuropsychologist who is being challenged by such a motion is sometimes offended by it. We urge neuropsychologists to set aside their personal feelings, accept the fact that the court provides a gatekeeping function, and focus their efforts on presenting the most reasonable scientific opinions.

As noted in the preceding chapter, there are essentially two standards considered by the courts at this time. The Frye standard is based on a 1923 federal Court of Appeals decision that dealt with the admissibility of polygraph evidence:

> Just when a scientific principle or discovery crosses the line between the experimental and demonstrable stages is difficult to define. Somewhere in this twilight zone the evidential force of the principle must be recognized, and while courts will go a long way in admitting expert testimony deduced from a well-recognized scientific principle or discovery, the thing from which the deduction is made must be sufficiently established to have gained general acceptance in the particular field in which it belongs (Frye, 1923).

This general acceptance test, in one form or another, continues to be the standard in many states.

In 1993 the United States Supreme Court rejected the Frye standard on the ground that it focused too narrowly on a single criterion, general acceptance. The Court felt that the standard for admissibility of a novel scientific theory should be based instead on the principles set forth in Rule 401, FRE, the rule governing admissibility generally. Taking that rule into consideration, as well as Rule 702, FRE, dealing with expert testimony in particular, the Court concluded that the focus of inquiry should be on "whether the reasoning or methodology underlying the testimony is scientifically valid" and "whether that reasoning and methodology properly can be applied to the facts in issue." Importantly, and many lawyers do not seem to appreciate this, the focus is not on the conclusions thereby generated, but "solely on principles and methodology" (Daubert, 1993).

The Supreme Court encouraged trial judges to be flexible in conducting this inquiry into scientific validity. It retained general acceptance as a consideration, but as only one of four factors it suggested be taken into consideration:

1. Falsifiability:
   Ordinarily, a key question to be answered in determining whether a theory or technique is scientific knowledge that will assist the trier of fact will

be whether it can be (and has been) tested. "Scientific methodology today is based on generating hypotheses and testing them to see if they can be falsified; indeed, this methodology is what distinguishes science from other fields of human inquiry" (Daubert, 1993).

2.   Peer review and publication;
      Another pertinent consideration is whether the theory or technique has been subjected to peer review and publication. Publication is not a *sine qua non* of admissibility.... But submission to the scrutiny of the scientific community is a component of good science in part because it increases the likelihood that substantive flaws in methodology will be detected.

3.   Known rate of error;
      Additionally, in the case of a particular scientific technique, the court ordinarily should consider the known or potential rate of error, and the existence and maintenance of standards controlling the technique's operations.

4.   General acceptance;
      Widespread acceptance can be an important factor in ruling particular evidence admissible, and "a known technique which has been able to attract only minimal support within the community" may properly be viewed with skepticism (Daubert, 1993).

The Daubert or Relevancy approach is used throughout the federal court system (Federal Judicial Center, 1994), and in many states, in one form or another.

We recommend that neuropsychologists become familiar with these two general standards and with the particular application of the standard used in the jurisdiction where they are serving as expert witnesses. Neuropsychologists should be aware that there are now a substantial number of court decisions applying these standards in specific cases. For example, the Arizona Court of Appeals applied a Frye standard in excluding testimony concerning brain mapping in a criminal case (Zimmerman, 1990). Neuropsychologists who become involved in a particular case would be well advised to learn about any such applications of these standards on the issues in which they are involved.

Presentation rules

Rules of procedure and rules of evidence govern the presentation of evidence at a trial. For example, Rule 611 of the Federal Rules of Evidence addresses the mode and order of interrogation of witnesses. Under that rule cross-examination of a witness is limited ordinarily to the subjects addressed to the witness during direct examination. In many state jurisdictions, however, cross-examination may be addressed to any subject, regardless of whether the subject was addressed during direct examination.

In addition to these formal rules, trial judges often have their own preferences about how evidence is presented in their courtroom. Some will allow the lawyers to present their respective cases however they want, so long as the lawyers comply with the applicable formal rules. Other judges want the lawyers to streamline their presentations to enhance jury attentiveness and expedite the trial process. Many judges encourage the use of demonstrative evidence, some to the point of installing computer technology in their courtroom and insisting the lawyers take advantage of it in their cases. Finally, some judges take an active role in the questioning of witnesses. This can involve as little as insisting that the witness answer questions in a particular way, but as much as directing questions to the witness. Neuropsychologists who are going to testify before a judge they have not seen previously should inquire about the formal rules applicable to the courtroom and also to the culture of that courtroom.

One recent innovation being tried in some jurisdictions is to allow jurors to ask questions. A common way for this to happen is after the lawyers finish their direct, cross and re-direct examinations the judge asks if any of the jurors have questions for the witness. Jurors who have questions write them on a sheet a paper. The judge reviews the questions with the lawyers and if they are relevant and otherwise proper the judge asks the witness the question. The judge may then allow the lawyers to follow-up on the subject matter with further direct and cross-examination. The obvious thinking behind this reform measure is to give the decision-maker an opportunity to learn about things the decision maker considers important in deciding the case. Properly prepared neuropsychologists should not be concerned about this kind of innovation, and other similar approaches likely to develop in the years ahead, since they should know the case well enough to address any questions related to their work.

Indeed, preparation is essential to testifying effectively, whether at a trial, in a deposition or in any other proceeding. The essential tasks for the neuropsychologist are to identify the questions one potentially may be asked to answer, determine what information is needed to answer those questions, obtain, assess and become familiar with the relevant data, and review the applicable scientific and technical literature. It certainly will be helpful as well to review testimony by other witnesses, usually in the form of deposition transcripts. As noted previously, anything the neuropsychologist reviews or discusses with anyone usually will be subject to disclosure to the other side. While some neuropsychologists may be deterred by this fact from reviewing materials or discussing the case, for fear of appearing to have been tainted, the neuropsychologist should seriously consider whether the benefits of being thoroughly prepared to testify outweigh the risks of jurors not trusting that there is a legitimate scientific basis for the expert's opinions.

Finally, it is important for the testifying neuropsychologist to remember that the audience, whether it is judge, jury or arbitrator, consists of lay persons, not psychologists or other health care professionals. The prudent witness never assumes that the audience is familiar with the subject matter until told or it becomes clear otherwise. Accordingly, testifying neuropsychologists should be alert to common misunderstandings and miscommunications between psychologists and nonpsychologists. We urge experts to define terms in plain language repeatedly. This is especially critical if the expert disagrees with the mainstream definition and has in mind a different referent than most other experts, including especially any other experts testifying in the case. Likewise, a great deal of misleading testimony arises when experts testify to generalities without clarifying that they are speaking in general or about different types of cases, and not referring to the particular patient in the case. Finally, the expert should be ready to provide illustrations, use demonstrative materials and even conduct demonstrations, if that will help convey to the audience the points the witness is making.

Conclusion

There is a profound need for objective, neutral observers who apply reliable and valid measures, admit the limits of what they know, and have sufficient independence to resist the pressures and incentives to distort findings or rationalize away their honest opinions. The future of forensic neuropsychology belongs to those who reason and offer opinions in a fair-minded way, drawing on pertinent empirical research and based on comprehensive, clean data. Consider to this end our summary in Table 1 of the distinction between the scientist and the pseudoscientist in forensic neuropsychology.

In forensic neuropsychology, the challenges to one's integrity will be substantial. We urge neuropsychologists to approach the forensic process with (1) a commitment to applying the basic principles of good scientific inquiry to their work, (2) a considered sense of their own professional abilities, including their strengths and limitations (3) a willingness to listen to opposing views with an open mind, (4) a recognition that they must make their methodology, analysis and conclusions understandable to a lay audience and (5) a determination to place the principles of honesty and fairness ahead of what is expedient.

We also suggest, in closing, that neuropsychologists who venture or are drawn into the forensic arena try to maintain a sense of humor. One cannot be involved in this field long without being exposed to everything from thoughtful wisdom to irresponsible deceit, and from incredible insights to outrageous insinuations. We counsel patience: The challenges of forensic neuropsychology will not be met in this lifetime.

Someone reportedly once asked Einstein why physicists had not made more advances, if they were so smart. Einstein is said to have answered, "because physics is hard". The same is true of assessing damage to and projecting the consequences of damage to the brain. Keep one's work in perspective – take the long view. When confronted with ethical dilemmas remember that reputations cannot be built in a day, but they can surely be ruined with swift dispatch. Avoid playing lawyer. Avoid becoming caught up in a team seeking to win at all costs. When testifying, remember that the jury and judge do not have the luxury of awaiting further research; they have the important responsibility to decide now, despite the missing information and imperfections of the existing data. Stay close to the data, search for the most plausible explanations, admit that which you do not know, and help the judge and jury when you can.

Table 1. Distinction Between Scientific and Pseudo-scientific Experts in Forensic Neuro-psychology.

Scientific expert	Pseudo-scientific expert
1. Open-minded, but skeptical of all experts regardless of personal characteristics, such as credentials, years of experience, publications, number of degrees, sources of degrees, awards, titles, association memberships.	1. Insinuates validity of opinions based on personal characteristcs such as credentials, source of degree, years of experience, organization memberships.
2. Views neuropsychology as a young research discipline with many issues yet to be resolved.	2. Views neuropsychology as sophisticated, with well established ecological validity and high intra disciplinary agreement on standards and methods
3. Describes procedures explicitly enough to permit replication and rebuttal, and encourages criticism and commentary from members of other disciplines.	3. Suppresses data, claims it is unethical to permit anyone besides members of the guild to evaluate and criticize neuropsychological tests or data, avoids disclosure of specific methodology, claims no one but insiders can understand data.
4. Openly explores alternative explanations for the data.	4. Presumes causation, minimizes investigation of other potential explanations, downgrades importance of alternative life stressors, claims only neuropsychological explanations apply to persons with possible brain injuries
5. Advocate of objectivity, evidence, reason, logic.	5. Advocate of plaintiff or defense.
6. Relies on empirical data interpreted with reason or logic	6. Relies on intuition, clinical experience, speculation, hunch, conjecture, imagination, what feels right.
7. Makes inferences from base rates, law of large numbers, descriptive data.	7. Generalized from small samples, anecdotal evidence, vivid illustrations, sensational data.
8. Welcomes reasoned criticism, questioning, cross-examination.	8. Resents and belittles cross-examination, overtly or covertly; treats criticism of neuropsychology as treason

Table 1 continues

Table 1 (continued).

9. Distinguishes between observation and inference, consensual/empirical data versus self-report data.

9. Blurs objective and subjective data, or denies the existence of any distinction between the two

10. Reads and relies on empirical research literature, admits limitations of knowledge

10. Relies on clinical experience in contradiction to well researched findings.

11. Attacks methodology of opposing experts.

11. Attacks person of opposing expert, makes ad hominem arguments.

12. Seeks explicit definition of alleged patterns and support for such allegations in peer-reviewed empirical literature.

12. Claims to see patterns which are not apparent to opposing expert, claims existence of signature profiles that prove causation, but cannot recall citations of independently replicated empirical demonstrations of their existence.

13. Draws attention to overall predictive power, consideration of all cells in table (true and false; positive and negative), points out that most sensitive test is sheer unfounded allegation.

13. Emphasizes sensitivity of tests; claims extreme accuracy

14. Expects temporal logic, dose-response logic, reasonable consistency of data.

14. Overlooks or makes excuses for unusual temporal relations, contradictory chronologies, and vague or implausible dose response allegations.

15. Responds to request for production of files by making materials available to the extent permitted by applicable laws and professional guidelines.

15. Responds to request for production of files and materials by seeking excuses to avoid disclosure and then by interpreting information requests as narrowly as possible to restrict access

16. Treats attorney inquiries objectively by focusing on substantive matters addressed and avoiding invitation to respond to matters personally or emotionally.

16. Tends to be defensive in responding to attorney challenges to work and conclusions.

17. Accepts outcome as product of a specific form of dispute resolution.

17. Views outcome as endorsement or rejection of science depending on whether it agrees or disagrees with the expert's position.

References

American Educational Research Association, American Psychological Association, and National Council on Measurement in Education (1985). *Standards for educational and psychological testing*. Washington, DC: American Psychological Association.

American Psychological Association (1992). Ethical principles of psychologists and code of conduct. *American Psychologist, 47*, 1597–1611.

American Psychological Association (1985). *Standards for educational and psychological testing*. Washington, DC: Author.

American Psychiatric Association (1994). *Diagnostic and Statistical Manual* (4th ed.). Washington, DC: Author.

Bersoff, D. N. (1995). *Ethical conflicts in psychology*. Washington, DC: American Psychological Association.

Barker, R.G. (1968). *Ecological psychology: Concepts and methods for studying the environment of human behavior*. Stanford, CA: Stanford University Press.

Binder, L.M., Rohling, M.L., & Larrabee, G.J. (1997). A review of mild head trauma. Part I: Meta-analytic review of neuropsychological studies. *Journal of Clinical and Experimental Neuropsychology, 19*, 421–431.

Binder, L.M. (1997). A review of mild head trauma. Part II: Clinical implications. *Journal of Clinical and Experimental Neuropsychology, 19*, 432–457.

Committee on Ethical Guidelines for Forensic Psychologists (1991). Specialty guidelines for forensic psychologists. *Law and Human Behavior, 15*, 655–665.

Daubert v. Merrill Dow Pharmaceuticals, Inc. 113 S.Ct. 2786 (1993).

Dunn, J.T., Lees-Haley, P.R., Brown, R.S., Williams, C.W., & English, L.T. (1995). Neurotoxic complaint base rates of personal injury claimants: Implications for neuropsychological assessment. *Journal of Clinical Psychology, 51*, 577–584.

Faust, D. (1991). Forensic neuropsychology: The art of practicing a science that does not yet exist. *Neuropsychology Review, 2*, 205–231.

Faust, D., Ziskin, J., & Hiers, J. B., Jr. (1991). *Brain damage claims: Coping with neuropsychological evidence*. Marina del Rey, CA: Law & Psychology Press.

Federal Judicial Center (1994). *Reference manual on scientific evidence*. Washington, DC: Federal Judicial Center.

Federal Rules of Civil Procedure (1997) 28 United States Code.

Federal Rules of Evidence (1997) 28 United States Code.

Fox, D.D., Lees-Haley, P.R., Earnest, K., & Dolezal-Wood. (1995a) Post-concussive symptoms: Base rates and etiology in psychiatric patients. *The Clinical Neuropsychologist, 9*, 89–92.

Fox, D.D., Lees-Haley, P.R., Earnest, K., & Dolezal-Wood. (1995b) Base rates of post-concussive symptoms in health maintenance organization patients and controls. *Neuropsychology, 9*, 606–611.

Frye v. United States, 293 F. 1013 (D.C. Cir. 1923).

Gouvier, W.D., Cubic, R., Jones, G., Brantley, P., & Cutlip, Q. (1992). Postconcussion symptoms and daily stress in normal and head-injured college populations. *Archives of Clinical Neuropsychology, 7*, 193–211.

Gouvier, W.D., Uddo-Crane, M., & Brown, L.M. (1988). Base rates of post-concussional symptoms. *Archives of Clinical Neuropsychology, 3*, 273–278.

Johnstone, B., & Farmer, J.E. (1997). Preparing neuropsychologists for the future: The need for additional training guidelines. *Archives of Clinical Neuropsychology, 12*, 523–530.

Kessler, R.C., McGonagle, K.A., Zhao, S., Nelson, C.B., Hughes, M., Eshleman, S., Wittchen, H., & Kendler, K.S. (1994). Lifetime and 12-month prevalence of DSM-III-R psychiatric disorders in the United States. *Archives of General Psychiatry, 51*, 8–19.

Lees-Haley, P.R., & Brown, R.S. (1993). Neuropsychological complaint base rates of 170 personal injury claimants. *Archives of Clinical Neuropsychology, 8*, 203–209.

Lees-Haley, P.R., Smith, H.H., Williams, C.W., & Dunn, J.T. (1995). Forensic neuropsychological test usage: An empirical survey. *Archives of Clinical Neuropsychology, 11*, 45–51.

Matarazzo, J.D. (1987). Validity of psychological assessment: From the clinic to the courtroom. *The Clinical Neuropsychologist, 1*, 307–314.

Matarazzo, J. (1990). Psychological assessment versus psychological testing. *The American Psychologist, 45*, 999–1017.

Ragge v MCA/Universal Studios, 165 F.R.D. 605 (C.D. Cal. 1995).

Regier, D.A., Boyd, J.H., Burke, J.D., Rae, D.S., Myers, J.K., Kramer, M., Robins, L.N., George, L.K., Karno, M., & Locke, B.Z. (1988). One-month prevalence of mental disorders in the United States. *Archives of General Psychiatry, 45*, 977–986.

Reitan, R. (1992). *Clinical neuropsychology: We've arrived but do we want to stay?* Paper presented to the American Psychological Association, Washington, DC.

Robins, L.N., Helzer, J.E., Weissman, M.M., Orvaschel, H., Gruenberg, E., Burke, J.D., & Regier, D.A. (1984). Lifetime prevalence of specific psychiatric disorders in three sites. *Archives of General Psychiatry, 41*, 949–967.

Rowe, P.M. (1997). The science of self report. *APS Observer, 10*, 3–38.

Satz, P., Zauch, K., McCleary, C., Light, R., Asarnow, R., & Becker, D. (1997). Mild head injury in children and adolescents: A review of studies (1970–1995). *Psychological Bulletin, 122*, 107–131.

State v. Zimmerman, 166 Ariz. 325, 802 P.2d 1024 (App. 1990)

Wetter, M.W., & Corrigan, S.K. (1995). Providing information to clients about psychological tests: A survey of attorneys' and law students' attitudes. *Professional Psychology: Research and Practice, 26*, 474–477.

Wilson, E.B. (1990). *An introduction to scientific research*. New York, NY: Dover.

Ziskin, J. (1995). *Coping with psychiatric and psychological testimony* (5th ed.). Marina del Rey, CA: Law and Psychology Press.

CONCLUSION:
TOWARD OBJECTIVE CLINICAL DECISION-MAKING IN ADVERSARIAL ASSESSMENTS

Jerry J. Sweet
Evanston Hospital, Evanston, Illinois, and Northwestern University Medical School, Chicago, Illinois

> *He that is possessed with a prejudice is possessed with a devil, and one of the worst kinds of devils, for it shuts out the truth, and often leads to ruinous error.* Tryon Edwards

In recent years, parties involved in adversarial proceedings have demonstrated considerable recognition of, and interest in, clinical neuropsychologists as professionals who can have a significant impact in these proceedings. This text on forensic neuropsychology is evidence of that strong recognition and interest. Whether related to personal injury, medical malpractice, competency, educational due process, disability status, or other similar context , clinical neuropsychologists as testifying experts, treating experts, or independent evaluators have been involved increasingly in adversarial evaluations. Within this context, objective information obtained by a professional who understands the nature and demands of the legal system is needed (e.g., Kreutzer, Harris-Marwitz, & Myers, 1990). There is little question that the strong scientist-practitioner foundations of clinical neuropsychology have provided essential empirical support for the development of the field, and, in particular, the development of clinical procedures that attract the attention of adversarial parties in need of relevant objective findings.

As we have seen within this text, the Daubert decision (*Daubert v. Merrell Dow Pharmaceuticals*, 1993) has established a new threshold for admission of information from experts. Although some may view this court ruling as a threat, clearly, no other healthcare field is better able, because of the strong connection between science and practice, to meet the consensual professional demands of the Daubert decision than clinical neuropsychology. Not only is there a strong empirical base in the field, but also a clear professional identity, as well as a series of definitions and professional guidelines. The reader is referred to the collection of seminal published readings from *The Clinical Neuropsychologist*

(TCN) (Adams & Rourke, 1992) for detailed information regarding definition, history, training, board-certification through the American Board of Professional Psychology (ABPP), use of testing assistants, and other important professional issues. A more recent publication in *TCN* by Eubanks (1997) documents the latest Division 40, Clinical Neuropsychology, American Psychological Association summary of important professional information pertinent to the field, including salient reference citations within the field.

Rather than be fearful of Daubert type standards, which will appear within forensic arenas and be refined at intervals across time, as scientist-practitioners we can embrace the associated opportunities and challenges toward the goal of attaining accountability and competent performance from expert witnesses in clinical neuropsychology. Readers interested in more detailed information than contained within the present text may wish to consult Heilbrun (in preparation) who will be devoting an entire text to discussion of the effects of the Daubert and other evidentiary rulings on forensic psychology and mental health assessments, in particular. In his text, Heilbrun provides a series of practice guidelines that can increase the probability of meeting current standards for expert and scientific evidence. Also of interest, the University of Michigan Law School maintains a continuously updated listing of Daubert and related rulings that affect expert and scientific evidence at a website that can be accessed via the internet. Known as "The Evidence Site", this site can be reached through *HTTP://www.law.umich.edu/*. Finally, Reed (1999) has thoroughly reviewed and provided details regarding the aplication of Daubert and other court rulings pertaining to admissability of expert testimony in each state. This informative article also summarizes all Federal Rules of Evidence.

As more is expected from professionals, there is an associated pressure to establish and attain individual standards. Perhaps because the field of clinical neuropsychology has adopted board-certification through ABPP as the clearest evidence of competence (Division 40 Executive Committee, 1989), or perhaps as a motivating force that increases willingness to undergo peer review, board-certified neuropsychologists engage in significantly more hours of neuropsychological practice per week *and* are involved in significantly more forensic activities than those who are not board-certified (Sweet, Moberg, & Westergaard, 1996). The number of board-certified clinical neuropsychologists is growing rapidly, and, in fact, clinical neuropsychology is the fastest growing of the various ABPP diplomate specialties.

As already indicated, changes in practice arenas typically bring new professional challenges and new professional issues that require attention. Along with increasingly frequent activity within adversarial proceedings is a commensurate demand for *objectivity* among the clinicians involved. Quite apart from having *objective* means of quantifying behavior is the recognition that *clinical inference* and *clinical decision-making* are, for every profession, the point at which degree of objectivity contained within an opinion may vary. For example, although magnetic resonance imaging (MRI) is widely recognized as the most *objective* non-invasive means of visualizing the myelin-coated 'white matter' of the brain, the meaning of any particular MRI image is only clear when a knowledgeable expert renders his or her *subjective* clinical judgement regarding the image. When expertise is not in question and an individual's expert opinion appears to deviate from consensual professional standards toward the side that has retained him or her, concerns may be raised regarding the individual's *objectivity* in rendering a particular opinion. To be clear, the issue of objectivity is beyond inter-rater reliability, a separate issue to which neuropsychology has always paid attention. (As an aside, one need only observe multiple medical specialists discussing a single patient's MRI, computed tomography, or x-ray films of whatever organ or body part is of interest, to realize that *all* specialists, not just psychologists, should collect data and be knowledgeable regarding inter-rater reliability per diagnostic procedure.) Recently, neuropsychologists have begun to articulate potential problems regarding objectivity within adversarial evaluations by clinical neuropsychologists (see discussions by Adams, 1997; van Gorp & McMullen, 1997; Guilmette & Hagan, 1997; McSweeny, 1997).

Despite the fact that actuarial approaches can be advantageous and may serve to increase objectivity (Dawes, Faust, & Meehl, 1989; Wedding & Faust, 1989), applicable formulae are not available for all clinical situations (Kleinmuntz, 1990). Therefore, clinicians, rather than formulae, must make most individual clinical and forensic decisions at present. With this reality in mind, the use of *decision aids* is among the suggested methods for increasing clinician objectivity and diagnostic accuracy in neuropsychological assessment (Garb & Schramke, 1996). In considering an explicit means of striving for increased objectivity and decreased bias in adversarial proceedings, Sweet and Moulthrop (1999) have offered a heuristic method that may prove useful to individual clinicians and to attorneys who work with neuropsychologists. Basically, this approach relies on self-examination questions employed by the clinician during his or her general work on a particular adversarial case and specifically pertaining to the construction and expression of the expert's opinion within a formal written report. For example, the clinician needs to consider: "Have I taken a position, in very similar cases, when retained by an attorney from one side that I did not take when retained by the opposite side?" and "Do I routinely apply the same decision-rules for establishing brain dysfunction no matter which side retains me?" and, with regard to a report, "Have I included statements that are at odds with the mainstream literature on this subject?", among other self-examination questions (Sweet & Moulthrop, 1999).

Additional strategies (described in greater detail in Borum, Otto, & Golding, 1993 and in Garb, 1998) for increasing clinician objectivity, and therefore decreasing bias and undue influences of attorneys in expert opinions, include: (1) noting the limitations of judgment tasks that are, at present, not well established, (2) documenting important case information in order to decrease reliance on memory (which may change across time), (3) gathering information in a comprehensive manner, when possible, (4) considering baserates and established diagnostic criteria, (5) following scientific standards of assessment, and (6) considering support for *and* against alternative hypotheses to the explanation given in a particular case.

Of course, when possible, reliance on empirically-based decision rules or actuarial formulae has been recommended repeatedly in the relevant literature.

When viewed from a dispassionate perspective, it is indisputable that objectivity can recede in the direction of bias favoring either side, plaintiff or defendant. In adversarial matters, righteousness and veridicality cannot be associated predictably with only one side, if examined across individual cases. Again, note that all professionals involved in rendering expert opinions in adversarial matters must bear the same burden of striving toward objectivity; clinical neuropsychology is no more vulnerable, in this regard, than any other field.

As noted throughout this text, much has transpired in recent years to increase forensic neuropsychological activities among qualified neuropsychology practitioners. These activities are a direct result of scientist-practitioner traditions within the sub-specialty, the same scientist-practitioner traditions that will undoubtedly herald future research and practice advances within this practice area.

References

Adams, K. (1997). Comment on ethical considerations in forensic neuropsychological consultation. *The Clinical Neuropsychologist, 11,* 294–295.

Adams, K., & Rourke, B. (Eds.) (1992). *The TCN guide to professional practice in clinical neuropsychology.* Lisse, Netherlands: Swets & Zeitlinger.

Borum, R. Otto, R., & Golding, S. (1993). Improving clinical judgment and decision making in forensic evaluation. *Journal of Psychiatry and Law, 21,* 35–76.

Daubert v. Merrell Dow Pharmaceuticals, 113 S. Ct. 2786 (1993).

Dawes, R., Faust, D., & Meehl, P. (1989). Clinical and actuarial judgement. *Science, 243,* 1668–1674.

Eubanks, J. (1997). Clinical neuropsychology summary information prepared by Division 40, Clinical Neuropsychology, American Psychological Association. *The Clinical Neuropsychologist, 11,* 77–80.

Garb, H. (1998). *Studying the clinician: Judgment research and psychological assessment.* Washington, DC: American Psychological Association.

Garb, H., & Schramke, C. (1996). Judgment research and neuropsychological assessment: A narrative review and meta-analyses. *Psychological Bulletin, 120,* 140–153.

Guilmette, T., & Hagan, L. (1997). Ethical considerations in forensic neuropsychological consultation. *The Clinical Neuropsychologist, 11,* 287–290.

Heilbrun, K. (1999). *Principles of forensic mental health assessment.* New York: Kluwers (formerly Plenum).

Kleinmuntz, B. (1990). Why we still use our heads instead of formulas: Toward an integrative approach. *Psychological Bulletin, 107,* 296–310.

Kreutzer, J., Harris-Marwitz, J., & Myers, S. (1990). Neuropsychological issues in litigation following traumatic brain injury. *Neuropsychology, 4,* 249–259.

McSweeny, A. J. (1997). Regarding ethics in neuropsychological consultation: A comment on Guilmette and Hagan. *The Clinical Neuropsychologist, 11,* 291–293.

Reed, J. (1999). Current status of the admissability of expert testimony after *Daubert* and *Joiner. Journal of Forensic Neuropsychology, 1,* 49–69.

Sweet, J., Moberg, P., & Westergaard, C. (1996). Five year follow-up survey of practices and beliefs of clinical neuropsychologists. *The Clinical Neuropsychologist, 10,* 202–221.

Sweet, J., & Moulthrop, M. (in press). Self-examination questions as a means of identifying bias in adversarial assessments. *Journal of Forensic Neuropsychology.*

van Gorp, W., & McMullen, W. (1997). Potential sources of bias in forensic neuropsy-
    chological evaluations. *The Clinical Neuropsychologist, 11,* 180–187.
Wedding, D., & Faust, D. (1989). Clinical judgment and decision making in neuropsy-
    chology. *Archives of Clinical Neuropsychology, 4,* 233–265.

AFTERWORD
FORENSIC NEUROPSYCHOLOGY: ADVANCES AND PROSPECTS

Byron P. Rourke

University of Windsor, Windsor, Ontario & Yale
University, New Haven, Connecticut

Overview of the text

The form, content, and unfolding of the issues presented in this work exemplify the elements and merits of the scientist-practitioner model of practice that guides the professional activities of the vast majority of clinical neuropsychologists (Rourke, 1995). There is probably no other applied discipline in psychology that is so committed to this model, and none that applies it more consistently and with good results than clinical neuropsychology.

As a case in point, the first chapters of this text concentrate on the psychometric underpinnings of our field. Issues ranging from reliability to base-rates are featured and discussed. These discussions provide the fundamental bases for the content issues addressed in the remaining chapters. Dealing with form first is as fundamental to the scientist-practitioner enterprise as is the construction of a home prior to its furnishing. And, just as a home becomes individualized by its furnishings, so too does a specialized area of practice become individualized as a result of its content. Without the basic form, neither type of individuation is possible. Therein lies the importance of this first section of the work.

As a case in point, Putnam and his colleagues focus on the details that arise in the context of forensic neuropsychological evaluations involving personal injury, such as traumatic brain injury. The sections of this chapter having to do with the estimation of premorbid cognitive and personality functioning are particularly important for their insightful and perspicacious analyses of the relevant dimensions and variables. An abundance of these issues, from actual assessment to base-rates of commonly reported neuropsychological symptoms, should be required reading for the forensic clinician.

Continuing in this vein, the contribution of Rankin and Adams is noteworthy for its combination of form and content. It follows a fairly standard path from interview through testing, interpretation, and the drawing of conclusions. Although one might prefer to protect objectivity and enhance validity through "blind analysis" of neuropsychological test results prior to examining history for significant moderator variables, their attempts at rigor in the process that they recommend are laudatory and compelling. Needless to say, this is what one would expect from the scientist-practitioner.

So too is the valiant attempt by Osmon to deal with the intellectual minefield of "executive functions." We learn from his assemblage of research that such functions have been hypothesized since the very early years of modern clinical neuropsychology, that they are rooted in both myths and theories of cortical functioning, and that the range of views held with regard to them is as broad as the gulf between comparative psychology and the clinical interview. In my view, it would be well, at present, to focus on the functional dimensions of such functions and to study their reliability and validity before testing hypotheses regarding their bases in brain activity. As Hebb (1949) so cogently observed, everything we know about the brain's relation to behavior started with the careful, systematic examination of behavior. At the very least, such a pursuit would discourage participation in the seemingly endless – and, so far, largely fruitless – debate regarding "frontal lobe tests".

Bieliauskas, in his chapter on the measurement of personality and emotional functioning, presents a cogent rationale for the importance of historical variables in the complete formulation regarding the current status of the forensic client. He points to the need for reliable and valid measurement of these dimensions, and highlights the insights that can be gained therefrom. His insistence on the need for psychologists who testify in an expert witness role to have a solid background in the foundations of knowledge germane to the profession is especially welcome. So too is his suggestion that psychologists who participate in such endeavors should obtain board certification through the American Board of Professional Psychology.

As a vehicle for capturing many of the foregoing dimensions that are relevant for forensic practice, I found the model presented by Kay to be of particular interest. His insistence on a multifactorial approach, rather than "yes-no" answers, to overly simplistic referral and/or trial questions is particularly refreshing. His discussions regarding the role of pain, anxiety, depression, posttraumatic stress disorder, and malingering are worthwhile adjuncts to some of the chapters that precede and follow his. The neuropsychologist who studies and reflects on all of these issues would be in a very good position to participate in the forensic process, from assessment through testimony at trial. Indeed, I would consider this chapter a kind of "primer" for the clinician who wishes to be involved in this complex undertaking.

The thorny issue of "ecological validity" is considered at some length by Sbordone and Guilmette. Their description of the many pitfalls and shortcomings involved in predicting everyday and vocational functioning from neuropsychological test results is particularly thorough. The authors do a good job of highlighting virtually all of the important issues in this area. That said, it is clear that an entire monograph could be devoted to these and related issues.

Sweet's presentation on malingering is the best summary of the relevant issues that I have read. It is thorough and succinct. The strategies for the detection of malingering should be required reading for all clinicians involved in forensic matters. The tests and methods suggested for promoting these strategies are state-of-the-art. This chapter is particularly notable for the order and rationality that the author brings to this complex issue.

The presentation by Rosenfeld and Ellwanger should be read in conjunction with that of Sweet. The use of cognitive psychophysiology in the detection of malingered cognitive deficit is very much in its infancy, but the potential for contributions to the problem of the detection of malingering seems promising. At the same time, one wonders just how successful these methods will be when strategies, such as those that effectively negate the relevance of "lie detection" methods, are employed by unscrupulous individuals. The principal problem with the detection of deception today is that those who are "guilty" are likely to be protected from the detection of deception, whereas those who are "innocent" run a risk of being judged as dissimulators principally as a result of the attention that they allocate to the material in question (Day & Rourke, 1974). This accentuates the need for considerable sophistication on the part of the investigator who employs any of these methodologies in search of "the truth". Also, there is the added complication that one would expect very few persons who are involved in litigation involving traumatic brain injury to submit to this procedure.

Ruff and Richardson address one of the very important classifications in forensic neuropsychological practice – "mild" traumatic brain injury. They are circumspect in their considerations of the implications that patients with this "diagnosis" present, and they are sufficiently prudent about the conclusions that can be arrived at on the basis of various dimensions of evidence. An even more thorough and negative analysis of the inadequacies of neuromedical measurement could be made, especially in this segment of the population that is, by the way, most often defined by neuromedical professionals, not by neuropsychologists. That said, the authors deal positively with the central issues faced by clinicians who encounter this "diagnosis" on a frequent basis.

"Neurotoxic tort" is the focus of Hartman's very interesting presentation. Although limited by the length afforded for a chapter in an edited book, Hartman does an admirable job of pointing out the principal issues involved in this very important, albeit murky, area of neuropsychological investigation. His sections on the "chain of inference" and, especially, "preparing for Daubert" are of particular relevance for the clinician.

The evaluation of young children for any purpose is a very complex endeavor. To help us with this task, we have an ever-widening scope and spectrum of relevant and sound neuropsychological investigations. Lorber and Yurk do an

admirable job of presenting and describing the scientific bases that have implications for the assessment of children, most of which have appeared only rather
recently. They also do an important service by providing the legal dimensions
and ramifications of these findings – especially regarding the provision of services – for brain-impaired children. This particular area of "forensic" service is
probably only fully appreciated by the practicing child-clinical/pediatric neuropsychologist. It should, however, be of considerable interest to all forensic
clinicians who identify themselves as providing services for families.

Do not go to court before reading Taylor's very interesting chapter on the
"legal environment." Although many of its explorations and explanations are
well known to the seasoned practitioner, it is particularly relevant for the relatively young practitioner, or for one who has come to our field in the relatively
recent past.

Lees-Haley and Cohen present an interesting chapter dealing with the neuropsychologist as expert witness. The middle sections of the chapter are of particular interest, especially the reflections about the admissibility of evidence, the
criteria for deciding the merits of one position or another, and the general "atmos-
phere" of court proceedings. Warning the uninitiated that the courtroom may be
full of sharks is interesting, even dramatic. But, in the final analysis, if the clinical
neuropsychologist simply tells the truth, most, if not all, of these concerns become
marginal or completely irrelevant. The final section of the chapter says as much.

Conclusions and future directions

Editor Sweet has assembled a very formidable group of scientist-practitioners
for presentations in this volume. They were aided by others involved in the
forensic practice, as appropriate.

Conclusions

The authors included in this work agree on a wide variety of matters. Some of
my conclusions, including those that are either explicitly or implicitly acknowledged by the authors, are as follows:

(1) Assessment procedures should be reliable and valid. Validity should
    include content, ecological, concurrent, and predictive dimensions.
    Although not addressed within the scope of the current text, construct
    validity might also be considered..

(2) Deficits antedating and post-dating injury, exposure, and so on are diffi-
    cult, though often possible, to determine.

(3) Deficits resulting from insult to the brain may be multifactorial, and these
    probably do interact with premorbid characteristics.

(4) Assessment procedures to deal with these multifactorial dimensions need
    to be multifaceted and comprehensive.

(5) Sometimes clients do not tell the truth. Sometimes they try to "fake bad."
    However, there are sufficient, reasonably reliable methods for addressing

issues surrounding malingering that, if applied in a clinically insightful context, deal in a judicious and reasonably adequate manner with potential malingering. That said, the potential for false positives and false negatives remains somewhat high.

(6)   The legal environment, wherein much of the work of the forensic neuropsychologist proceeds, is anything but uncomplicated. Indeed, knowledge about this environment is essential if the clinician is to perform in a prudent and responsible manner within it. At the same time, it is also important to realize how simple our role is, that is , to tell the truth as we see it.

Future directions

The following are some directions that might prove beneficial for expanding our understanding of, and work within, this field:

(1)   Rigorous and independent investigation of the effects of neurotoxins should be pursued on a number of fronts. Efforts to do so are well under way (e.g., White & Krengel, 1995). The results of such studies should put us in a much better position to evaluate the complex neuropsychological interactions that obtain in this area.

(2)   A valid perspective vis-a-vis the long-term implications of pediatric head injury has yet to be developed. For example, it may come as a surprise that there is little or no relationship between severity of injury and psychosocial outcome (Butler, Rourke, Fuerst, & Fisk, 1997). More generally, the complex nature of pediatric head injury is coming to light only recently (e.g., Ewing-Cobbs, Fletcher, & Levin, 1995).

(3)   The more general contributions of electrophysiological studies to our understanding of relevant issues of neuropsychological development in children (e.g., Stelmack, Rourke, & van der Vlugt, 1995) should yield fruit with respect to our capacity to assess them and, in consequence, provide evidence that would be helpful in the forensic arena.

(4)   More attention needs to be given to issues of competency assessments in persons of all ages. This is especially important for those who suffer from dementing diseases of adulthood. Appreciation of the heterogeneity evident in some of these diseases (e.g., Fisher et al., 1997) would be highly desirable.

(5)   Future volumes of this sort should devote some space to forensic implications and ramifications of "recovered memories," and attentional deficit disorder. At present, these and related matters constitute a veritable minefield for the forensic clinician.

(6)   A code of ethics and professional practice specifically geared to the clinical practice of neuropsychology vis-a-vis forensic matters would be particularly desirable. Efforts to address such issues are well under way (e.g., Adams, 1997). Related to this is the issue of competence to carry out forensic activities – a dimension addressed by some of the authors.

Final comment

The investigators who have authored this work have offered solid and worthwhile contributions to our understanding of the role of the neuropsychologist in the forensic milieu. What is especially noteworthy is the large extent of virtual unanimity regarding many important issues in the field. Among other things, this attests to the continuing contribution of the scientist-practitioner model to this and other domains of our profession (Rourke, 1995). And it seems clear that the continued application of this model will serve to insure the further expansion of the knowledge base so critical for professional practice in these domains.

References

Adams, K. M. (1997). Comment on ethical considerations in forensic neuropsychological consultation. *The Clinical Neuropsychologist, 11*, 294–295.

Butler, K., Rourke, B. P., Fuerst, D. R., & Fisk, J. L. (1997). A typology of psychosocial functioning in pediatric closed-head injury. *Child Neuropsychology, 3*, 98–133.

Day, D. A., & Rourke, B. P. (1974). The role of attention in "lie-detection." *Canadian Journal of Behavioural Science, 3*, 270–276.

Ewing-Cobbs, L., Fletcher, J. M., & Levin, H. S. (1995). Traumatic brain injury. In B. P. Rourke (Ed.), *Syndrome of nonverbal learning disabilities: Neurodevelopmental manifestations* (pp. 433–459). New York: Guilford Press.

Fisher, N. J., Rourke, B. P., Bieliauskas, L. A., Giordani, B., Berent, S., & Foster, N. L. (1997). Unmasking the heterogeneity of Alzheimer's disease: Case studies of individuals from distinct neuropsychological subgroups. *Journal of Clinical and Experimental Neuropsychology, 19*, 713–754.

Hebb, D. O. (1949). *The organization of behavior.* New York: John Wiley.

Rourke, B. P. (1995). The science of practice and the practice of science: The scientist-practitioner model in clinical neuropsychology. *Canadian Psychology, 36*, 259–287.

Stelmack, R. M., Rourke, B. P., & van der Vlugt, H. (1995). Intelligence, learning disabilities, and event-related potentials. *Developmental Neuropsychology, 11*, 445–465.

White, R. F., & Krengel, M. (1995). Toxicant-induced encephalopathy. In B. P. Rourke (Ed.), *Syndrome of nonverbal learning disabilities: Neurodevelopmental manifestations* (pp. 460–475). New York: Guilford Press.

ABOUT THE EDITOR AND CONTRIBUTORS

The Editor

Jerry J. Sweet, Ph.D., ABPP-CN, CL is Director of the Neuropsychology Service at Evanston Hospital, Evanston, IL and Associate Director of Clinical Training in the Psychology Department of Northwestern University. He is Associate Professor of Clinical Psychiatry and Behavioral Sciences at Northwestern University Medical School and Adjunct Associate Professor in the Department of Psychology at Northwestern University. He is board-certified in clinical neuropsychology and in clinical psychology by the American Board of Professional Psychology (ABPP). He is a Fellow of the Division of Clinical Neuropsychology and the Division of Clinical Psychology of the American Psychological Association and is also a Fellow of the National Academy of Neuropsychology. Dr. Sweet co-edited *Handbook of Clinical Psychology in Medical Settings* and co-authored *Psychological Assessment in Medical Settings*, and has published numerous book chapters and peer-reviewed research studies on a variety of topics. He is founding Associate Editor of *Journal of Clinical Psychology in Medical Settings* and currently serves on the editorial boards of *The Clinical Neuropsychologist*, *Archives of Clinical Neuropsychology*, *Journal of Forensic Neuropsychology*, *Journal of Clinical Psychology* and the internet periodical *Journal of Credibility Assessment and Witness Psychology*. Dr. Sweet also serves on the Board of Directors of the American Academy of Clinical Neuropsychology and the Board of Directors of the Association of Postdoctoral Programs in Clinical Neuropsychology.

The Contributors

Russell L. Adams, Ph.D., ABPP-CN,CL is Professor, Department of Psychiatry and Behavioral Sciences, at the University of Oklahoma Health Sciences Center in Oklahoma City, Oklahoma, where he is currently Director of the Adult Neuropsychology Laboratory (since 1978), Director of Postdoctoral Training in Neuropsychology, and Director of the Clinical Psychology Internship Training Program. He serves on journal editorial boards, including *The Clinical Neuropsychologist* and *Archives of Clinical Neuropsychology*. Dr. Adams is board-certified in clinical neuropsychology and in clinical psychology by the American Board of Professional Psychology (ABPP). He is a Fellow of the Division of Clinical Psychology and the Division of Clinical Neuropsychology of the American Psychological Association and is also a Fellow of the National Academy of Neuropsychology.

Stanley Berent, Ph.D., ABPP-CN,CL is currently Professor and Chief of Psychology in the Department of Psychiatry, Director of the Neuropsychology Division, and Co-Director of the Neurobehavioral Toxicology Program at the University of Michigan Medical Center in Ann Arbor, Michigan. He holds board certification in clinical psychology and clinical neuropsychology from the American Board of Professional Psychology (ABPP). Dr. Berent initiated formal programs in neuropsychological testing at both the University of Michigan Medical Center and at the Ann Arbor VAMC. In his clinical practice, teaching and research, Dr. Berent is active in the measurement of behavior and other physiological function in neurological and other medical conditions, including the interplay between cognitive function and psychopathology, and the contributions of environmental neurotoxins to disorder and disease. His work is reflected in approximately 50 externally funded projects, 2 books, 15 book chapters and over 200 publications in scientific journals.

Linas Bieliauskas, Ph.D., ABPP-CN, CL is associate professor in the Departments of Psychiatry and Psychology at the University of Michigan, where he is also coordinator of postdoctoral training in the Neuropsychology Program, and staff psychologist at the Ann Arbor V.A. Medical Center. He is board certified in both Clinical Psychology and Clinical Neuropsychology by the American Board of Professional Psychology (ABPP) and is a fellow of the Divisions of Clinical Health Psychology and Clinical Neuropsychology of the American Psychological Association (APA), and a fellow of the Michigan Psychological Association. He is past president (1992–1993) of the International Neuropsychological Society, past president (1997–1998) of the Division of Clinical Neuropsychology within APA, and Executive Director for the American Board of Clinical Neuropsychology. Dr. Bieliauskas is the Co-Editor of *Aging, Neuropsychology, and Cognition*, was Associate Editor (1997–1998) and is currently on the editorial board of *Journal of Clinical and Experimental Neuropsychology*, was Associate Editor (1987–1994) and is currently on the editorial board of *The Clinical Neuropsychologist*, and is also currently on the editorial board of *Journal of Clinical Psychology in Medical Settings*. His primary research interests include cognitive and affective changes due to normal and abnormal aging and in predictors of treatment response in patients with chronic pain. He has published widely on the relationship between depression and cognitive change in dementing conditions, the implications of neuropsychological test performance for prediction of disease effects and critical daily functioning in the elderly, and on relationships between psychological factors and physical illness and disease.

Joel W. Ellwanger, Ph.D., is currently a member of the research staff at the Geriatric Psychiatry Research Center at San Diego VA Medical Center, which is affiliated with the University of California, San Diego. He previously completed an NIMH research fellowship at the San Diego VA Medical Center. Dr. Ellwanger is a psychophysiologist whose primary interests and numerous research publications have related to malingering, adult psychosis, and aging.

Wm. Drew Gouvier, Ph.D., is Associate Professor in the Department of Psychology at Louisiana State University. He previously held a faculty position in the Department of Rehabilitation Medicine at the University of Alabama at Birmingham. He is a Fellow of the National Academy of Neuropsychology, and served as the 1995 and 1996 convention program chair for that organization. Dr. Gouvier serves on the editorial boards of *Archives of Clinical Neuropsychology* and *Journal of Forensic Neuropsychology*. He has published extensively on head injury and a variety of assessment topics.

Thomas J. Guilmette, Ph.D., ABPP-CN is Assistant Professor in the Department of Psychology at Providence College in Providence, RI. He is also the Director of Neuropsychology at Southern New England Rehabilitation Center at St. Joseph Hospital for Specialty Care in Providence,RI. Dr. Guilmette is a Clinical Assistant Professor, Department of Psychiatry and Human Behavior, Brown University School of Medicine. He is board-certified in clinical neuropsychology by the American Board of Professional Psychology (ABPP). Dr. Guilmette serves on the editorial board of *Journal of Forensic Neuropsychology*.

Paul R. Lees-Haley, Ph.D., ABPP-FN is in private practice in Woodland Hills (Los Angeles) California, specializing in the evaluation and treatment of trauma victims. He is board-certified in forensic psychology by the American Board of Professional Psychology (ABPP) and board-certified by the American Board of Vocational Experts. Dr. Lees-Haley is on the editorial boards of *Journal of Forensic Neuropsychology*, *Trial Science Reporter*, *E&F Vanguard Series*, *Hippocrates' Lantern*, *California Psychologist*, and *Claims*. He is co-editor, with D. Price, of *The Insurer's Handbook of Psychological Injury Claims*, author of *The Last Minute Guide to Psychological and Neuropsychological Testing* and *Pseudoscientific Mumbo Jumbo*, and co–author, with W. Kizorek, of *Psychological Claims Investigation*. Dr. Lees-Haley received the Nelson Butters award for Research Contributions to Clinical Neuropsychology from the National Academy of Neuropsychology. He has 160 professional publications and is a Woodrow Wilson Fellow.

David E. Hartman, Ph.D., ABPN is the Director of the Rush Human Performance Laboratory at the Isaac Ray Center of Rush Presbyterian-St. Luke's Medical Center and is a private practice forensic and medical neuropsychologist in Chicago, Illinois. Dr. Hartman is an Assistant Professor at Rush University and is on the editorial boards of *Neuropsychology Review*, *Archives of Clinical*

Neuropsychology, and *Journal of Forensic Neuropsychology.* He is the author of *Neuropsychological Toxicology,* 2nd edition. Dr. Hartman is board-certified in clinical neuropsychology by the American Board of Professional Neuropsychology (ABPN). He is a Fellow of the National Academy of Neuropsychology.

Thomas Kay, Ph.D. is in private practice in New York City, is Assistant Professor of Clinical Rehabilitation Medicine at New York University Medical Center (where he has been since 1982), and adjunct professor of Psychology (Neuropsychology) at the City University of New York (Queens College). He served from 1987 to 1993 as Director of Research for the NYU Medical Center Research and Training Center on Head Injury and Stroke, and as Assistant Department Head for Psychology Research within the Rusk Institute of Rehabilitation Medicine. He has served on the boards of a number of professional organizations, including the National Head Injury Foundation and the New York Neuropsychology Group. He was the recipient of the 1990 Clinical Service Award from the National Head Injury Foundation. Dr. Kay has participated in a number of grant funded educational and research projects, and has published and lectured extensively, primarily in the areas of mild traumatic brain injury, return to work after head injury, the impact of head injury on the family system, and rehabilitation. He is chief author of the Head Injury Family Interview, a structured interview for persons with head injury and their families. Publications include a Federally funded educational monograph and videotape entitled "The Unseen Injury: Minor Head Trauma" (distributed by the Brain Injury Association), a co-edited text entitled "Head Injury: A Family Matter", and a co-authored article providing a proposed "Definition of mild traumatic brain injury" (developed within the American Congress of Rehabilitation Medicine) in the *Journal of Head Trauma Rehabilitation.*

John E. Kurtz, Ph.D. is an Assistant Professor in Psychology at Villanova University and a licensed psychologist in the Commonwealth of Pennsylvania. His research interests include personality change and development in adulthood and alternative methods for psychological assessment. In addition to teaching graduate and undergraduate courses in psychological testing, clinical psychology, and personality theory, Dr. Kurtz directs the Psychology Internship Program for undergraduate majors at Villanova.

Rudy Lorber, Ph.D., ABPP-SC, BP is an Assistant Professor in the Department of Pediatrics of Northwestern University Medical School. He is in private practice in Highland Park, Illinois. Dr. Lorber has a special interest in advocacy for educational services for learning disabled and neurologically impaired children. He is board-certified in school psychology and in behavioral psychology by the American Board of Professional Psychology (ABPP). From 1992 through 1997, Dr. Lorber was a Level 1 Impartial Due Process Hearing Officer for the Illinois State Board of Education. He is on the professional staff of Evanston Hospital, Evanston, IL, in the Departments of Psychiatry and Pediatrics.

Manfred J. Meier, Ph.D., ABPP-CN was strongly influenced at the University of Wisconsin-Madison by Harry F. Harlow in early neuropsychology, Samuel H. Friedman in the clinical psychology of neurological and psychiatric disorders, and David A. Grant in psychological measurement. He subsequently joined Starke R. Hathaway at the University of Minnesota and established one of the early neuropsychological laboratories, which he directed, along with the health psychology clinic. A Research Career Development Award from what is now the National Institute of Neurological Disorders and Stroke provided opportunity for research and supervisory responsibility for psychology interns and post-doctoral fellows in neuropsychology. Before retiring as director of related clinical training and research programs at Minnesota in 1995, Professor Meier received an Outstanding Contributors to Knowledge Award from the Professional Directorate of the American Psychological Association (1992), and served as Presidents of the Minnesota Society of Neurological Sciences (1969–1970), the International Neuropsychological Society (1980–1981), APA Division of Clinical Neuropsychology (40: 1984–1985), and American Board of Clinical Neuropsychology (1982–1988). Dr. Meier is board-certified in clinical neuropsychology by the American Board of Professional Psychology (ABPP). He has published numerous professional writings on various topics, including the history of American clinical neuropsychology.

David C. Osmon, Ph.D., ABPP-CN is an Associate Professor in the Department of Psychology at the University of Wisconsin-Milwaukee (UWM). Dr. Osmon is currently Director of Clinical Training in the APA-approved doctoral program at UWM and teaches neuropsychological theory and assessment courses, as well as the introduction to clinical psychology course. He is board-certified in clinical neuropsychology by the American Board of Professional Psychology (ABPP). He has maintained a private practice in clinical neuropsychology since 1980, working in a variety of settings (hospital-based service, forensic, outpatient office setting, inpatient rehabilitation unit, long-term residential facility for developmentally disabled patients). Dr. Osmon has published over thirty articles and chapters, and a book, all on various aspects of neuropsychology, including assessment, structural and functional imaging, neuropathological correlates of cognitive impairments, and the structure of cognition.

Steven H. Putnam, Ph.D. is a staff clinical neuropsychologist at the Detroit Medical Center's outpatient neurorehabilitation program. He serves as Assistant Professor of Physical Medicine and Rehabilitation in the Wayne State University School of Medicine, Adjunct Assistant Professor in the Psychology Department, and Associate Graduate Faculty in the Wayne State University Graduate School. At present he serves on the editorial boards of *The Clinical Neuropsychologist*, *Journal of Head Trauma Rehabilitation*, *Journal of Forensic Neuropsychology*, and *Advances in Medical Psychotherapy*. He has served on the Program Committee for the scientific conferences of the International Neuropsychological Society and is a regular presenter at the annual meetings of the American

Psychological Association, International Neuropsychological Society, National Academy of Neuropsychology, and the MMPI symposia. He has published a number of papers pertaining to neuropsychological and personality assessment following traumatic brain injury, professional practices of neuropsychologists, and the evaluation of rehabilitation treatment efficacy following TBI.

Eugene J. Rankin, Ph.D., is Director of the Department of Neuropsychology at Iowa Methodist Medical Center/Younker Rehabilitation Center in Des Moines, Iowa. Until recently he worked at Emmanuel Rehabilitation Center in Omaha, NE and was Adjunct Assistant Clinical Professor at the University of Nebraska-Lincoln, where he provided clinical supervision in neuropsychology to clinical psychology graduate students.

Ann M. Richardson, Esq., has worked exclusively in the area of personal injury law since 1981. Currently a solo practitioner in San Francisco, California, she specializes in cases involving closed head injury, sexual harassment and discrimination, premises liability, psychological malpractice, and automobile accident. She is a graduate of the California Western School of Law, where she was a staff writer for the *International Law Journal* and Editor of California Western's *Law Review*. Ms. Richardson is on the Board of Directors of the Consumer Attorneys of California.

Joseph H., Ricker, Ph.D., ABPP-CN,RP is Director of Training and clinical neuropsychologist in the Department of Rehabilitation Psychology and Neuropsychology at the Rehabilitation Institute of Michigan. He is an Assistant Professor in the Department of Physical Medicine & Rehabilitation and the Department of Neurology at Wayne State University School of Medicine. Dr. Ricker is board-certified in both Clinical Neuropsychology and Rehabilitation Psychology by the American Board of Professional Psychology (ABPP). He has published and presented in the area of neuropsychological assessment of neurologic rehabilitation populations, particularly individuals with brain injury and stroke. Dr. Ricker is a member of the editorial board of the *Journal of Head Trauma Rehabilitation*, and regularly serves as a reviewer for other neuropsychology and medical journals. He is currently editing a text on differential diagnosis in adult neuropsychological assessment. Dr. Ricker has been active in professional organizations, such as Division of Clinical Neuropsychology of the American Psychological Association, within which he served as a member of the division's scientific program committee from 1995–1998, as the Co-Chair in 1999, and as the Chair in 2000.

J. Peter Rosenfeld, Ph.D., is Professor in the Psychology Department of Northwestern University, Evanston, IL., where he also has held appointments in the Department of Psychiatry and Behavioral Sciences and Department of Neurobiology and Physiology. Dr. Rosenfeld was President of the Association for Applied Psychophysiology and Biofeedback in 1991–1992 and served on the

National Institute of Health Neurological Sciences study section in 1985–1987. He currently serves on the editorial boards of *Applied Psychophysiology and Biofeedback*, *Journal of Credibility Assessment and Witness Psychology*, *Journal of Neurotherapy*, and *International Journal of Rehabilitation and Health* and reviews articles for numerous other journals. He has published extensively on a wide range of psychophysiological assessment procedures. Dr. Rosenfeld is also a licensed clinical psychologist and certified biofeedback practitioner, with a practice in Evanston, IL.

Scott R. Ross, Ph.D. is an Assistant Professor of Psychology at DePauw University. His published work has focused on contextual aspects of psychological assessment and the predictive utility of neuropsychological measures. Currently, he is engaged in an ongoing program of research concerned with the relationship between individual differences in personality and performance on traditional neuropsychological tests

Byron P. Rourke, Ph.D., ABPP-CN, FRSC is Professor of Psychology and a University Professor at the University of Windsor, and a member of the faculty of the School of Medicine, Yale University. He has served as president of the INS, the Division of Clinical Neuropsychology of the APA, and the American Board of Clinical Neuropsychology. He is Co-Editor and co-founder of *The Clinical Neuropsychologist*. He co-founded the *Journal of Clinical and Experimental Neuropsychology*, and served as its Co-Editor for many years. In addition he co-founded and served as editor for *Child Neuropsychology*. He has been elected a Fellow of the Royal Society of Canada. He received the Distinguished Contributions to Psychology as a Profession Award of the Canadian Psychological Association (CPA) for 1994. According to the CPA Awards Committee, Dr. Rourke's research into Nonverbal Learning Disabilities (NLD) has a "direct, positive, and pervasive influence on clinical practice and has significantly enhanced knowledge in the field". Dr. Rourke is board certified in Clinical Neuropsychology by the American Board of Professional Psychology.

Ronald M. Ruff, Ph.D., ABPP-RP, is Director of Neurobehavioral Rehabilitation Services at St. Mary's Medical Center in San Francisco, California, and is an Associate Adjunct Professor of Neurosurgery and Psychiatry at the University of California, San Francisco. He is board-certified in rehabilitation psychology by the American Board of Professional Psychology (ABPP). He is a Fellow of the Division of Clinical Neuropsychology of the American Psychological Association and the National Academy of Neuropsychology. Dr. Ruff has published over 80 articles and book chapters, most related to traumatic brain injury and cognitive rehabilitation, and has lectured extensively on mild traumatic brain injury. He has been a participant in major data collections of mild, moderate, and severe brain injury populations extensively on topics related to traumatic brain injury, including the National Traumatic Coma Data Bank, a 7-year project funded by NIH. Dr. Ruff was the founder and served as board member and President of the San Diego Head Injury Foundation. In addition, he has authored a number of neuropsychological tests, including the Ruff Figural Fluency Test, the Ruff 2 & 7 Selective Attention Test, and the Ruff-Light Trail Learning Test. Dr. Ruff has been a peer reviewer for most journals in the field of neuropsychology, and is on the editorial board of *Applied Neuropsychology*.

Robert J. Sbordone, Ph.D., ABPP-CN, ABPN, maintains a private practice in Irvine, California. Dr. Sbordone is board-certified in clinical neuropsychology by the American Board of Professional Psychology (ABPP) and the American Board of Professional Neuropsychology (ABPN). He has authored or edited more than 100 publications, including *Neuropsychology for the Attorney*; *Disorders of Executive Functions: Civil and Criminal Law Applications* (with H. Hall, Ph.D.); *The Ecological Validity of Neuropsychological Testing* (with C. Long, Ph.D.); and *Forensic Neuropsychology* (in preparation with A. Purisch, Ph.D. and L. Cohen, Esq.) He is a Fellow of the National Academy of Neuropsychology. Dr. Sbordone serves on the Board of Directors of the American Board of Professional Neuropsychology and on the editorial board of *Neurolaw*.

Christine L. Swartz, Ph.D. attended Bates College in Lewiston, Maine, where she received a Bachelor of Arts degree in Psychology and Russian. After completing a pre-professional psychology training program at the Devereux Foundation in Devon, Pennsylvania, she entered the graduate program in clinical psychology at Saint Louis University and received the Doctor of Philosophy degree in psychology. She completed an APA-approved internship in Child Clinical Psychology at the University of Rochester in New York and is currently completing a post-doctoral fellowship in pediatric neuropsychology at the University of Michigan. Dr. Swartz's areas of interest include pediatric neuropsychology and toxicology.

J. Sherrod Taylor, J.D., is a senior partner in the law firm of Taylor, Harp and Callier in Columbus, Georgia. A graduate of the University of Georgia School of Law, he serves on the adjunct faculty of the Emory University School of Medicine. He is the author of *Neurolaw: Brain and Spinal Cord Injuries* (ATLA Press/Clark Boardman Callaghan, 1997) and Editor-in-Chief of *The Neurolaw Letter*, a monthly periodical for legal and health professionals. Mr. Taylor has published extensively on the interface of neuropsychology and the law.

Helene Yurk, Ph.D., is Director, Pediatric Neuropsychology Service and Associate Director, Clinical Psychology Predoctoral Internship. Dr. Yurk is an Assistant Professor Professor of Clinical Psychiatry at the University of Chicago Medical Center. Specialty interests and experience include neuropsychological investigation of the relationship between chronic medical disease and quality of life, particularly school and social functioning, in children. She has extensive experience with children who have kidney and liver disease, as well as children with Systemic Lupus Erythematosus.

CONTRIBUTORS ADDRESS LIST

Russell L. Adams
Dept. of Psychiatry & Behavioral Sciences
University of Oklahoma Health Sciences
Center
P.O. Box 26901
Oklahoma City, OK 73190-3048

Stanley Berent
University of Michigan Medical Center
1500 East Medical Center Drive
Ann Arbor, MI 48109-0840

Linas A. Bieliauskas
Psychology Service (116B)
VA Medical Center
2215 Fuller Road
Ann Arbor, MI 48105

Larry J. Cohen
3200 North Central Avenue
Suite 1800
Phoenix, AZ 85012

Joel W. Ellwanger
VA Medical Center
3350 La Jolla Village Drive
San Diego, CA 92161

Wm. Drew Gouvier
Louisiana State University
Psychology Department
Baton Rouge, LA 70803-5501

Thomas J. Guilmette
Department of Psychology
Providence College
Providence, RI 02918

David E. Hartman
Forensic and Medical Neuropsychology
401 E. Illinois St, Suite 320
Chicago, IL 60611

Thomas Kay
333 E. 34th St., Suite 1N
New York, NY 10016

John E. Kurtz
Department of Psychology
Villanova University
Villanova, PA 19085-1699

Paul R. Lees-Haley
Lees-Haley Psychological Corporation
21331 Costanso Street
Woodland Hills (Los Angeles), CA 91364

Rudy Lorber
Lake Shore Neuropsychological Services
600 Central Avenue, #228
Highland Park, IL 60035

Manfred J. Meier
101 Trail W10601, Box 10
Bruce, WI 54819-0010

David C. Osmon
Department of Psychology
University of Wisconsin-Milwaukee
P.O. Box 413
Milwaukee, WI 53201

Steven H. Putnam
Detroit Medical Center-Rehabilitation
Institute of Michigan
42005 W. Twelve Mile Road
Novi, MI 48377-3113

Eugene J. Rankin
Department of Neuropsychology
Iowa Methodist Medical Center/Younker
Rehabilitation Center
1200 Pleasant Street
Des Moines, IA 50309-1453

Joseph H. Ricker
Rehabilitation Institute of Michigan
261 Mack Blvd., Suite 555
Detroit, MI 48201

Ann M. Richardson
One Sansome, 19th Floor
San Francisco, CA 94104

J. Peter Rosenfeld
Psychology Department
Northwestern University
102 Swift Hall
Evanston, IL 60208

Scott R. Ross
DePauw University
7 E. Larrabee Street
103C Harrison Hall
Greencastle, IN 46135

Byron P. Rourke
Department of Psychology
University of Windsor
Windsor, Ontario
Canada N9B 3P4

Ronald M. Ruff
Neurobehavioral Rehabilitation
St. Mary's Medical Center
450 Stanyan Street
San Francisco, CA 94117-1079

Robert J. Sbordone
7700 Irvine Center Drive, Suite 750
Irvine, CA 92618

Christine Swartz
NeuroBehavioral Resources, Inc.
325 East Summit
Ann Arbor, MI 48104

Jerry J. Sweet
Neuropsychology Service
Evanston Hospital
2650 Ridge Avenue
Evanston, IL 60201-1718

J. Sherrod Taylor
c/o Taylor & Harp
The Corporate Center, Suite 900
Columbus, GA 31902

Helene Yurk
Department of Psychiatry
University of Chicago
5841 South Maryland Avenue, MC 3077
Chicago, IL 60637

SUBJECT INDEX

AUTHOR INDEX

Mac and Mc are treated as the same

STUDIES ON NEUROPSYCHOLOGY, DEVELOPMENT, AND COGNITION

1. *Fundamentals of Functional Brain Imaging: A Guide to the Methods and their Applications to Psychology and Behavioral Neuroscience.* Andrew C. Papanicolaou
1998. ISBN 90 265 1528 6 (hardback)

2. *Forensic Neuropsychology: Fundamentals and Practice.* Jerry J. Sweet (ed.)
1999. ISBN 90 265 1544 8 (hardback)

3. *Neuropsychological Differential Diagnosis.* Zakzanis, Leach, and Kaplan
1999. ISBN 90 265 1552 9 (hardback)